A Structured Approach to Business Forecasting

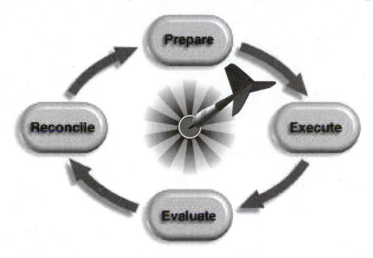

www.duxbury.com

www.duxbury.com is the World Wide Web site for Thomson Brooks/Cole and is your direct source to dozens of online resources.

At www.duxbury.com you can find out about supplements, demonstration software, and student resources. You can also send email to many of our authors and preview new publications and exciting new technologies.

www.duxbury.com

Changing the way the world learns®

THOMSON TITLES OF RELATED INTEREST

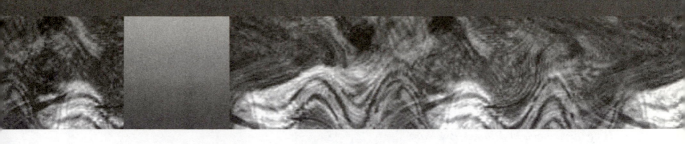

FORECASTING
Practice and Process
for Demand Management

Hans Levenbach

Delphus, Inc.

James P. Cleary

Formerly of Lucent, Inc.

THOMSON

BROOKS/COLE ™

Australia • Canada • Mexico • Singapore • Spain
United Kingdom • United States

THOMSON

BROOKS/COLE

Forecasting: Practice and Process for Demand Management
Hans Levenbach and James P. Cleary

Publisher and Executive Editor: *Curt Hinrichs*
Assistant Editor: *Ann Day*
Editorial Assistant: *Fiona Chong*
Technology Project Manager: *Earl Perry*
Executive Marketing Manager: *Tom Ziolkowski*
Executive Marketing Communications Manager:
   Nathaniel Bergson-Michelson
Project Manager, Editorial Production: *Kelsey McGee*
Art Director: *Lee Friedman*
Print Buyer: *Rebecca Cross*

Permissions Editor: *Bob Kauser*
Production Service: *The Book Company: Dustine Friedman*
Text Designer: *Andrew Ogus*
Copy Editor: *Julie F. Nemer*
Illustrator: *Scientific Illustrators*
Cover Designer: Gia Giasullo/Studio eg
Cover Image: Ann Monn@Royalty–Free/Corbis
Cover Printer: *Phoenix Color Corp*
Compositor: *Cadmus Professional Communications*
Printer: *R.R. Donnelley, Crawfordsville*

For more information about our products, contact us at:
Thomson Learning Academic Resource Center
1-800-423-0563
For permission to use material from this text or product, submit a request online at
http://www.thomsonrights.com.
Any additional questions about permissions can be submitted by e-mail to
thomsonrights@thomson.com

Library of Congress Control Number: 2004117231
Student Edition: ISBN 0-534-26268-6

Thomson Higher Education
10 Davis Drive
Belmont, CA 94002-3098
USA

Asia (including India)
Thomson Learning
5 Shenton Way
#01-01 UIC Building
Singapore 068808

Australia/New Zealand
Thomson Learning Australia
102 Dodds Street
Southbank, Victoria 3006
Australia

Canada
Thomson Nelson
1120 Birchmount Road
Toronto, Ontario M1K 5G4
Canada

UK/Europe/Middle East/Africa
Thomson Learning
High Holborn House
50–51 Bedford Row
London WC1R 4LR
United Kingdom

Latin America
Thomson Learning
Seneca, 53
Colonia Polanco
11560 Mexico
D.F. Mexico

Spain (including Portugal)
Thomson Paraninfo
Calle Magallanes, 25
28015 Madrid, Spain

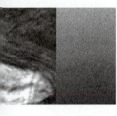

To my precious wife, Suzanne, and our wonderful children Jody, Phill, and Amy and adorable grandchildren Jordan, Daphne, and Ivan
— H.L.

To my dearest wife, Lee, and our children Ken and Beth and grandson Eli
— J.P.C.

HANS LEVENBACH is president of Delphus Inc., a software-development and sales-consulting firm specializing in decision analysis, decision support, forecasting, and operations management. He started his career at AT&T Bell Laboratories as an applied statistician, participating in forecasting projects and developing analytical support systems for the Bell Operating Companies. During his many years of teaching, he served as adjunct professor in the Business Schools of Columbia University and New York University. In 1996, he was a visiting professor at the Stern School of Business at NYU. In his professional life, he served as president, treasurer, and board member of the International Institute of Forecasters (IIF). He is a member of INFORMS, ASA, and an elected member of ISI. In June 2003 he was elected Fellow of the IIF. Levenbach graduated from Acadia University (Canada) with a degree in physics and mathematics and received the MS in electrical engineering from Queen's University (Canada) and MA and PhD in mathematical statistics from the University of Toronto. He is involved in all aspects of Delphus and particularly enjoys working with forecasting practitioners and software developers.

JAMES CLEARY has held numerous leadership positions in marketing and finance at AT&T, Lucent Technologies, Avaya, and New York Telephone Company. Cleary developed courses in time-series and econometric modeling for AT&T, taught an Industrial Engineering course at Columbia for a year, as well as a managerial course about the function of forecasting. Recently, he developed market-sizing and quantified business-case value propositions that demonstrate the product benefits from the perspectives of end users, service providers, and manufacturers. Cleary directed the business management functions of strategic planning, business planning, forecasting, and results analysis. He also led a forecast improvement team that improved the accuracy of customer-team demand forecasts. At AT&T, Cleary was the director of market research and analysis in the business communications services unit; his financial positions included CFO of data communications services and CFO of computer systems.

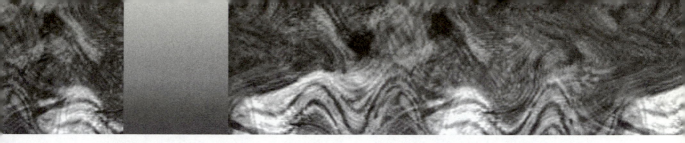

Brief Contents

Part 1

Introducing the Forecasting Process xxxi

  1 – Forecasting as a Structured Process 1

  2 – Classifying Forecasting Techniques 27

Part 2

Exploring Time Series 65

  3 – Data Exploration for Forecasting 67

  4 – Characteristics of Time Series 117

  5 – Assessing Accuracy of Forecasts 161

Part 3

Forecasting the Aggregate 194

  6 – Dealing with Seasonal Fluctuations 195

  7 – Forecasting the Business Environment 235

Part 4

Applying Bottom-Up Techniques 277

  8 – The Exponential Smoothing Method 279

  9 – Disaggregate Product-Demand Forecasting 321

Part 5

Forecasting Models 356

  10 – Creating and Analyzing Causal Forecasting Models 357

  11 – Linear Regression Analysis 391

  12 – Forecasting with Regression Models 427

  13 – Building ARIMA Models: The Box-Jenkins Approach 465

  14 – Forecasting with ARIMA Models 511

Part 6

Improving Forecasting Effectiveness 541

  15 – Selecting the Final Forecast Number 543

  16 – Implementing the Forecasting Process 577

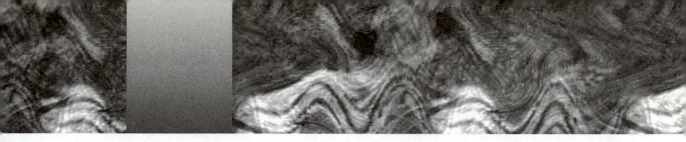

Contents

Preface xxiii

Part 1
Introducing the Forecasting Process xxxi

Chapter 1   Forecasting as a Structured Process   1

1.1   Inside the Crystal Ball   2
Forecasting Defined   2
What Is Demand Forecasting?   2
Learning from Actual Examples of Forecasting   4
What Are Forecasting Models?   6

1.2   Is Forecasting Worthwhile?   10
Role of the Forecaster   10
Who Are the End Users of Forecasts?   10

1.3   Creating a Structured Forecasting Process   11
Stages in the Forecasting Process   11
Defining User Needs, Forecastable Items, and the Forecaster's Resources   12
Identifying Factors Likely to Affect Changes in Demand   12
A Car-Buying Analogy   13
Analysis of Data Sources   14
Judging the Quality of Data   15
Evaluating Forecasting Methods   17

1.4   Establishing an Effective Demand Forecasting Strategy   18

Summary   20
References   21
Problems   21

CASE 1A   Production of Ice Cream   23

CASE 1B   Forecasting Peak Power Demand in a Large Metropolitan Area   23

CASE 1C   Energy Forecasting   24

CASE 1D   A Perspective on the Automotive Industry   24

CASE 1E   College Textbook Publishing   25

CASE 1F   The Textile Industry   25

Chapter 2   Classifying Forecasting Techniques   27

2.1   Selecting a Forecasting Technique   28
Qualitative Methods   28
Quantitative Methods   31
Statistical Forecasting   33

2.2   A Life-Cycle Perspective   35

2.3   Market Research   37

2.4   New Product Introductions   38

2.5   Promotions and Special Events   38

2.6   Sales Force Composites and Customer Collaboration   39

2.7   Neural Nets for Forecasting   39
Rule-Based Forecasts   40

2.8   Execution: Projecting Historical Patterns—A Prototypical Forecasting Application 40
Forecasting with Moving Averages 42
Fit versus Forecast Errors 47
Weighting with Recent History 48
Judgment and Modeling Expertise 50
A Multimethod Approach 52

2.9   How to Forecast with Weighted Averages 52
Simple Smoothing as a Weighted Average 53
Choosing the Smoothing Weight 55
Forecasting with Limited Data 55
Evaluating Forecasting Performance 55

Summary 57

References 58

Problems 59

CASE 2A   The Delphi Method 61
CASE 2B   Input-Output Analysis 61
CASE 2C   Dynamic Systems Modeling 62
CASE 2D   Focus Group Research 62
CASE 2E   Promotions: Are They Really Worth It? 63
CASE 2F   Technological Forecasting 63

Part 2
Exploring Time Series 65

Chapter 3   **Data Exploration for Forecasting 67**

3.1   Exploring Data 68
Time Plots 69
Scatter Diagrams 71
Linear Association 74

3.2   Creating Data Summaries 76
Typical Values 76
Variability 78
Tabulating Frequencies 80

3.3   Displaying Data Summaries 84
Stem-and-Leaf Display 84
Box Plots 85
Quantile-Quantile Plots 86

3.4   Serially Correlated Data 88
Autocorrelation 88
Sample Autocorrelation Functions 92
Detecting Autocorrelation 93

3.5   What Does Normality Have to Do with It? 96
Determining Precision with Standard Errors 99
Establishing Confidence and Prediction Intervals 99

3.6   The Need for Nontraditional Methods 102
Insuring against Unusual Values 102
M-Estimators 103

Summary 105

References 106

Problems 107

CASE 3A   Production of Ice Cream 109
CASE 3B   Twin Rivers 109
CASE 3C   Estimating Promotion Lifts Using Running Medians 110
CASE 3D   Domestic Automobile Production 111

Appendix 3A: The Need for Robustness in Correlation 112
A Robust Correlation Coefficient 112

Appendix 3B: Comparing Estimation Techniques 113
Unbiasedness 114
Consistency 114
Asymptotic Efficiency 115

Chapter 4   **Characteristics of Time Series 117**

4.1   Visualizing Components in a Time Series 118
Trends and Cycles 120
Seasonality 121
Irregular or Random Fluctuations 124
Weekly Patterns 126
Trading-Day Patterns 127

4.2   A First Look at Trend and Seasonality 128
Exploring Components of Variation 129
Day-of-the-Week-Effect 135

4.3   What Is Stationarity? 136
Detecting Nonstationarity with
Correlograms 137
Removing Nonstationarity Using
Differencing 139
Logarithmic Transformations with
Nonstationary Data 142

4.4   Classifying Trends 145

4.5   How to Detect Trends 146
The Basics of Trend Analysis 147
Building the Trend Analysis Spread
Sheet 148

Summary 148

References 149

Problems 149

CASE 4A   Production of Ice Cream 154

CASE 4B   Demand for Air Travel 154

**CASE 4C   Domestic Automobile
Production 155**

**CASE 4D   Sales and Advertising of a
Weight-Control Product 155**

**Appendix 4A A Two-Way Table
Decomposition 156**
Contribution of Trend and Seasonal
Effects 158
Interpreting the Residual Table 158

**Chapter 5   Assessing Accuracy of
Forecasts 161**

**5.1   The Need to Measure Forecast
Accuracy 161**

5.2   Analyzing Forecast Errors 161
Lack of Bias 162
What Is an Acceptable Precision? 163

5.3   Ways to Evaluate Accuracy 166
The Fit Period versus the Test
Period 166
Goodness-of-Fit versus Forecast
Accuracy 167
Item-Level versus Aggregate
Performance 168
Absolute Errors versus Squared
Errors 168

**5.4   Measures of Forecast
Accuracy 169**
Measures of Bias 169
Measures of Precision 170

**5.5   Comparing with Naïve
Techniques 173**
NAÏVE_1 Technique 174
Relative Error Measures 175

5.6   Tracking Tools 176
Ladder Charts 176
Prediction-Realization Diagram 177
Prediction Intervals for Time Series
Models 178
Prediction Interval as a Percent
Miss 180
Prediction Intervals as Early Warning
Signals 180
Trigg Tracking Signals 183

5.7   How to Monitor Forecasts 183
Quick and Dirty Control 184
The Tracking Signal in Action 185
Adapting the Tracking Signal to
Other Spreadsheets 187

Summary 188

References 189

Problems 189

CASE 5A   Production of Ice Cream 191

CASE 5B   Demand for Air Travel 191

CASE 5C   Demand for Bicycles 192

Part 3

Forecasting the Aggregate 194

**Chapter 6   Dealing with Seasonal
Fluctuations 195**

6.1   Seasonal Influences 196
Removing Seasonality by
Differencing 197
Seasonal Decomposition 200
Uses of Seasonal Adjustment 201

**6.2   The Ratio-to-Moving-Average
Method 202**
Step 1: Trading-Day Adjustment 202
Step 2: Calculating a Centered
Moving Average 203
Step 3: Trend-Cycle and Seasonal
Irregular Ratios 203
Step 4: Seasonally Adjusted Data 205

**6.3   Multiplicative and Additive
Seasonal Decompositions 205**
Decomposition of Monthly Data 205

Decomposition of Quarterly Data 208
Seasonal Decomposition of Weekly
Point-of-Sale Data 211

**6.4   Census Seasonal Adjustment
Method 215**
Why Use the X-11/X-12
Programs? 215
The X-11/ X12 Programs 217
Program Output 218
A Forecast Using X-12 220

6.5   Resistant Smoothing 221

**6.6   How To Detect Seasonal
Cycles – Formal-Wear Rental
Revenue 225**
The Basics of Seasonal
Analysis 225

Summary 227

References 228

Problems 228

CASE 6A   Production of
Ice Cream 232

CASE 6B   Demand for Air Travel 232

CASE 6C   Demand for Apartment
Rental Units 232

CASE 6D   Domestic Automobile
Production 233

CASE 6E   Sale of Twin Screw Extruders
to the Food Industry 234

Chapter 7   **Forecasting the Business
Environment 235**

**7.1   Forecasting with Economic
Indicators 236**
Macroeconomic Demand
Analysis 236
Origin of Leading Indicators 236
Use of Leading Indicators 237
Composite Indicators 243
Reverse Trend Adjustment of the
Leading Indicators 244
Sources of Indicators 245
Selecting Indicators 245

**7.2   Trend-Cycle Forecasting with
Turning Points 247**
Pressures Analysis 247
Ten-Step procedure for Making a
Turning-Point Analysis 249

Preparing a Cycle Forecast for
Revenues 250
Alternative Approaches to
Turning-Point Forecasting 253

7.3   Using Elasticities 253
Determinants of Demand 254
The Price Elasticity 254
Price Elasticity and Revenue 256
Cross-Elasticity 257
Other Demand Elasticities 258
Estimating Elasticities 258

**7.4   Econometrics and Business
Forecasting 260**
Uses of Econometric Models 261
Types of Econometric Models 262
A Recursive System 262
Some Pitfalls 263

**7.5   Using Pressures to Analyze
Business Cycles 265**
1/12 Pressures 266
3/12 Pressures 266
12/12 Pressures 267

Summary 268

References 268

Problems 269

CASE 7A   Production of Ice Cream 271

CASE 7B   Top-Down Forecast of
Integrated Circuit
Production 271

CASE 7C   Demand for Bicycles 273

CASE 7D   Sales of Domestic
Automobiles 274

CASE 7E   Export of Japanese
Automobiles 275

Part 4

Applying Bottom-Up Techniques 277

Chapter 8   **The Exponential Smoothing
Method 279**

**8.1   What Is Exponential Smoothing?
280**

8.2   Smoothing Weights 281
Equally Weighted Average 281

Exponentially Decaying Weights 282
Simple Exponential Smoothing 282

8.3 Types of Smoothing Techniques 286

8.4 Smoothing Levels and Constant Change 287

8.5 Damped and Exponential Trends 292
Prediction Limits for Exponential Smoothing Models 298

8.6 Seasonal Models 301

8.7 Handling Special Events with Smoothing Models 307
Event Adjustments for Outliers 308
Extensions of Event Modeling 309

Summary 310

References 310

Problems 311

CASE 8A Production of Ice Cream 313

CASE 8B Demand for Apartment Rental Units 313

CASE 8C Energy Forecasting 314

CASE 8D A Perspective of the Automotive Industry—Japanese Automobile Production 315

CASE 8E The Demand for Chicken 317

CASE 8F Sale of Twin Screw Extruders to the Food Industry 317

Appendix 8A 318
Model Formulations 318
Notation 318
Nonseasonal Techniques 318
Additive-Seasonal Techniques 319
Multiplicative-Seasonal Techniques 319

Chapter 9 Disaggregate Product-Demand Forecasting 321

9.1 Forecasting for the Supply Chain 322
Systems for the Supply Chain Pipeline 323
Operating Lead Times 325
What Is Demand Management? 325

Distributional Resource Planning: A Time-Phased Planned Order Forecast 326

9.2 A Framework for an Integrated Demand Forecasting System 328
A System Architecture 329
Dimensions of Demand 329
Role of Planning 331
Reconciling Cross-Functional Forecasting Processes 332
Market Intelligence and Judgmental Overrides 333
Analyzing Demand Variability 334

9.3 Automated Statistical Forecasting 337
Selecting Models Visually 338
Automatic Method Selection 341
Searching for Optimal Smoothing Procedures 343
Error-minimization Criteria 344
Searching for Optimal Smoothing Weights 344
Starting Values 345

9.4 Disaggregate Product-Demand Forecasting Checklist 346

9.5 How to Create a Time-Phased Replenishment Plan 347
Basic Distribution Resource Planning 347

Summary 348

References 349

Problems 349

CASE 9A Demand for Ice Cream 351

CASE 9B Demand for Spare Ribs 351

CASE 9C Forecasting for Inventory Control 354

Part 5
Forecasting Models

Chapter 10 Creating and Analyzing Causal Forecasting Models 357

10.1 A Model-Building Strategy 358

10.2 What Are Regression Models? 359

The Regression Curve 360
A Simple Linear Model 361
The Least-Squares Assumption 362
Linear Regression: One Explanatory
Variable 363

**10.3   Creating Multiple Linear
Regression Models 363**
Some Examples 365

**10.4   Learning from Residual
Patterns 369**
A Run Test for Randomness 372
Nonrandom Patterns 374
Graphical Aids 376
Identifying Unusual Patterns 377

**10.5   Validating Preliminary Modeling
Assumptions 378**
Transformations 379
Achieving Additivity 379

Summary 382

References 382

Problems 383

CASE 10A   Demand for Ice Cream 386

CASE 10B   Demand for Air Travel 386

**CASE 10C   Forecasting Bicycle Demand
387**

**CASE 10D   Forecasting Foreign Imports
387**

**CASE 10E   The Commercial Real Estate
Market: Office Occupancy
Rates 388**

**CASE 10F   Demand for Apartment
Rental Units 388**

Appendix 10A Achieving Linearity
389

Chapter 11   Linear Regression Analysis 391

11.1   Graphing Relationships 391
What Is a Linear Relationship? 395

**11.2   Creating and Interpreting
Output 395**
The Precision of the Estimated
Regression 395
R^2 Statistic 397
Interchanging the Role of Y
and X 400
Linear Correlation 401

**11.3   Making Inferences about Model
Parameters 402**
The Normality Assumption in
Regression 402
Important Distribution Results 403
Significance of Regression
Coefficients 404
Inferences from Summary
Statistics 404
An Incremental F Test 406

11.4   Autocorrelation Correction 407
First-Order Autocorrelation 407
Testing for Serial Correlation 408
Adjusting for Serial Correlation 409
Causal Regression with Prediction
Intervals 414

Summary 415

References 416

Problems 416

**CASE 11A   Demand for (Methylene
Dipheylene Diisocyanate) 421**

CASE 11B   Demand for Air Travel 422

CASE 11C   Energy Forecasting 422

**CASE 11D   New York Mets Home
Attendance 423**

**CASE 11E   Japanese Automobile
Export 425**

CASE 11F   Twin Rivers 425

Appendix 11A A Robust Correlation
Coefficient 425

**Chapter 12   Forecasting with Regression
Models 427**

**12.1   Multiple Linear Regression
Analysis 428**

**12.2   Assessing Model
Adequacy 428**
Collinearity Due to Trends 429
Overfitting 430
Extrapolation 430
Outliers 430
Multicollinearity 431
Invalid Assumptions 432
Application: Relating Business
Telephone Demand to Nonfarm
Employment 432

12.3 Selecting Variables 434
Regression by Stages 435
Application: Building a Cross-
Sectional Model for Additional
Telephone-Line Development 436

**12.4 Indicators for Qualitative
Variables 440**
Use of Indicator Variables 440
Qualitative Factors 440
Dummy Variables for Different
Slopes and Intercepts 442
Measuring Discontinuities 442
Adjusting for Seasonal Effects 443
Eliminating the Effects of Outliers 444

12.5 Analyzing Residuals 445

**12.6 The Need for Robustness in
Regression 446**
Why Robust Regression? 446
M-Estimators 447
Calculating M-Estimates 448

12.7 Multiple Regression Checklist 450

**12.8 How to Forecast with
Qualitative Variables 451**
Modeling with a Single Qualitative
Variable 451
Modeling with Two Qualitative
Variables 452
Modeling with Three Qualitative
Variables 454

Summary 455

References 456

Problems 456

**CASE 12A Forecasting Market Price for
Residential Housing 459**

CASE 12B The Demand for Chicken 460

CASE 12C Energy Forecasting 461

**CASE 12D Predicting Academic
Performance 462**

**CASE 12E The Commercial Real Estate
Market: Office Occupancy
Rates 464**

**Chapter 13 Building ARIMA Models:
The Box-Jenkins Approach 465**

**13.1 Why Use ARIMA Models for
Forecasting? 466**

**13.2 The Linear Filter Model as a
Black Box 467**

13.3 A Model-Building Strategy 468

**13.4 Identification: Interpreting
Autocorrelation and Partial
Autocorrelation Functions 469**
Autocorrelation and Partial
Autocorrelation Functions 470
The Mixed ARMA Process 474
Invertibility and Stationarity 476
Seasonal ARMA Process 476

**13.5 Identifying Nonseasonal ARIMA
Models 477**
Identification Steps 478

**13.6 Estimation: Fitting Models to
Data 478**

**13.7 Diagnostic Checking: Validating
Model Adequacy 484**
Overfitting 485
Chi-Squared Test 485
Periodogram Analysis 486

**13.8 Implementing Nonseasonal
ARIMA Models 486**
Index of Consumer Sentiment 487
Seasonally Adjusted U.S. Money
Supply 490
The FRB Index of Industrial
Production 492

**13.9 Identifying Seasonal ARIMA
Models 494**

**13.10 Implementing Seasonal ARIMA
Models 496**
Preliminary Data Analysis 496
Model Summary 499
Some Forecast Test Results 500

13.11 ARIMA Modeling Checklist 502

Summary 505

References 505

Problems 506

**CASE 13A Demand for Ice
Cream 508**

**CASE 13B Demand for Apartment
Rental Units 508**

CASE 13C Energy Forecasting 508

**CASE 13D U.S. Automobile Production
509**

CASE 13E   Demand for Chicken 509

CASE 13F   ARIMA Models for Ballpark
Attendance of a Sports Team
510

Chapter 14   Forecasting with ARIMA
Models 511

14.1   ARIMA Models
for Forecasting 512

14.2   Models for Forecasting
Stationary Time Series 512
Creating a Stationary Time
Series 512
White Noise and the Autoregressive
Moving Average Model 513
One-Period Ahead Forecasts 515
l-Step-Ahead Forecasts 517
The Black-Box Representation 518

14.3   Models for Nonstationary
Time Series 519
Forecast Profile for ARMA (1, *q*)
Models 520
Forecast Profile for an ARIMA
(0, 1, 1) Model 520
Forecast Profile for an ARIMA
(1, 1, 1) Model 521
Three Kinds of Trend Models 521
A Comparison of an ARIMA (0, 1, 0)
Model and a Straight-Line
Model 522

14.4   Seasonal ARIMA Models 525
A Multiplicative Seasonal ARIMA
Model 525

14.5   Forecast Probability
Limits 528
Probability Limits for an MA(2)
Model 528
Probability Limits for ARIMA
Models 530

14.6   ARIMA Forecasting Checklist
531

Summary 531

References 532

Problems 532

CASE 14A   Production of
Ice Cream 534

CASE 14B   Demand for Rental Units
534

CASE 14C   Energy Forecasting 535

CASE 14D   U.S. Automobile Production
535

CASE 14E   The Demand for Chicken 536

CASE 14F   Forecasting Ballpark
Attendance for a Sports
Team 536

Appendix 14A   Expressing ARIMA Models in
Compact Form 537

Appendix 14B   Forecast Error and Forecast
Variance for ARIMA
Models 539

Part 6
Improving Forecasting Effectiveness 541

Chapter 15   Selecting the Final Forecast
Number 543

15.1   Preparing Forecast Scenarios
544
Combining Forecasts
and Methods 545
Averaging Forecasts 545

15.2   Establishing Credibility 546
Setting Down Basic Facts —
Forecast-Data Analysis
and Review 546
Causes of Change 547
Analyzing Forecast Errors 547
Factors Affecting Future
Demand 548
Creating the Final Forecast 548
Verifying Reasonableness 549
Selecting a Final Forecast
Number 550
Role of Judgment 551

15.3   Using Forecasting Simulations
552
Choosing the Holdout Period 552
Fixed-Origin Simulations 554
Rolling-Origin Simulations 555

Analyzing Simulation Errors by
Lead Time 556

15.4   **Designing Forecasting
Simulations 557**
Single versus Multiple Fit Periods
558
Updating versus Recalibrating 559
Dealing with Short Time Series 559

15.5   **Reconciling Sales Force and
Customer Inputs 559**

15.6   **Gaining Acceptance from
Management 560**
Forecast Presentation and Approval
560
The Forecast Package 561
Forecast Presentations 562

15.7   **The Forecaster's Checklist 563**
Summary 564
References 564
Problems 565

CASE 15A   **Demand for Ice Cream 573**

CASE 15B   **Demand for Apartment Rental
Units 573**

CASE 15C   **Energy Forecasting 573**

CASE 15D   **U.S. Automobile
Production 574**

CASE 15E   **Demand for Chicken 574**

CASE 15F   **Combining Forecasts 575**

Chapter 16   **Implementing the Forecasting
Process 577**

16.1   **PEERing into the Future: A
Framework for Process
Improvement 577**
Prepare 578
Execute 579
Evaluate 580
Reconcile 586

16.2   **An Implementation
Checklist 590**
Selecting Overall Goals 590
Obtaining Adequate Resources 591
Defining Data 592
Forecast Data Management 592
Selecting Forecasting Software 593
Training 593
Coordinating Modeling Efforts 593
Documenting for Future Reference
594
Presenting Models to Management
594

16.3   **Using Virtual Forecasting Ser-
vices 594**
Economic and Demographic
Forecasts 595
Database Management 595
Training Seminars 596

16.4   **The Forecast Manager's
Checklists 596**
Forecast Standards 597
Implementation 597
Software Selection 598
Summary 600
References 601

CASE 16A   **Demand for Ice Cream 602**

CASE 16B   **Demand for Apartment Rental
Units 602**

CASE 16C   **Energy Forecasting 602**

CASE 16D   **U.S. Automobile Production
603**

CASE 16E   **Demand for Chicken 603**

Glossary 605

Index 613

Preface

Over the past couple of decades, the use of computerized and improved statistical forecasting methods has greatly enhanced the productivity and effectiveness of forecasting in business, government, and the private sectors of society. This development is in part due to the uncertain and changing nature in competitive markets, global economic expansions, financial objectives, shifting demographics, and operational environments facing a business enterprise.

The clear need for improved planning to reduce costs and enhance customer satisfaction in manufacturing companies, for instance, has increased the desire to apply better forecasting approaches to the planning and management of change in the supply chain. Fortunately for practicing forecasters, computer-based techniques have greatly simplified the way they do their work. Ready access to data sources, spreadsheet modeling, and sophisticated quantitative methods have given rise to a wide variety of data-intensive techniques that are readily applied in a relatively short time at a reasonable cost. Still, a forecast practitioner can easily be overwhelmed by a plethora of forecasting techniques that are not readily understood. Moreover, the manager or end user of the forecasting process has been offered little guidance in how to make effective and appropriate use of these powerful (often inadequately documented) techniques in real-world situations.

BUSINESS FORECASTING: A DATA-DRIVEN PROCESS

Up to recent times, business forecasting was most closely linked to economic and financial thinking. During the 1980s, however, economic forecasting suffered from a lack of credibility, media ridicule, and shortcomings in accuracy goals. Nowadays, the meaning of business forecasting has broadened considerably to include forecasting demand throughout a supply chain from supplier of raw material to consumer of finished goods. Called demand forecasting, it generally attempts to predict future customer demand for a company's goods and services. This new focus is nowadays directed more to forecasting the disaggregated elements of product demand for supplying warehouses, distributors, channels, and consumers than to economic- and financial-driven aggregates. Demand forecasting, within the context of this book, means that the firm predicts the *right* amount of the *right* product to be in the *right* place at the *right* time for the *right* price, which is one of the underpinnings of what is now known as demand forecasting and replenishment planning for the supply chain.

By presenting a unified and practical orientation to the subject, we aim to prepare a practicing forecaster or student learning about forecasting to become a productive professional. To this end, we present the most widely accepted, currently practiced quantitative methods in business forecasting. The principal unifying theme of this book is the presentation of forecasting as a process rather than a series of disconnected techniques. A further unifying theme is our constant emphasis on the role and importance of looking at data, or data analysis, as practiced by scientists and statisticians. The problems, examples, and computer exercises included herein are also consistent with our goal of preparing the reader for the immediate practice of business forecasting as a data-driven process.

WHAT IS NEW?

This book is a complete revision of our 1984 book *The Modern Forecaster* and its predecessors *The Beginning Forecaster* and *The Professional Forecaster*. Although basic principles underlying the forecasting process have not changed, a sea change has occurred to forecasting in the business environment from top-down macroeconomic forecasting to bottom-up operational forecasting. This has meant a realignment of major topics as well as the introduction of new material pertaining to bottom-up demand forecasting.

SCOPE

In this book, we have emphasized the following:

- Establishment of a process for effective forecasting. Specific methods and techniques are presented within the context of the overall process.

- Selection of the forecasting and analytical techniques most appropriate for any given problem. The methods discussed, many representing the current state of the art, are the ones that have proved to be most useful and reliable to practicing forecasters.

- Preliminary analysis of historical data before attempting to build models. Computer-generated graphical displays enable you to see in a picture what you might otherwise have to glean from a spreadsheet or stack of computer "green-sheets."

- Performance of diagnostic analysis on fit and forecast errors. To determine what the unexplained variation might tell us about the adequacy of the model, we emphasize the importance and usefulness of displaying residual patterns and diagnostics. Residuals are emphasized throughout as essential in all phases of an effective model-building effort.

- Use of robust and resistant methods to complement traditional methods. Experience with a wide variety of realistic applications has convinced us that data are rarely well behaved enough for the direct application of standard statistical assumptions. The robust/resistant methods produce results that are less subject to departures from conventional assumptions and to the distortions caused by a few outlying data values. By comparing traditional and robust results, the practitioner is in a better position to decide which are most appropriate for the problem at hand.

- Refocusing the attention of practitioners away from the mechanistic execution of computer software and toward a greater understanding of data and the processes generating data.

This book describes a number of basic, well-established, and proven forecasting methods that are applicable to a wide variety of real-world business applications. The practicing forecaster will find that the techniques explained in this book provide preliminary models for measuring improvements resulting from building increasingly complex models. Likewise, managers and end users of forecasts will find in this book a comprehensive treatment of how to evaluate basic forecasting approaches. In addition, the material offers a guide to using, interpreting, and communicating practical forecasting results.

For the practicing forecaster and researcher the book will be of interest because it extends the basic principles to meet the need of the experienced forecaster. The development of the book progresses in a natural fashion from the basic, most widely used techniques to the more sophisticated, less practiced methods. In this progression, the book includes up-to-date statistical forecasting tools in exploratory data analysis, elements of robust/resistant estimation, regression diagnostics, and state-space models.

We analyze and forecast variables by emphasizing basic forecasting techniques. We begin the analysis with traditional approaches and follow them with resistant (those that safeguard against unusual values) and robust (those that safeguard against departures from classical statistical assumptions) alternatives to the same problems. More advanced techniques, including the ARIMA (autoregressive, integrated, moving average) models based on the Box-Jenkins methodology and some dynamic regression and econometric modeling with multiple variables, are considered as well. However, some of the more esoteric methods, such as neural networks, vector autoregression, and GARCH, are not included because they appear to be more relevant in applications to finance than demand planning.

Many examples are drawn from the experience of the authors as practicing forecasters, teachers, and consultants in industry. Our experience suggests that modern business forecasting applications contain a common thread independent of the particular field or industry. That is, the characteristics of the data and the modeling steps required are vital to the understanding of any forecasting method. However, as we have stressed throughout the book, the context of the business problem must not be forgotten; it plays a vital role throughout the forecasting endeavor. Data sets from a variety of sources have also been used throughout to make certain points or illustrate a particular technique.

COVERAGE

Our experience suggests that, in practice, the failure of many forecasting efforts begins with flaws in the quality and handling of data rather than in the lack of modeling sophistication. Thus, our objective has been to place greater emphasis on data-analytic methodology (much of it intuitive and graphical) as a key to improved forecasting.

A number of forecasting methods useful to students of forecasting are not covered in great detail in this book. The omitted methods are typically used when quantitative data are scarce or nonexistent. As an example, the whole area known as technological forecasting, which requires grounding in probabilistic (in contrast to data-analytic) statistical concepts, is not treated. On the other hand, new-product forecasting, for which adequate data are rarely available, is new in this edition. Because this book deals with exploratory data analysis along with confirmatory modeling, we have emphasized techniques for which a reasonable amount of data is available or can be collected.

The focus and emphasis on formal statistical methods in most forecasting books are rooted in the days of limited computing power for empirical work. Some of the materials in these texts are mathematically elegant; others are designed to make it easy for the instructor to provide packaged lectures, problem exercises, and examination questions. Our experience in the corporate world suggests that statistical theory tends to be overemphasized at the expense of simpler methods. Although not grounded as firmly in theory, simpler methods can frequently do as well as, and at times surprisingly better than, their complex cousins. The computer has made it feasible to warehouse lots of relevant data and perform complex analytical calculations in a flash. We are now able to effectively analyze ever-larger amounts of data, much of this through graphical means and data mining techniques from data warehouses/marts, in shorter time frames than before. The availability of relevant data, simple paradigms, and the experience of individual forecasters need to be more balanced than ever before.

Economics, mathematics, and mathematical statistics have provided much of the formal underpinnings and rationale in forecasting practice before the widespread availability of desktop computing power. As a result, certain tools, such as the Durbin-Watson test, which uses statistical hypothesis testing theory, have become popular "must-have" techniques in any forecaster's arsenal, despite the fact that readily produced graphical alternatives (such as correlograms) have become much more informative. Much of hypothesis testing, in fact, is not really required for business forecasting because confidence intervals give, for all practical purposes, identical results and are closer to the business realities. In this book, hypothesis-testing theory is intentionally de-emphasized in favor of confidence interval estimation. We understand this goes along with some recent trends occurring in statistics curricula for MBA students.

SOFTWARE

Although many companies have implemented formal forecasting systems for demand planning, their usefulness in handling small problems and classroom exercises is still quite limited. A wide range of statistical forecasting packages and modules built into integrated supply chain management (SCM) and enterprise resource planning (ERP) systems can be found with the same functionality to run the most widely used forecasting techniques. However, using systems rather than packages in a course may not be cost-effective or simple enough to support. In fact, most practitioners widely use the ubiquitous Excel spreadsheet software in the Microsoft Office system to carry out much of their daily analytical work (including forecasting). Hence, we provide an easy-to-use Excel add-in to help students carry out the exercises and examples contained in this book.

It is highly desirable, both as student and practitioner, to have an operational understanding of the Excel spreadsheet software; but it is not required. With almost equal ease, we can use IBM's Lotus 1-2-3 or Correl's Quattro Pro for performing these spreadsheet functions. However, Excel is the market leader, by far, so we provide our illustrations and software compatibility only with Excel.

COURSEWARE

The material in this book can be used for turnkey courses and seminars for enhancing the technical skills of forecasters performing demand forecasting functions in corporate environments. We suggest four courses here: Structured Approach to Demand Forecasting (SADF), Demand Forecasting and Market Analysis (DFMA), Time Series and Smoothing Techniques (TSST), and Data Analysis and Modeling Demand (DAMD).

A. Structured Approach to Demand Forecasting (SADF)

Target population: This course is intended for entry-level forecasters and managers involved in market and demand forecasting. This course is also recommended for forecast users in sales, marketing, budgeting, human resources, and operational organizations who require an appreciation in the use numerical forecasting tools.

Description: SABF is designed to provide hands-on skills for dealing effectively with the principles and techniques of data analysis, graphical presentation, and interpretation of forecasting methods and results. The course focuses on those forecasting methods for products and services that have become the most widely accepted and prominently used by forecasters in industry. Topics appear in Chapters 1–3, 4 (Sections 4.1 and 4.2), 5 (Section 5.1), 7 (Sections 7.1 and 7.2), 8 (Section 8.1), 9 Sections (9.1–9.3), 15, and 16.

B. Demand Forecasting and Market Analysis (DFMA)

Target population: This course is intended for forecasters and managers with limited experience and background in quantitative analyses. This course is also recommended for forecast users in sales, marketing, budgeting, human resources, and operational organizations who require a sound foundation in the use quantitative forecasting tools.

Description: DFMA is designed to provide basic statistical and analytical skills for understanding the economic and empirical foundations on which product/service-demand and market estimation is based. This course focuses on skill-based techniques for market analysis, new product forecasting, and demand analysis for products and services through all stages of their life cycle. Topics appear in Chapters 1, 3 (Sections 3.1–3.3), 5 (Sections 5.1–5.3), 7, 9 (Sections 9.1–9.3), and 15.

C. Time Series and Smoothing Techniques (TSST)

Target population: This course is intended for experienced forecasters and forecast managers with some background and experience in quantitative analyses. This course is also recommended for analysts in sales, marketing, budgeting, human resources, and operational organizations who require a sound foundation in the use of quantitative forecasting tools for item-level forecasting.

Description: TSST is designed to provide the intermediate statistical skills for understanding the theoretical and empirical foundations on which automatic forecasting methods are based. This course focuses on skill-based techniques for historical data analysis, visualization, uncertainty analysis, modeling diagnostics, and presentation of time series occurring in disaggregate forecasting applications. Topics appear in Chapters 1, 2 (Section 2.9), 3 (Sections 3.1–3.4), 4 (Section 4.1), 5 (Sections 5.1–5.3), 6 (Sections 6.1–6.2), 8, 9 (Sections 9.1–9.3), and 10–15.

D. Data Analysis and Modeling Demand (DAMD)

Target population: This course is intended for experienced forecasters and analysts with some background and experience in quantitative analyses. This course is also recommended for managers in sales, marketing, budgeting, human resources, and operational organizations who require a sound foundation in the use of statistical modeling tools for macro-level forecasting applications.

Description: DAMD is designed to provide the enhanced statistical skills for understanding the theoretical and empirical foundations on which data analysis and statistical forecasting models are based. This course focuses on skill-based techniques for data analysis, visualization, chance distributions, regression/ econometric modeling, residual analysis, and presentation of results occurring in a broad range of aggregate forecasting applications. Topics appear in Chapters 1, 3, 4, 7, 10–12, and 15.

For example, a certification curriculum for forecast practitioners might follow these course sequences: for a beginning forecaster/forecast manager, A-B; for an intermediate forecaster, A-B-C, A-B-D; and for an advanced forecaster, A-B-C-D. Other options for using this material include:

- *Train-the-trainer services*. These enable organizations to achieve a quick, effective start-up of training to provide timely services for new hires, organizational changes, or redeployment of resources.
- *Customized training*. Courses can be adapted to provide specific modules addressing the needs of distinct audiences involved in forecasting: power users, casual users of forecasts, managerial users, and new hires.
- *Training an installed base on software upgrades*. This provides continuity of use of forecasting software systems adopted by forecasting organizations.
- *On-site training*. The same courses can be held at a corporate training facility.

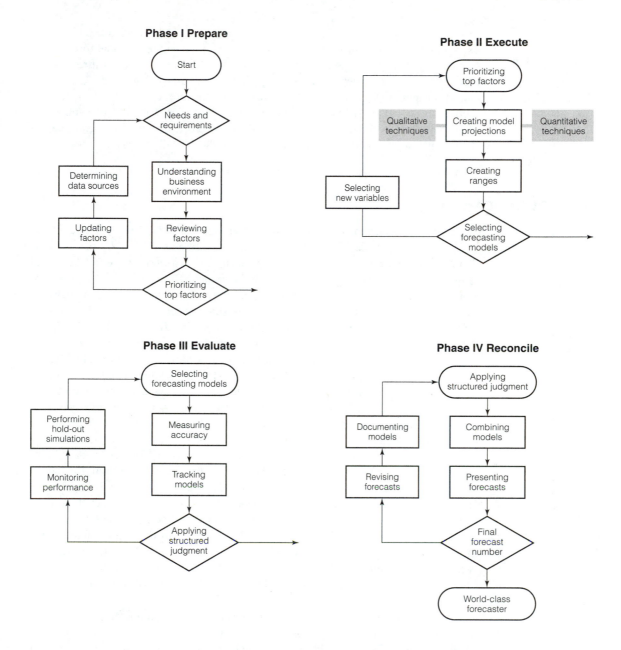

Phase I Prepare

Start → Needs and requirements → Understanding business environment → Reviewing factors → Prioritizing top factors

Determining data sources → Updating factors

Phase II Execute

Prioritizing top factors → Creating model projections → Creating ranges → Selecting forecasting models

Qualitative techniques · Quantitative techniques

Selecting new variables

Phase III Evaluate

Selecting forecasting models → Measuring accuracy → Tracking models → Applying structured judgment

Performing hold-out simulations → Monitoring performance

Phase IV Reconcile

Applying structured judgment → Combining models → Presenting forecasts → Final forecast number → World-class forecaster

Documenting models → Revising forecasts

ORGANIZATION

This book is divided into seven parts. Part 1 comprises two chapters on how to start making a forecast, introducing the forecasting process along with a classification of forecasting techniques and the use of the moving average as the prototypical projection technique. The three chapters of Part 2 deal with basic statistical concepts of forecasting and measuring forecasting accuracy. The next two chapters in Part 3 concern seasonal decomposition, top-down market-based forecasting, and econometric techniques. In Part 4, two chapters cover disaggregate item-level forecasting and exponential smoothing, which plays a key role in bottom-up operational forecasting. Forecasting with models is the subject of the five chapters of Part 5. Linear regression techniques are used primarily for short-term forecasting applications with causal factors. Chapters 15 and 16 deal with the comprehensive Box-Jenkins methodology for the ARIMA family of linear models. Part 6 treats the examination of management needs in acquiring excellence in forecasting performance, dealing with delivering the final forecast, and improving the overall forecasting process.

ACKNOWLEDGMENTS

We have very much appreciated the support and encouragement from a number of our colleagues. In particular, Len Tashman (professor emeritus, University of Vermont) contributed valuable material on forecast accuracy measurement, simulation, and implementing exponential smoothing methods. Everette S. Gardner, Jr. (University of Houston) was instrumental in providing the material for most of the spreadsheet exhibits illustrating forecasting methods within spreadsheets. We are thankful to Pete Weber (IBM, retired) for his creation of the problems in Chapter 15. In addition, we thank a number of forecasting experts who reviewed individual chapters and provided valuable feedback: Estela Bee Dagum (University of Bologna), Anne Koehler (Miami University), Keith Ord (Georgetown University), Herman Stekler (George Washington University), and Tom Yokum (San Angela State). For the case studies and some examples, we have benefited from the term projects submitted by a number of students in the Industrial Forecasting course that one of the authors taught for a decade at Columbia University. For their fine work, we thank the following student contributors: Jaideep Singh Bajaj, Tom Baker, Isabelle Boccon-Gibod, Murli Rama Chandran, Jihn-Fang Chang, Hung W. Char, Steven P. Choy, Uri Cohen, Ziad M. Dalloul, Ratnadeep R. Damle, Demetrios P. Demetriou, Richard Drobner, David Fischer, Jeffrey A. Gerwin, Jeffrey A. Golding, Erika M. Hamizar, Karim Hatoum, Kenneth Hattem, Tony C. Hom, Kiyofumi Ichikawa, Jeremy S. Kagan, Jonathan Koerner, Anand R. Krishna, Antony Kurniawan, Keng-Peng Lee, Douglas Leone, David E. Levy, Jenny Lew, Stephen Minnig, Abraham Reifer, Javier Rodriguez, Jeong Taek Ryu, Marcello Sciota, Allen Shalitsky, Philip Sheih, Ajit V. Shirodkar, Kenneth Silber, Ganapathy S. Sivan, Menelaos Tassopoulos, J. C. Tshishimbi, Takayuki Uchida, Sharon Varnelas, Yos Watganai, Kenneth Yu, Guiyu Zhou, and Houlin Zhou.

No book project of this significance could succeed without the diligent efforts of many people. We very much appreciate the contributions of Brooks/Cole-Duxbury Publisher/Editor Curt Hinrichs, Development Editor Cheryll Linthicum, Assistant Editor Ann Day, Editorial Assistant Fiona Chong, Executive Marketing Communications Manager Nathaniel Bergson-Michelson, and Technology Project Manager Earl Perry. We are grateful also for the work of a number of people outside the Thomson Higher Education umbrella: Dustine Friedman of The Book Company, Martha Ghent, Julie F. Nemer, Andrew Ogus, Walt Paczkowski, Scientific Illustrators, and Erin Taylor.

Introducing the Forecasting Process

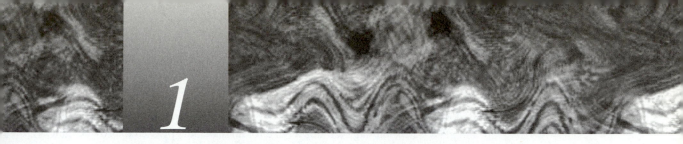

1

Forecasting as a Structured Process

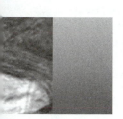

"The earth is degenerating these days. Bribery and corruption abound. Children no longer mind parents. Every man wants to write a book, and it is evident that the end of the world is approaching fast."

ATTRIBUTED TO AN ANCIENT TEXT

AS YOU BEGIN TO READ THIS BOOK, you may find it helpful to keep the following in mind:

- A grasp of economics, mathematics, and statistics, although necessary for the forecaster, will not in itself ensure successful forecasting.

- For the best results, apply such knowledge within a sound framework—a forecasting process.

- Following a sound process, which describes the sequence of activities to be followed, can reduce the chances of inadvertently overlooking a key step.

- The omission of a key step, whether deliberate or inadvertent, can jeopardize a forecaster's credibility, and credibility is a forecaster's livelihood.

This chapter describes

- What a forecasting process is
- Why it is a worthwhile approach in the business forecasting profession
- How, when, and by whom forecasting is done
- The systematic steps in the forecasting process

1.1   INSIDE THE CRYSTAL BALL

A wise person once said that he who lives by the crystal ball soon learns to eat ground glass. The same sage left this advice for all managers pressed to provide their corporate bosses with projections: Give them a number or give them a date, but never give them both. Unfortunately for those in the business of forecasting, the **demand** for products and services must provide both numbers and a time line.

Forecasting Defined

The simplest definition of forecasting is a process that has as its objective the prediction of future **events** or conditions. More precisely, forecasting attempts to predict change in the presence of uncertainty. Forecasting is all about change and chance. If future events represented only a quantifiable change from historical events, future events or chance conditions could be readily predicted through quantitative projections of historical trends into the future. Methodologies that are used to describe historical events with mathematical equations (or models) for the purpose of predicting future events are classified as quantitative **projection techniques.** However, there is much more to forecasting than projecting past trends.

 Forecasting is a process that has as its objective the prediction of future events or conditions.

Experience and intuitive reasoning quickly reveal that future events or chance conditions are not solely a function of historical trends. Even familiar abstractions such as **trend, cycle,** and **seasonality,** although extremely useful to business forecasters, cannot be completely relied on when it comes to predicting future events. In addition, in the commercial world goods and services are bought by individuals for innumerable reasons. Therefore, business forecasting must include other ingredients to complement quantitative projection techniques.

A forecast is not an end product but rather an input to the planning process. A forecast provides *advice* to planners and decision makers as to what will happen under an assumed set of circumstances. Often a forecast is a prediction of future values of one or more **variables** under "business as usual" conditions. In planning activities, this is often referred to as the status quo or the baseline. Forecasts are also required for a variety of "what if" situations and for the formulation of business plans to alter base-case projections that have proved unsatisfactory.

What Is Demand Forecasting?

Forecasting for demand management generally attempts to predict future customer demand for a firm's goods and services. For some time, this process has been closely linked to **macroeconomics.** In recent years, economic forecasting has suffered from a lack of credibility, media ridicule, and shortcomings in accuracy.

Business journals and newspaper articles frequently comment on the inability of economic forecasters to predict recessions: In 1996, *Fortune* magazine stated, "The biggest problem with economic forecasters is that they generally can't tell us what we most want to know."

More recently, demand forecasting has become much more focused on the disaggregate elements of product demand for supplying warehouses, distributors, channels, and consumers on economic and financially driven aggregates.

Demand forecasting is the process of predicting future customer demand for a firm's goods and services.

It is generally recognized that accurate forecasts are necessary and provide significant improvements in manufacturing, distribution, and the operations of retail firms. Over time, the scope of business forecasting has broadened to include forecasting more detailed, micro elements of the demand for goods to supply warehouses, distributors, brokers, channels, accounts, and consumers. Demand-driven forecasting of the right amount of the right product in the right place at the right time is one of the underpinnings of demand forecasting for the **supply chain.** In Chapter 9, we describe the demand forecasting process as an integral part of a supply chain process.

The starting point for the forecasting process is to identify all the things that are needed to put a forecast together. These are inputs; typical inputs are finding sources of **data** about the item to be forecast; obtaining information about external conditions, that is, about factors in the environment influencing a forecast; determining the needs of the user of the forecast; gathering the human and financial resources required to produce a forecast; and listing projection techniques. These are inputs not only to the forecasting process but also to the forecaster's judgment, which is applied throughout the process. The forecasting process also requires knowledge about the outputs of the process: formatting the output of the final product, presenting the forecast to the **forecast users,** and evaluating the forecast on an ongoing basis.

Once forecasting needs have been identified, a data-gathering network capable of continuously providing pertinent information about market conditions must be established. The data that have been gathered are then placed into some form of **database** or **data warehouse** for ease of analysis. Data gathering and analysis can be very time consuming and should both precede and follow the production of the forecast.

The forecast user will generally specify the format of the forecast output and consult with the forecaster about the kinds of analyses and/or variables that should be considered.

The end product of the forecasting process is clearly the forecast itself. A forecast should not be considered permanent or never changing. The dynamic nature of any market (e.g., consumer demands for goods and services) dictates that the forecasting process be revisable and repeatable at some future time. Because the value of any forecast is based on the degree to which it can provide advice in a decision-making process, the view of a market and its demands on a company within that marketplace (as expressed in terms of a forecast) must be current to be useful.

Learning from Actual Examples of Forecasting

As a unifying thread throughout this book, we frequently examine practical forecasting problems drawn from our broad experience: as practitioners in forecasting in industry, as consultants to forecasting organizations in government and commercial business settings, and as instructors to managers and students interested in learning about business forecasting. Where appropriate, we also use **time series** (data about changes through time) from real-world sources to illustrate forecasting methods and to compare or contrast results.

A time series is a set of chronologically ordered points of historical data, such as the sales revenue received by month or units shipped by week for an extended period.

The forecasting problems that we borrow from our industrial experience arise from the requirement for accurate, timely, and reliable forecasts of demand, sales revenues, product shipments, and services; throughout the book we develop forecasts of these items from actual data under realistic assumptions.

The telecommunications industry illustrates a number of considerations common to many market-based forecasting applications (Exhibit 1.1a). The market that generates telephone toll revenues may be viewed, in part, as the number of business telephones from which calls can be made. Toll messages (calls) are regarded as the *quantity of service* rendered (or product sold). The correspondence between revenues and messages is not one to one because additional factors, such as the distance between the parties, time of day, and duration of calls, cause variation in the revenue per message.

In general, the overall state of the economy, as measured by an economic indicator such as nonfarm employment, is known to influence the demand for business telephone service (Exhibit 1.1b). Different measures of economic activity, such as interest rates, industrial production, the unemployment rate, gross domestic product (GDP), volume of imports versus exports, and inflation rates, have special significance in other industries to help determine the size of some market at a designated time.

The revenue-quantity relationship, in the most general sense, is similar to that encountered in forecasting revenues from passenger-miles of transportation, mortgage commitments from housing starts, expenditures for goods and services purchased during tourism travel, tax revenues from retail sales, and revenues from

EXHIBIT 1.1 Sales-Revenue Forecasting Problem for Tele-communications

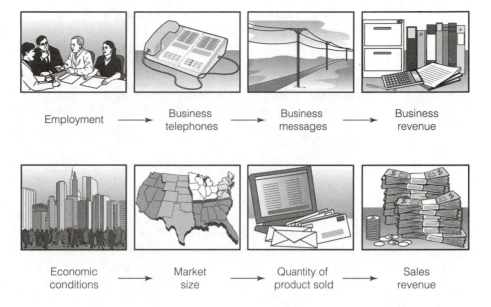

Employment ⟶ Business telephones ⟶ Business messages ⟶ Business revenue

Economic conditions ⟶ Market size ⟶ Quantity of product sold ⟶ Sales revenue

barrels of crude oil after refining. In each instance, the revenue depends on the mix of the products sold. However, for financial planning purposes, very accurate aggregated revenue forecasts can be derived without the necessity of forecasting every product or product combination and multiplying that by the sales price.

In the retail industry, department store sales may be influenced by a number of regional economic variables such as the **consumer price index,** average weekly earnings, and the unemployment rate. Retailers may also feel that the number of shopping days between Thanksgiving and Christmas has a major impact on the Christmas holiday sales volume, so their needs tend to be expressed by accurate disaggregated unit forecasts.

Market planning and forecasting at electric utilities require demand and energy models, where demand refers to the level of electricity consumption at a particular time and energy refers to the level of total use of electricity over a given period of time. Residential electricity consumption is highly influenced by weather, economic, and demographic factors. Weather influences are measured by heating degree-days and cooling degree-days. The economic factors used are price and disposable income, and the demographic influences include size and age of dwelling, age of family residents, number and type of electrical appliances, and type of space- and water-heating equipment.

The consumer goods industry provides a somewhat broader sales and operational forecasting application, in which inventory, bills of material, routings, lead times, and customer orders must be accurately forecasted in a timely way before schedules and plans can be effectively established. Sales and operational forecasting incorporates the business plan, sales plan, production plan, and marketing plans into one information source. Detailed forecasts are prepared as inputs for planning inventory, establishing customer service and determining production loads. They must be created at a *disaggregated* level in order to account for the product and customer detail required for manufacturing operations.

Sales and operational forecasting involves the marketing, sales, production, and financial plans to determine the disaggregated forecasts of product demand or services.

Exhibit 1.2 depicts a comprehensive view of a packaged-goods producer. The **manufacturer** produces a product for export; direct sales to consumers, the government, and the military; and sales to an extensive network of retailers. A grocery wholesaler or co-op retailer might distribute the product to supermarkets, grocery stores, and warehouse stores. Other distributors sell the product to chain drug stores, discount mass merchandisers, and variety stores. The entities being forecast are often product groupings segmented by geography (sales region or market zone) and customer-specific categories (warehouses, channels, or accounts).

What Are Forecasting Models?

A **forecasting model** is a *job aid* for forecasters: It creates a simplified representation of reality. The forecaster tries to include in the representation those factors that are critical and to exclude those that are not. This process of stripping away the nonessential and concentrating on the essential is the essence of forecast modeling.

A forecasting model is a simplified representation of reality for making projections.

Although abstract, models permit the forecaster to estimate the effects of important future events or trends. In the telecommunications industry, for example, there are thousands of reasons why subscribers want their telephones connected or disconnected or why they place calls over a network. It is beyond the scope of the forecaster to deal with all these reasons. Therefore, a forecaster attempts to distill these many influences down to a limited number of the most pertinent factors. In particular, consider a forecasting model for telephone demand in a major metropolitan area such as Detroit, Michigan; the model might look like Exhibit 1.3. This model assumes that the automobile industry creates jobs for people, who then buy homes or rent apartments and want telecommunications services. The telephone forecaster's job, for instance, is to determine the relationships among employment levels, household growth, land use, and telephone demand.

Mathematical equations are used to develop models that represent a real-world situation. For the telephone model, such an equation might take the form

EXHIBIT 1.2
Comprehensive View
of a Packaged-Goods
Producer: U.S.
Confectionary

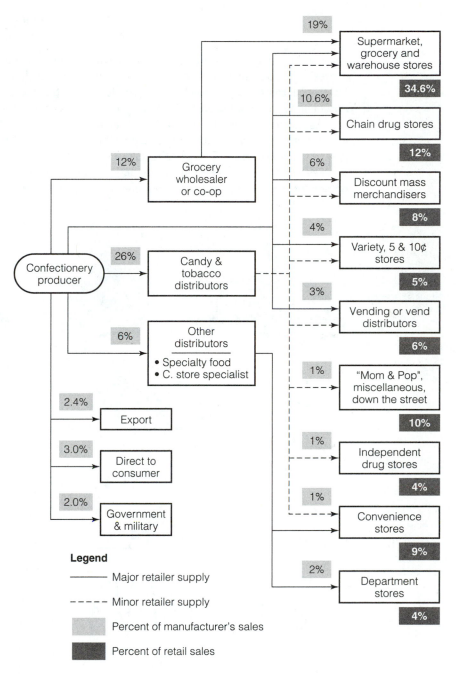

Legend

——— Major retailer supply

- - - - Minor retailer supply

Percent of manufacturer's sales

Percent of retail sales

$$\text{Telephone demand} = b_0 + b_1 \text{ (Number of employees)}$$
$$+ b_2 \text{ (Number of housing starts)}$$

where b_0, b_1, and b_2 are coefficients determined from historical data. Models such as these simplify the analysis of some problems, but, of course, do not account for all the factors that cause people to behave as they do. Notice that the model summarized in the equation does not include information on the prices of other goods and services.

As another example, consider tourism demand forecasting. International tourism has grown very rapidly over the past few decades and has become a major part of the global trade. Tourism demand measures a visitor's use of a quantity of a good or service; such measures commonly found in **tourism forecasting** include number of visitors to a destination, number of transportation passengers, and amount

EXHIBIT 1.3 A Forecasting model: The Demand for Telephone Service in Detroit

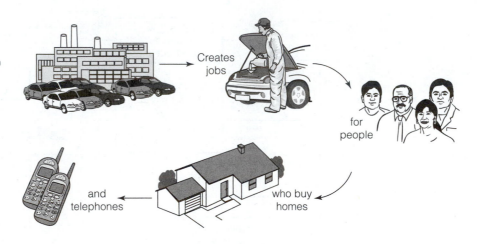

Creates jobs

for people

and telephones

who buy homes

EXHIBIT 1.4 Forecasting Monthly Tourism Demand

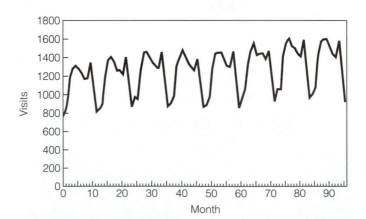

of tourism expenditures. Some factors that are known to affect tourism demand include personal disposable income, travel costs, natural and human-made disasters, and weather (Exhibit 1.4).

There is a trade-off between simplicity and completeness in every model-building effort. Multiequation causal models (Chapter 10) are commonly used to approximate the relationships between retail consumption and its drivers: price, advertising spending, coupons, competitive influences, and seasonality (Exhibit 1.5). On the premise that there is a strong relationship between consumer purchases and factory shipments, a related causal model around factory shipments would include among its drivers retail consumption, merchandising, trade allowances, and promotional lift variables.

Modeling and projection techniques are tangible and structured, just like the forecasting process. Projection techniques can produce quantifiable and reproducible results. As the analytical engine of a forecasting model, these techniques provide the basis for understanding **forecast error** impacts. Models perform similar tasks regardless of the data they use; although some inputs to the forecasting process depend on the nature of the given situation, projection techniques do not. For this reason, the forecaster must exercise sound judgment in selecting and using the projection techniques for any given forecast. Through a systematic process of elimination, the forecaster can identify those projection techniques that will provide the greatest assistance in the development of the forecast output.

A forecasting technique is a systematic procedure for producing and analyzing forecasts.

In this book, we analyze and forecast variables by emphasizing basic forecasting techniques. We begin the analysis with traditional approaches and follow them with **resistant** (those that safeguard against unusual values) and **robust** (those that safeguard

EXHIBIT 1.5 Forecasting Weekly Shipments of Canned Beverage Product as a Function of Retail Consumption

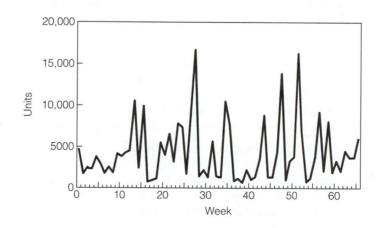

against departures from classical modeling assumptions) alternatives to the same problems. More advanced techniques, including the autoregressive, integrated, **moving average** models (ARIMA) based on the Box-Jenkins methodology, dynamic **regression,** and some econometric modeling with multiple variables and equations, will be considered as well.

1.2 Is Forecasting Worthwhile?

The process of forecasting is not an exact science; it is more like an art form. As with any worthwhile art form, the forecasting process is definitive and systematic and is supported by a set of special tools and techniques that are dependent on human judgment and intuition.

Role of the Forecaster

The forecaster is an advisor. The completed forecast must meet the requirements of the end user in terms of timeliness, format, methodology, and presentation. In the forecaster–end user relationship, the end user is knowledgeable about the environment surrounding the problem and variables that should be considered. The forecaster is knowledgeable about the forecasting process and specific forecasting methods most appropriate for the problem. In larger firms, the volume and complexity of required forecasts are usually sufficient to support a full-time, well-trained forecasting staff.

 The forecaster is an advisor, not just a producer of numbers.

Who Are the End Users of Forecasts?

The diversity of business activities creates work for many kinds of planners, or end users of forecasts, each with a special set of problems. The problems may be viewed in terms of a business's function and the time horizon for that function.

- *Executive managers* are concerned with current performance but even more concerned with future direction—strategic planning. In which markets should the business operate over the next 5–10 years? An executive manager must identify and analyze key trends and forces that may affect the formulation and execution of strategies, including economic trends, technological developments, political climates, market conditions, and assessment of potential competitors.

- *Financial managers* are concerned with financial planning, for which they need short-term (1–3 months), medium-term (up to 24 months), and long-term (more than 2 years) forecasts. For example, cash flow projections are needed to negotiate lines of credit in the short term and estimates for capital investment are needed for planning in the long term.

■ *Sales and marketing managers* are concerned with short- and long-term forecasts of demand for products and services. Forecasting methods suited to products and services have existed for some time. In forecasting a **new product,** these methods are applicable if analogous products exist or if careful market trials can be conducted. The demand for the product can then be related to the economic or demographic profiles of the people in the market areas. These relationships can then be used to predict the product's acceptance and profitability in other areas having their own economic and demographic characteristics.

■ *Planners of competitive strategies* use forecasting techniques to forecast the total market—for example, total gasoline consumption, passenger-miles of traffic between cities, automobile purchases by size (sedan, compact, subcompact), or computer storage requirements. Given the total market, each firm within it will then estimate its **market share** on the basis of product differences, price, advertising, quality of service, market coverage (including the size of the sales force), geography, and other factors specific to the market for the product or service. In many cases, market share is also estimated by using quantitative modeling approaches.

■ *Production and inventory managers* are generally concerned with very short-term forecasts (days or weeks). Production managers use forecasts to plan raw material and capacity requirements and schedule resources for manufacturing. In inventory management, exponential smoothing models find extensive application. (This important technique is like a weighted **moving average,** in which the most current data are given the greatest weight.) For extremely complex inventory systems, these models can produce many forecasts, which are closely monitored for unusual deviations between estimated and actual inventories. Sometimes deviations can be interpreted as event-driven and can be modeled to alter demand projections. Large deviations are flagged as exceptions for future scrutiny and reevaluation of the forecast-generating model.

1.3 CREATING A STRUCTURED FORECASTING PROCESS

Suppose that you are in charge of making the forecast of the demand for a product or service for your company for the next few weeks, months, or even years. How do you begin to plan your work? The specific operations that must be performed in these and other stages of the forecasting process are diagrammed in a **flowchart** inside the front cover of this book. Each chapter highlights the flowchart operations relevant to that chapter while emphasizing the iterative nature of the forecasting process.

Stages in the Forecasting Process

The four key stages by which forecasting is done are:

Preparation. Of primary importance when we prepare a forecast is that better forecasts result when the proper process has been meticulously followed. At this stage, we try to identify and understand the context in which the forecast is to be developed.

Execution. Once the forecasting context has been established, we can turn to the execution stage. A systematic execution of a forecasting methodology leads to a better understanding of the factors that influence the demand for a product or service. The forecaster who has a good handle on demographic, economic, political, land-use, competitive, and pricing considerations will develop an expertise in making the best possible forecasts of the demand for a company's products and services.

Evaluation. The forecasting cycle is typically an iterative process. Once the forecasting models are built, we still need to turn our attention to the evaluation stage. How well have the models performed in the past? The process of forecasting focuses attention on evaluating forecasts and using the right methodology for a given forecast. For example, do not use short-term methods for long-term forecasts.

Reconciliation. During the forecasting cycle, we could be making changes to the models, projections, and assumptions behind our forecasts. But in the end we need to come up with a final forecast, essentially a number or set of numbers on which the company can make its future plans. Instead of focusing first on the numbers they hope will result from a forecast, forecast managers and users must reconcile their planning approaches so that the most likely methodology will produce accurate forecasts. Selecting the right forecasting methodology is the focus of the next chapter.

Together these stages make up the PEER methodology.

The first two steps of the Preparation stage involve defining the **parameters** that will govern the forecast and making first choices among alternative projection techniques (see flowchart). These considerations help the forecaster answer the question: Can cost-effective and timely forecasts be provided to assist planners or managers in making their decisions?

Defining User Needs, Forecastable Items, and the Forecaster's Resources

First, a forecaster identifies the forecast users and their information needs. For example, toll revenue forecasts are needed by a telecom business to determine the expected net income and return on investment for a base case. You want to be certain too that you have an understanding of which products or services should be measured in your forecast.

Next, a forecaster's own practical needs must be recognized; if they are overlooked, the quality of the forecast will be diminished. So you should consider your time, administrative support needed, expenses for computerized forecast production, and transportation for field visits.

Identifying Factors Likely to Affect Changes in Demand

A forecaster also needs information about the business environment in which a company operates: Which factors have affected the demand for a product or service in the past and are likely to affect the demand in the future? For example, in the consumer goods industry, the demand for a product is forecast along with a measure of the effect a change in the price of the product will have on its demand. Or the forecaster may need to consider demographic, economic, and market factors; factors such as income, market potential, fashion, and consumer habits are usually an integral part of a formal demand theory.

- *Income* measures a consumer's ability to pay for a company's goods or services. The price of a company's goods or services and the prices of its competitors are certainly important.

- The *market potential* represents the total market for products or services being forecast. This might be the number of households or business establishments.

- *Fashion* and *consumer habit* are crucial because innovation and change create new products and services, thus causing people's tastes and habits to change. These changes must be monitored. For example, the introduction of air transportation caused people to change travel habits; the resulting impact on the railroad industry was tremendous. Also, the introduction of computers has impacted people's work habits.

A beginning forecaster can develop a simple demand theory without having to build complex models. For example, you may be required to project the sales of a product or service per household. A total **sales forecast** can be obtained by multiplying the forecast of this ratio by an independent forecast of the number of households. In this way, an important relationship can be modeled that uses relatively simple forecasting methods. This gives a first approximation, which can provide valuable and timely information to decision makers.

A beginning forecaster can develop a simple demand theory without having to build complex models.

In addition to these demand factors, supply considerations should also be taken into account. In forecasting public services (such as the residential telephone or power utility in some areas), it is important to recognize that a corporate charter requires a company to serve customer demand. Its management does not have the option of meeting only a part of the demand. In competitive industries, where this is not so, the forecaster and the forecast user must evaluate the interaction of demand and supply before arriving at the final forecast.

A Car-Buying Analogy

At first glance, the first two steps of the forecasting process as depicted in the flow-chart may appear somewhat abstract, so consider as an analogy the process of buying a new car.

The first step in defining a car buyer's problem (definition of the item) is seemingly straightforward—a new car. It may later turn out that other alternatives (a used car or public transportation) provide more appropriate solutions to the real problem; the potential car buyer does not overlook this possibility, but wants to do some preliminary checking to see if the assumption that a car is needed will hold up.

Therefore, the potential buyer defines the car's users and their requirements. In this example, the users are family members and their needs will be determined as part of the forecasting process. The needs primarily relate to the use to which the car will be put: The car might be needed for commuting, for family chores, for vacation trips, or for teenager transportation. The intended use will strongly influence the selection of the type of car to be purchased.

Potential constraints also need to be considered; these include family size and the family's financial limitations (including money for a down payment, the availability of financing, insurance costs, and maintenance and operating costs). To help in a basic understanding of these, the potential buyer might read publications relating to new car quality, consult books, search the Internet, talk to friends about their experiences, and go to several dealers to discuss prices and terms of sale. This leads to a listing of alternative solutions.

The problem-definition step concludes with a determination of the costs versus benefits of the alternative solutions. What this modest forecast teaches the potential car buyer is to look for solutions in which the benefits exceed the costs. But has the car buyer been sufficiently accurate in measuring the costs and benefits?

 There are numerous alternative ways of generating any forecast.

Analysis of Data Sources

Because all forecasting methods require data, the forecaster analyzes the availability of data from both external (outside a business or industry) and internal (within the company or its industry) sources. For example, one potential source of **internal data** is a corporate **data warehouse,** which normally contains a rich history of revenues, expenses, capital expenditures, product sales, shipments, prices, and marketing programs.

The availability of **external data** is improving rapidly. Most of the required demographic data (age, race, sex, households, and so forth), forecasts of **economic indicators**, and related variables can be readily obtained from computerized data sources and from industry and government publications. A partial listing of data sources and services for forecasting is given in DeLurgio (1998, App. A) and Makridakis and Wheelwright (1987, Chap. 29). Another overall reference for data sources at the national and international levels is the *Statistical Yearbook*, published annually by the United Nations.

With the explosion of Internet websites, potential sources of valuable data are becoming limitless. Here are just a few of the many **Uniform Resource Locator (URL)** addresses that you might find useful for business forecasting. Some of these sites are also linked to data sources provided by universities, governments, and other organizations.

International Institute of Forecasters	**www.ms.ic.ac.uk/iif/index.htm**
Centre for Forecasting, Lancaster University	**www.lancs.ac.uk/users/mansch/ manageme/research/forecast.htm**
Time Series Data Library	**www.personal.buseco.monash.edu.au/ ~hyndman/forecasting/**
Forecasting principles resource—Scott Armstrong	**www-marketing.wharton. upenn.edu/ forecast/data.html**
Notre Dame—Business Forecasting Course	**www.nd.edu/~keating/forecastinghome/**
The Federal Statistics Web page—links to many federal agencies in the United States	**www.fedstats.gov**

Judging the Quality of Data

The analysis of data for forecasting purposes requires a careful consideration of the quality of data sources. There are several criteria that can be applied to data to determine their appropriateness for modeling.

- **Accuracy.** Proper care must be taken that the required data are collected from a reliable source with proper attention given to accuracy. **Survey data** exemplify the need to ensure accurate data. Survey data are collected by government agencies and research firms from questionnaires and interviews to determine the future plans of consumers and businesses. The **consumer confidence index** in Exhibit 1.6 is the result of a survey made by the **Conference Board.** These data have certain limitations because they reflect only the respondents' anticipation (what they expect others to do) or expectations (what they themselves plan to do), not firm commitments. Nevertheless, such information may be regarded as a valuable aid to forecasting demand directly or as an indication of the state of consumer confidence concerning the economic outlook.

EXHIBIT 1.6 Time Plot of the Consumer Confidence Index

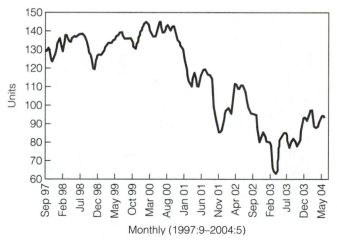

(*Source*: The Conference Board)

EXHIBIT 1.7 Time Plot of the Yearly INDPRO from the Federal Reserve Board (IP, industrial production)

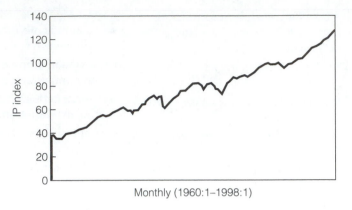

Monthly (1960:1–1998:1)

(*Source*: Board of Governors of the Federal Reserve System)

- ■ **Conformity.** The data must adequately represent the phenomenon for which it is being used. If the data purport to represent economic activity, the data should show upswings and downswings in accordance with past historical **business cycle** fluctuations. Data that are too smooth or too erratic may not adequately reflect the patterns desired for modeling. The Federal Reserve Board **Index of Industrial Production** (INDPRO) (Exhibit 1.7) is an example of a cyclical indicator of the economy. It is evident that the data are consistent with historical expansions and contractions (see Chapter 7). The INDPRO index measures changes in the physical volume or quantity of the output of manufacturers, mineral suppliers, and electric and gas utilities. The index does not cover production on farms, in the construction industry, in transportation, or in various trade and service industries. Since the **Federal Reserve Board** first introduced the INDPRO index in 1920, it has been revised from time to time to take into account the growing

EXHIBIT 1.8 Time Plot of Monthly New Housing Units Started in the United States (seasonally adjusted annual rates)

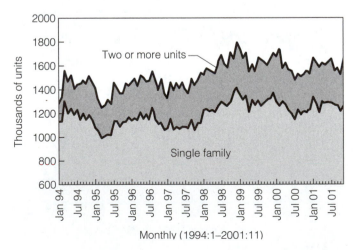

Monthly (1994:1–2001:11)

(*Source*: www.census.gov)

EXHIBIT 1.9 Time Plot of Monthly Changes in Telephone Access Lines

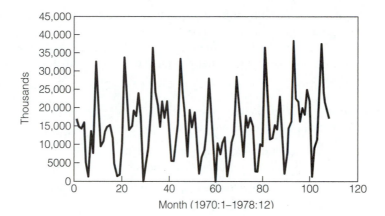

complexity of the economy, the availability of more data, improvements in statistical processing techniques, and refinements in methods of analysis.

- **Timeliness.** It takes time to collect data. Data collected, summarized, and published on a timely basis are of greatest value to the forecaster. Often preliminary data are available first, so that the time delay before the data are declared official may become a significant factor. Demographic data may fall into this category for many users. The monthly housing starts data shown in Exhibit 1.8 are demographic data reported by contractors and builders for use by government and private industry. Such external data are, of course, subject to adjustment because of data-collection delays and reporting inaccuracies.

- **Consistency.** Data must be consistent throughout the period of their use. When definitions change, adjustments need to be made in order to retain the logical consistency in the historical patterns. The monthly change in telephone access lines (Exhibit 1.9) is an example of internal data that is available to a company forecaster of telecommunications demand. It shows a consistent pattern. If the data pattern shows abrupt level changes or unusual variation, then a forecaster should check into how the data are defined. Definitions may vary because of changes in the structure of an organization, accounting procedures, or product and service definitions. As part of the forecasting process, throughout this book various kinds of internal data are related to economic/demographic indicators, survey data, and other external data in a number of examples and spreadsheet exhibits.

A model or technique based on historical data will be no better than the quality of its source.

Evaluating Forecasting Methods

The forecaster has now reached the end of the preparation phase in the forecasting process. After executing the forecasting methods (Chapters 3, 4, and 6–14), the forecaster evaluates each method. This forecast model evaluation step is called *diagnostic checking* and often results in modifications of the initial models until acceptable models are obtained.

A variety of test statistics and graphical analyses are reviewed to decide whether the model is acceptable. After passing these tests, predictions are generated from the models.

Reconciling Final Forecasts

After completing the evaluation of forecasting methods:

- Informed judgment, important throughout the process, is used to select the forecast values from among the possible candidates.

- Estimates are made about the reliability of the forecast and a presentation is made to gain the acceptance of the forecast.

- The forecasts are monitored to ensure their continuing relevance and forecast changes are proposed when necessary.

1.4 ESTABLISHING AN EFFECTIVE DEMAND FORECASTING STRATEGY

Throughout the book, we discuss many methods, tools, and techniques used in business forecasting. However, you should not lose sight of the forest for the trees—successful approaches need to be assimilated within an overall process. We believe there are ten maxims that must be kept in mind to produce a successful forecasting process; we have incorporated these into the PEER process.

Prepare

1. **Avoid surprise.** Involve management. Management tends to avoid dealing with forecasting issues because of an aversion to change and uncertainty. Forecasting is all about change and chance. In dealing with uncertainty in the external environment, in the instability of organizational structure, or in changing user needs, forecasters must remain in constant dialog with their clients.

2. **Understand the big picture.** Understand that forecasts can be wrong. Forecasting tends to become a high priority with management only during times of crises and unexpected changes in the business environment. Dealing with changing market realities and uncertainties and confronted by competitive pressures, management should view the future as subject to a broad range of dissimilar influences. Forecasting must be accepted as an evolutionary process responsive to and molded by change. It does not deal with static, status quo situations.

3. **Practice modeling.** Incorporate intuition. Models help management learn quickly from the past without relying exclusively on trial-and-error approaches. Forecasters must learn how to better incorporate their intuition into the formal, **quantitative forecasting** tools. Usually, we tend to interact with mathematical and statistical models through software to build our forecasting models. As

expert systems become more commonplace, the need for intuitive insight may link judgment and quantitative modeling more closely as a unified process.

Execute

4. **Run the numbers.** Don't bet on them! When asked to predict the thickness of an ordinary sheet of paper when folded 40 times, most of us will miss the answer by orders of magnitude. Most will offer an answer between 3 inches and 10 feet. The actual answer is on the order of the distance between Earth and the moon! In business applications, arguing about expectations can be counterproductive. For example, 30 years ago *Time Magazine* publisher Henry Luce was quoted as saying, "By 1980 all power (electric, atomic, solar) is likely to be costless."

5. **Remember that all models are wrong.** Validate quantitative methods empirically. **Models** allow for effective tracking of **forecast performance** against expectations. Some of the methods are elegant; others can be misleading. Empirical evidence suggests that for industrial applications the theory is much overrated and simpler methods can do as well. With modern computing power, we can effectively analyze large amounts of data, much of this through efficient graphical means in shorter time frames than before. The computer has made it feasible to warehouse lots of relevant data and perform complex analytical calculations in a flash. The availability of relevant data, simple paradigms, and the experience of individual forecasters may be more balanced than ever before. As the renowned statistician George P. Box once noted, "All models are wrong, but some are useful."

Evaluate

6. **Aim for accuracy.** Pursue an attainable goal. It is *not* certainty that forecasters want; it is an understanding of alternatives and possibilities. Forecasts should simply be recognized as having ambiguities and uncertainties. The key is to identify a model with the most appropriate **forecast profile** over the **forecast horizon.**

7. **Training.** Establish a training program. Somehow, institutionalized expertise has a way of falling through the cracks in a rapidly changing environment because new forecasters have not maintained continuity with the best practices of experienced forecasters long on the job. Likewise, what a forecaster needs to know today may become obsolete in the future. Companies need to maintain a training program for forecasters. Just as the management process is subject to change, so is forecasting.

Reconcile

8. **Cooperate.** Achieve consensus on assumptions and rationale. Forecasts are based on assumptions about the economy, demographics, and industry and market forces. Recognize differences among common assumptions, not just the numbers! Forecasters must strive toward a consensus on assumptions and rationale rather than a confrontation over what the right number is.

9. **Collaborate.** Enhance forecasting knowledge in management, the salesforce, and customers. Forecast users and providers can absorb and use tools that work for them. They will apply forecasting techniques, no matter how sophisticated, in their daily tasks if they provide them with tangible benefits. With more and better data, the value of enhanced forecasting knowledge and methods will soon become evident to forecast users and providers, giving them access to better forecasting information.

10. **Communicate.** Sell a forecast as advice. The forecast should be useful and provide information that is relevant to the decision maker. Advice that is useful will get management approval, not wow the administration with technical know-how, sophisticated modeling output, or excessive wit.

SUMMARY

This chapter has emphasized the need to think of forecasting as a process. Forecasting is a **structured process** that produces a specific output, namely advice about the future. The purpose of the forecasting process is to identify and evaluate systematically all factors, which are most likely to affect the course of future events and to produce a realistic view of the future. Because the future is not completely predictable, the systematic structure of the forecasting process establishes the foundation on which the most important ingredient (human judgment and intuition) is based. At first glance, such a process may seem inefficient and interminable; but in practice you will discover that a reasonable course will often become apparent, especially with experience.

Problem definition is probably the most critical phase of any forecasting project. It is necessary in this stage to define what is to be done and to establish the criteria for successful completion of the project or forecast. The four PEER stages of the forecasting process are independent of the item to be forecast and the input parameters. It is essential to agree on the required outputs, time, and money to be devoted to solving a problem, the resources that will be made available, the time when an answer is required, and, in view of these, the level of accuracy that may be achievable. Data analysis, forecasting, and model-building should only begin after these kinds of agreements have been reached. If the prospects for reasonably accurate forecasts are good, the forecaster proceeds to the next step of the process.

 A properly trained forecaster is one who does the right things in the correct sequence.

REFERENCES

DeLurgio, S. (1998). *Forecasting Principles and Applications*. New York: Irwin/McGraw-Hill.

Makridakis, S., and S C. Wheelwright, eds. (1987). *The Handbook of Forecasting, A Manager's Guide*. 2nd ed. New York: John Wiley & Sons.

PROBLEMS

1.1   Referring to Exhibits 1.1 and 1.2, suggest some factors (drivers of demand) for the following products/services within their respective market or industry sectors. Provide a rationale for your choices.

  a. Bottled beverage for a soft drink distributor

  b. Infant formula for health-product manufacturer

  c. Disk drives for a computer hardware distributor

  d. CD players for an electronics manufacturer

  e. Car wax for a consumer products manufacturer

  f. Farm machinery for a durable goods manufacturer

  g. Computer chips for for a semiconductor component manufacturer

  h. Costume jewelry for a fashion/cosmetics distributor

  i. Frozen foods for a retail chain

  j. Hospital admissions for a health-care management firm

  k. Admissions for a college or university system

  l. Service calls for a cable company

1.2   Referring to Exhibit 1.3, construct a descriptive, graphical model for the products within a region or customer location for the appropriate industry sector in Problem 1.1.

1.3   Producers of consumer products, such as automobiles and beer, could make forecasts for their industries with aggregate projections for the quarterly employment and unemployment time series. Because there is variation in consumer-purchasing patterns for automobiles and beer based on age, race, and sex differentials, how would you attempt to make predictions that are more accurate for these companies?

1.4   As the economy goes into a recession, consumers' ability to buy goods and services declines. As measured by real personal income (constant dollars), this indicator is the amount of income a person has to spend before deducting personal income tax payments. During the 1980s, much of the growth in personal income came from transfer payments (nonwage payments by government to individuals), which include unemployment compensation, Social Security benefits, welfare payments, and other income redistribution schemes. Contrast the effects that a large increase in personal income arising from augmented transfer payments would have on the beer and automobile sales.

1.5   Recognizing the need to differentiate between domestic and import sales of automobiles, how might a business forecaster go about selecting the appropriate variables to analyze?

1.6   Beer producers manufacture premium and low-priced beers as well as a variety of differentiated products to broaden their market. Given its non-durable nature and its relative cheapness, suggest what primary factors would drive the level of beer production.

1.7   Both automobiles and beer are subject to intense advertising. Discuss the relative impact on sales volatility and market share in each industry over the course of a business cycle.

1.8   The automobiles industry is a direct user of steel for car bodies. To assess the future course of output for the steel industry, what type of factors would a business forecaster need to consider?

1.9   For the following steel-consuming industries, suggest some economic time series a business analyst would need to consider for forecasting purposes: (a) railroads, (b) oil and gas construction, (c) automotive, (d) mining, (e) shipbuilding, (f) machinery, (g) appliances, and (h) furniture.

1.10 Financial forecasters consider forecasts of firms' future corporate earnings to be important in valuing their common shares. Unfortunately, earnings changes have been shown to be largely unsystematic or simply a matter of chance. In predicting quarterly and annual earnings for a company, what factors might you want to consider to help improve these forecasts?

1.11 Financial investors use accounting data and financial ratios for predicting corporate failures and bond ratings. What type of variables do you suggest for predicting corporate bankruptcies?

1.12 The future path of interest rates is important to (1) investment managers of pension funds, trusts, and other accounts; (2) real-estate development companies, mortgage bankers, and management companies; and (3) nonfinancial corporations in every major sector of the economy.

    a. Discuss how the forecaster must take into account both individual industry conditions, such as housing, and the specific needs of the investment decision maker.

    b. How are interest rates related to the demand for money as measured by, say, M1 (the measure most narrowly defined by the Federal Reserve Board, this consists of the most liquid forms of money, namely currency and checkable deposits).

    c. In terms of the generic forecasting problem in Exhibit 1.1, describe the elements involved in forecasting interest rates. What are the markets?

1.13 Assume that a forecaster is attempting to analyze the national industry sales of cross-country skis. Suggest several factors that are clearly related to ski sales, and discuss the rationale as to how these factors could be used to forecast ski sales.

1.14 An aircraft-engine company with multinational sales is making a forecast for jet-engine sales.

    a. Provide a set of economic, political, and technological assumptions that the company forecaster should consider in order to prepare the study.

    b. List the types of supply and demand data the forecasting group might be interested in using.

    c. Suggest a number of economic indicators that may be relevant in the study for correlation analysis.

    d. Without using formal statistical calculations, how can you most effectively display such correlations graphically?

1.15 Forecasting is an integral part of the planning process. A firm's sales forecast for the coming year is pessimistic, suggesting a 10% decline from the current year. Management reviews these data and decides that two courses of action are possible: (1) prepare for the expected decline in sales and (2) attempt to offset the expected decline in sales. Discuss the steps required for each course of action.

1.16 In the car-buying analogy, buying a new car is likened to the selection of appropriate forecasting techniques. Make a list of desirable and undesirable product features that would characterize a new automobile (positives: seat belts, driver trunk release, etc.). For each feature, assign a score (0–5, low–high) indicating how you would rate each one in a buying decision.

1.17 Many economic indicators have been successfully used to determine the current or future state of industry. For the following factors, describe which types of decision maker/forecast user would be most interested in a given factor and how important each factor will be in the ultimate decision making process.

Factor/Indicator	Decision Maker
New business formation	Capital banker
Manufacturing orders	R&D management
Dow Jones index	Pension fund manager
Housing starts	Manufacture of refrigerators
Building contracts	Marketing sales
Management-wholesaler	Building supplies

1.18 To forecast tourism demand we need to establish measures of tourism activity. For the following activities, suggest an appropriate unit of measure of demand.

Activity Measured	Measure of Demand
Number of people traveling away from home	
Groups of people traveling away from home together	
Total nights visitors spend away from home	
Distance traveled while away from home	
Total money spent purchasing goods and services related to the trip	

Source: Frechtling, D.C. (1996). *Practical Tourism Forecasting.* Oxford: Butterworth-Heinemann.

CASE 1A PRODUCTION OF ICE CREAM

An entrepreneur has designed an ice cream plant to be built in an urban region of the country. Based on the recent performance of super-premium (gourmet) ice cream, she feels that if the business can capture 7% of the super-premium market, it will yield a monthly internal rate of return of 21.5%. The entrepreneur has designed the daily production requirements, equipment, worker-power, and facility requirements based on these sales targets.

Recent recessionary trends have concerned her, and now she would like to backtrack and see what the outlook of the ice cream industry is before she begins her venture. The entrepreneur requests information on the ice cream industry that would indicate whether this venture is a wise investment.

(1) Create a structured forecasting process in the form of a flowchart for the entrepreneur of what she should consider before embarking on the forecasting project.

(2) Discuss the demand factors and consumption trends she might want to investigate, such as price, income, demographics, advertising, health and nutrition trends, and convenience foods. Are there others?

CASE 1B FORECASTING PEAK POWER DEMAND IN A LARGE METROPOLITAN AREA

For energy utilities to operate in an efficient manner, it is necessary to forecast power consumption in the areas they are servicing. There are two types of demand of interest: total consumption and peak demand. The need to forecast total consumption arises because the utility needs to know how much power to produce while keeping labor hours and the amount of power bought from other sources to a minimum. Peak demand forecasting deals with knowing what the public's maximum draw is and when it occurs. The chief engineer wants to create a forecasting course for his engineering staff and would like you to gather the basic background information.

(1) Considering the horizon of the forecast, what would be some of the key factors that could determine total consumption and peak demand?

(2) What is the market for power in a large metropolitan area, such as New York City?

CASE 1C ENERGY FORECASTING

Energy is indispensable to the functioning of every industry and to the everyday life of every individual, and, therefore, it is considered as a national security issue. The demand for energy is based on the demand for products and services that require energy input. The main areas of energy use are the residential, commercial, industrial, transportation, and public utilities sectors. The supply of energy is derived from several sources, including oil, coal, gas, hydropower, nuclear, and miscellaneous (solar, wind, geothermal, synthetic fuel, etc.). An economic research firm is preparing a White Paper to be placed on its website and has enlisted your assistance in the analysis of data sources and the quality of the data. You have learned that there are several criteria that can be applied to data to determine their appropriateness for modeling. Using these criteria, discuss the applicability of energy data for input to a 10-year forecast of the energy balance, including consumption and energy production. Industry trade organizations, such as the Independent Petroleum Association of America (IPAA) and Organization for Petroleum Exporting Countries (OPEC), are a good place to start a search for data source.

CASE 1D A PERSPECTIVE ON THE AUTOMOTIVE INDUSTRY

The automotive industry has been a major factor in the worldwide growth and economic prosperity of industrialized nations such as the United States. The motor vehicle affects the lives of almost everyone around the world. It has also been associated with dramatic improvements in production techniques and engineering skills, which in turn have made possible innovative developments in other industries. From time to time, the underpinnings of what makes an industry tick change, so the CEO of an automotive parts supplier has enlisted your help in looking for such transitions. The CEO has handed you a 1988 report on the world car industry, the last time such a study was undertaken by his company. "It is time to take another look" is his suggestion to you.

This 1988 report states that automobile sales in the United States will continue to mirror general economic conditions, especially changes in disposable personal income. One favorable U.S. demographic trend, however, is the rapid increase projected for the 30- to 54-year-old age bracket, the major car-buying years. Also influencing the demand are (1) auto prices in relation to all consumer prices; (2) the availability, cost, and average maturity of automobile credit; (3) the used-car market and its effect on trade-in allowances; (4) the style, engineering, safety, and quality enhancements that hasten the obsolescence of existing vehicles; and (5) the cost and availability of gasoline.

(1) Investigate the current state of the automotive industry and consider whether these factors need to be updated, modified, or eliminated. Create a list of factors, and using your own judgment, assign an importance to each one on a scale of 1 to 5. Then, for a list of current factors, decide whether the importance of each factor has changed in importance in a positive, negative, or neutral direction.

(2) Create a flowchart of a modeling process you may want to follow to forecast the demand for automobiles over the next decade.

CASE 1E COLLEGE TEXTBOOK PUBLISHING

The college textbook publishing industry is a particularly challenging area for forecasting. College textbooks are highly polished products requiring years of preparation and are geared for a very limited yet noncaptive market. Accurately forecasting the potential demand for each book seems crucial. As a first assignment, it might be useful for you to provide a book publisher with some assistance in forecasting the textbook industry.

(1) Using a source such as the Association of American Publishers, find out some facts about the publishing industry: How many publishers are there in the college textbook industry? How many big, medium-size, and small publishing houses are there? What are the sales revenues of textbooks in each category over the past few years?

(2) Describe the typical **lead time** for a textbook from conception to placement in bookstores.

(3) Assuming a textbook can last 15 years going through three editions, sketch the lifecycle of a college textbook as it passes through its editions. Graph relative sales versus time.

(4) Develop some key factors influencing the global demand for college textbooks.

(5) Who will use this forecast and what are their needs?

(6) Can you imagine other industries with marketing, sales, financial, and production environment similar to the college textbook publishing industry?

(7) Create a flowchart for a structured forecasting process for this industry.

CASE 1F THE TEXTILE INDUSTRY

The textile industry started in the United States during the Industrial Revolution in the late 1800s as the country began to develop manufacturing independence to meet the needs of an industrial fast-changing society. During the 1980s, the U.S. dominance in this global industry changed rapidly as Far Eastern countries began developing greater efficiencies and more cost-effective manufacturing plants to meet the needs of the international textile community.

(1) Trace the development of the textile industry in the United States over the past 100 years or so, indicating the key periods when the industry was going through major changes.

(2) Develop up to seven key factors driving the demand for textiles during each one of these phases.

(3) Explain which factors might have resulted in a quantitative analysis if data gathering and computing resources had been available.

(4) Develop a flowchart for creating a modeling program to make forecasts for a textile manufacturer in the United States (or your country) today.

2

Classifying Forecasting Techniques

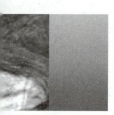

"Things should be made as simple as they are, but not simpler."
ALBERT EINSTEIN

THIS CHAPTER PROVIDES AN OVERVIEW of the most widely used forecasting techniques available for solving business-forecasting problems. One of the first things you will need to put a forecasting model together is a listing of projection techniques. We describe a way of classifying projection techniques into qualitative and quantitative methods. Whatever the technique, you need to start the selection process with

- A statement of the forecasting problem in terms of the stages of a product/service life cycle
- An economic theory stating what changes will affect demand for a product or service
- A gathering of market intelligence from field sales forecasters, market research studies, and competitive analyses
- A listing of plans for new products and special events or promotions

The most common application of a projection technique involves some form of smoothing to diminish or highlight an important aspect of the data. Familiarity with the moving average helps to motivate the basic ideas behind a smoothing technique for forecasting. In addition, a moving average produces a "typical" forecast that is widely accepted for its practical, intuitive appeal.

2.1 SELECTING A FORECASTING TECHNIQUE

Many different forecasting techniques are available, ranging from elementary smoothing methods to more complex ARIMA and econometric models. How are we to select the best one for a particular situation? Most of these methods have a sound mathematical basis to them, but what good are they without proper preparation? How can we expect to derive accurate forecasts before we have asked the right questions? Delving straight into the formulas is no substitute for a careful examination of the pros and cons of what is available.

Forecasting techniques can be classified as either qualitative or quantitative. This distinction, however, may have no bearing on the accuracy of the forecast achievable by a particular method.

Quantitative methods are characterized by a rigorous data acquisition procedure along with a mechanical application of techniques. Qualitative methods may lack rigorous data acquisition and involve techniques that are more intuitive.

Qualitative Methods

Qualitative methods provide the framework within which techniques (including forms of quantitative analyses, such as decision trees and linear programming) are brought to bear on a particular problem. The objective of a qualitative method is to bring together in a logical, unbiased, and systematic way all information and judgments that relate to the factors of interest. These methods use human judgment and rating schemes to turn qualitative information into numeric estimates.

Qualitative methods are most commonly used in forecasting something about which the amount, type, and quality of historical data are limited.

Familiar qualitative methods include the panel of consensus, Delphi method, market research (focus groups), surveys, visionary forecasts, and historical analogies. Treatments of these subjects may be found in forecasting textbooks listed at the end of the chapter. An assessment of textbooks dealing with forecasting methods can be found in Cox and Loomis (2001). A brief description of some of these qualitative techniques follows.

Panel Consensus

Perhaps the most widely practiced qualitative method, a panel consensus can be as simple as having managers sit around a conference table and decide collectively on the forecast for a product or service. Bringing executives from various business disciplines together increases the amount of relevant information available to the decision makers. A further advantage of the approach is the speed with which forecasts can be obtained, particularly in the absence of complete historical or market data. This advantage may be offset by the lack of accountability for the forecast.

Also, the typical problems of group dynamics become apparent here and are compounded by the relative rank of the executives. Unfortunately, the person with the best insight may not have sufficient weight to sway the whole group decision.

Delphi Method

This method is used to obtain the consensus of a panel of experts about a problem or issue. The Delphi method (Linstone and Turoff, 1975; Rowe and Wright, 1999) attempts to avoid the possible negative aspects associated with group dynamics (e.g., suppression of minority opinions, domination by strong individuals who may be incorrect, unwillingness to change public positions, and bandwagon effects). Therefore, instead of bringing these experts together in a debating forum, the Delphi method relies on the distribution of questionnaires to the experts with an admonishment not to discuss the problem among themselves. They may not know who the other members of the panel are, and they are not provided with individual opinions or estimates.

The initial questionnaire may be used to state the problem and to obtain preliminary estimates and reasons or assumptions behind them. The responses are then summarized and fed back to the panel. Members with widely differing estimates are asked to review the responses and, if appropriate, revise their estimates. Through several iterations it may be possible to refine the differences among experts to a usable range of opinion. However, there is no attempt to force an expert to accept the majority opinion. If an expert feels strongly about another position and can articulate it persuasively, the method provides a range of opinion, which may be desirable in conditions of high uncertainty.

Criticisms of the Delphi method include questions about panel members' true level of expertise, the clarity (or outright vagueness) of questionnaires, and the reliability of forecasts.

Historical Analog

This method uses the history of similar products as a reasonable guide in situations such as the introduction of a new product. For example, the introduction of digital television into households can be related to the earlier introduction of color television; perhaps the type of growth curve is comparable here. The depletion of natural resources may be viewed similarly. Wood burning was replaced by coal, which was replaced by oil. As oil resources are eventually depleted, and if nuclear power continues to face problems, then solar technology could become a serious energy alternative.

Historical analogs may also be useful in the shorter term when a new product replaces and improves on its predecessor. For example, each new generation of computers can be evaluated in terms of price and performance relative to the existing market. Comparing the current improvements in price and performance with previous new product introductions, given their rate of price and performance improvement, can suggest the appropriate introduction or replacement rate.

Surveys

Business surveys have been widely used throughout the world to measure economic movements such as manufacturing production in a country. The variables used in these surveys are typically qualitative in nature with only a few responses possible, such as Larger, Smaller, and Unchanged. Some examples of variables used in a manufacturing production survey include volume of production, production capacity, prices, orders, purchases, and time of deliveries (Bergstrom, 1995). The responses are then further calibrated into barometer-type series in which the difference between larger and smaller responses is summarized. The resulting series are reported by central statistical agencies for use by business economists and managers to get a pulse of the overall economy in relation to their own particular industry sector.

Visionary Technological Forecasting

This approach offers a variety of techniques that attempt to predict future technological trends. Often a set of "S" curves are constructed from data representing factors such as speed, efficiency, horsepower, and density to predict the characteristics of the next generation of technological products. For example, the capacity of a memory chip to store a given number of bits of information can be plotted over time (often using semilogarithmic scales). By extrapolating this growth curve, the forecaster in effect predicts the next breakthrough. Similarly, the constant dollar cost per chip can be plotted and extrapolated. Because there are relatively few data values for most items being forecast, significant judgment is required and assumptions must be developed and evaluated. There are physical and theoretical limits to certain factors, such as speed not exceeding the speed of light and efficiency not exceeding a certain value. Several recent books dealing with technological forecasting include Martino (1993), Makridakis (1990), Porter et al. (1991), and Twiss (1992).

Morphological Research

This method attempts to identify all possible parameters that may be part of the solution to a problem. A (multidimensional) box is created showing all possible combinations of parameters. Each possibility is then individually evaluated. By determining the number of parameters by which the proposed technology differs from present technology, we can evaluate which breakthroughs are most likely to occur.

Role-Playing

In a role-playing scenario, several panel members are assigned the role of the competitor. (One of the potential drawbacks of using the Delphi and panel consensus techniques for forecasting demand is that the competitor is typically not represented.) Several panel members are made responsible for developing information about the competitor and for creating competitor strategies and reaction plans. In a simulated forecasting session, assumptions developed by the home team are challenged by the competition. The separation of roles may allow a greater range of possibilities to be explored and more realistic assumptions to be developed than would otherwise occur.

Decision Trees

Decision trees are used to help decide on a course of action from a set of alternative actions that could be taken. Alternative actions are based on selected criteria such as maximization of expected revenues and minimization of expected costs. The method uses probability theory to assess the odds for the alternatives. In most cases, however, the probability assessments are subjective in nature and cannot be tested for validity. Decision trees are frequently used in making pricing and product-planning decisions and for developing hedging policies to protect against future currency changes in international financing arrangements.

Consider a situation in which a firm is deciding how to respond to a published request for bids for 1000 units of a nonstandard product. The firm's managers believe there is a 30% chance of winning the contract with a bid of $1000 per unit and a 70% chance of losing the contract to a competitor. A win would result in $1 million in revenue (1000 units $\times$ $1000/unit). At a price of $750 per unit, the probability of a win is expected to be 60%. A win of $750 would result in $750,000 in revenue.

If the decision is made to go with a bid of $1000/unit, the expected value is equal to the probability of a win (0.3) multiplied by the revenue ($1 million) plus the probability of loss (0.7) multiplied by the revenue ($0), or $300,000. Similarly, for the alternative $750 bid the expected value is (0.6) ($750,000) + (0.4) ($0), or $450,000. If the managers' expected probabilities are correct, a lower bid would yield more revenue.

The profit margin for the alternative bid is smaller but the probability of winning the bid is substantially increased. If a firm has little available capacity, a $1000 bid might be appropriate. If it wins the bid, the job will be very profitable. If they lose the bid, they still have plenty of business. A firm with a smaller backlog of orders on hand may be more interested in keeping the volumes up to help maintain revenues and operating efficiencies.

Quantitative Methods

If appropriate and sufficient data are available, then quantitative methods can be employed. Quantitative methods can be classified into two more categories: statistical and deterministic.

 Quantitative methods are often classified into statistical and deterministic approaches.

- **Statistical (stochastic) methods.** These methods focus entirely on patterns, pattern changes, and disturbances caused by random influences. This book extensively treats quantitative techniques, including moving averages (this chapter), summary statistics (Chapter 3), time series decomposition (Chapter 4), and **exponential smoothing** (Chapter 8). These projection techniques can be viewed as basically self-driven because forecasts are derived purely from historical influences. Linear regression modeling (Chapters 10–12) and ARIMA models (Chapters 13–14) are among the more sophisticated and most widely used forecasting methods. Multiequation econometric models, **multivariate** dynamic models, and state-space models, also fall in this category, but are beyond the scope of this book.

- **Deterministic methods.** These methods incorporate the identification and explicit determination of relationships between the factor being forecast and other influencing factors. Deterministic techniques include anticipation surveys, growth curves, **leading indicators,** and **input-output tables.** Economic indicators are discussed in Chapter 7 in connection with business cycle forecasting.

EXHIBIT 2.1   Comparison of Analysis and Forecasting Techniques

		Qualitative				
		Delphi method	Market research	Panel con-sensus	Visionary forecast	Historical analoge
Pattern of data that can be recognized and handled	Horizontal Trend Seasonal Cyclical			Not applicable		
Minimum data requirements				Not applicable		
Time horizon that is most appropriate	Short term (0–3 mo)		×	×		
	Medium term (3 mo–2 yr)	×	×	×	×	×
	Long term (2 yr or more)	×	×	×	×	×
Accuracy[a]	Predicting patterns	5	5	5	5	5
	Predicting turning points	4	6	3	2	3
Applicability[a]	Time required to obtain forecast	4	8	4	3	5
	Ease of understanding and interpreting the results	8	9	8	8	9
Computer costs[a]	Development Storage requirements Running			Not applicable		

(*Source*: Adapted from Chambers, Mullick, and Smith, 1974, pp. 63–70; Makridakis and Wheelwright, 1980, pp. 292–93)
[a]On a scale of 0 to 10, with 0 being the least applicable.
NA, not applicable; TCSI, trend, cycle, seasonal, irregular.

Input-Output Analysis

This method was developed by W. Leonfief (1986) as a method for quantifying relationships among various sectors of the economy. This forecasting approach, generally used for long-range forecasts, can be used to answer one or more of the following questions: What is happening in the economy or industry sector? What is important about different aspects of the economy or industry sectors? How should we look at the economy or industry sectors? How should we look at changes in the economy or industry sectors?

Dynamic Systems Modeling

This branch of modeling involves building evaluation models that replicate how systems operate and how decisions are made. In a business environment, the management models the flows of orders, materials, finished goods, and revenues and subsystems are developed for functional areas such as marketing/selling, pricing, installation/maintenance, research, product development, and manufacturing. The information and operational **feedback** systems are also modeled. The objective might be to evaluate alternative policies to determine the combination of policies and strategies that will result in growth in assets employed and profitability.

| | Quantitative | | | | | | | | | |
| Statistical | | | | | | | Deterministic | | | |
Summary statistics	Moving average	Exponential smoothing	ARIMA (BOX-JENKINS)	TCSI decomposition (Shiskin X-11)	Trend projections	Regression model	Econometric model	Intention-to-buy anticipation survey	Input-output model	Leading indicator
×	×	×	×	×	×	×	×	×	×	×
×	×	×	×	×	×	×	×	×	×	×
×	×	×	×	×		×	×			
×			×	×		×	×			
5 points	5–10 points	3 points	3 yr by mo	5 yr by mo	5 points	4 yr by mo	4 yr by mo	2 yr by mo	>1000	5 yr by mo
×	×	×	×	×	×	×	×	×		×
×	×	×	×	×	×	×	×	×	×	×
×					×	×	×		×	
2	2	3	2	7	4	8	2	2	2	2
NA	2	2	6	8	1	5	7	8	0	5
1	1	1	7	5	4	6	9	5	10	3
10	9	7	5	7	8	8	4	10	3	10
0	1	1	8	6	3	5	8	NA	10	4
4	1	1	7	8	6	7	9	NA	10	2
1	1	1	9	7	3	6	8	NA	10	NA

The equations that describe the system are not based on correlation studies; rather, they are descriptive in nature. For example, the number of salespeople this month equals the number last month plus new hires minus losses. Equations are then developed describing how hires and losses are determined. If an individual salesperson can sell a given amount of product, the desired sales force equals the desired total sales divided by the quota per salesperson. Hires are initiated when the actual sales force size falls below the desired level.

In a similar manner, a set of equations is developed that represents the behavior of the system or business. **Parameters** are established and the model is exercised using an evaluation language incorporated in computer software. A properly developed model should be able to simulate past behavior and provide insights into strategies that can improve the performance of the system.

Statistical Forecasting

Within statistical techniques (Exhibit 2.1), there are essentially two approaches. The first approach is best illustrated by time series decomposition. The primary assumption on which this methodology is based is that the historical data can be decomposed into several unobservable components, such as trend, seasonality, cycle, and irregularity, and that these components can then be analyzed and projected into the future on an individual basis. A self-driven forecast is then obtained by combining the projections for the components.

A **decomposition method** is an approach to forecasting that regards a time series in terms of a number of unobservable components, such as trend, cycle, seasonality, and irregularity.

An underlying assumption made in a time series approach is that the factors that caused demand in the past will persist into the future. Time series analysis then helps to identify trends in the data and the growth rates of these trends. For instance, the prime determinant of trend for many consumer products is growth in numbers of households.

Time series analysis can also help identify and explain cyclical patterns repeating in the data roughly every two, three, or more years—commonly referred to as the business cycle. A cycle is usually irregular in depth and duration and tends to correspond to changes in economic expansions and contractions.

Trend, seasonality, and cycle are only abstractions of reality. These concepts help us think about how to structure data and models for them.

In Chapter 7, we show how these concepts can be effectively used to make a turning-point analysis and forecast. Other uses of time series analysis include inventory forecasts dealing with daily or weekly shipments of units over short-term sales cycles or lead times, sales forecasts dealing with dollar-based volumes on a monthly to annual basis (this also includes seasonality, which is related to weather and human customs), and economic forecasts dealing with quarterly and annual time series.

A second group of time series techniques includes the model-based approaches associated with the Box-Jenkins and econometric modeling methodologies. Their theoretical foundations are primarily statistical and economic in nature and do not necessarily assume that the underlying historical data can be represented by separately identifiable components; rather, the data have an overall representation in which the components are generally not separately identified or modeled.

A model-based approach to forecasting represents the situation usually in terms of mathematical equations.

The econometric approach may be viewed as a cause-effect approach. Its purpose is to identify the drivers responsible for demand. The econometric models of the U.S. economy, for example, are very sophisticated and represent one extreme of econometric modeling. These models are built to depict the essential quantitative relationships that determine output, income, employment, and prices.

It is general practice in econometric modeling to remove only the seasonal influence in the data prior to modeling. The trend and cyclical movements in the data should be explicable by using economic and demographic theory. The Detroit model, discussed in Chapter 1, is an example of how an econometric system is used in the telecommunications industry. The growth in revenues might be analyzed, projected, and related to business telephones in service, a measure that is related to the level of employment. It is not necessarily assumed that the drivers that caused demand in the past will persist in the future; rather, the factors believed to cause demand are identified and forecast separately.

There is often a finer distinction made within the model-based approaches: (1) the Box-Jenkins methodology versus (2) econometric methods. Although there are similarities in their mathematical foundation, these two model-based approaches offer significant practical differences in the way relationships among variables are constructed and in interpreting model parameters.

As part of a final selection, each technique must be rated by the forecaster in terms of its general reliability and applicability to the problem at hand, relative value in terms of effectiveness as compared to other appropriate techniques, and relative performance (accuracy) level. With selection criteria established, the forecaster can proceed to produce a list of potentially useful extrapolative techniques (Exhibit 2.1). An understanding of the data and operating conditions is the forecaster's primary input now. This knowledge must, however, be supplemented by a thorough knowledge of the techniques themselves.

2.2 A LIFE-CYCLE PERSPECTIVE

Exhibit 2.2 provides broad recommendations for forecasting methods related to a product's life cycle. These techniques are described in Armstrong (1985, 2001); Chambers, Mullick, and Smith (1974); DeLurgio (1998); Mullick et al. (1987); and Murdick and Georgoff (1986).

The life cycle of a successful product is characterized by a number of stages: product development, product introduction, rapid growth, mature stage, fall-off, and abandonment.

Product development. The life cycle begins with the decisions made and actions taken before the product is introduced. The feasibility and marketability of a new product will depend on the availability of the appropriate technology, **R&D** funding, product designs, resource allocation, and business strategies. The techniques used to forecast a product's future in the early stages are included in the Delphi method, market research related to the characteristics of

EXHIBIT 2.2 The Life Cycle of a Product

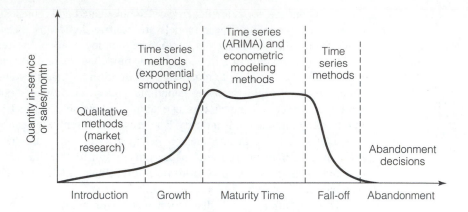

the market and consumers' willingness to pay, panel consensus, visionary forecasts, historical analogs, decision trees, and other methods that can be applied with little or no historical data about the product concept.

Product introduction. Product introduction begins the next phase of the **product life cycle.** Quite often, supply limitations, **pipeline** backups, and lack of customer awareness result in only a gradual buildup of the quantity of products sold. High marketing costs will often result in net losses. High failure rates may also place increased demands on manufacturing test and repair resources. A period of time will be required for the product to gain acceptance in the marketplace. At this stage, qualitative methods, such as Delphi and panel consensus, and market-research estimates may be preferred.

Rapid growth. The next stage in the life of a successful product is rapid growth. The product fills a need not otherwise met in the marketplace, or its price-performance characteristics are superior to its competitors, and it is adopted rapidly—faster, in fact, than would be accounted for by average growth in market or economic conditions. The growth period is characterized by increasing sales, and profits begin to rise. With historical data at hand, statistical (time series) methods become applicable. If fewer than 18 months of historical data are available, exponential smoothing techniques may be tried. For longer stretches of data, ARIMA models are preferable. If 24 months of data are available, research shows that the forecasting performance of ARIMA models become superior to exponential smoothing models (see, for example, Granger and Newbold, 1986; Makridakis and Hibon, 1979).

Mature stage. As the product enters a mature stage or steady state, competition is intense and marginal competitors will usually drop out of the market. During the maturity phase, intense competition coupled with increased price reductions and advertising costs will squeeze profits. Cost-reduction efforts focused on the reduction of material costs can be employed to increase profits. With adequate data available on sales, prices, economic factors, market size, and so forth, econometric modeling techniques can be applied. These techniques offer an explanatory capability that the time series methods lack. Whether this is an important consideration depends on the circumstances that have created the need for the forecast; in the mature phase, a product is frequently modified or its price adjusted to maintain its competitiveness.

Fall-off. As new technology or other competitive products with superior price-performance characteristics also enter the market, the demand for the mature product falls off at a rate greater than can be attributed to economic or market size considerations. Consideration is given to whether the product will become obsolete or be enhanced in the near future. Enhancing an existing product will extend the maturity phase at a much lower cost than designing a new product. Once again, time series techniques may be more responsive in projecting the rapid decline of the product's sales.

Abandonment. In the final stage of the life cycle, the product is about to be abandoned. It is no longer profitable, and its past history may not help in determining when it will be withdrawn. Financial considerations or plans to introduce a new product to customers will determine the product's fate. The pruning of losing products is a strategy employed by corporate turnaround specialists to get losing companies back on track.

2.3 MARKET RESEARCH

Market research is often conducted to determine market potential, market share, desirable or unfavorable product attributes, responses to changes in price or terms and conditions, customer preferences, and key factors that customers consider important in deciding among a variety of purchase or lease products from alternative vendors.

The American Marketing Association defines market research as the systematic gathering, recording, and analyzing of data about problems in the marketing of goods and services.

The old adage "a problem well defined is half-solved" is especially true in market research. In the early stages of a market research project, the objectives may be fairly well defined but the nature of the specific questions to be asked may be less well defined. Often small focus groups (three to ten potential respondents, randomly selected) are brought together to discuss the questions. A moderator directs the sessions. By asking specific questions and directing the discussion, the moderator determines whether the respondents understand the questions, believe other issues are important (leading to additional questions), and whether the project should continue. The moderator then makes recommendations for fieldwork: questionnaire development, testing, and surveying.

It is important to remember that the results of market research do not predict the future. Instead, market research studies may be used to identify the characteristics of existing users (income, age, product's life cycle position, nature of business, number of establishments, etc.). They may also be possible to estimate market potentials by gathering similar data about customers or potential customers (Clifton et al., 1991). An example of this approach using regression models is illustrated in Chapter 12 in which variables are selected based on previous market research studies combined with available data sources.

2.4 NEW PRODUCT INTRODUCTIONS

As competitive pressures mount, companies are driven to introduce new products at an ever-increasing pace. It is not uncommon to see a 20–40% turnover of product assortment in many companies. Promotional and new product forecasts are initially based on the analogous product history of similar products and finalized by consensus with input from marketing, sales, finance, distribution, and manufacturing. Other studies, such as intent-to-purchase market research studies may also be used to firm up the forecast. For the initial period, a borrowed model from an analogous product can be used for forecasting. When approximately six values of historical data become available, smoothing and time series methods become effective for short-term trending.

Many products have direct predecessors—products that have become outdated, gone out of style, or become unprofitable. Demand for a new product can be estimated by inheriting the historical pattern of its predecessor and using exponential smoothing procedures for forecasting. For products that have no predecessors or analogs, an estimate of demand can be correlated to a forecast of the overall size of the product line. Then, by projecting a mix ratio for the new product, we can obtain a historical profile that can be used with a smoothing model for updates and monitoring.

2.5 PROMOTIONS AND SPECIAL EVENTS

Manufacturers spend much time and effort trying to find ways to market their products. What does a manufacturer hope to gain from a promotion? Trade promotions are an important part of a manufacturer's plan to increase profitability and sales to retailers. Promotions are widely used in a number of industries to stimulate additional demand for a product or service. In the consumer goods industry, where promotional campaigns are used to maintain or gain additional market share, companies spend a significant amount of advertising money to beat out the competition. As a result, sales demand patterns for promoted products can create very wide swings in the data.

The effect of a promotion introduces several stages in addition to the normal pattern of demand. Before an announced promotional campaign, the demand for a product might fall below expected levels due to the drop in demand as consumers anticipate the advantage of a promotional price. During the promoted period, a promotional blip may result from the increased demand for the product along with a pull-ahead amount for the demand that would have been there after the promotion ended. After the close of a promotional campaign, demand might drop again below normal levels for a short period as the **forward buy** dissipates and demand returns to normal patterns. These patterns differ across promotions in type, intensity, and duration.

2.6 Sales Force Composites and Customer Collaboration

In a typical forecasting situation, forecasts from a field sales force may need to be reconciled with a centralized, administrative forecast. In this process, information flowing from the field includes specific customer developments (key accounts), pricing strategies, and market information. Feedback from the central staff typically deals with relevant economic developments, the competitive environment, corporate promotional plans, capacity constraints, and new product information. Although tactical in nature, the intuitive judgment of the field sales force plays a critical role in formulating a consensus forecast for **production planning** and in setting sales quotas.

The sales force composite is the compilation of estimates from the salespeople adjusted for expected changes in demand and presumed biases.

To gain additional cost savings and efficiencies, some companies and their customers collaborate in the forecasting and planning process by sharing information and co-managing processes. For instance, a consumer goods manufacturer can collaborate with a retailer to share supply and demand forecasts for certain product lines. The retailer develops demand forecasts for the product line in its stores and the manufacturer produces to meet that demand; this can benefit both sides of the partnership in increased sales and market share.

2.7 Neural Nets for Forecasting

The **artificial neural network** (ANN) is a means of processing complex data when a problem is not fully understood or well specified. Many forecasting problems suffer from noisy and incomplete data, which do not lend themselves to formal modeling approaches. Neural nets are inspired by the architecture of the human brain, which processes complex data using interconnected neurons and processing paths; these interconnections are simulated in a computer program. Each neuron takes many input signals and then, based on an internal weighting system, produces a single output that is typically sent as input to another neuron. The neurons are tightly integrated and organized into different layers.

A neural net, or artificial neural network, is a processing architecture derived from models of neuron interconnections of the human brain.

In a forecasting application, the input layer can receive multiple inputs (e.g., the drivers of the forecasted item) and the output layer produces the projections of the item of interest. Usually one or more hidden layers are sandwiched in between the two. This structure makes it impossible to predict or know the exact flow of the data, but allows the system to learn and generalize from experience. A neural net is useful because it can be exposed to large amounts of data and discover patterns and relationships within them (see Chatfield, 1993; Faraway and Chatfield, 1998; Zhang and Wu, 1998).

Rule-Based Forecasts

Rule-based forecasting employs dozens of empirical rules to model a time series for forecasting. These rules are distilled from published empirical research, surveys of professional forecasters, and tape-recorded sessions with forecasting experts. The result of a rule-based forecasting procedure is a combined forecasting model. A number of extrapolative procedures are fit to a time series, such as a random walk, a least-squares trend line, or an exponential smoothing model. At each forecast lead time, the methods' projections are averaged by a set of rules that determines how to give weights to the various components of the combined model.

A rule-based forecast uses empirical rules to model a time series for forecasting.

For example, a possible rule is that the weight assigned to the random walk component of the combined model is raised from a base of 20% if recent trends depart from the global trend, there are shifts detected in the level of the series, or the series is considered suspicious in that it seems to have undergone a recent change in pattern.

A test using the M1 competition data (contains 1001 monthly, quarterly, and annual series that have been used as "wind-tunnel" data for testing extrapolation methods in the 1982 M-competition; see www.marketing.wharton.upenn.edu/forecast/data.html) and a rule-based forecast resulted in significantly more accurate forecasts than obtained by **combining forecasts** with simple averaging of individual smoothing methods experts (Collopy and Armstrong, 1992; Armstrong, 2001). Incorporating business judgment into the extrapolations can further enhance rule-based procedures, something that a simple smoothing technique is unable to do directly.

2.8 EXECUTION: PROJECTING HISTORICAL PATTERNS— A PROTOTYPICAL FORECASTING APPLICATION

It may come as a surprise to many, but most forecasting practitioners use simple spreadsheets to create forecasts for the products and services in their company. Why is this so? Computing power is plentiful and forecasting packages and sophisticated techniques have been around for a long time. It turns out that simple techniques do quite well; more sophisticated techniques often do not do any better, at least not

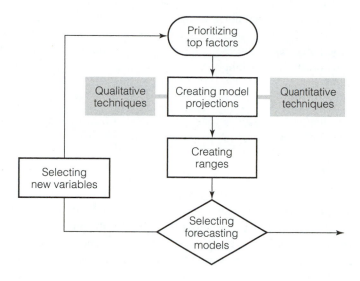

enough to justify the personal and financial investments required to get the added benefits. This may change over time, but perhaps not as quickly as some might expect. In the remainder of this chapter, we introduce the most basic techniques that have stood the test of time and are routinely applied in many forecasting situations. Moreover, this allows us to introduce the basic concept of smoothing that underlies so many of these techniques.

Many extrapolative techniques make use of some form of smoothing of historical data. Most often, smoothing refers to a procedure of taking weighted sums of data in order to smooth out very short-term irregularities. To smooth a time series, in which the effect of seasonality and the impact of any irregular variation need to be minimized or eliminated, we must perform some type of smoothing operation. To forecasters, smoothed data reveal information about secular trends (those of long duration) and economic cycles, which need to be understood and projected using a set of assumptions about the future. Smoothed data are also required in econometric modeling in order to have meaningful estimates of changes for many economic time series.

A smoothing technique is a procedure that uses weighted sums of data to create a projection or to smooth irregularities in data.

Smoothed data are easy to interpret. A plot of a smoothed employment series or of a smoothed industrial production series shows something about the state of the economy. Percentage changes, which are calculated from transformations or reexpressions of smoothed data, are readily interpreted by laypersons and corporate managers as saying something about the changes going on in the business or economy. Percentages based on unadjusted series, on the other hand, behave too erratic to be practically meaningful.

A transformation is a mathematical operation to enhance the analysis, display, and interpretation of data.

If we can interpret the differences between unadjusted and smoothed historical data, then we might be able to model this in quantitative (statistical) terms. If these differences are less important or irrelevant for the purposes of a given analysis, then we use smoothing to remove them. In macroeconomics, the seasonal component is often removed, and the notions of trend-cycle and turning point come about from analyzing a smoothed version of the resulting data (see Chapter 7).

At times, unadjusted data and smoothed data may indicate movement in opposite directions because of a technical feature in the smoothing method. As forecasters gain experience, these anomalies can be properly interpreted and their meaning adequately communicated to the forecast users. It is also worth noting that certain sophisticated time series models, known as ARIMA models (described in Chapters 14 and 15) do not require smoothed data as inputs, yet can produce excellent forecasting results.

Forecasting with Moving Averages

We have already stated that time series analysis helps us identify and explain any regularly recurring or systematic patterns in data owing to trends and seasonality. The most popular technique for revealing patterns in time series over the short term is based on averaging a sliding window of values over time, known as the simple moving average. This technique is used to smooth historical data in which the effect of seasonality and the impact of randomness can be minimized or eliminated. The result is a smoothed value that can be used as a typical, self-driven forecast for the next period or the immediate future. In doing so, we realize that too much smoothing will cause a delayed or even unnoticed change in direction in the data. Therefore, we need to use discretion and match the degree of smoothing to each particular application.

A simple moving average is a smoothing technique in which each value carries the same weight.

The most rudimentary smoother using moving averages is the (unweighted) simple moving average, in which each value of the historical data carries the same weight in the smoothing calculation. Because these moving averages are used so frequently in practice, we will give them specific names, similar to a variable name in a software program. It will also help to distinguish them from other kinds of moving averages. For n values of a set of historical data $\{Y_1, \ldots, Y_n\}$, the three-term moving average 3_MAVG_t has the formula

$$3\_MAVG_t = (Y_t + Y_{t-1} + Y_{t-2})/3$$

In general, the p-term moving average is given by

$$P\_MAVG_t = (Y_t + Y_{t-1} + Y_{t-2} + \cdots + Y_{t-(p-1)})/p$$

where Y_t denotes the actual (observed) value at time t, and p is the number of values included in the average. This moving average technique is so basic and important in forecasting that we give it the name P_MAVG_T, a P-term simple moving average forecast starting from the time period $t=T$.

As long as no trending is expected in the immediate future or no seasonality is present, this is an effective, readily understood, and practical approach for carrying out a typical forecast without the need for explanatory factors. A forecast made at a specific point in time $t=T$ for the *next* period is obtained by setting the forecast equal to the value of the moving average at time $t=T$:

$$Y_T(1) = P\_MAVG_T$$

The one-period-ahead forecast $Y_t(1)$ is a moving average based on a simple average of the current period's value Y_t and the previous $t-(p-1)$ values.

The forecast m steps ahead is the m-period-ahead forecast, given by a repetition of the one-period-ahead forecast:

$$Y_T(m) = Y_T(1)$$

The set of values $\{Y_T(1), Y_T(2), Y_T(3), Y_T(4), \ldots, Y_T(m)\}$ is called a forecast profile; it produces a constant-level forecast, in this case. Thus, the moving average technique can only produce a constant-level profile for any historical pattern. If level forecasts are expected, this may be an appropriate approach to use.

The forecast profile is the set of m-period-ahead forecasts produced by a forecasting technique.

Exhibit 2.3 shows a comparison of a three-period moving average and a 12-period moving average on a time series with 29 annual values. The forecast profile is a constant level in both situations. However, because the 12-period moving average (12_MAVG) contains more terms in the averaging calculation than the three-period

EXHIBIT 2.3 Comparison of a Three-Period and 12-Period Moving Average on an Annual Housing Starts Series	Datum Numbers	Housing Starts (1963–1991)	3_MAVG	12_MAVG
	1	1603.2		
	2	1528.8		
	3	1472.8	1534.9	
	4	1164.9	1388.8	
	5	1291.6	1309.8 ·	
	6	1507.6	1321.4	
	7	1466.8	1422.0	
	8	1433.6	1469.3	
	9	2052.2	1650.9	
	10	2356.6	1947.5	
	11	2045.3	2151.4	
	12	1337.7	1913.2	1605.1
	13	1160.4	1514.5	1568.2
	14	1537.5	1345.2	1568.9
	15	1987.1	1561.7	1611.8
	16	2020.3	1848.3	1683.1
	17	1745.1	1917.5	1720.9
	18	1292.2	1685.9	1702.9
	19	1084.2	1373.8	1671.0
	20	1062.2	1146.2	1640.1
	21	1703.0	1283.1	1611.0
	22	1749.5	1504.9	1560.4
	23	1741.8	1731.4	1535.1
	24	1805.4	1765.6	1574.1
	25	1620.5	1722.6	1612.4
	26	1488.1	1638.0	1608.3
	27	1376.1	1494.9	1557.4
	28	1192.7	1352.3	1488.4
	29	1014.5	1194.4	1427.5
	30		1194.4	1427.5
	31		1194.4	1427.5
	32		1194.4	1427.5
	33		1194.4	1427.5

(*Source*: Dielman, 1996, Ex 3.4 HSTARTS3.DAT)

moving average (3_MAVG), the former gives a smoother fitted curve through the historical data (Exhibit 2.4). The typical forecast (=1427.5) appears to be around the center of the data, whereas the forecasts for shorter-term moving average (=1194.4) start closer to the last value (#29 = 1014.5).

The three-period moving average tends to follow the cyclical pattern of the data, but will **lag** behind the peaks and troughs in business cycles. However, as forecasting models, all moving averages produce level (no trend) forecasts. Hence, the moving average is only useful for short-term forecasting in which a trend or seasonal patterns are not anticipated in the period of the forecast horizon.

Moving averages are also used to remove seasonal variation with the ratio to moving average method (Chapter 6). As we show in that **seasonal adjustment** procedure, a 12-period moving average smooths out a seasonal pattern in a *monthly*

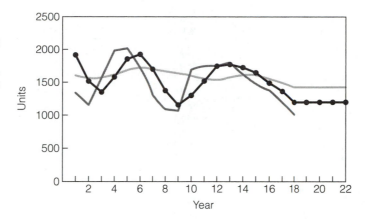

EXHIBIT 2.4 Time Plots of a Three-Period and 12-Period Moving Average on 29 Annual Values of a Housing Starts Series

time series. Likewise, a four-period moving average smooths out a seasonal pattern in *quarterly* series. In these smoothing applications, the period of the moving average plays an important role.

Seasonality is a regularly recurring or systematic yearly variation in a time series. A seasonal adjustment is an analytical procedure designed to remove and adjust for seasonality in the data.

The larger the value of p, the greater the smoothing effect. A measure of the average age for data smoothed by a moving average is given by $(p+1)/2$. Thus, monthly data, smoothed by a 12_MAVG has an average age of 5.5. The average age is defined in Chapter 8, where we show how it relates to the smoothing parameter for simple exponential smoothing.

Although moving averages are appropriate for projecting level series, it is important to understand the effect that a moving average has on a trending time series. In the construction industry, for instance, the prime determinant of trend in housing starts is mortgage rates.

A trend is the basic tendency of a measured variable to grow or decline over a long period.

Exhibit 2.5 shows the calculations for the three-period and 12-period moving average of an annual, trending mortgage rate series. Once again, the longer-term moving average (12_MAVG) leads to constant-level forecasts (=10.99) close to the overall average of the data, whereas the constant-level forecasts for the

EXHIBIT 2.5	Data Number (annual)	Mortgage Rates (1963–1991)	3_MAVG	12_MAVG
Comparison of a Three-Period and	1	5.80	—	—
12-Period Moving	2	5.75	—	—
Average on an	3	5.74	5.76	—
Annual Trending	4	6.14	5.88	—
Mortgage Rate Series	5	6.33	6.07	—
	6	6.83	6.43	—
	7	7.66	6.94	—
	8	8.27	7.59	—
	9	7.59	7.84	—
	10	7.45	7.77	—
	11	7.78	7.61	—
	12	8.71	7.98	7.00
	13	8.75	8.41	7.25
	14	8.77	8.74	7.50
	15	8.80	8.77	7.76
	16	9.33	8.97	8.02
	17	10.49	9.54	8.37
	18	12.26	10.69	8.82
	19	14.13	12.29	9.36
	20	14.49	13.63	9.88
	21	12.11	13.58	10.26
	22	11.88	12.83	10.63
	23	11.09	11.69	10.90
	24	9.74	10.90	10.99
	25	8.94	9.92	11.00
	26	8.83	9.17	11.01
	27	9.77	9.18	11.09
	28	9.68	9.43	11.12
	29	9.01	9.49	10.99
	30		9.49	10.99
	31		9.49	10.99
	32		9.49	10.99
	33		9.49	10.99

(*Source*: Dielman, 1996, Ex 3.4 HSTARTS3.DAT)

shorter-term moving average (= 9.49) starts closer to the last data value (#29 = 9.01). In Exhibit 2.6, we can see that moving averages lag behind the historical data, more so with the longer-term moving average. The extrapolations are level forecasts, which generally means that moving averages do not make effective forecasting tools for trending data.

In the treatment of ARIMA time series models in Chapters 13 and 14, the term *moving average* appears again, but in a different context with a somewhat different meaning.

Moving medians can be used instead of moving averages. The main advantage of a moving median compared to moving average smoothing is that outliers influence its results less. Thus, if there are outliers in the data, median smoothing typically produces smoother curves than moving average–based smoothing on the same moving window

EXHIBIT 2.6 Time
Plots of a Three-
Period and 12-Period
Moving Average on
an Annual Mortgage
Rate Series

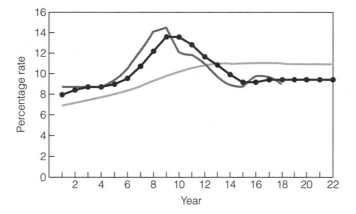

width. The main disadvantages of median smoothing are that in the absence of clear outliers it may produce more "jagged" curves than moving average and that it does not allow
for weighting (Velleman and Hoaglin, 1981). Moving medians can be used effectively to
separate or estimate a promotional impact from a baseline forecast. Moving averages
have a tendency, in this situation, to follow the promotion peaks and troughs, which
reduces the effectiveness of separating the promotion effect from the baseline series.

Moving median smoothing, compared to moving average smoothing, is less influenced by outliers.

Fit versus Forecast Errors

In practice, you need to make a careful distinction between fit errors and forecast
errors. For a one-period-ahead forecast, the forecast error is given by $Y_{T+1} - Y_T(1)$ or,
in general, $Y_{T+m} - Y_T (m)$ for the m-period-ahead forecast error. Using the data in
Exhibit 2.3, we see in Exhibit 2.7 that there should be an important distinction made
between fit errors and forecast errors; they are not the same. We need to be aware of
different time **origins** T and the horizons (lead times).

A fit error is the difference between the actual value and the fitted value, calculated
over the historical period of the time series. A forecast error is the difference between the actual value and the forecast, calculated in the forecast period.

How do we assess the adequacy of this kind of forecast? One way is to compare
the moving average forecasts with a simpler, naïve forecast, which simply uses the
current period's value for the next period's forecast: $Y_t(1) = Y_t$. This naïve forecast is

EXHIBIT 2.7 Fit
Error versus Forecast
Error for 3_MAVG of
Annual Mortgage
Rates

	Fit Errors	$Y_{T+1}-F_T(1)$	$Y_{T+2}-F_T(2)$	$Y_{T+3}-F_T(3)$
$T=4$	$6.14-5.88$	$6.33-5.88$	$6.83-5.88$	$7.66-5.88$
$T=5$	$6.33-6.07$	$6.83-6.07$	$7.66-6.07$	$8.27-6.07$
$T=6$	$6.83-6.43$	$7.66-6.43$	$8.27-6.43$	$7.59-6.43$

(Exhibit 2.5; $n=29$)

called NAÏVE_1 (or NF1) and serves as a benchmark for level forecast profiles. The naïve forecast also serves as an important theoretical starting point for forecasting price levels and interest rates in financial applications.

A naïve forecast NF1 uses the current period's actual value for the next period's forecast.

Exhibit 2.8 shows a comparison of the forecast errors found in Exhibit 2.7 with the corresponding NF1 errors. Because the data are trending in this stretch of the series, the NF1 performs better than a one-period-ahead moving average forecast. Similar calculations can be made for the other forecast horizons. In short, a moving average is not appropriate for projecting trending (nonstationary) data. We discuss the evaluation of forecast errors in more detail in Chapter 5.

Weighting with Recent History

In the previous section, we introduce the simple (unweighted) moving average, in which each value of the data carries the *same* weight in the smoothing calculation. For forecasting very short-term demand based on weekly or daily data, the most recent historical period is usually the most informative. Hence, it is not uncommon to simply average the demand from the most recent past few periods as a typical measure of the next period's demand.

There are more flexible ways to use a moving average for forecasting. In the case of retail sales, we could use a weighted moving average to place more emphasis on end-of-week sales than on beginning-of-week sales. In connection with most seasonal adjustment, adaptive filtering, and outlier correction procedures, the weighted moving average also plays an important role. In another variation, the recent values

EXHIBIT 2.8
NAÏVE_1 versus
3_MAVG Forecast
Errors for Annual
Mortgage Rates

Time origin	NAÏVE_1	One-Period-Ahead $Y_{T+1}-F_T(1)$	Two-Period-Ahead $Y_{T+2}-F_T(2)$	Three-Period-Ahead $Y_{T+3}-F_T(3)$
$T=4$	0.40	0.45	0.95	1.78
$T=5$	0.19	0.76	1.59	2.20
$T=6$	0.50	1.23	1.84	1.16

(Exhibit 2.5; $n=29$)

are weighted more heavily than past values using exponentially declining weights; this forms the basis of the exponential smoothing method discussed in Chapter 8.

Simple exponential smoothing is a weighted average procedure using exponentially declining weights.

Recall that, for n values of a time series $\{Y_1, \ldots, Y_n\}$, the p-term (unweighted) moving average is given by

$$P\_MAVG_t = (Y_t + Y_{t-1} + Y_{t-2} + \cdots + Y_{t-(p-1)})/p$$

Now, a p-term-weighted moving average has the formula

$$P\_WMAVG_t = (w_1 Y_t + w_2 Y_{t-1} + w_3 Y_{t-2} + \cdots + w_p Y_{t-(p-1)})$$

where Y_t is the value of the historical data at time t and w_p is the weight assigned to $Y_{t-(p-1)}$. The w values are known as weights and must sum to unity. Should the weights be positive and not sum to unity, however, then the weighted moving average must be divided by the sum of the weights (see Exhibit 2.9 for an example of a weighted average calculation). Because of its importance, we use the notation P_WMAVG_t for the P-term weighted moving average. Thus, the P-term simple moving average P_MAVG_t is a special case in which the w values are all equal to 1.

The forecast, made at time $t = T$ for the next period $T+1$, is known as the one-period-ahead forecast $Y_T(1)$ and, like the simple moving average, is defined by the latest smoothed value:

$$Y_T(1) = P\_WMAVG_T$$

and an m-period ahead forecast $Y_T(m)$ is assumed to be the same smoothed value m periods ahead: $Y_T(m) = Y_T(1)$. In other words, a moving average forecast is based on a weighted or unweighted moving average of the current periods value Y_t and the previous $t - (p - 1)$ values. From the formula, it is clear that the forecast profile $\{Y_T(1), Y_T(2) \ldots Y_T(m)\}$ is a level forecast made at time $t = T$ for m periods ahead. Hence, the value of the moving average method lies in situations in which forecasts are expected to be level and nontrending, at least in the very short term.

EXHIBIT 2.9
Weighted Average
Calculation

Value	Weight	(Value) × (Weight)
1.14	0.96	1.0944
0.00	1.00	0.0000
1.24	0.95	1.1780
2.08	0.86	1.7888
2.18	0.85	1.8530
Total	4.62	5.9142

Weighted average $= 5.9142/4.62 = 1.28$

Some caution should be exercised in considering moving averages as a smoothing and forecasting procedure because:

■ Isolated outlying values may cause undue distortion of the smoothed series.

■ Extreme values at the most current time will probably create unrealistic levels in the forecasts.

■ Cyclical peaks and troughs are rarely followed smoothly by the procedure.

Extrapolative techniques are used during the forecasting process to describe the historical behavior of a time series in simplistic, mathematical terms and then, using that same mathematical model, to predict the future characteristics of the data. By using more than one extrapolative technique, the forecaster can be reasonably sure of avoiding biases, which are inherent in using any single extrapolative technique.

Judgment and Modeling Expertise

When describing complex reality in terms of simplified models, such as moving averages, clearly no single model can be considered universally adaptable to any given forecasting situation. The assumptions and theories on which the extrapolative techniques are based limit their appropriateness and reliability. The forecaster should be careful to avoid using techniques for which the data characteristics do not match the assumptions of the method.

Later, when dealing with linear regression models (Chapter 11), we develop a relationship between the annual housing starts data and the annual mortgage rates. Intuitively, these two series should be related in the sense that in the construction industry mortgage rates influence housing starts. On the surface, there does not appear to be much similarity when comparing the basic time plots (Exhibits 2.4 and 2.6). However, after applying some business and modeling logic, we uncover a potentially useful, predictive relationship between the two variables.

Exhibits 2.10 and 2.11 depict a monthly series for a pharmaceutical product over a 54-month period as well as the year-over-year percentage changes in units sold (shipped) during the same period in the prior year. Exhibit 2.10 is dominated by a declining trend, whereas the annual changes are quite volatile. If the trend in the original data were predominantly linear, the annual changes would fluctuate about a constant level. Exhibit 2.11 is dominated by large fluctuations around a constant level of approximately −800. Could there be any economic, demographic, or political influences causing these patterns to change direction? Are the apparent seasonal influences of an additive or multiplicative nature? What projection techniques should be used in planning a forecast of these data?

Bear in mind first that a greater number of techniques are appropriate for the time horizon 1 year ahead than are appropriate for 2-year-ahead forecasts. As we approach forecasts 2 or more years ahead, moving average, exponential smoothing, Box-Jenkins, and time series decomposition methods become less applicable.

Also apparent is that more techniques handle trending data than handle cyclical data. If we assume a turning point will occur in the second year, the moving

EXHIBIT 2.10 Time Plot of a Monthly Units Shipped for a Pharmaceutical Product over a 54-month Period

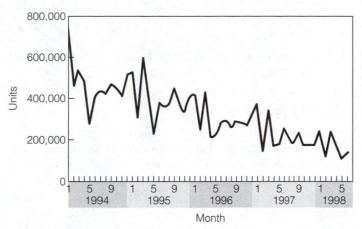

(last demand, 7/98)

average, exponential smoothing, and trend projection techniques are no longer applicable.

In terms of accuracy of a forecast for the 1-year-ahead horizon, the ARIMA, econometric, regression, and time series decomposition techniques appear to be the most promising. If there is a turning point in the second year, univariate ARIMA models should not be considered.

If we consider time constraints and the desire to present an easily understood method, the regression and time series decomposition approaches turn out to be quite valuable, followed by trend projection and exponential smoothing for 1-year-ahead forecasts. The time series decomposition and linear regression methods should also be considered for 2-year-ahead forecasts. Different conclusions might result, however, when:

- Shorter time horizons are involved.

- Data gathering and computational costs are important.

EXHIBIT 2.11 Time Plot of Year-by-Year Changes in the Data Plotted in Exhibit 2.10

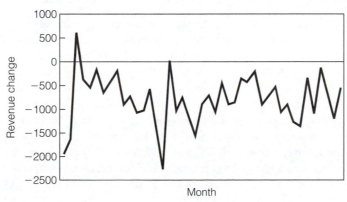

Source: (Exhibit 2.10)

- Accuracy requirements are less stringent.
- Time is not a constraint.
- The ease of understanding and explaining forecast methods is extremely important.

A Multimethod Approach

The purpose of using more than one technique is to ensure that the forecasting approach will be as flexible as possible and that the forecaster's judgment (which is so critical to the forecasting process) is not overly dependent on one particular extrapolative technique. It is not uncommon to see forecasters develop a preference for one forecasting technique over another and then use that technique almost exclusively, even in a new situation. Such a preference can become easily established because of the highly specialized nature of some of the techniques.

One of the lasting myths about forecasting is that complex models are more accurate than simple models. Some forecasters uncritically prefer the most sophisticated statistical techniques that can be found. The accuracy of an extrapolative technique, however, is not necessarily a direct function of the degree of its sophistication. In many cases, this tendency can greatly reduce the effectiveness of a forecasting model because complex models may become unbelievable when unexpected pattern changes occur in the time series. A simpler model, on the other hand, may remain relatively unaffected by the change.

We recommend that two or more extrapolative techniques be used every time to describe the historical behavior of the data and to predict future behavior. In essence, this allows us to evaluate alternative views of the future. It is also necessary to provide a risk level, in terms of forecast probability limits, associated with each alternative. A comparison can be made of the alternative views of the future, hence increasing the chances that the selected forecast level is reasonable.

2.9 HOW TO FORECAST WITH WEIGHTED AVERAGES

Surveys show that many business forecasters produce forecasts using some variation of weighted average smoothing, known as simple smoothing. Simple smoothing is the familiar name for simple exponential smoothing, which is covered in Chapter 8. There are at least two reasons for the popularity of these smoothing models. First, the computations are relatively simple, especially in a spreadsheet. Second, smoothing models are quite accurate and compare very favorably to more complex forecasting models, especially for a volatile series such as the demand for a repair part.

As with moving average models, an important assumption behind simple smoothing is that no consistent trend or pattern of growth exists. Thus, the aim in simple smoothing is to make a typical forecast—an estimate of the level or average value of the data. This estimate is used as a short-range forecast and is continuously adjusted to keep pace with changes. Simple smoothing is easy to automate, so the model is widely used in forecasting service parts for manufactured products when there are a large number of forecasts to process on a regular schedule. Simple smoothing is a standard tool in inventory systems, where numerous demand forecasts must be updated on a monthly or quarterly basis. Another common application of simple smoothing is to generate short-range projections of sales, costs, and expenses for budgets (Winters, 1960).

Simple Smoothing as a Weighted Average

To illustrate simple smoothing, consider the demand for a repair part. In Exhibit 2.12, the connected line represents historical demand, and the large dots are forecasts. The initial forecast in the graph, for the second month of Year 1, was made at the end of the first month of that year. The one-period-ahead forecast for the third month of Year 1 was made at the end of the second month, and so on. At the end of Year 2, a number of forecasts had been made for number of Year 3. Because the model was given no historical information beyond Year 2, further adjustments to the one-period-ahead forecasts cannot be made. No trend is assumed, so the forecasts are a horizontal straight line into the future.

Simple smoothing tracks the changes in demand by continuously adjusting the forecasts in a direction opposite to that of the forecast error. When a forecast is too high, the next forecast is reduced. When a forecast is too low, the next forecast is increased. If we get lucky and a forecast is perfect, the next forecast is unchanged. Exhibit 2.13 shows the spreadsheet for updating a weighted average at the end of each month. Historical demand is shown in column Data, with fitted demand in column Fcst. The WEIGHT that controls the updating process is 0.13 (see later discussion for how to choose the best weight). Forecasts are computed by calculating a weighted sum of the data:

$$\text{Weighted sum} = \text{Actual data} + (\text{Weight} \times \text{Previous weighted sum})$$

EXHIBIT 2.12
Simple Smoothing
for Demand of a
Repair Part

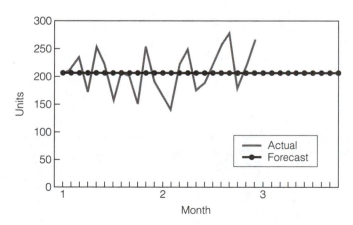

(*n*=22)

EXHIBIT 2.13 Monthly Demand of a Repair Part

SIMPLE EXPONENTIAL SMOOTHING

								WEIGHT	MSE
TITLE1:	SIMPLE EXPONENTIAL SMOOTHING			WEIGHT: 0.00					
TITLE2:	Demand for Repair Part			MSE: 1491.2507					
X-AXIS:	MONTH								1491.2507
Y-AXIS:	UNITS							0.00	1502.0870
								0.01	1515.0717
Year	Mon	Per	Data	Fcst	Error	Error^2		0.02	1527.1356
1	1	1	206	206.0	0.0	0.0		0.03	1538.3285
	1	1	206	206.0	0.0	0.0		0.04	1548.7201
	2	2	212	206.0	6.0	36.0		0.05	1558.3939
	3	3	235	206.8	28.2	796.4		0.06	1567.4421
	4	4	170	210.4	−40.4	1636.1		0.07	1575.9615
	5	5	255	205.2	49.8	2481.0		0.08	1584.0492
	6	6	222	211.7	10.3	106.8		0.09	1591.8003
	7	7	156	213.0	−57.0	3250.0		0.10	1599.3053
	8	8	206	205.6	0.4	0.2			
	9	9	200	205.7	−5.7	31.9			
	10	10	150	204.9	−54.9	3015.7			
	11	11	256	197.8	58.2	3390.0			
	12	12	192	205.3	−13.3	178.1			
2	1	13	168	203.6	−35.6	1268.1			
	2	14	140	199.0	−59.0	3478.8			
	3	15	220	191.3	28.7	822.9			
	4	16	248	195.0	53.0	2804.4			
	5	17	174	201.9	−27.9	779.9			
	6	18	188	198.3	−10.3	106.0			
	7	19	214	197.0	17.0	290.4			
	8	20	255	199.2	55.8	3116.6			
	9	21	279	206.4	72.6	5266.2			
	10	22	178	215.9	−37.9	1433.8			
	11	23	214	210.9	3.1	9.3			
	12	24	266	211.3	54.7	2987.7			
3	1	25	162	218.4	−56.4	3186.1			
	2	26	275	211.1	63.9	4082.2			
	3	27	280	219.4	60.6	3670.7			
	4	28	227	227.3	−0.3	0.1			
	5	29	234	227.3	6.7	45.5			
	6	30	288	228.1	59.9	3584.5			
	7	31	226	235.9	−9.9	98.3			
	8	32	200	234.6	−34.6	1198.8			
	9	33	279	230.1	48.9	2389.0			
	10	34	196	236.5	−40.5	1638.4			
	11	35	246	231.2	14.8	218.6			
	12	36	293	233.1	59.9	3583.5788			

(Fcst, forecast; MSE, mean squared error; Mon, month; Per, period)

Forecast errors, or actual demand minus forecasts, are shown in the column labeled Error. Again, forecasts that are too high cause a reduction in the next forecast, whereas forecasts that are too low cause an increase.

Exhibit 2.14 shows an alternative way to compute a one-step-ahead forecast for the first month of Year 3. The historical demand is listed in the column labeled Data and the individual weights assigned to each past data point are computed in column Weight. A graph of these declining weights is shown in Exhibit 2.15.

Choosing the Smoothing Weight

So far, we have taken the smoothing weight 0.13 in Exhibit 2.14 on faith. Now let us look at how this weight was selected. The objective in choosing a smoothing weight is to minimize the sum of squared (one-period-ahead) forecast errors. The reason for squaring is to emphasize the large errors. In most businesses, people can live with a series of small forecast errors, but large errors are extremely disruptive.

A graph of the sum of squared errors versus the smoothing weight is U-shaped and often flat near the minimum sum of squared errors, indicating that several smoothing weights will give the same results. In this case, any weight in the range 0.125 to 0.135 will give similar performance. Weights are usually chosen from the range 0 to 1. We can make the list of weights in the data table as detailed as we like, although there is usually not much to gain by using an increment smaller than 0.05.

Forecasting with Limited Data

Exhibit 2.13 is designed to work with any number of **observations.** We can even get a forecast from one or two observations. To illustrate, suppose that we have only one demand value available for the first month of Year 1. The forecasting formulas automatically repeat that value as the forecast for the second month of Year 1. This is the only reasonable thing to do with one historical data point. When the actual demand for the second month of Year 1 comes in, the formulas automatically update the next forecast according to the forecast error in month 1 of Year 1. When the third actual demand comes in, another adjustment is made, and so on. More data are always better in forecasting, but this spreadsheet does the job when we have no choice but to forecast from a few data points.

Evaluating Forecasting Performance

By withholding the last 12 months of demand in Year 3, we can rerun the analysis and investigate the performance of simple smoothing over the holdout period. With the 24 observations, the mean squared error (MSE) = 1502 with a smoothing weight of 0.0. The MSE is calculated by averaging the squared errors. The forecast for the holdout period is level 206 based on the simple smooth model. The forecast for the NAÏVE_1 model is the last observed value (= 266) for month 12 of Year 2. With level forecasts for periods 25–36, Exhibit 2.16 displays forecast errors for the two models. Based on the MSE calculated over the holdout period, it appears that the simple smoothing model has not done a very good job of forecasting Year 3. Had we used a smoothing parameter of 0.13, the forecasts would be 218, the MSE over the **fit period** would be larger (=1621, vs. 1502), and the MSE over the forecast period would be reduced (2206 vs. 2930) but still worse than the NAÏVE_1 model in this particular situation. In practice,

EXHIBIT 2.14
Smoothing Weight
for Monthly Demand
of a Repair Part

ANALYSIS OF WEIGHTS ON PAST DATA
TITLE1: WEIGHTS ON PAST DATA
TITLE2: SIMPLE EXPONENTIAL SMOOTHING
X-AXIS: PERIOD
Y-AXIS: WEIGHT

SMOOTHING WEIGHT = 0.13
NBR. OF PERIODS = 36

Year	Mon	Per	Data	WEIGHT	DATA*WEIGHT
		1	206	0.006648	1.369495
1	1	1	206	0.000993	0.204637
	2	2	212	0.001142	0.242066
	3	3	235	0.001312	0.308423
	4	4	170	0.001509	0.256453
	5	5	255	0.001734	0.442161
	6	6	222	0.001993	0.442460
	7	7	156	0.002291	0.357377
	8	8	206	0.002633	0.542437
	9	9	200	0.003027	0.605331
	10	10	150	0.003479	0.521837
	11	11	256	0.003999	1.023681
	12	12	192	0.004596	0.882484
2	1	13	168	0.005283	0.887555
	2	14	140	0.006072	0.850149
	3	15	220	0.006980	1.535573
	4	16	248	0.008023	1.989666
	5	17	174	0.009222	1.604569
	6	18	188	0.010600	1.992727
	7	19	214	0.012183	2.607260
	8	20	255	0.014004	3.571014
	9	21	279	0.016097	4.490931
	10	22	178	0.018502	3.293312
	11	23	214	0.021266	4.551006
	12	24	266	0.024444	6.502136
3	1	25	162	0.028097	4.551663
	2	26	275	0.032295	8.881137
	3	27	280	0.037121	10.393807
	4	28	227	0.042668	9.685526
	5	29	234	0.049043	11.476091
	6	30	288	0.056371	16.234965
	7	31	226	0.064795	14.643607
	8	32	200	0.074477	14.895338
	9	33	279	0.085605	23.883904
	10	34	196	0.098397	19.285812
	11	35	246	0.113100	27.822600
	12	36	293	0.130000	38.090000
	SUM			**1**	**240.9191895**

(n = 36; Mon, month; NBR, number; Per, period)

EXHIBIT 2.15 Graph of Weights of Past Data for Repair Parts

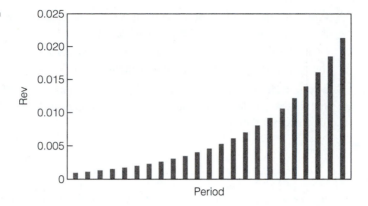

we would be making repeated forecasts at different time origins, in which case the simple smoothing model might very well outperform the NAÏVE_1 model on average. It should also be evident that a single measure of performance is not sufficient.

SUMMARY

Understanding traditional methods is fundamental to the selection of the most appropriate forecasting technique. To succeed, the forecaster must

1. Perform a general analysis on the time series
2. Perform a screening procedure that reduces the list of all available extrapolative techniques to a list of those extrapolative techniques that are capable of handling the data in question

EXHIBIT 2.16 NAÏVE_1 Model Forecast Errors versus Simple Smoothing Model Forecast Errors for Demand for Repair Data

Time Origin, $T = 24$	Actual Demand	Simple Smoothing Forecast (206) Error	NAÏVE_1 Forecast (266) Error
25	162	−44	−104
26	275	69	9
27	280	74	14
28	227	21	−39
29	234	28	−32
30	288	82	22
31	226	20	−40
32	200	−6	−66
33	279	73	13
34	196	−10	−70
35	246	40	−20
36	293	87	27
MSE		2930	2190

(Exhibit 2.13; $n = 24$, $m = 12$)

3. Perform a detailed examination of the techniques that are still considered to be the most appropriate for the given situation

4. Make the final selection of two or more techniques that are considered to be the most appropriate for the given situation.

The next few chapters discuss ways to improve data analysis skills, build and evaluate quantitative models, and enhance the management of these analytical functions and the forecasts they lead to.

REFERENCES

Armstrong, J. S. (1985). *Long-Range Forecasting: From Crystal Ball to Computer*. New York: John Wiley & Sons.

Armstrong, J. S. (2001). *Principles of Forecasting: A Handbook for Researchers and Practitioners*. Boston: Kluwer Academic.

Bergstrom, R. (1995). The relationship between manufacturing production and different business survey series in Sweden 1968–1992. *Int. J. Forecasting* 11, 379–93.

Chambers, J. C., S. K. Mullick, and D. D. Smith (1974). *An Executive's Guide to Forecasting*. New York: John Wiley & Sons.

Chatfield, C. (1993). Neural networks: Forecasting breakthrough or passing fad? *Int. J. Forecasting* 9, 1–3.

Clifton, P., H. Nguyen, and S. Nutt (1991). *Market Research Using Forecasting in Business*. Stoneham, MA: Butterworth and Heinemann.

Collopy, F., and S. Armstrong (1992). Rule-based forecasting: Development and validation of an expert systems approach to combining time series extrapolations. *Manage. Sci.* 38, 1394–414.

Cox, J. E., and D. G. Loomis (2001). Diffusion of forecasting principles through books. In *Principles of Forecasting: A Handbook for Researchers and Practitioners*, edited by S. Armstrong, 633–49. Boston: Kluwer Academic.

DeLurgio, S. A. (1998). *Forecasting Principles and Applications*. New York: Irwin/McGraw-Hill.

Dielman, T. E. (1996). *Applied Regression Analysis for Business and Economics*. 2nd ed. Belmont, CA: Wadsworth.

Faraway, J., and C. Chatfield (1998). Time series forecasting with neural networks: A comparative study using the airline data. *Appl. Statist.* 47, 231–50.

Granger, C. W. J., and P. Newbold (1986). *Forecasting Economic Time Series*. 2nd ed. Orlando, FL: Academic Press.

Leontief, W. (1986). *Input/Output Economics*. 2nd ed. New York: Oxford University Press.

Linstone, H., and M. Turoff (1975). *The Delphi Method: Techniques and Applications*. Reading, MA: Addison-Wesley.

Makridakis, S. (1990). *Forecasting, Planning and Strategy for the 21st Century*. New York: Free Press.

Makridakis, S., and M. Hibon (1979). Accuracy of forecasting: An empirical investigation. *J. Royal Statist. Assoc. A* 142, 97–145.

Makridakis, S., and S. C. Wheelwright (1980). *Forecasting Methods for Management*. 3rd ed. New York: John Wiley & Sons.

Martino, J. P. (1993). *Technological Forecasting for Decision-Making*. 3rd ed. New York: McGraw-Hill.

Mullick, S. K., et. al. (1987). Life-cycle forecasting, In *The Handbook of Forecasting*, edited by S. Makridakis and S. C. Wheelwright, Chap. 19. New York: John Wiley & Sons.

Murdick, R. G., and D. M. Georgoff (1986). How to choose the best technique—or combination of techniques—to help solve your particular forecasting dilemma. *Harvard Business Rev.* 64, 110–20.

Porter, A. L., A. T. Roper, T. W. Mason, F. A. Rossini, and J. Banks (1991). *Forecasting and Management of Technology*. New York: John Wiley & Sons.

Rowe, G., and G. Wright (1999). The Delphi technique as a forecasting tool: Issues and analysis. *Int. J. Forecasting* 15, 353–75.

Twiss, B. C. (1992). *Forecasting for Technologists and Engineers: A Practical Guide for Better Decisions*. London: Peter Peringrinus.

Velleman, P. F., and D. C. Hoaglin (1981). *Applications, Basics, and Computing of Exploratory Data Analysis*. Boston: Duxbury Press.

Winters, P. R. (1960). Forecasting sales by exponentially weighted moving averages. *Manage. Sci.* 6, 324–42.

Zhang, G., B. E. Patuwo, and M. Y. Hu (1998). Forecasting with artificial neural networks: The state of the art. *Int. J. Forecasting* 14, 35–62.

PROBLEMS

2.1 Consider a projection technique for a new product that is based on a projection of sales per month (10% per month). Assume limited sales data as given in the following table.

Actual Sales per Month (cumulative)	Forecasted Percentage	Year-End Projection
100	8.33	1200
140	16.16	866
230	25.00	920
282	33.32	846
360	41.65	864

Forecasted percentage assumes 1/12 of total yearly sales are made each month. What are some of the deficiencies of this method? Consider the viewpoint of customers and salespeople in different geographical regions, store preferences, and seasonality.

2.2 Cash flow forecasting requires information about the level of cash flowing into a firm per month. Using a visual model describe the flow of future cash in terms of variables whose levels are already known at the time the forecast is made.

2.3 An important distinction made among types of forecasting models concerns whether they are causal or noncausal (causal models are those that have as their basis a set of causal relationships). For the following forecasting situations, provide your reasons for choosing to adopt a causal or noncausal modeling approach.

   a. A chart analysis of the stock exchange

   b. Sales of produce in supermarkets

   c. The acceptability of a new product in the market

   d. The demand for an existing product or service

2.4 The p-term moving average is given by
$p\text{-MAVG}_t = (Y_t + Y_{t-1} + Y_{t-2} + \cdots + Y_{t-(p-1)})/p$.
Show that $p\text{-MAVG}_t = p\text{-MAVG}_{t-1} + \{$Correction term$\}$. If the observation Y_{t-p} is assumed to equal the moving average value $p\text{-MAVG}_{t-1}$, find a new expression for $p\text{-MAVG}_t$ in terms of the $p\text{-MAVG}_{t-1}$ and Y_t only. Interpret the formula.

2.5 Show that an updated one-period-ahead forecast for a p-term moving average $p\text{-MAVG}_t$ is F_{T+1}
$(1) = F_T(1) + (Y_{t+1} - Y_{t-p+1})/p$, so that when p is large, a p-term moving average results in one-step-ahead forecasts that do not change very much.

2.6 For the following set of data {200, 135, 195, 198, 310, 175, 155, 130, 220, 277, 235}, determine the one-period-ahead forecasts for a 3_MAVG$_t$ and a 5_MAVG$_t$.

   a. Plot the observed data with the one-period-ahead forecasts.

   b. Contrast the smoothness of the forecast patterns.

   c. Compare the forecasts for period 12 with an overall average of the 11 numbers.

2.7 Repeat the questions in Problem 2.6 replacing the 8th value (130) with 600. Interpret the impact of an outlier on the MAVG one-period-ahead forecasts.

2.8 Repeat the questions in Problem 2.6 adding 400 to each of the last four values. Interpret the impact of a level-shift on the MAVG one-period-ahead forecasts.

2.9 A Naïve_1 forecast uses the current period's value for the next period's forecast: $F_T(1) = Y_T$. Refer to the housing starts in Exhibit 2.3 and the mortgage rates in Exhibit 2.4.

   a. What are the first four Naïve_1 forecasts $F_{29}(1), F_{29}(2), F_{29}(3),$ and $F_{29}(4)$?

   b. What are the first four Naïve_1 forecasts $F_9(1), F_9(2), F_9(3),$ and $F_9(4)$?

   c. Contrast your findings in parts (a) and (b).

   d. What does a forecast profile look like for a Naïve_1 technique?

2.10 For the housing starts data in Exhibit 2.3, apply a six-period moving average to create a 4-year-ahead forecast using the first 25 data values (i.e., determine $F_{25}(1), F_{25}(2), F_{25}(3),$ and $F_{25}(4)$). Create the corresponding Naïve_1 forecasts (see Problem 5), and contrast the corresponding forecast profiles for each technique.

2.11 For the mortgage rates data (Exhibit 2.4) create a 5-year-ahead forecast using a five-period moving average from the first 24 data values (i.e., determine $F_{24}(1), F_{24}(2), \ldots, F_{24}(5)$). Create the corresponding Naïve_1 forecasts (see Problem 2.5).

2.12 A forecast error is defined as Actual value minus Forecast $= Y_{T+1} - F_T(1)$ for the one-period-ahead forecast or, in general, $Y_{T+m} - F_T(m)$ for the m-step-ahead forecast error. For the housing starts forecasts in Problem 2.6, determine the forecast errors. Then determine the forecast errors for the mortgage rates data in Problem 2.7. Compare the performance of the two models in the two examples using a time plot. Make a graphical assessment of the accuracy of the two methods.

2.13 Using a three-period moving average, determine the following ten one-period-ahead forecasts for the housing starts in Exhibit 2.3: $F_{19}(1)$, $F_{20}(1)$, ..., $F_{28}(1)$. Find the corresponding forecast errors $e_{19}(1)$, $e_{20}(1)$, ..., $e_{28}(1)$. Compare forecasting accuracy with the Naïve model.

   a. Calculate the average of the ten forecast errors.

   b. Calculate the average of the ten absolute values of the forecast errors.

   c. Calculate the average of the ten squared forecast errors.

   d. Contrast the two methods of forecasting.

2.14 Repeat Problem 2.9 for the mortgage rates in Exhibit 2.6.

2.15 Sales have been absolutely flat at 100 units per week for the past 15 weeks. Contrast the Naïve_1 forecast method with the three-term moving average forecast for this situation.

2.16 As in Problem 2.15, sales have been absolutely flat at 100 units per week except for week 16 (4% under), week 17 (20% over), and week 18 (3% under) due to a promotion. The forecaster is interested in making a business as usual or baseline forecast for weeks 19 through 21.

   a. Use a three-period moving average to make the forecasts.

   b. As an alternative method, prepare a three-period moving median forecast, in which the median (middle value of the three ordered values) is used instead of the average.

   c. Contrast the two forecasting techniques.

CASE 2A THE DELPHI METHOD

Your company is interested in conducting a survey to determine the feasibility of introducing a product into a new industry. Budget limitations for travel prohibit bringing a panel of experts together for multiple meetings, so you recommend using the Delphi method instead. You assure your management that you can achieve accurate results provided the company adheres to a carefully controlled process for conducting the survey questionnaires. For instance, you caution your management that participants are never to engage in face-to-face discussions in order to prevent any member from dominating the entire group's responses and, thus, the anonymity of participants and the controlled communication of the Delphi method will help eliminate any distorting bandwagon effects. To prepare yourself for conducting a Delphi study, you research previous studies to learn about the strengths and pitfalls of an actual study. You discover that the areas of primary concern are (1) the specification of the study's objectives, (2) the study's area of interest, and (3) the study's methodology.

Find a study in the literature (e.g., Brock, J. L. (1978). *A Forecast for the Grocery Retailing Industry in the 1980's*. Ann Arbor, MI: UMI Research Press).

(1) Create a flow diagram of how the study was conducted.

(2) Describe the event statements (questions) used in the questionnaires.

(3) Rank the events by the degree of consensus in the various rounds.

(4) Summarize the strengths and weaknesses encountered in the study.

CASE 2B INPUT-OUTPUT ANALYSIS

Your management has expressed interest in analyzing the sectors in your firm's industry. You have read that input-output analysis attempts to quantify relationships among various sectors in the economy. Can this approach also be applied to analyzing industry sectors? For instance, consider an ink manufacturer that wants to project the effects of a new type of computer technology on the demand for ink. Although ink does not have a direct demand associated with computers, ink is used in printers by firms manufacturing, selling, and servicing computers.

You quickly volunteer to find out, thinking that this might improve your chances for a promotion. In your first attempt, you find out that the first practical use of input-output theory occurred shortly after World War II. President Roosevelt asked the U.S. Labor Department to assess the probable effect on the American economy of the impending transition from war to peacetime. A static input-output model was constructed on the basis of a matrix of input coefficients derived from the 1939 model of the U.S. economy. Although most economic experts predicted a slump in steel, usually considered a war industry, the input-output calculations led to the opposite conclusion. The substitution of a normal peacetime demand vector for the wartime vector of final demand led to the determination that demand for steel, and consequently employment in the steel industry, would rise sharply. Subsequent developments have demonstrated that this conclusion was indeed correct. However, you would like to find a more current study using the methodology. In the literature, find a study and:

(1) Clearly state the objectives and underlying assumptions.

(2) Describe the structure of the model.

(3) Show a sample input-output table.

(4) Summarize the pertinent results and implications of the study.

CASE 2C DYNAMIC SYSTEMS MODELING

System dynamics is an investigation of the information feedback characteristics of systems and the use of models for the design of improved systems and guiding policy. This is a broad statement, and to assist your management to understand the potential value of the systems modeling method in your business, you offer to do some research.

(1) What are the basic steps of a system dynamics approach?

(2) What are the key characteristics and assumptions that go into such a feedback system?

(3) How are decisions formulated in the model?

(4) What are the outputs and how are they used in practice?

(5) Find an example of a systems dynamics model in the literature.

CASE 2D FOCUS GROUP RESEARCH

The VP of Market Research has placed a call to you after she heard about your taking a forecasting course at the local university. She is requesting that you perform an in-depth, necessarily qualitative study of the growing senior citizens market. You jump at the opportunity with a great deal of enthusiasm, even though your knowledge is very limited at this stage of your career. Nevertheless, you will find a way to complete the assignment in a timely fashion. To get a handle on this approach, you decide to put together a concise memorandum on the salient features of a focus group research project. Using various sources in the literature,

(1) Determine the benefits of a focus group study.

(2) What should be the composition of the group?

(3) What are the roles of the moderator?

(4) How do focus groups differ from individual in-depth interviews?

(5) What are some disadvantages of this approach?

(6) Contrast three approaches to conducting focus group interviews: (a) exploratory, (b) clinical, and (c) phenomenological.

CASE 2E PROMOTIONS: ARE THEY REALLY WORTH IT? (SECTION 2.5 CONT.)

Manufacturers, distributors, and retailers use promotions for their own purposes. These entities have both similar and conflicting interests that can affect the promotion. A promotion can be favorable to the manufacturer but not the retailer, and vice versa. For instance, a retailer can be forced to overbuy to avoid being undersold. Often, a retailer will not pass the promotion on to the end consumer, which has a potentially negative impact on the manufacturer. Consequently, when you find out in a meeting with the VP of Production that a proposed trade deal could have serious negative impact on your company's business, what can you do about it? You know knowledge is power, so you decide to get some good information about promotion modeling in front of the VP, so that the organization can study the potential impact of a promotion before one is implemented. From the literature,

(1) Identify a model developed to study the profitability and sales effects of promotions.

(2) Discuss how the consumer and retailer (or retailer and manufacturer) affect the promotions.

(3) Consider what the initiator of the promotion expects from a promotion.

(4) Describe the different types of promotions (e.g., discounts, incentives, and advertising).

CASE 2F TECHNOLOGICAL FORECASTING

Technological forecasting can be described as a prediction about the future characteristics of useful machines, procedures, or techniques. The VP of Corporate Planning has come to you for your knowledge of forecasting techniques acquired through your interest in reading about the subject. Although the VP does not have a particular problem in mind at this time, he would like to learn more about the trend extrapolation techniques used in this approach, as well as about the hazards of technological forecasting.

(1) Prepare a survey of the trend extrapolation (growth curve) methodology.

(2) Provide a real-world illustration from the literature.

(3) Summarize some of the key advantages and disadvantages of the approach.

Exploring Time Series

3

Data Exploration for Forecasting

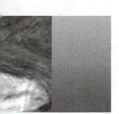

"Time series analysis consists of all the techniques that, when applied to time series data, yield, at least sometimes, either insight or knowledge, AND everything that helps us choose or understand these procedures"
JOHN W. TUKEY, 1980

AS A STATISTICAL METHODOLOGY, much data analysis in forecasting is informal. It is important to realize that:

- An understanding of historical data in forecasting demand will be enhanced when we can identify key patterns in a time series.

- Analyzing data is part of the forecasting process. For example, when data contain trends, contain seasonal patterns, or have outliers, the use of some commonly used forecasting techniques such as moving averages is inappropriate.

- Statistical summaries are beneficial in describing the shape or distribution of data patterns and forecast errors.

- Assuming unrealistic distributions for forecast errors can lead to misleading results when assessing forecasting performance. Most testing procedures implicitly assume that the data follow a **normal distribution,** and this may not be the case in reality.

- Describing the association or mutual dependence between values within the same data at different time periods, called autocorrelation analysis, is a key aspect of any projection technique using historical data.

This chapter deals with the statistical basis for data analysis in forecasting. As in the modeling process, we find that

■   Data analysis is open-ended and iterative in nature.

■   The steps may not always be clearly defined.

■   The nature of the process depends on what information is revealed at various stages. At any given stage, various possibilities may arise, some of which will need to be explored separately.

3.1 EXPLORING DATA

When Princeton University wanted to undertake a study of the paired uses of electricity and gas in townhouses, it contacted the residents of Twin Rivers, a nearby planned community in New Jersey. One of the authors of this book was a resident there at the time and participated in this study, not realizing that some new analysis techniques used on the data would eventually be published in the ground-breaking book *Exploratory Data Analysis* (Tukey, 1977). The purpose of the study, during a winter in the mid-1970s, was to examine differences in energy use and make comparisons with structural aspects of the 152 individual townhouses and the behavioral aspects of their inhabitants. As a statistician, I took great delight in being a participant and was intrigued to later looking at the results and the data from the study. The data were gathered automatically through a special device that was hooked up to the telephones and the energy sources in the home. There were questions to be answered periodically about our lifestyle, the details of which have long escaped my memory. Nevertheless, some novel uses of graphing techniques with schematic data plots using these data can be found throughout this book. These techniques, new at the time, have now become a familiar part of many business statistics books.

Studying the patterns in the data improves the forecaster's chances of successfully modeling data for forecasting applications. Through exploratory data analysis (EDA), a forecaster can start the important task of finding indicators of demand that are generally quantitative in nature. Tukey (1978, p. 1) also likens EDA to detective work: "A detective investigating a crime needs both tools and understanding. If he has no fingerprint powder, he will fail to find fingerprints on most surfaces. If he does not understand where the criminal is likely to have put his fingers, he will not look in the right places." A planned forecasting and modeling effort that does not include provisions for exploratory data analysis often misses the most interesting and important results; but it is only a first step, not the whole story (Hoaglin et al., 1983; Tukey, 1977; Velleman and Hoaglin, 1981).

Exploratory data analysis means looking at data, absorbing what the data are suggesting, and using various summaries and display methods to gain insight into the process generating the data.

Many business forecasting books describe a variety of classical ways to summarize data. An entertaining yet informative cartoon guide covering these is Gonick and Smith (1993). For example, the familiar histogram is widely used in practice. In addition, there are a number of lesser-known techniques that are especially useful in analyzing large quantities of data that have become accessible as a result of the increased flexibility in data management and computer processing. Because of their potential value to forecasting, we describe them in some detail.

The most commonly used graphical displays for analyzing time series are the **histogram, time plot, scatter diagram,** and **autocorrelation plot.** However, stem-and-leaf diagrams, box plots, and quantile-quantile (Q-Q) plots can be quite informative, are readily produced on a computer, and are becoming very useful in forecasting.

Why are graphical displays useful in forecasting? A graphical display is often easier to interpret than the tabular form of the same data. Graphical displays are flexible in their ability to reveal alternative structures present in data or to show relationships among variables. A wise choice in the scale of a graphical display can also make the difference between seeing something important in the data or missing it altogether. For example, rates of growth and changing rates of growth tend to be easier to interpret from graphs with a logarithmic scale than with an arithmetic scale.

Graphics software has given us virtually unlimited power to display data but not much guidance on how to display data most effectively. Although forecasters have always used graphics for analysis and presentation, the principles for effective graphical design in statistical analyses are still evolving (Cleveland, 1994; Robbins, 2005; Schmid, 1983; Tufte, 1990, 2000).

Time Plots

Time plots (or sequence plots) are graphs that show values arranged sequentially in time. If data are recorded at equal time intervals, the corresponding values must be plotted at equally spaced time intervals. These time intervals may be hours, days, weeks, months, quarters, or years.

A time plot is a graph in which the data values are shown sequentially in time.

Time plots provide a useful initial step in the forecast modeling process. Many macroeconomic variables, such as the nonfarm employment (NFRM), industrial production (IP), and U.S. GDP, are dominated by a strong trend, so time plots offer an opportunity to make visual comparisons of their growth patterns.

When a time series is reexpressed (transformed) into another form more useful for a particular analysis, the time plot often shows why a new form may give better results. When analyzing trending data, for example, the percentage changes of a given time series reveal growth rates over time. Exhibit 3.1 shows a plot of the seasonally adjusted U.S. Nonfarm Employment (NFRM). The establishment payroll survey, the Current Employment Statistics (CES) survey, is based on a sample of 400,000 business

EXHIBIT 3.1
Time Plot of
U.S. NFRM,
Seasonally Adjusted

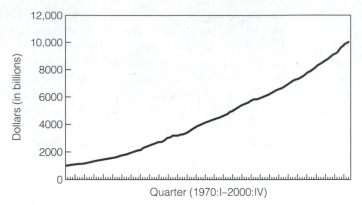

Quarter (1970:I–2000:IV)

(*Source*: U.S. Dept. of Labor, Bureau of Labor Statistics, www.bls.gov)

establishments nationwide. According to the Bureau of Labor Statistics website (www.bls.gov), "the CES Employment is the total number of persons on establishment payrolls employed full or part time who received pay for any part of the pay period which includes the 12th day of the month. Temporary and intermittent employees are included, as are any workers who are on paid sick leave, on paid holiday, or who work during only part of the specified pay period. A striking worker who only works a small portion of the survey period, and is paid, would be included as employed under the CES definitions. Persons on the payroll of more than one establishment are counted in each establishment. Data exclude proprietors, self-employed, unpaid family or volunteer workers, farm workers, and domestic workers."

Exhibit 3.2 shows the percentage changes from previous months. Clearly, trend has been essentially removed in this graph and attention can be focused on the employment growth (or decline).

On the other hand, when financial forecasters analyze rates of return, they use trends in prices for much shorter periods (granularity is defined in terms of days,

EXHIBIT 3.2 Time
Plot of Percentage
Change from the
Previous Month for
NFRM

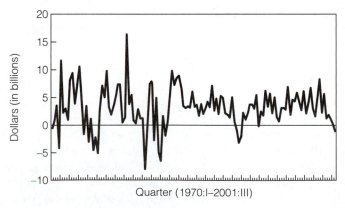

Quarter (1970:I–2001:III)

(*Source*: U.S. Dept. of Labor, Bureau of Labor Statistics, www.bls.gov)

EXHIBIT 3.3 Time
Plot of Monthly
Values of the FSDJ:
(a) Original Scale;
(b) Logarithmic Scale

(a)

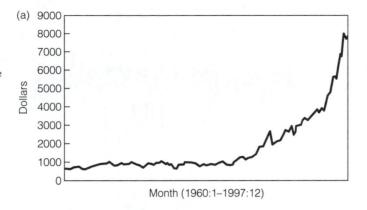

(b)

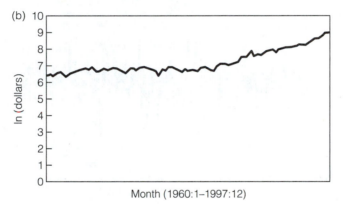

hours, and minutes). Although the Dow Jones Industrial Average of Stock Prices (FSDJ) has been increasing for many years (Exhibit 3.3a), trends are not easily analyzed by period-over-period changes in short periods because of the granularity of the data. When patterns of growth appear to be trending strongly, we need to look at the logarithms of the data and note how linear (like a straight line) the resulting pattern appears (Exhibit 3.3b). For linear patterns on a logarithmic scale, changes (or rates of return) closely resemble percentage changes calculated from the original price data. When we look at the percentage changes of FSDJ in Exhibit 3.4a, the pattern is identical to the pattern of changes in logged FSDJ (Exhibit 3.4b), even though the time plot shows too much volatility to suggest relatively constant growth rates in the original pattern.

Scatter Diagrams

When the values of one time series (or variable) are paired with corresponding values of a related time series (or variable), a relationship between the variables can be depicted in a scatter diagram. One variable is plotted on the horizontal scale and the other variable is plotted on the vertical scale. Such a plot is a valuable tool for studying the relationship between two or more sets of variables.

EXHIBIT 3.4 Time
Plot of Percentage
Changes in Monthly
Values of the FSDJ:
(a) Plot Derived from
Exhibit 3.3a; (b) Plot
Derived from Exhibit
3.3b

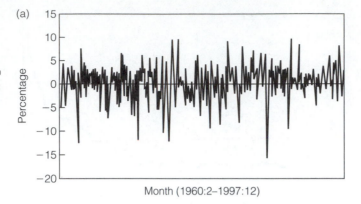

(a)

Month (1960:2–1997:12)

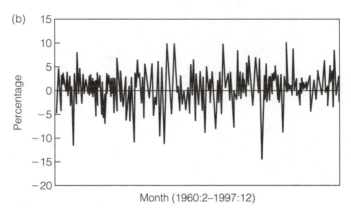

(b)

Month (1960:2–1997:12)

A scatter diagram is a plot in which paired values of two variables are plotted on the same diagram.

The scatter diagram in Exhibit 3.5 relates annual housing starts and mortgage rate data for a 29-year period. This is the same data we discussed in Chapter 2, where moving average forecasts for annual housing starts and mortgage rates were introduced (Exhibits 2.3 and 2.5). Forecasts of housing starts are used for planning for expansions or cutbacks within the construction industry. They are also used for forecasts of goods and services used by the home buyer, such as refrigerators and telephone access lines. Here it is more difficult to suggest a simple forecasting relationship because of the wider dispersion of points.

As a prelude to linear regression analysis (finding the line that best fits the points of the scatter diagram; see Chapters 11 and 12), we may search for variables that are related to one another. If a variable has been plotted against another variable that is dependent on it, we may see a functional relationship between the two variables. As part of the modeling process, scatter diagrams can suggest if certain relationships among variables can be assumed to be linear on the basis of physical,

EXHIBIT 3.5 Scatter Diagram of Annual Housing Starts versus Mortgage Rates for a 29-Year Period

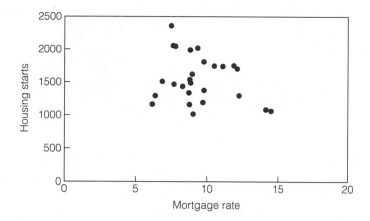

(*Source:* Exhibits 2.3 and 2.5)

economic, or even intuitive hypotheses. After regression analysis, scatter diagrams play a diagnostic role in the graphical analysis of **residual series** to help verify that assumptions are reasonable and that a proposed statistical model provides a good fit to the data.

Whenever the scatter diagram does not suggest a clear relationship between the variables, it might be useful to look at transformations of these variables as well (see Chapter 10). For annual (trending) data, it often makes sense to look at the relationship between changes or growth rates (percentage changes) of variables to be forecast. Exhibit 3.6 shows the relationship between the period-over-period changes in the housing starts and mortgage rates variables. It is evident that there is a strong negative relationship in the respective changes of these variables, as shown by the clear downward pattern. For forecasting annual housing starts, this suggests that modeling the relationship between the changes in the variables (Exhibit 3.6) might be more promising for forecasting housing starts than a model based on the original

EXHIBIT 3.6 Scatter Diagram of Changes in Annual Housing Starts versus Changes in Annual Mortgage Rates for a 29-Year Period

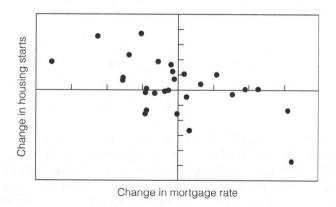

(*Source:* Exhibits 2.3 and 2.5)

levels (Exhibit 3.5). The two approaches imply different model structures and hence can lead to quite different forecasts (see Chapter 12).

A transformation is a mathematical operation to enhance the analysis, display, and interpretation of data.

Linear Association

When two series have a strong positive association, the scatter diagram has a scatter of points along a line of positive slope. A negative association shows up as a scatter of points along a line with negative slope. A measure of such linear association between a pair of variables Y and X is given by the Pearson product moment correlation coefficient r, where r is an averaging formula using the sample mean and sample **standard deviation** of the two variables, respectively. Although spreadsheet and statistical packages routinely calculate r, it is useful to view it as the result of an averaging process, namely, of the product of the standardized variables Y and X: Average{(Standardized Y_t) · (Standardized X_t)}, with a divisor of $(n - 1)$ instead of n. When a variable is standardized, it has a zero mean and a unit standard deviation, which is useful for making comparisons and correlations between variables that have very different sizes or scales of measurement.

A standardized value is obtained by subtracting the sample mean from the data and dividing by the sample standard deviation.

Exhibit 3.7 shows a spreadsheet calculation for obtaining the product moment correlation between the annual housing starts and mortgage rates. The coefficient can vary between $+1$ and -1, so that $r = -0.20$ suggests a weak negative association between housing starts and mortgage rates.

The product moment correlation coefficient is a measure of linear association between two variables.

Although the product moment correlation coefficient for the housing starts versus mortgage rates data is only about -0.20 (Exhibit 3.7), the correlation coefficient for the respective annual change in these variables turns out to be -0.57. Both are negative, as expected, but the latter reflects a much stronger linear association. This suggests that the strength of the relationship between housing starts and mortgage rates lies in their respective growth rates, not the levels.

EXHIBIT 3.7	Housing Starts	Mortgage Rates	Col 1 × Col 2
Calculation of the	0.168898	−1.384396	−0.233822
Product Moment	−0.051812	−1.405769	0.072836
Correlation Coef-	−0.217938	−1.410043	0.307302
ficient between	−1.131335	−1.239058	1.401789
Annual Housing	−0.755474	−1.157839	0.874718
Starts and Mortgage	−0.114703	−0.944107	0.108292
Rates	−0.235737	−0.589311	0.138923
	−0.334226	−0.328558	0.109813
	1.500873	−0.619234	−0.929391
	2.403886	−0.679079	−1.632429
	1.480404	−0.538016	−0.796480
	−0.618717	−0.140474	0.086913
	−1.144684	−0.123375	0.141226
	−0.026003	−0.114826	0.002986
	1.307751	−0.102002	−0.133393
	1.406240	0.124554	0.175153
	0.589850	0.620413	0.365951
	−0.753694	1.377025	−1.037857
	−1.370734	2.176384	−2.983244
	−1.435998	2.330271	−3.346265
	0.464958	1.312906	0.610447
	0.602902	1.214589	0.732279
	0.580060	0.876892	0.508650
	0.768732	0.299815	0.230477
	0.220219	−0.042157	−0.009284
	−0.172550	−0.089178	0.015388
	−0.504802	0.312639	−0.157821
	−1.048865	0.274167	−0.287564
	−1.577502	−0.012234	0.019300
	0.000000	0.000000	$r = -0.201611$
	1.000000	1.000000	

(*Source:* Exhibits 2.3 and 2.5)

The Correlation Matrix

The correlation matrix gives a representation of the degree of linear association or correlation when there are more than one pair of variables. This matrix is an array of all sample correlation coefficients between pairs of variables. Exhibit 3.8 shows a correlation matrix for the toll revenues (REV), toll messages (MSG), business telephones (BMT), and nonfarm employment (NFRM) in the telecommunications forecasting example used in this book (see the CD).

A correlation matrix gives a representation of the degree of linear association or correlation when there are more than one pair of variables.

The diagonal of a correlation matrix consists of 1s because each variable is perfectly correlated with itself. The variables are numbered 1–4, and each appears as a

EXHIBIT 3.8		1	2	3	4
Correlation Matrix of	1 REV	1			
Ordinary Cor-	2 MSG	0.963	1		
relations for the Four	3 BMT	0.765	0.772	1	
Time Series in the	4 NFRM	−0.128	−0.105	0.015	

Correlation Matrix of Ordinary Correlations for the Four Time Series in the Telecommunications Example (BMT, business telephones; MSG, toll messages; NFRM, nonfarm employment; REV, toll revenues)

(*Source*: Levenbach and Cleary, 1984)

row and a column. At the intersection of each row and column is the correlation coefficient relating the row variable to the column variable. For example, the coefficient of correlation between toll revenues and toll messages is very high ($= 0.963$). The business telephones have a positive correlation with toll messages ($= 0.772$) and a low correlation with nonfarm employment ($= 0.015$). The negative but very small correlations of the nonfarm employment data with revenues and messages are not intuitively satisfying; however, the low values suggest that these results are probably not significant.

In practice, it turns out that looking at correlations among transformed values of the variables can also be very useful. In particular, the period-to-period changes (month-to-month, year-over-year, etc.) and percentage changes are most frequently applied to time series as a prelude to modeling. Because there will be other types of correlation measures introduced in this text, we will use *ordinary correlation coefficient* interchangeably with the *product moment correlation coefficient*.

3.2 CREATING DATA SUMMARIES

To get a better understanding of patterns and structure in data, we extract some essential features from the data and summarize them with a few meaningful numbers. These summary measures are called statistics. Summary statistics are commonly used for simply describing some aspect of the data that needs to be highlighted for a particular application, such as measures of central tendency or dispersion; summarizing salient features of a frequency distribution, such as percentiles; comparing two or more frequency distributions; and confirming an analysis, such as calculating prediction intervals on forecasts and determining confidence limits for model parameters.

Typical Values

The location, or central tendency, of a data set is the center of the data when they are arranged in order of size; this is the typical or average measurement. The most commonly used measure of central tendency is the familiar arithmetic mean or sample average (the sum of n data values $\{Y_1, Y_2, \ldots, Y_n\}$ divided by n). In Chapter 2, we encounter this statistic when we define the moving average of p successive values $\{Y_t, Y_{t-1}, \ldots, Y_{t-(p-1)}\}$ of a time series by

$$(Y_t + Y_{t-1} + Y_{t-2} + \cdots + Y_{t-(p-1)}) / p$$

where Y_t is the actual value at time t, and p is the number of values included in the average. This is the quantity **P_MAVG** that we introduced as a prototypical projection technique in Chapter 2. Another familiar measure of central tendency is the median. The median is the middle value when the data are arranged in order from lowest to highest.

The reason that no single measure of location is always the best is that each provides its own perspective and insights, and in practice outliers or unusual values can seriously distort the representativeness of certain statistics, such as the mean.

 The central tendency or location of a set of data is a middle, typical value in the data set.

Outliers

Unusual or atypical values, called outliers, occur quite frequently in data used by forecasters. Summary statistics based on the averaging process, most notably the arithmetic mean and standard deviation, are quite sensitive to outliers. A statistic is said to be resistant if a change in a small fraction of the data will not produce a large distortion of a total calculated value—if it is resistant to weird or unusual values. Statistics based on the median or arithmetic mean of truncated data appear to be much more resistant to outliers. The arithmetic mean is clearly not resistant because its value can be changed by arbitrarily increasing only one of its terms. The median, on the other hand, is quite resistant. A seminal study on measures of location and their properties under varying realistic assumptions of distribution is the Princeton Robustness Study (Andrews et al., 1972); however, sufficient attention has not been given to data distributions encountered in forecasting.

 An outlier is an unusual or atypical value in a data set.

When we deal with forecast accuracy (Chapter 5), we obtain averages of numbers based on forecast errors (squared errors, absolute errors, etc.). To properly interpret a measure of forecast accuracy, we must also be sensitive to the role of unusual values in these calculations. For example, consider the following set of numbers, representing 12 absolute percentage errors of monthly forecasts made over 1 year:

$$\{1.1, 1.6, 4.7, 2.1, 3.1, 32.7, 5.8, 2.6, 4.8, 1.9, 3.7, 2.6\}$$

By ranking the data from smallest to largest, we obtain the ordered set:

$$\{1.1, 1.6, 1.9, 2.1, 2.6, 2.6, 3.1, 3.7, 4.7, 4.8, 5.8, 32.7\}$$

Over the 12 monthly periods, the mean absolute percentage error (MAPE = 5.6) can be viewed as a typical percentage error for a month. The median absolute percentage error (MdAPE = [2.6 + 3.1]/2 = 2.9) is an outlier-resistant measure and gives quite a different answer. Note that the arithmetic mean has been severely distorted by the outlying value 32.7. The arithmetic mean of the numbers when we exclude the outlier is 3.1 and appears to be much more representative of the data, like the MdAPE. Overall, an average absolute percentage error for the 12 months is more likely to be around 3% per month than around 6% per month, and this would be more typically used to summarize the year's forecast performance.

What should forecasters do in practice? Although the arithmetic mean is the conventional estimator of location, forecasters should not accept it uncritically. As our example illustrates, one outlier can have an undue effect on the arithmetic mean and pull an estimate of the bulk of data away from its representative value. It is always best to calculate and compare multiple measures for the same quantity to be estimated.

The Trimmed Mean

One simple way to make the arithmetic mean less sensitive to outliers is first to delete, or trim, a proportion of the data from each end and then calculate the arithmetic mean of the remaining numbers. Such a statistic is called a trimmed mean. The midmean, for example, is a 25% trimmed mean because 25% of the data values have been trimmed from each end (i.e., the mean is taken of the values between the 25th and 75th percentiles). The deletion of values is based on their order, but the deleted values are not necessarily the extreme values. For the sample of 12 percentage errors in our example, the midmean is (2.1 + 2.6 + 2.6 + 3.1 + 3.7 + 4.7)/6 = 3.1, again a fairly typical value.

Variability

In addition to measures of location, certain measures of variability (also called measures of spread or dispersion) are also useful. Some commonly used examples are the range and standard deviation and variations of these. The range is simply the difference between the largest and smallest value in a set of data. To measure variability we need to consider how far away numbers are from their typical value, such as the arithmetic mean or median. Then, to summarize variability in the data, we can square or take absolute values of these deviations and seek a typical value for these deviations. This is a measure of spread or dispersion.

Spread or variability in a data set is a measure of dispersion around the measure of central tendency.

The (Sample) Standard Deviation

The most familiar measure is the sample standard deviation, which is the square root of the (sample) **variance.** For a sample of values $\{Y_1, Y_2, \ldots, Y_n\}$, the sample variance is the sum of squared deviations from the sample mean divided by $n - 1$.

Although the standard deviation is by far the most familiar measure of variability, it is not always the most effective. Like the sample mean, it can be misleading when there are outliers. Taking, instead, the absolute deviations from the mean and averaging the result, is better because a large absolute deviation is not as extreme as a squared deviation. This gives rise to the mean absolute deviation (MAD) statistic. However, because it is an arithmetic mean, the MAD statistic may still not be sufficiently resistant to outliers.

A sample standard deviation is a conventional measure of spread.

Median Absolute Deviation

To ensure greater sensitivity to outliers, the median absolute deviation (MdAD) is a viable alternative to the sample standard deviation and MAD. The calculation of a MdAD is illustrated in Exhibit 3.9. The data are the numbers used earlier to explain the median and arithmetic mean. The MdAD is calculated by first sorting the data from smallest to largest (column 2) and picking the median, which is $(2.6 + 3.1)/2 = 2.85$, because there is an even number of data values. Absolute deviations from the median are calculated next (column 3), and the result is reranked (column 4). The midvalue of the latter set of numbers is the MdAD, $(0.95 + 1.25)/2 = 1.1$.

The Interquartile Difference

The interquartile difference (IQD) is the difference between the 75th and 25th percentile positions in the ordered data set. For our sample of absolute percentage errors {1.1, 1.6, 4.7, 2.1, 3.1, 32.7, 5.8, 2.6, 4.8, 1.9, 3.7, 2.6}, the 25th percentile position

---

EXHIBIT 3.9
Calculation of the Median Absolute Deviation (MdAD)

Data (n)	Data (sorted)	Deviations from the Median	Absolute Deviations from the Median
1.1	1.1	−1.75	0.25
1.6	1.6	−1.25	0.25
4.7	1.9	−0.95	0.25
2.1	2.1	−0.75	0.75
3.1	2.6	−0.25	0.85
32.7	2.6 Median = 2.85	−0.25	0.95 MdAD = 1.1
5.8	3.1	0.25	1.25
2.6	3.7	0.85	1.75
4.8	4.7	1.85	1.85
1.9	4.8	1.95	1.95
3.7	5.8	2.95	2.95
2.6	32.7	29.85	29.85

1st quartile $= \dfrac{1}{4}(n + 1) = 13/4 = 3.25$

Median $= (2.6 + 3.1)/2 = 2.85$

Midmean $= (2.1 + 2.6 + 2.6 + 3.1 + 3.7 + 4.7)/6 = 3.13$

3rd quartile $= \dfrac{3}{4}(n + 1) = 39/4 = 9.75$

is $(n + 1)/4 = 13/4 = 3.25$, that is, between the third and fourth smallest values. The 75th percentile is $3(n + 1)/4 = 9.75$, between the ninth and tenth values in the ranked data. We can scale MdAD and IQD by dividing by 0.6745 and 1.349, respectively, giving the unbiased median absolute deviation (UmdAD) and the unbiased interquartile difference (UIQD). The divisors used for scaling are empirically determined so that, for normally distributed data, this scaling makes these measures good approximations of their theoretical counterpart σ (the population standard deviation) if the number of observations can be assumed to be very large.

The MAD, MdAD, and IQD are other useful measures of spread.

Tabulating Frequencies

Suppose that we have been asked to predict the average travel time between a central service center and customers' locations for a cable-company installer (Exhibit 3.10). These data, which are collected during the same period, are called **cross-sectional.** They differ from time series data, which are collected in different periods. The following questions are to be answered: What is the most typical travel time (i.e., where are most of the data concentrated)? How much variability is there in the data? Are there any extreme or unusual values—values that do not seem to fit? Can the overall behavior of the data be described?

Typical Values and Variability

For this data set, the average travel time is 12.4 minutes and the standard deviation is 8 minutes. Scanning the data, it is apparent that there are some unusually large travel times. This tends to inflate both the arithmetic mean and the standard deviation. The median travel time is 10 minutes, which is sufficiently different from the mean to warrant some further attention to the data.

An alternative method is to look at the median along with a resistant scale statistic. As noted before, an approximate unbiased scale statistic is UMdAD = MdAD/0.6745 = 5.2. Recall that the divisor 0.6745 is used because the result is approximately equal to the standard deviation of a normal distribution, assuming the number of observations is large and the data are actually a random sample from a normal distribution with (population) variance σ^2. An alternative interpretation of a typical travel time is the median $\pm$ UMdAD, that is, 10 minutes, give or take 5 minutes, which is quite different from the standard approach.

The UMdAD is an unbiased measure of spread comparable to a standard deviation for data that is normally distributed.

Service Order Number	Time (min)	Service Order Number	Time (min)
1	11	26	9
2	10	27	11
3	23	28	10
4	7	29	16
5	9	30	8
6	13	31	8
7	14	32	7
8	4	33	10
9	4	34	6
10	3	35	10
11	23	36	6
12	9	37	6
13	7	38	6
14	14	39	22
15	10	40	18
16	5	41	11
17	21	42	8
18	30	43	12
19	4	44	11
20	13	45	40
21	11	46	27
22	6	47	17
23	32	48	3
24	10	49	20
25	11	50	12

EXHIBIT 3.10   Travel Times from a Central Service Center to a Customer's Location for 50 Service Orders

Unusual Values

The values that do not seem to fit can be filtered out by a simple rule. In the terminology of Tukey (1977), a step is defined as 1.5 times the IQD. Inner fences are values one step above the 75th percentile and one step below the 25th percentile:

$$\text{Lower inner fence} = \text{25th percentile} - 1.5 \text{ IQD}$$

$$\text{Upper inner fence} = \text{75th percentile} + 1.5 \text{ IQD}$$

Outer fences are values two steps above and below these percentiles:

$$\text{Lower outer fence} = \text{25th percentile} - 3.0 \text{ IQD}$$

$$\text{Upper outer fence} = \text{75th percentile} + 3.0 \text{ IQD}$$

A value outside an inner fence can be an outlier, but a value outside an outer fence is much more likely to be one. The choice of 1.5 is arbitrary, but it seems to work well in practice. For the service time data (Exhibit 3.10), we find that

$$\text{Lower inner fence} = 7 - 1.5(7.5) = -4.25$$

$$\text{Upper inner fence} = 7 + 1.5(7.5) = 18.25$$

and

$$\text{Lower outer fence} = 7 - 3.0 \, (7.5) = -15.5$$

$$\text{Upper outer fence} = 7 + 3.0 \, (7.5) = 29.5$$

According to our rule, there are three large outliers (30, 32, and 40). Alternatively, we can also calculate the median ± 3 UMdAD, which gives $(10 - 16$ and $10 + 16)$ and find the same three outliers plus an additional one (27).

Overall Behavior of the Data

It is apparent that the way the data are displayed (Exhibit 3.10) is not very enlightening. To describe the overall behavior of the data, we may decide to condense the raw data by placing the numbers into cells or classes. These classes can be either numerical or attributive (e.g., designated as either a "commercial" or "residential" service order) in nature. Our data can be easily grouped into numerical classes of 3-minute intervals (Exhibit 3.11).

The procedure for counting the number of occurrences of a given characteristic in a grouping of data gives rise to frequencies. These frequencies, when considered as fractions, can be displayed as a relative frequency distribution (Exhibit 3.11). The relative percentage of total number of orders contained in each class suggests that 9–11 minutes is the typical or most frequent travel time. Most of the observations are clustered within 3–14 minutes. The one observation falling in 39–41 minutes looks very unusual, and all values greater than 26 minutes look suspect because they are so far from the apparent average. Perhaps there were extenuating circumstances that caused these travel times to be so great. But how does the analyst determine if some of the data are truly unusual?

EXHIBIT 3.11
A Frequency Distribution for the 50 Service-Order Travel Data

Interval (min)	Number of Service Orders	Relative Percentage of Orders	Cumulative Percentage of Orders
3–5	6	12	12
6–8	11	22	34
9–11	15	30	64
12–14	6	12	76
15–17	2	4	80
18–20	2	4	84
21–23	4	8	92
24–26	0	0	92
27–29	1	2	94
30–32	2	4	98
33–35	0	0	98
36–38	0	0	98
39–41	1	2	100
	Total = 50	Sum = 100	

(*Source:* Exhibit 3.10)

A grouping interval must be selected before tallying the data; this interval will depend on the range of variation, the number of data values, and the palatability of the display to the user. In Exhibit 3.11, the intervals have a width of 3 minutes. An interval of 1 or 2 minutes would have resulted in a long table without providing added information. These groupings should be uniquely defined so that there is no ambiguity about which cell a given tally belongs in. The relative frequencies are then determined by counting the number of data values in a cell and dividing by the total number of data values recorded. In the first cell, for example, there are six service orders out of 50, giving a relative frequency of 12%. This information is then appropriately summarized as a relative frequency distribution in the third column.

The last column in Exhibit 3.11 is another relative frequency distribution, the cumulative percentage of orders. Once the 21- to 23-minute interval is reached, 92% of the observations have been counted. This reinforces the suspect nature of the remaining observations. A relative frequency distribution should be displayed as a cumulative frequency distribution when we are interested in quantities such as the proportion of items below (or above) a given standard value or when we are investigating whether a distribution follows some particular mathematical form (see the section on quantile-quantile plots).

The graph of the frequency distribution can take the form of a histogram (Exhibit 3.12). The bins in the histogram are intervals suitably chosen to make a meaningful distribution of frequencies. A good rule of thumb is to select from 8 to 15 equally spaced intervals. In a histogram, data are plotted as bars rather than as a single graph line. Exhibit 3.11 shows that there are a very few long travel times (more than 25 minutes) and that the times around the median (10 minutes) are the most typical. The distribution is said to have a tail skewed to the right because a number of values are far above the median. However, the shape of the histogram can be sensitive to the choice of class intervals. It is often desirable to try out several different class intervals to be certain that the results appear reasonable.

EXHIBIT 3.12 A Histogram Plot of the Frequency Distribution for 50 Service-Order Travel Data

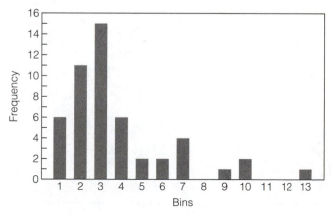

(*Source:* Exhibit 3.10)

3.3 DISPLAYING DATA SUMMARIES

Stem-and-Leaf Displays

Stem-and-leaf displays are useful because they show inherent groupings; show asymmetrical trailing off, data going farther in one direction than another; highlight unexpected values; show approximately where the values are centered; and show approximately how widely the values are spread.

Exhibit 3.13 shows the stem-and-leaf display for the service-order travel data in Exhibit 3.10. The stem is the vertical column; its divisions are multiples of 10. The leaves are the horizontal rows of numbers; each of the numbers in a leaf represents a unit digit. For example, the first number is 3 (service order number 10 in Exhibit 3.10). The tens digit is 0 and the unit digit is 3. By entering each number in this manner and ordering the numbers, we get a visual impression of the distribution of the data as well as an ordering of the data.

> A stem-and-leaf diagram is a device for depicting frequencies as well as actual values in a simple display.

Frequent values stand out (e.g., 10 and 11) as do atypical values and values that are absent or missing. For example, there are no values in the 33–39 range. Notice that the leaves have been ordered. The right-hand column provides a count that is useful as a check that all the values are entered. A cumulative count will also turn out to be useful in the quick calculation of certain statistics, such as the interquartile difference IQD.

When the data are concentrated, as these are, it is often desirable to split each tens unit of the stem into two ranges (0–4 and 5–9). This is illustrated in Exhibit 3.14. The absence of data in the 35–39 range becomes more apparent. Additional information can be added by using symbols to identify various qualities of data that may be helpful in understanding differences in data. For example, a circle might be

EXHIBIT 3.13	Tens Digit	Unit Digit	Number of Orders
Stem-and-Leaf	0	33444566666777888999	20
Display of the	1	00000111111223344678	21
Service-Order Travel	2	012337	6
Data	3	02	2
	4	0	1
	5		0
	6		0
	7		0
	8		0
	9		0
			Total = 50

(*Source:* Exhibit 3.10)

used to indicate that a travel value represented the first trip of the day (which usually takes longer than others because it begins at the service company's garage).

Choosing the number of lines L for a stem-and-leaf display is analogous to determining the number of intervals or the interval width for a histogram. A useful formula for the maximum L when n is less than 100 is $L = 2n^{1/2}$, for n greater than 100 the formula $L = 10 \log_{10} n$ is recommended, where L is taken as the largest integer not exceeding the right-hand side of the expression (Hoaglin et al., 1983). For our example, $N = 50$, so we have $2(50)^{1/2} = 14.1$. Thus a ten-line stem-and-leaf display for the service-order travel data seems satisfactory.

Box Plots

Percentiles

When describing a frequency distribution, the pth percentile is the value that exceeds $p\%$ of the data. The median is the 50th percentile; that is, it is the value that exceeds 50% of the data. For example, the frequency distribution shows that the median of the service-order travel data is 10 minutes. The distribution does not have a symmetrical shape, so it is important to summarize the distribution with more than one percentile.

 Percentiles describe a frequency distribution. The median is the 50th percentile.

Although percentiles can be used to describe a frequency distribution, it is often desirable to summarize a distribution with the smallest set of values possible. Rather than displaying a sequence of percentiles, we can use a box plot, which is a five-number summary of a frequency distribution (Tukey, 1977). A box plot concisely depicts the median, the upper and lower quartiles (the 75th and 25th percentiles), and the two extremes of any group of data.

EXHIBIT 3.14	Range	Unit Digit	Number of Orders
Stem-and-Leaf	0–4	33444	5
Display with Split	5–9	566666777888999	15
Stem	10–14	000000111111223344	18
	15–19	678	3
	20–24	01233	5
	25–29	7	1
	30–34	02	2
	35–39		0
	40–44	0	1
	45–49		0
			Total = 50

Median = (25th + 26th observation)/2 = (10 + 10)/2 = 10

Source: Exhibit 3.10

A box plot is a display based on a five-number summary of a frequency distribution, consisting of the median, upper and lower quartiles, and two extreme values in a data set.

A box plot for the service-order travel data is shown in Exhibit 3.15. Fifty percent of the data values are tightly grouped (these are depicted by the box, which includes all data falling between the 25th and 75th percentiles; these quartiles are also called hinges). The upper tail appears longer than the lower tail. The median value is slightly lower than the midrange of the box. The upper extreme value appears to be an outlier because it is so far away from the bulk of the data. Because there may be values along the whisker (the vertical line), this conclusion needs to be confirmed by additional analyses. This plot gives a surprising amount of information for such a simple display.

The simple box plot summarizes the distribution in terms of five quantities. In addition, the distance between the quartiles is the IQD, a measure of dispersion. Missing from this plot is the number of values in the distribution, which affects the reliability of the estimate of the median. To indicate sample size, a box plot can be modified so that the depth of a notch in the box is proportional to the square root of the sample size (McGill et al., 1978).

Even with this change, a single comprehensive box plot may not be enough by itself. One weakness is its inability to identify or discern data from two different populations. Exhibit 3.16 shows box plots for the service-order travel times for three categories of telephones. It shows that there are really two distributions: orders for nonkey telephone sets and orders for key telephone sets (a key telephone set is linked to multiple telephone numbers). A single box plot would mask these differences.

Quantile-Quantile Plots

When the quantiles (or percentiles) of one distribution are plotted in a scatter diagram against the quantiles of a second distribution, we get a quantile-quantile (Q-Q) plot. If two data sets have the same probability distribution, the Q-Q plot is linear.

EXHIBIT 3.15 Box Plot of the Service-Order Travel Data

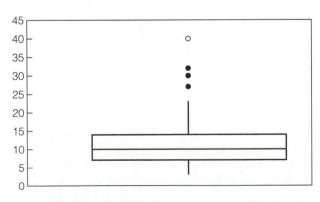

Source: Exhibit 3.10

EXHIBIT 3.16 Box Plots for the Service-Order Travel Times for Three Categories of Telephones

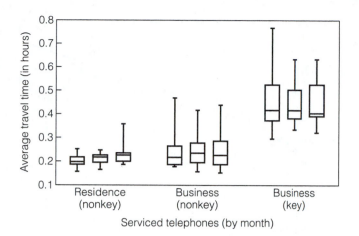

For example, the quantiles of an empirical data set can be compared to the quantiles of the standard normal distribution ($\mu = 0$, $\sigma = 1$) to test for normality. Such a display is known as a normal probability plot.

 The quantile-quantile (Q-Q) plot is used to determine if two data sets have the same probability distribution.

Exhibit 3.17 shows some normal probability plots for situations in which the data are not normally distributed. When the distribution of the data is right-skewed, the highest observations tend to be far from the center and the lowest observations tend to be close to the center. Thus, the highest and lowest observations both exceed the expected normal deviates. For left-skewed data distributions, the highest and

EXHIBIT 3.17 Normal Probability Plots in which the Data Are Not Normally Distributed

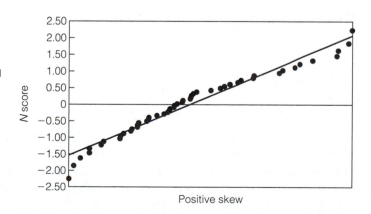

EXHIBIT 3.18
Normal Probability
Plot for the Service-
Order Travel Data

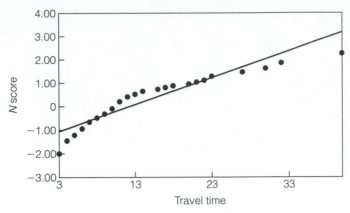

(*Source:* Exhibit 3.10)

lowest observations tend to be less than expected normal deviates. The pattern weaves around the line in the case of light- and heavy-tailed data distributions.

A normal probability plot for the service-order travel data (Exhibit 3.10) is shown in Exhibit 3.18. This plot shows an upper tail in the empirical distribution that is much longer than that of the normal distribution. The tails of many data sets are longer than those of the normal distribution, but generally by less than in this example.

3.4 SERIALLY CORRELATED DATA

When data are ordered sequentially, as in a time series, data sequences one or more time periods apart are often statistically related. As a result of the chronological ordering of historical data, most time series exhibit a mutual dependence of successive values, called serial correlation. For example, in a trending series there is a close association between successive periods because adjacent values follow a very similar pattern. In particular, for a straight-line trend, successive changes are constant. This effect can also be shown among values that are not successive. For example, when we look at the pattern of highs and lows in a monthly time series over a number of years, we often note a relationship of values 4, 6, or 12 periods apart. More generally, this is known as autocorrelation.

Most time series exhibit a mutual dependence of successive values, known as serial correlation.

Autocorrelation

An objective of autocorrelation analysis is to describe correlation in time series. Autocorrelations are also used frequently in the analysis of residuals from a forecasting model. In this situation, the forecaster seeks an absence of autocorrelation, namely a purely random pattern in the residuals. We will also frequently use autocorrelation analysis in the identification of ARIMA models for the Box-Jenkins model-building strategy (Chapters 13 and 14). It also plays a significant role in the analysis of residuals in forecasting models (Chapter 10) and ARIMA modeling applications.

Through autocorrelation analysis, we measure the effects of mutual dependence in values of a time series.

The calculation of the first-order **autocorrelation coefficient** is analogous to calculating a product moment correlation coefficient r between Y_t and Y_{t-1} (Exhibit 3.7). For the mortgage rate date, this is done by creating two columns in a spreadsheet where the first column is the original data and the second column is a shifted (one row down) version of the first, when Y_{t-1} is positioned adjacent to Y_t (Exhibit 3.19). The first difference is the original rate minus the lagged rate.

EXHIBIT 3.19	Mortgage Rates	Lagged 1	First Difference
Annual Mortgage	5.8		
Rates, Rates Time-	5.75	5.8	−0.05
Lagged One Period,	5.74	5.75	−0.01
and First Differences	6.14	5.74	0.4
	6.33	6.14	0.19
	6.83	6.33	0.5
	7.66	6.83	0.83
	8.27	7.66	0.61
	7.59	8.27	−0.68
	7.45	7.59	−0.14
	7.78	7.45	0.33
	8.71	7.78	0.93
	8.75	8.71	0.04
	8.77	8.75	0.02
	8.8	8.77	0.03
	9.33	8.8	0.53
	10.49	9.33	1.16
	12.26	10.49	1.77
	14.13	12.26	1.87
	14.49	14.13	0.36
	12.11	14.49	−2.38
	11.88	12.11	−0.23
	11.09	11.88	−0.79
	9.74	11.09	−1.35
	8.94	9.74	−0.8
	8.83	8.94	−0.11
	9.77	8.83	0.94
	9.68	9.77	−0.09
	9.01	9.68	−0.67
			Mean = 0.114643
			SD = 0.884913

To calculate the first-order autocorrelation coefficient (Exhibit 3.20), we first create the standardized values of the mortgage rates and the shifted values. Then, we place the product of the 28 common values in the next column. Finally, we simply calculate the arithmetic mean of the values in the last column to obtain an estimate of first-order autocorrelation coefficient, r_1. The standard deviation of column 1 = 2.34 (based on 28 values) and standard deviation of column 2 = 2.42 (based on 28 values); the two column averages are 9.16 and 9.02. This gives $r_1 = 0.90$. The coefficient can take on values between $+1$ and -1, so $r_1 = 0.9$ suggests a strong positive association between mortgage rates and a lagged 1 version of itself (Exhibit 3.21).

In practice, the formula used in software packages for autocorrelation at lag 1 has some technical differences (a factor $\sqrt{n/(n-1)}$ but yields the same interpretation from the data. We introduce a more commonly used formula later in this chapter when dealing with autocorrelation plots.

 A time series is said to be lagged if it is a time-shifted version of itself.

EXHIBIT 3.20
Calculation of First-Order Autocorrelation Coefficient of Annual Mortgage Rates

Mortgage Rates	Lagged Mortgage Rates	Standardized Mortgage Rates	Standardized Lagged Mortgage Rates	Column 3 × Column 4
5.75	5.80	−1.4572	−1.3265	1.93
5.74	5.75	−1.4615	−1.3472	1.97
6.14	5.74	−1.2905	−1.3513	1.74
6.33	6.14	−1.2093	−1.1863	1.43
6.83	6.33	−0.9956	−1.1079	1.10
7.66	6.83	−0.6409	−0.9017	0.58
8.27	7.66	−0.3802	−0.5593	0.21
7.59	8.27	−0.6708	−0.3077	0.21
7.45	7.59	−0.7307	−0.5882	0.43
7.78	7.45	−0.5896	−0.6459	0.38
8.71	7.78	−0.1922	−0.5098	0.10
8.75	8.71	−0.1751	−0.1262	0.02
8.77	8.75	−0.1665	−0.1097	0.02
8.80	8.77	−0.1537	−0.1014	0.02
9.33	8.80	0.0728	−0.0891	−0.01
10.49	9.33	0.5686	0.1296	0.07
12.26	10.49	1.3250	0.6080	0.81
14.13	12.26	2.1242	1.3381	2.84
14.49	14.13	2.2781	2.1095	4.81
12.11	14.49	1.2609	2.2580	2.85
11.88	12.11	1.1626	1.2763	1.48
11.09	11.88	0.8250	1.1814	0.97
9.74	11.09	0.2480	0.8555	0.21
8.94	9.74	−0.0939	0.2987	−0.03
8.83	8.94	−0.1409	−0.0313	0.00
9.77	8.83	0.2609	−0.0767	−0.02
9.68	9.77	0.2224	0.3110	0.07
Mean = 9.16	9.02	0	0	0.90
SD = 2.34	2.42	1	1	

(*Source:* Exhibit 3.19)

EXHIBIT 3.21
Scatter Diagram of
Mortgage Rates ver-
sus Lagged 1 Values

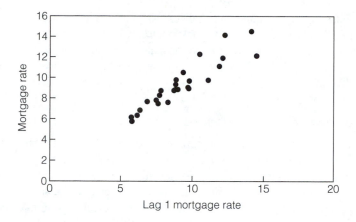

(*Source:* Exhibit 3.19)

For annual mortgage rates, the changes in successive values (known as first differences) are seen to vary around zero (Exhibit 3.19). The Lagged 1 column shows the mortgage rates shifted down one row and the First Difference column is the difference between Mortgage Rates column and Lagged 1 column. The mean and standard deviation of the First Differences column are 0.11 and 0.88, respectively.

 First differences are changes in successive values in a time series.

In Exhibit 3.6, we depict a strong relationship between annual changes in housing starts and annual changes in mortgage rates. To use these data as a leading indicator for forecasting housing starts, we want to show that there is a strong relationship between changes in housing starts and the *lagged* changes in mortgage rates so that the 1-year-ahead housing start forecast can be based on actual (observed) mortgage rates. In Exhibit 3.22, we compare the scatter of Exhibit 3.6 with the corresponding scatter

EXHIBIT 3.22
Scatter Diagram of
Changes in Housing
Starts versus
Changes in Mortgage
Rates Lagged One
Period

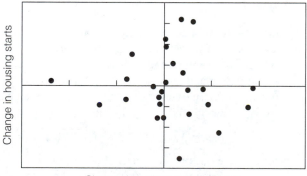

Source: Exhibit 3.19

with *lagged* mortgage rate changes. As expected, there appears to be a negative relationship between the two variables, but not a strong one. The correlation coefficient is only −0.12 compared to the original correlation between the contemporaneous changes (−0.57). This lack of linear association is clearly evident in the scatter diagram.

Sample Autocorrelation Functions

In autocorrelation analysis, a time series is related to lagged versions of itself. To illustrate the association between successive values in the mortgage rate series, we have depicted the mortgage rate column and Lagged 1 column in Exhibit 3.19 as a scatter diagram in Exhibit 3.21. There is evidence of a strong positive (almost linear) association between the values in the two columns of data. This dependency on past values is used to measure the strength of the relationship between the values of a time series and its values one or more time periods away. We quantify this mutual dependence in a time series with a set of autocorrelation coefficients.

Autocorrelation coefficients measure the degree of association between lagged values of a time series.

A useful way to visualize the effect of autocorrelation is to display the sample autocorrelation function, known as the ordinary correlogram. The ordinary correlogram is determined by correlating a time series with versions of itself in which the dimension of time has been shifted. This yields a set of coefficients, which can be plotted successively, corresponding to each time shift or lag. The ordinary or autocorrelogram is simply a plot of the estimated autocorrelation coefficients.

An (ordinary) correlogram is a display of the sample autocorrelation coefficients.

The interpretation of a correlogram is more an art than a science and requires substantial experience. Correlograms are commonly used for time series with the following characteristics: pure randomness, low-order serial correlation, trend, seasonality, and alternating and rapidly changing fluctuations.

Given n values of a time series $\{Y_t, t = 1, 2, \ldots, n\}$, a commonly used formula for the first sample autocorrelation coefficient r_1 is

$$r_1 = \sum (Y_t - \overline{Y})(Y_{t-1} - \overline{Y}) / \sum (Y_t - \overline{Y})^2$$

where $\overline{Y}$ is the average (arithmetic mean) of the n values of the time series. The subscript t in the value Y_t denotes time; thus, Y_{t-1} is the value of the series one period earlier.

In a similar fashion, the formula for the sample autocorrelation coefficient r_j between values of the time series separated by j periods is

$$r_j = c_j/c_0 \quad \text{for } j = 0, 1, 2, 3, \ldots, k$$

where

$$c_j = \frac{1}{n} \sum_{t=1}^{t=j+1} (Y_t - \bar{Y})(Y_{t-j} - \bar{Y})$$

For large n, r_k is approximately normally distributed with a mean of 0 and variance of $1/n$. Thus, approximate 95% confidence limits for autocorrelation coefficients can be plotted at $\pm 1.96 \sqrt{n}$. (We show in the next section how confidence limits are determined.) The observed values of r_j that fall outside these limits suggest that the corresponding autocorrelation coefficient is significantly different from 0 at the 5% level. If the first 20 values of the correlogram are plotted, we might expect one significant value even if the r_j values are random.

In practice, the number of r_j values calculated for a correlogram varies with the length of the series and the length of the seasonal cycle. Generally, for a 12-month seasonal series, approximately 36–60 monthly values are appropriate. No more than approximately $n/4$ to $n/3$ correlations should be calculated for a series of length n. Otherwise, there will not be enough terms in the calculation for higher lags for any inferences to be meaningful. It is desirable that $n > 50$.

Detecting Autocorrelation

Correlograms should be interpreted, with caution, in terms of the patterns of autocorrelation coefficients more than the behavior of individual values of the coefficients.

A purely **random time series** is one in which there is no time dependence between values any number of time periods (lags) apart and thus no systematic pattern that can be exploited for forecasting. Thus, autocorrelation coefficients for a purely random time series should be 0 at all lags, except at lag 0. At lag 0, the time series is perfectly related to itself and has the maximum value 1. For a sampled random series, a correlogram shows sample estimates of autocorrelations of the data. Hence, the correlogram of a random series is 1 at lag 0 and nearly 0 for all lags different from lag 0. Hence, the correlogram shown in Exhibit 3.23 is unity at lag 0 and

EXHIBIT 3.23
Correlogram of a
Random Time Series

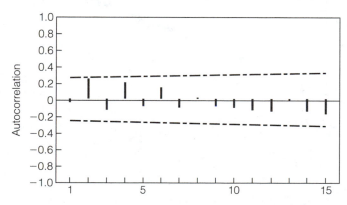

very close to 0 for all nonzero values k of the lag (there are 15 of these plotted on the abscissa.

A random time series is one in which there is no time dependence between values that are any number of time periods apart.

Note that a random series is not necessarily normally distributed. A normal probability plot of the 60 numbers does not support a normal distribution assumption for the data (Exhibit 3.24a), but, in fact, the random series was generated from a lognormal distribution (Exhibit 3.24b).

EXHIBIT 3.24
(a) Normal
Probability Plot
of a Random
(Nonnormal) Series;
(b) Lognormal
Probability Plot of
the Random Series

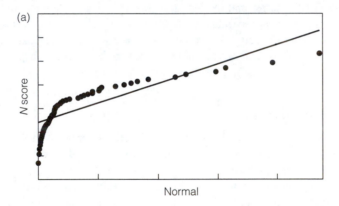

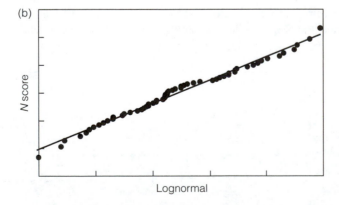

As we see in the next chapter, patterns found in a correlogram can be used to analyze associated patterns in the basic data, such as seasonal periods, which is useful for the specification of ARIMA time series models (Chapter 13).

Patterns of autocorrelation coefficients can give insight into the structure or characteristics of a time series.

This association between two different time series can be quantified with a set of cross-correlation coefficients. The cross-correlogram displays the correlations between lagged versions of two time series. Cross-correlations play a major role in estimating **transfer function**s for ARIMA time series models and dynamic regression models (Pankratz, 1991), but are beyond the scope of this book.

Low-Order Autocorrelation

Many time series, even after a certain amount of **differencing,** exhibit short-term correlations (or memory) in their pattern. Thus, autocorrelation at shorter lags is greater than autocorrelation at longer lags.

For a time series having low-order dependence, a correlogram shows a decaying pattern.

Exhibit 3.25a shows a correlogram of the annual housing starts data, and Exhibit 3.25b shows a correlogram of the annual mortgage rate series. Note that the spikes tend to get progressively smaller. Once inside the bounds (roughly, $2 / \sqrt{n}$, $n = 29$), the spikes cannot be interpreted as significant. In this case, the trending in the data induces a memory pattern, and the correlogram corroborates this. Thus, a correlogram can help identify structure in the data, but the pattern is not unique to the individual time series.

Periodic Cycles

Other time series have a tendency to show periodicities in a fairly regular way. This means that correlograms for these series also show a similar **periodicity.** A series such as the changes in the telephone access lines (Exhibit 1.9) is a very special kind of periodic series in which the highs and lows correspond to a monthly cycle (known as seasonality). It is interesting to note that correlograms of seasonal data have their very own cyclical pattern; notice the high autocorrelation at lag 12 and at multiples of 12 for the access line-gain data displayed in Exhibit 3.26. The spike at lag 12 and to a lesser extent at lags that are multiples of 12 (24, 36, etc.) is characteristic of a monthly seasonal pattern. The pattern of positive values in the correlogram of housing starts (Exhibit 3.27), in contrast to the cyclic positive and negative values of the telephone access line correlogram, is indicative of the relative strength of a trend in the housing starts data; the predominantly seasonal pattern in the housing starts is

EXHIBIT 3.25 (a) Correlograms of the Annual Housing Starts; (b) Correlograms of the Annual Mortgage Rate Data

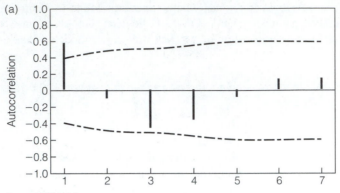

Source: Exhibit 2.3

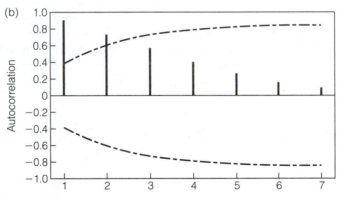

Source: Exhibit 3.19

also reflected. The monthly sale of a pharmaceutical product (shown in Exhibit 2.10) is an example of a seasonal time series in which the seasonal peaks and troughs are not consistent throughout the historical period. Consequently, its correlogram in Exhibit 3.28 lacks the definitive pattern of a seasonal time series and will be more difficult to model.

 A time series with periodic cycles has a correlogram showing a similar periodicity.

3.5 WHAT DOES NORMALITY HAVE TO DO WITH IT?

Statistical analyses usually proceed along the following lines:

1. Postulate a probability model, including unknown parameters, for a situation involving uncertainty.

2. Use data to estimate the unknown parameters in the model.

3. Plug the estimated parameters into the model and do calculations.

EXHIBIT 3.26
Correlogram for the
Monthly Change in
Telephone Access
Lines

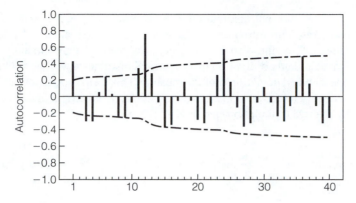

(*Source:* Exhibit 1.9)

EXHIBIT 3.27
Correlogram for the
Monthly Housing
Starts

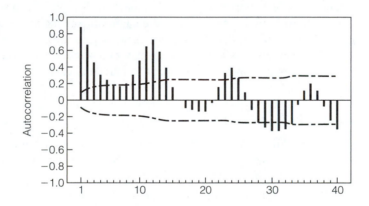

(*Source:* www.census.org.gov/const/C20/startsua.xls)

EXHIBIT 3.28
Correlogram for the
Monthly Sales of a
Pharmaceutical
Product

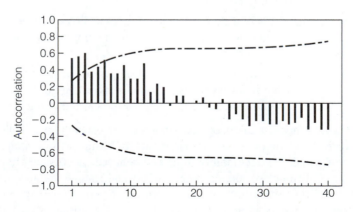

(*Source:* Exhibit 2.10)

For a normal distribution with (population) mean μ and standard deviation σ, we know from introductory statistics that the interval $(\mu - 1.64\sigma, \mu + 1.64\sigma)$ contains 90% of the distribution. Similarly, there is approximately a 68% probability that a random observation from a normal distribution will be between plus and minus one standard deviation σ of the mean μ. This well-known normal distribution, which is deeply rooted in statistical theory, looks symmetrical and is completely specified by the two (population) parameters (μ, σ).

 In inferential statistics, probability models are used to make statements about the probability that a certain portion of the data will fall within a specified range.

One reason why forecasters use histograms is to get a sense of how closely historical demand, forecast errors, or model residuals resemble a normal probability distribution. For example, we might assume that daily changes in stock prices follow a normal distribution. Using historical data to estimate the mean and standard deviation of the assumed distribution, we then use the model to predict future price changes. When we look at 50 daily changes of a stock, we are considering a sample of size 50 from a hypothetical population of all daily changes of the stock (Granger, 1993).

The normality assumption is widely used among practitioners and is justified primarily for the following reasons:

- Observed data are often represented reasonably well by a normal distribution. This can be verified by the use of empirical frequency distributions or various normal probability plotting techniques.

- When the data are not normally distributed, it is theoretically possible to find a transformation of the data that renders the distribution normal. Although this is not always practical, sometimes a very simple transformation (such as taking the logarithm or square root of the data) results in residuals that appear approximately normal.

- In the words of Tukey, "Practice dictates a choice between what can be done and should be done. In the absence of anything better, normality usually implies what can be done."

- Fortunately, the normality assumption permits us to apply a very extensive (although not always realistic), often simple, and quite elegant set of statistical tests of significance to a multitude of forecasting problems.

Much statistical theory used in forecasting is built on a mathematical foundation that is not often realized in the practice of business forecasting. For determining a typical value and variability from well-behaved data (usually implying they are normally distributed), the arithmetic mean and sample variance can be shown to be "best" (i.e., the estimators have optimal properties; see Appendix 3B) if the data can be viewed as a random sample from the normal distribution. Least-squares techniques are not only not "best," but their results can be very misleading when the data deviate even a little from normality assumptions. Because of this, it is important to have forecasting procedures that are robust against data that would otherwise distort results.

 When dealing with confidence intervals and hypothesis testing, correlation and least-squares regression analysis generally require an underlying normal distribution for the inferential assumptions.

Determining Precision with Standard Errors

In statistics, the sample has to be the product of a well-specified process, such as a random sample, and this assumption is difficult to meet with business time series. Every sample statistic from a random sample is a random variable. To answer the question of how we can assess the uncertainty in the sample average of daily changes in stock prices, we need to examine the sampling distribution of the sample average. If we are **sampling** from a population, this sampling distribution has a (population) mean and standard deviation (called the standard error) that depend directly on the population values of the underlying distribution of the random sample. Business statistics textbooks tell us that according to the Central Limit Theorem, regardless of the underlying population, the distribution of the sample average tends toward a normal distribution as the sample size becomes large. As a practical rule of thumb, a sample size of 30 is considered to be large enough to assume a normal distribution for the sampling distribution.

 A random sample from a population is a set of randomly selected observations from that population.

Establishing Confidence and Prediction Intervals

Statements of uncertainty are frequently required when we make forecasts or determine the reliability of parameter estimates in a forecasting model. These statements are based on assumptions about the data-generating process—usually involving a normality assumption. Typical statements of uncertainty may take the form of confidence intervals, as in "I want to be 95% sure that in repeated samples of a test, the true value of the slope parameter in the regression model is within certain bounds"; or prediction intervals, as in "I want to be 95% sure that in a future sample, the forecast will fall within certain bounds."

 Statements of uncertainty about forecasts are made with prediction intervals.

You generally need normal distribution theory to obtain confidence limits for parameter values in a regression model. We can get an indication of the accuracy of estimation from the confidence interval, which consists of a lower and upper limit within which we are fairly certain that the unknown population parameter (e.g., the mean) is located. To construct a confidence interval, we use the sample statistic, the standard error of the sample statistic, and a *t* factor (or statistic). The (population) standard deviation of a sample statistic is called its standard error and the *t* factor is determined from the level of confidence desired. For example, for 95% confidence (two-sided) and a sample size of 12, the *t* factor is 2.2 (found in the table for the *t* distribution). For sample sizes greater than 30, the *t* factor is 2.0 to the first decimal place. Hence, for practical purposes, we use $t = 2$ whenever we construct a 95% confidence interval for a model parameter. The confidence interval takes the general form:

$$\{\text{Statistic}\} \pm t \text{ factor} \cdot \{\text{Standard error of the statistic}\}$$

The particular numbers used in the formula can be found in the typical computer output of a statistical analysis.

Statements of uncertainty about parameters in models are made with confidence intervals.

Prediction intervals and confidence intervals make statements about unknown quantities. Their validity is based on a process of repeated sampling, which give rise to probability statements for an observed interval. If the interval is wide, we have high uncertainty even if all our assumptions are correct. On the other hand, if the interval is narrow, we have low uncertainty only if assumptions hold.

A prediction interval for a forecast, also called an interval forecast, is a formal statement about the limits of forecast error. The margin of error determines a lower and upper limit for the prediction interval. A prediction interval typically comprises two components: (1) an estimate of the forecast error variance (or its square root, the standard error) and (2) a multiplier for the standard error, derived to ensure that the interval is wide enough to meet the specified probability. The appropriate multiplier depends on the assumed probability distribution for the forecast errors. Standard practice is to make the assumption that forecast errors are normally distributed about a mean of zero. In a normal distribution, a multiplier of 1.64 generates a prediction interval wide enough to encompass 90% of the possible error values and 1.96 is the factor for a 95% interval.

A prediction interval expresses with a defined probability how confident we can be, based on our forecasting equation, that the future will materialize within a margin of error around the forecast.

Exhibit 3.29 shows the 95% confidence intervals for the population mean and median of the service-order travel data (Exhibit 3.10). The data ranges between

EXHIBIT 3.29
(a) Box Plot of
Service-Order Travel
Data (Exhibit 3.10);
95% Confidence
Intervals for
(b) Population Mean
(μ) and (c) Median

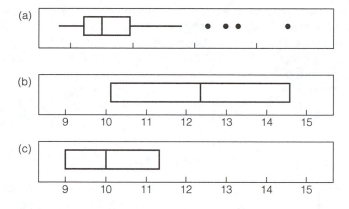

3 and 40, the median is 10, and the mean is 12.36. Because the data are right-skewed, the confidence interval for the mean is centered to the right of the center of the confidence interval for the median.

Exhibit 3.30 illustrates the prediction intervals resulting from fitting a linear trend, exponential smoothing procedure to a time series of IBM annual earnings (1944–84). The forecast origin is 1984, and forecasts were generated for 1985–89. None of the 90% prediction in intervals (not shown) contains the actual values that occurred for 4 years after the 1985 forecast.

A 99% prediction interval for the 1-year-ahead forecast means that "We can be 99% certain those 1985 earnings will fall between $49.9 and $53.5 billion." The equivalent statement in terms of forecasting error is "We can be 99% certain that our forecast error for next year's earnings will not exceed $1.15 billion." Note that 1.15 is the distance from the 1985 forecast to either the lower or upper limit. Unless the original data have been transformed, prediction intervals will be symmetrical about a point forecast if normality is assumed.

The appropriate multiplier for a 99% prediction interval is necessarily larger than that for a 90% interval. As shown in Exhibit 3.30, for each year the distance from the lower to the upper limit is greater. For 1985, we can be sure with 99% confidence that earnings will fall between $49.9 and $53.5 billion. So, except for this one-step-ahead forecast, we can be certain that the chosen model is inappropriate for these data.

EXHIBIT 3.30
Approximate 99%
Prediction Interval
for IBM Annual
Earnings (in billions
of dollars)

Year	Lower Limit	Forecast	Upper Limit	Actual
1985	49.9	51.7	53.5	50.1
1986	53.8	57.4	61.1	51.2
1987	57.2	63.2	69.2	54.2
1988	61.1	68.9	76.8	59.7
1989	64.3	74.7	85	62.7

Average forecast error = 1.80
Average percent error = 18.9%

There are prediction interval calculations for linear regression models, ARIMA, and exponential smoothing models. Indeed, for exponential smoothing alone, a number of different approaches have been employed. See, for example, Chatfield (1993) for a thorough discussion of the issues surrounding the calculation of forecast intervals in practice.

Prediction interval calculations are specific to the method employed.

3.6 THE NEED FOR NONTRADITIONAL METHODS

The need for a nontraditional approach to **estimation** is motivated by two problems. First, a forecaster never has an accurate knowledge of the true underlying distribution of the random errors in a model. Second, even slight deviations from a strict parametric model can give rise to poor statistical performance for classical (i.e., associated with the ordinary least-squares method) estimation techniques. Estimators that are less sensitive to these factors are often referred to as *robust*.

There appear to be many meanings of the word *robustness* in modern statistical literature (Hoaglin et al., 1983). In the context of estimation, robustness of efficiency means that parameter estimates are highly efficient not only under idealized (usually normal) conditions but also under a wide class of nonstandard circumstances. The Princeton Robustness Study (Andrews et al., 1972) was an early effort to analyze systematically this concept for estimates of location. Estimates that have robust efficiency are often very resistant to outliers. This can be a very valuable consideration because real-life data are frequently nonnormal and possess hard-to-detect outlying observations.

The use of robust or resistant methods is illustrated by Mallows (1979), who describes several analyses of large data sets (more than 1000 observations) in which robustness considerations have proved relevant.

Insuring against Unusual Values

We have already pointed out that outliers can distort certain estimators; a robust or resistant procedure must produce estimates that are not seriously affected by the presence of a few outliers. Thus, some robust estimators are designed to be resistant to unusual values—they give less weight to observations that stray from the bulk of the data.

In the data set {1.1, 1.6, 4.7, 2.1, 3.1, 32.7, 5.8, 2.6, 4.8, 1.9, 3.7, 2.6} that we used earlier in the chapter, the standard deviation is 8.6, the MdAD is 1.1, and the IQD is 2.75. For normally distributed data, the standard deviation can be approximated by dividing the MdAD by 0.6745 (= 1.63) or by dividing the IQD by 1.35 (= 2.04). We can determine outlier boundaries by calculating the mean plus and minus three standard deviations; for the upper limit this is $5.55 + (3)(8.6) = 31.35$, which almost encompasses the suspected outlier (= 32.7). This is because the calculation of the standard deviation is itself distorted by the outlier because it gives equal weight to all observations. However, if we consider the alternative means of determining outlier boundaries, the median plus and minus three times UMdAD, we get $2.9 + 3(1.63) = 7.79$ for the upper limit. Now 32.7 is clearly shown to be very unusual and to be far away from the bulk of the data!

Large data sets with numerous unusual values are prevalent in planning systems for forecasting replenishments of SKUs (stock-keeping units) for inventory, production, and distribution planning applications (see Chapter 9).

M-Estimators

The M-estimation method can be used to reduce automatically the effect of outliers by giving them a reduced weight when we compute the typical value of the data. The method is based on an estimator that makes repeated use of residuals in an iterative procedure. This estimator, called an M-estimator, is the maximum likelihood estimate (a good statistical property to possess) for the location parameter of a heavy-tailed distribution (Huber, 1964). Basically, the Huber distribution behaves like a normal distribution in the middle range and like an exponential distribution in the tails. Thus, the bulk of the data appears normally distributed, but there is a greater chance of having extreme observations in the sample.

A Numerical Example

We will work out an example of calculating an M-estimate based on a small set of (ordered) data corresponding to 15 percentage errors on weekly forecasts: $\{-67, -48, 6, 8, 14, 16, 23, 24, 28, 29, 41, 49, 56, 60, 75\}$.

The first step in data analysis is to investigate the data graphically. Note there are a couple of extremely low values, -67 and -48. We might then prepare a Q-Q plot. If the data are normally distributed, they should lie along a straight line. The Q-Q plot can also be used to get a quick-and-dirty robust estimate of the (μ, σ) parameters of the normal distribution. By eyeballing a straight line through the bulk of the points on the Q-Q plot, we can determine that the Y intercept at $X = 0$ and the slope of the line correspond to estimates of μ and σ, respectively.

The plot (Exhibit 3.31) demonstrates that the two smallest data values are indeed extreme, indicating that the normality assumption is not valid. On the other

EXHIBIT 3.31
The Q-Q Plot for the Example Data Set (mean = 20.93; Standard Deviation = 36.46)

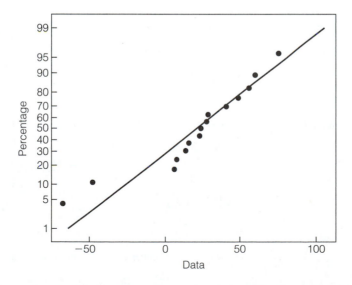

hand, the ordinary least-squares estimates, calculated from the data, are $\mu = 20.93$ and $\sigma = 37.74$, which could be used to superimpose a straight line with intercept 20.93 and slope 37.74 on the Q-Q plot. As can be seen from Exhibit 3.31, the straight line determined from the least-squares estimates does not represent the bulk of the data on the Q-Q plot very well.

Exhibits 3.32 and 3.33 show the calculations for the Huber M-estimator of location and its corresponding standard error for our data set $\{-67, -48, 6, 8, 14, 16, 23, 24, 28, 29, 41, 49, 56, 60, 75\}$. The median of the data is 24. Column 2 is the difference between the raw data and the median. The median absolute deviation is 17; it is the median of the absolute value of this column (the eighth largest absolute value in column 2). An approximate unbiased scale statistic is the UMdAD = MdAD/0.6745 = 25.2.

EXHIBIT 3.32
Calculations of the Huber M-Estimator of Location ($K = 2$, and $\theta_0 = 24$, median)

$Y_{(i)}$	$Y_{(i)} - \theta_0$	w_{I1}^2	$Y_{(i)} - \theta_1$	w_{I2}^2
-67	-91	50.4/91	-92.13	50.8/92.13
-48	-72	50.4/72	-73.13	50.8/73.13
6	-18	1	-19.13	1
8	-16	1	-17.13	1
14	-10	1	-11.13	1
16	-8	1	-9.13	1
23	-1	1	-2.13	1
24	0	1	-1.13	1
28	4	1	2.87	1
29	5	1	3.87	1
41	17	1	15.87	1
49	25	1	23.87	1
56	32	1	30.87	1
60	36	1	34.87	1
75	51	50.4/51	49.87	1

EXHIBIT 3.33
Calculations for the Huber M-Estimator of the Standard Error ($K = 2$ and $\theta_0 = 24$, median)

Iteration 0 $\theta_0 = \text{Median} = 24$

Iteration 1 $s = \text{MdAD} / 0.6745 = 17/0.6745 = 25.20$

$Ks = 50.40$

$\theta_1 = \left[\sum W_i^2 y_i / \sum W_i^2 \right] = 357.38 / 14.24 = 25.13$

Iteration 2 $s = \text{MdAD} / 0.6745 = 17.13 / 0.6745 = 25.40$

$Ks = 50.80$

$\theta_2 = \left[\sum W_i^2 y_i / \sum W_i^2 \right] = 358.72 / 14.25 = 25.17$

$V(\theta) = \left[\sum W_i^4 (y_i - \theta_2)^2 / (n^*)^2 \right]^{1/2}$

$= \left[\sum W_i^4 (y_i - 25.17)^2 / (13)^2 \right]^{1/2} = 8.28$

where $n^* = $ number of observations receiving full weight

The results for the mean, the 0.25-trimmed mean (midmean), and the Huber M-estimator are listed in the following table.

	Mean	Midmean	M-Estimator
θ	20.93	25.78	25.17
$V(\theta)$	9.74	7.37	8.28

On a technical note, $V(\theta)$ is the estimated standard deviation (standard error) of the *location estimator* and should not be confused with the estimated standard deviation of the *sample values*.

We have now shown several times that the arithmetic mean (and hence any moving average) is very sensitive to outliers (extreme values) in the data. Both the trimmed mean and Huber M-estimator provide estimates that are less sensitive to these extreme values. The standard error of the M-estimator is somewhat smaller than of the arithmetic mean. However, the standard error for the midmean is significantly less. This is somewhat expected because computations for the midmean begin by trimming (Winsorizing) 40% of the data, this 40% being the most extreme values (three from each end). However, these values still have on some effect on (but less than on **least-squares**) the Huber M-estimator because their associated weight is not zero. The M-estimator appears again in regression analysis (Chapter 12) to weight extreme values so that they cannot unduly distort estimates in regression relationships.

SUMMARY

This chapter has stressed the importance of basic statistical tools for all phases of **model building**—exploratory as well as confirmatory. A number of summary measures are considered: statistics, such as the mean, median, and trimmed mean, used to describe central tendency; measures of dispersion, such as the standard deviation, median absolute deviation, and range; box plots, which are a convenient way of visually displaying a five-number summary of a distribution; correlation analysis, which is a useful tool for looking for linear associations between pairs of variables; and correlograms, which are used to characterize patterns in data, such as trend and periodic cycles.

Correlograms and their interpretations play a major role in ARIMA time series modeling (see Chapters 13 and 14). They are used in the analysis of residuals from regression models. As part of the diagnostic checking process, correlograms can be useful in validating the assumptions that residuals are randomly distributed (see Chapter 10).

Stem-and-leaf displays show groupings of data values, distribution or spread of the data, and asymmetrical trailing off at the low or high end. Central tendency, missing values, and unusual values can also be identified with this display. The display is particularly useful for examining residuals when a model is fit to a data set. The Q-Q plot is used to determine whether a data set, especially residuals from a regression model, follows a normal distribution. If it does, hypothesis testing can be carried out or confidence limits can be constructed.

Summaries are necessary to quantify information inherent in the shape or distribution of data. We plot frequency distributions to show what percentage of the total number of data each interval contains, and cumulative frequencies or percentiles can be used to determine whether a data set follows a normal probability distribution.

When assumptions are not satisfied, a statistical analysis can give misleading results and inappropriate forecasting models. In addition, confidence intervals for the parameters may be stated too conservatively. When the data analysis suggests the data are nonnormal with possible outliers, robust or resistant methods should be considered as a way of dealing with the estimation and modeling problems. Trimmed means and M-estimates are two examples of robust procedures that complement the usual (least-squares) procedures. When they are in agreement, they should be reported together. When substantial differences exist between the two analyses, the data should be examined more thoroughly for outliers or bad data values. Even if you use robust techniques, you should still plot the data for a thorough examination of it.

What we desire, in practice, are robust procedures that are resistant to nonnormal tails in the data distribution, so that they give rise to estimates that are much better than those based on normality.

Correlation analysis is a useful tool for looking for linear associations between pairs of variables, but it cannot replace thoughtful and thorough data analysis on the part of the forecaste. Similarly, variables that are otherwise highly correlated may appear to have a low correlation coefficient because of the presence of outliers (Appendix 3A).

REFERENCES

Andrews, D. E., P. J. Bickel, E. R. Hampel, P. J. Huber, W. H. Rogers, and J. W. Tukey (1972). *Robust Estimates of Location: Survey and Advances*. Princeton, NJ: Princeton University Press.

Belsley, D. A., E. Kuh, and R. E. Welsch (1980). *Regression Diagnostics: Identifying Influential Data and Sources of Collinearity*. New York: John Wiley and Sons.

Chatfield, C. (1993). Calculating interval forecasts (with discussion). *J. Bus. Econ. Stat.* 11, 121–44.

Cleveland, W. S. (1994). *Visualizing Data*. Summit, NJ: Hobart Press.

Cryer, J. D., and R. B. Miller (1994). *Statistics for Business*. Belmont, CA: Wadsworth.

DeLurgio, S. (1998). *Forecasting Principles and Applications*. New York: Irwin/McGraw-Hill.

Devlin, S. J., R. Gnanadesikan, and J. R. Kettenring (1975). Robust estimation and outlier detection with correlation coefficients. *Biometrika* 62, 531–45.

Dielman, T. E. (1996). *Applied Regression Analysis for Business and Economics*. 2nd ed. Belmont, CA: Wadsworth.

Draper, N. R., and H. Smith (1998). *Applied Regression Analysis*. 3rd ed. New York: John Wiley and Sons.

Gonick, L., and W. Smith (1993). *A Cartoon Guide to Statistics*. New York: HarperCollins Publishers.

Granger, C. W. J. (1993). Forecasting stock prices: Lessons for forecasters. *Int J. Forecasting* 8, 3–13.

Hanke, J. E., and A. G. Reitsch (1998). *Business Forecasting*. 6th ed. Englewood Cliffs, NJ: Prentice Hall.

Hoaglin, D. C., F. Mosteller, and J. W. Tukey (1983). *Understanding Robust and Exploratory Data Analysis*. New York: John Wiley and Sons.

Huber, P. J. (1964). Robust estimation of a location parameter. *Ann. Math. Stat.* 35, 73–101.

Huber, P. J. (1997). Speculations on the path of statistics. In *The Practice of Data Analysis: Essays in Honor of J. W. Tukey,* edited by D. R. Brillinger, L. T. Fernholz, and S. Morgenthaler. Princeton, NJ: Princeton University Press.

Levenbach, H., and J. P. Cleary (1984). *The Modern Forecaster.* Belmont, CA: Wadsworth.

Makridakis, S., S. C. Wheelwright, and R. J. Hyndman (1998). *Forecasting Methods and Applications.* 3rd ed. New York: John Wiley and Sons.

Mallows, C. L. (1979). Robust methods—Some examples of their use. *Am. Statist.* 33, 179–84.

McGill, R. J., J. W. Tukey, and W. A. Larsen (1978). Variation of box plots. *Am. Statist.* 32, 12–16.

Menzefricke, U. (1995). *Statistics for Managers.* Belmont, CA: Wadsworth.

Pankratz, A. (1991). *Forecasting with Dynamic Regression Models.* New York: John Wiley and Sons.

Robbins, N. B. (2005). *Creating More Effective Graphs.* New York: John Wiley and Sons.

Schmid, C. F. (1983). *Statistical Graphics: Design Principles and Practices.* New York: John Wiley and Sons.

Tryfos, P. (1998). *Methods for Business Analysis and Forecasting: Text and Cases.* New York: John Wiley and Sons.

Tufte, E. (1990). *Envisioning Information.* Cheshire, CT: Graphics Press.

Tufte, E. (2000). *The Visual Display of Quantitative Information.* 2nd. ed. Cheshire, CT: Graphics Press.

Tukey, J. W. (1962). The future of data analysis. *Ann. Math. Statist.* 33, 1–67.

Tukey, J. W. (1977). *Exploratory Data Analysis.* Reading, MA: Addison Wesley.

Velleman, P. E., and D. C. Hoaglin (1981). *Applications, Basics, and Computing for Exploratory Data Analysis.* North Scituate, MA: Duxbury Press.

Wilson, J. H., and B. Keating (1998). *Business Forecasting.* New York: Irwin McGraw-Hill.

PROBLEMS

3.1. For the annual mortgage rate data in Exhibit 2.3:

   a. Determine a suitable stem-and-leaf display.

   b. Calculate a 15% trimmed mean (or midmean) and compare with the median. What does the midmean tell us about the symmetry of the data?

   c. Find the 75th (upper-quartile) and 25th (lower-quartile) percentiles, and calculate the IQD. Also determine the extreme values for the data.

   d. Find the median, and construct a box plot using the five-number summary.

   e. Define QL − 1.5 IQD and QU + 1.5 IQD as the outlier cutoffs, where QL and QU denote the lower and upper quartiles, respectively. Determine the outlier(s).

   f. Calculate the standard error for the 15% trimmed mean.

   g. Calculate a Huber M-estimator of location using $K = 1.5$ for two iterations, and summarize the results.

   h. Compare your results in (g) with the mean and trimmed mean.

3.2. Repeat Problem 3.1 for one or more of the following data sets:

   a. Waiting times for hospital booking (HOSPITAL, Menzefricke, 1995, 47)

   b. Savings and Loan rates of return (S&LRETURN, Dielman, 1996, Table 5.7)

3.3. Repeat Problem 3.1 using the data set {−67, −48, 6, 8, 14, 16, 23, 24, 28, 29, 41, 49, 56, 60, 75}.

3.4. Compare the results for Problem 3.3 with those for the data set {1.1, 1.6, 4.7, 2.1, 3.1, 32.7, 5.8, 2.6, 4.8, 1.9, 3.7, 2.6}, in the chapter. How well do the midmean, median, and, mean agree among themselves for the two data sets?

3.5. For the monthly toll revenue data (TOLLREV) on the CD, plot and interpret the data as a quarterly and annual time series. By inspection determine:

   a. the trend pattern

   b. the nature of the seasonal variation

   c. the cyclical variation

3.6. For the monthly toll message data (MSG) on the CD create an annual time series. Using the annual data, make a stem-and-leaf diagram.

   a. Create several measures of location (mean, median, and midmean).

b. Determine several measures of scale (standard deviation and UMdAD). Do these measures suggest the presence of unusual data values? Which measure do you recommend be used in this case?

c. Create a box plot. In what range does the middle 50% of the data lie?

d. Create a normal probability plot. Is it reasonable to assume a normal distribution? If so, how do you estimate the mean and variance? How does this compare with your answer to (a) and (b)?

3.7   Repeat Problem 3.6 for one or more of the following monthly time series on the CD:

a. Unemployment rate data (UNEMP, Dielman, 1996, Table 4.7)

b. Silver prices data (SILVER, Dielman, 1996, Table 4.8)

c. Prime rate data (PRIME4, Dielman, 1996, Table 4.14)

d. Retail sales for U.S. retail sales stores (RETAIL, Hanke and Reitsch, 1998)

e. Wheat exports (SHIPMENT, Dielman, 1996, Table 4.9)

f. Shipments of spirits (DSHIP, Tryfos, 1998, Table 6.23)

g. Retail sales in a region (RSALES, Tryfos, 1998, Table 6.25)

h. Australian beer production (AUSBEER, Makridakis et al., 1998, Table 2.2)

i. International airline passenger miles (AIRLINE, Box, Jenkins, and Reinsel, 1994, Series G)

j. Australian electricity production (ELECTRIC, Makridakis et al., 1998, Fig. 7-9)

k. French industry sales for paper products (PAPER, Makridakis et al., 1998, Fig. 7-20)

l. Shipments of French pollution equipment (POLLUTE, Makridakis et al., 1998, Table 7-5)

m. Sales of recreational vehicles (WINNEBAG, Cryer and Miller, 1994, p. 742)

n. Sales of lumber (LUMBER, DeLurgio, 1998, p. 34)

o. Sales of consumer electronics (ELECT, DeLurgio, 1998, p. 34)

3.8. Repeat Problem 3.7 for one or more of the following quarterly time series on the CD:

a. Sales for Outboard Marine (OMC, Hanke and Reitsch, 1998, Table 4.5)

b. Shoe store sales (SHOES, Tryfos, 1998, Table 6.5)

c. Domestic car sales (DCS, Wilson and Keating, 1998, Table 1-5)

d. Sales of The GAP (GAP, Wilson and Keating, 1998, p. 30)

e. Air passengers on domestic flights (PASSAIR, DeLurgio, 1998, p. 33)

3.9. Create a scatter diagram of monthly toll revenues (TOLLREV) versus monthly toll messages (MSG). Interpret the pattern and make an educated guess about the value of the correlation coefficient; then calculate it precisely.

3.10. Determine the correlograms of the toll revenue (TOLLREV) and toll message (MSG) series for the (a) monthly, (b) quarterly, and (c) annual versions of the data. Interpret the patterns in the correlogram in terms of individual spikes as well as the overall pattern.

3.11. For the annual versions of TOLLREV and MSG data, create new differenced time series $Z_t = Y_t - Y_{t-1}$ in a spreadsheet. Then:

a. Plot the differenced data for each series.

b. Calculate the corresponding correlograms and interpret spikes and overall pattern.

c. Create the probability plots and interpret similarities and differences in these plots for the two differenced series.

3.12. Repeat Problem 3.11 for the quarterly versions of TOLLREV and MSG, creating the differenced series by calculating $Z_t = Y_t - Y_{t-4}$ in a spreadsheet for these data before proceeding with parts (a), (b), and (c).

CASE 3A PRODUCTION OF ICE CREAM (CASE 1A CONT.)

The ice cream entrepreneur has asked you to give her some insight into the shipment patterns by performing some basic data analysis. The table that follows contains almost 6 years of monthly shipments of ice cream (in thousands of gallons):

Month	YR01	YR02	YR03	YR04	YR05	YR06
Jan	61,977	61,968	61,366	56,652	57,467	51,997
Feb	60,814	66,002	70,455	65,458	60,805	56,299
Mar	74,369	76,598	81,295	79,739	75,110	68,633
Apr	76,066	79,638	82,088	74,887	68,867	67,056
May	86,254	87,764	85,814	81,544	80,433	74,556
Jun	90,227	92,460	97,512	94,009	85,781	78,412
Jul	95,945	96,038	97,074	88,657	81,090	83,585
Aug	90,040	91,054	88,818	89,554	81,523	76,473
Sep	77,097	77,938	79,383	74,416	67,102	—
Oct	70,511	73,243	66,631	63,071	61,310	—
Nov	59,541	60,291	59,749	57,467	57,785	—
Dec	58,608	60,603	61,213	56,068	53,886	—

(1) Show the annual variations by constructing box plots side by side for each of the first 5 years. How do you interpret the display?

(2) Calculate the monthly averages and comment on the pattern of these averages. Do you need to consider alternative measures of location (i.e., any unusual values)?

(3) Make a time plot, and discuss the pattern.

(4) Construct the correlogram, and interpret the pattern in the autocorrelations.

(5) Calculate and plot the 12-term centered moving average along with the data. What effect does that have on the pattern?

(6) Create a simplistic projection for September through December of YR06 with a 12-term moving average. Does it make sense?

(7) Summarize your key findings in one or two paragraphs.

CASE 3B TWIN RIVERS

We introduce this chapter with the Twin Rivers data from Tukey's *Exploratory Data Analysis* (1977, Chap. 8). The file **EI_GaUse.xls** on the CD contains the 152 (EIUse, GaUse) pairs (winter energy consumption in therms; IEUse = therms of electricity; GaUse = therms of gas; 1 therm = ~30 kilowatt hours and exactly 100,000 BTU).

(1) Explore a potential relationship between EIUse and GaUse.

   (a) Create a scatter diagram, and comment on whether this display appears informative to you.

   (b) Sort the EIUse column of data into approximately ten groupings (horizontal cuts). You may find that the first and last two EIUse groupings should be wider than the middle groupings to balance the amount of data in each group.

- Using the medians of the groups as an anchor, create multiple box plots of GaUse. Use the medians of EIUse groups on the horizontal axis and display the box plots side by side
- Can you now detect more of a pattern than you might have seen in a scatter diagram? What do you think the next steps should be?
- If you have plotted the group's median (GaUse) inside each box plot, you may want to smooth them and look for a trend in the medians. (For smoothing patterns, see Chapter 2.)

(2) Does there appear to be a connection between electric and gas uses for these townhouses? (At this stage, we are satisfied just to gain some insight. We do some confirmatory analysis on this same data in Chapter 11 dealing with regression analysis.)

(3) Review the scatter diagram again; are there townhouses that appear unusually high or low compared to the rest? How might this influence your sense of the stretch in the scatter?

CASE 3C ESTIMATING PROMOTION LIFTS USING RUNNING MEDIANS (SECTION 2.5 AND CASE 2E CONT.)

The Marketing VP is in dire need of estimates of a promotion's impact, apart from the business-as-usual baseline forecast. You are provided with start and end dates in the pattern of historical promotions, and you decide to use running medians as a filter to separate out its effects. The following series represents a constant level baseline ($=10$) with a promotion that lasts three periods {10, 10, 10, 10, 10, 10, 10, 7, 5, 9, 12, 20, 17, 8, 6, 7, 10, 10, 10, 10, 10, . . .}.

(1) Separately, use a three-period moving average (3_MAVG) and a five-period moving average (5_MAVG) on the data and calculate the residual as $10 -$ Fit, in each case. What is the estimated lift for three periods?

(2) Repeat (1), for a three- and five-period moving median.

(3) Plot the original data and the two smoothed series. Provide a qualitative assessment of these two procedures for removing the constant level from the original data.

(4) Repeat the procedure by running the 3_MAVG and 5_MAVG smoothers in succession before calculating the lift. Do the same for the moving medians. Contrast your findings.

CASE 3D DOMESTIC AUTOMOBILE PRODUCTION (CASE 1D CONT.)

The U.S. automobile market used to be one of the largest and most profitable in the world. Over time, U.S. manufacturers lost market share to the Japanese. Then, the Japanese started to manufacture automobiles in the United States in the hopes of reducing their exposure to exchange rate fluctuations. In addition, there are other countries marketing and manufacturing automobiles in the United States now. In this environment, it is extremely difficult to forecast the demand for domestic automobiles. Your assignment as an analyst for an industry association is to first understand the underlying data by producing various data summaries and graphical displays. A table of U.S. automobile production by month from 1977 to 1985, which includes a recession in the early 1980s, follows (from *Automotive News*, 1982 and 1986 *Market Data Book* issues).

Month	1977	1978	1979	1980	1981	1982	1983	1984	1985
Jan	712,548	688,096	786,839	531,670	459,862	266,858	452,949	670,687	701,921
Feb	700,399	689,355	738,808	624,301	482,098	311,926	511,981	695,171	673,582
Mar	936,870	905,041	889,775	643,200	623,571	455,808	589,394	783,807	747,433
Apr	819,154	875,652	723,248	580,855	653,306	474,222	549,322	672,200	757,752
May	880,640	929,094	935,198	527,353	666,389	499,242	627,517	712,979	776,811
Jun	956,064	872,012	826,678	534,085	712,524	546,807	692,046	684,381	694,122
Jul	694,436	524,766	583,765	417,038	518,456	436,644	473,239	524,872	618,537
Aug	514,643	512,073	445,349	312,758	397,334	353,656	536,421	557,060	550,469
Sep	757,266	741,768	622,851	534,480	457,731	425,047	663,210	541,792	635,329
Oct	876,044	915,978	783,843	675,001	535,129	415,197	717,972	701,686	760,970
Nov	792,743	862,127	627,346	562,380	418,948	404,542	684,263	678,927	668,083
Dec	652,867	637,336	454,669	473,764	354,697	383,921	614,038	554,159	541,031

(1) Plot the monthly time series in two ways: (a) as a time plot by month from 1977 to 1985 and (b) as a set of time plots in which the horizontal axis is Jan–Dec. What is the value in creating the second plot?

(2) Create and interpret a time plot and correlogram of the first-differenced data. Create and interpret a time plot and correlogram of seasonal (order 12) differenced data. Repeat for data after taking first difference and seasonal difference of order 12. Look for evidence of regularity in patterns that you may be able to describe. (a) Which set of differenced data do you think is closest to a set of random numbers? (b) For this series, create a stem-and-leaf diagram and a Q-Q plot. Is there much evidence for or against normality in the random series?

(3) Repeat (1) and (2) for the quarterly series. Are your conclusions any different? If so, how?

APPENDIX 3A: THE NEED FOR ROBUSTNESS IN CORRELATION

As in the case of the arithmetic mean and variance, the uncritical use of the sample product moment correlation coefficient r can be quite misleading because it can be very sensitive to certain outliers and data that are not normally distributed.

A Robust Correlation Coefficient

For a more robust assessment, we desire a correlation coefficient that is less sensitive to outliers than the product moment correlation coefficient r. One example of a robust estimator of correlation, called r^*(SSD), is based on the standardized sums and differences of two variables, say X and Y (Devlin et al., 1975). The formula is given by

$$r^* \text{ (SSD)} = (V_+^* - V_-^*) / (V_+^* + V_-^*)$$

where V_+^* and V_-^* are robust variances of a sum vector Z_1 and a difference vector Z_2. The calculations are not complicated, and we offer the derivation and justification for this formula in Appendix 11A.

Exhibit 3.34 shows how to calculate r^* for ten pairs of numbers by using the median and MdAD as robust estimates of location and scale, respectively. The calculation shows that $r^* = 0.97$, compared to the product moment correlation coefficient $r = 0.80$. At first sight, it might appear that the second observation for Y ($Y_2 = -134$) is the likely culprit. With the (Y_2, X_2)-pair removed, $r^* = 0.95$ and $r = 0.76$. But the real influential point is the last observation; when this pair is removed, the resulting nine pairs of numbers yield the same values for r^* and r (=0.96). Hence, r clearly understates the degree of linear association between Y and X; the measure r^* is more representative of the bulk of the data than r.

EXHIBIT 3.34
Calculation of a
Robust Estimate r^*
for Ten Pairs of
Numbers

Y	X	$\dot{Y} = (Y - \tilde{Y})/S_Y^*$	$\dot{X} = (X - \tilde{X})/S_X^*$	$Z_1 = \dot{Y} + \dot{X}$	$Z_2 = \dot{Y} - \dot{X}$
73	36	0.89	1.03	1.92	−0.14
−134	−35	−3.43	−2.62	−6.05	−0.81
23	18	−0.16	0.1	−0.06	−0.26
86	22	1.16	0.31	1.47	0.85
−62	−21	−1.93	−1.9	−3.83	−0.03
38	16	0.16	0	0.16	0.16
−18	−11	−1.01	−1.38	−2.39	0.37
−22	−3	−1.09	−0.97	−2.06	−0.12
44	16	0.28	0	0.28	0.28
78	100	0.99	4.31	5.3	−3.32

$\tilde{Y} = \text{Median}(Y) = 30.5$ $S_X^* = \text{MdAD}(X) = 19.5$
$S_Y^* = \text{MdAD}(Y) = 48.0$
$\text{MdAD}(Z_1) = 2.00$ $\text{MdAD}(Z_2) = 0.255$
$r^*(\text{SSD}) = (22 - 0.255^2) / (22 + 0.255^2) = 0.968$
(Product moment correlation coefficient $r = 0.80$)

We have already noted the impact an outlier (properly defined) can have on the calculation of a correlation coefficient. Exhibit 3.35 shows correlations for the telecommunications example; the robust correlation matrix for the percentage changes in toll revenue (PCTREV), toll message (PCTMSG), business telephones (PCTBMT), and nonfarm employment series (PCTNFRM) can be compared with the matrix of ordinary product moment correlation coefficients. For comparison, the ordinary product moment correlation coefficients are shown in brackets underneath the corresponding robust estimates. It appears that most product moment correlation coefficients may be somewhat understated. This may be due to the presence of one or more discrepant values lying along the off-diagonal (from top left to bottom right) direction in the scatter plots for the variables.

If you need to look at many pairs of correlations, it is often useful to compute each pairwise correlation in two ways—first with a sample product moment correlation coefficient r and then with robust correlation coefficient r^*. Any large differences between these coefficients implies the presence of outliers distorting the true estimate of correlation. In a plot of r versus r^*, the points close to the $45°$ line ($r = r^*$) would imply that there are no outliers adversely affecting r in the data; on the other hand, the presence of larger deviations suggests that you should look for outliers. These pairs of variables could then be assessed for their influence on the regression analysis (Belsley, Kuh, and Welsch, 1980; Cryer and Miller, 1994; Draper and Smith, 1998).

EXHIBIT 3.35

Matrix of Robust and Ordinary Correlation Coefficients for the Telecommunications data (ordinary correlation coefficients in brackets)

	1	2	3	4
1 PCTREV	1			
	[1.00]			
2 PCTMSG	0.76	1		
	[0.67]	[1.00]		
3 PCTBMT	0.61	0.46	1	
	[0.50]	[0.55]	[1.00]	
4 PCTNFRM	0.67	0.81	0.64	1
	[0.63]	[0.70]	[0.60]	[1.00]

APPENDIX 3B: COMPARING ESTIMATION TECHNIQUES

A typical estimation procedure goes as follows. First, obtain data from a random experiment. Then, construct a model that relates the data to the physical quantities of interests (parameters) through an error structure. From this model, apply some theory to obtain the estimators. These estimators will be expressed as some function of the original raw data. The important point is that the estimators are themselves random variables. Therefore, each estimator should have a probability distribution. The shape of the distribution of an estimator (where it is centered, how it is concentrated, and so on) will tell you essentially everything about this estimator. In a way, unbiasedness, consistency, and asymptotic efficiency can be viewed as technical ways of describing the desirable shapes of the distributions of estimators.

In selecting a reasonable estimator, one goal is to be able to test how the estimates differ from the true (unknown) parameters of the regression line. One set of criteria for choosing the estimates is that they possess certain desirable properties. Among those, unbiasedness, consistency, efficiency, and minimum mean-squared error are the most often mentioned in discussions about comparative estimation techniques. Although these concepts may have limited direct consequences on forecasting, they are nevertheless of considerable value in regression analysis and econometric forecasting, so the practitioner does well to have a familiarity with them.

Strictly speaking, unbiasedness, consistency, and asymptotic efficiency are properties of estimators of a (real-value) parameter. You may occasionally run into an unbiased test, a consistent estimator of a vector-value parameter, or an asymptotically efficient ranking procedure, but they are really generalizations of the same concept when the estimation of a real-value parameter is involved. Therefore, we discuss here only the basic forms of these concepts.

Unbiasedness

Exhibit 3.36 describes the distribution of an estimator δ of the unknown parameter θ. If you place a wedge along the x axis, you can see that, at some point, the wedge will balance the density function. This balancing point is the expected value of the estimator δ, denoted by $E(\delta)$. An estimator δ is an unbiased estimator of θ if this balancing point happens to be θ; in symbols, $E(\delta) = \theta$.

Unbiasedness is a very restrictive property to require of an estimator. It is a convenient property to have if it comes naturally in the theory. For example, the least-squares criterion in regression leads to unbiased estimators.

Consistency

As more and more observations are accumulated, the estimators should become better and better. Consistency is merely a formal statement of this property. Let δ_n denote the estimator of θ based on n observations. Exhibit 3.37 shows the distribution of δ_n, for $n = 10$ and $n = 100$. If as $n \to \infty$ (in some formally defined way) δ_n approaches θ, then δ_n is said to be a consistent estimator of θ. In Exhibit 3.37, this means the density function is much narrower for $n = 100$ than for $n = 10$. It is a lot easier to have a typical δ_{100} close to θ than to have a typical δ_{10} close to θ.

EXHIBIT 3.36
Unbiasedness of an
Estimator

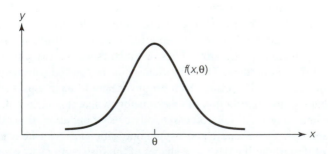

y

$f(x, \theta)$

θ

x

EXHIBIT 3.37
Consistency of an
Estimator

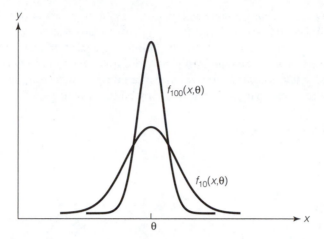

$f_{100}(x, \theta)$

$f_{10}(x, \theta)$

Asymptotic Efficiency

If the density function of δ_n becomes narrower as n increases, then δ_n is likely to be consistent. Some other estimator, say d_n of θ may have the same property, however. How do we compare δ_n and d_n? Let $g_n(x; \theta)$ denote the density of d_n. If for large values of n at least, $f_n(x; \theta)$ is always narrower than $g_n(x; \theta)$, then δ_n is likely to be closer to θ than d_n. When this is so, δ_n is said to be asymptotically (this means as $n \to \infty$) more *efficient* than d_n. If an estimator δ_n beats or ties every other d_n in this sense, δ_n is said to be an asymptotically efficient estimator of θ. Because the spread of the distribution of δ_n is usually measured by Var (δ_n), δ_n is said to be asymptotically efficient if

$$[\text{Var}(\delta_n) \,/\, \text{Var}(d_n)] \leq 1$$

for all other d_n and all large values of n.

An estimate or statistic is unbiased if its mean (expected) value is equal to the true value. An unbiased estimate is efficient if its variance is smaller than the variance of any other estimate. An estimate is consistent if it comes close in some sense to the true value of the parameter θ, as the sample size becomes arbitrarily large.

In summary, we find that a robust correlation coefficient, r^*, serves as an alternative to the sample product moment correlation coefficient; the differences between the two kinds of coefficients could imply the presence of outliers. If both conventional and robust methods yield similar results, we can quote conventional results with an added degree of confidence because classical assumptions are likely to be reasonable for these data. If the results differ significantly, we need to dig deeper into the data source to find the reasons for the difference.

When choosing among estimation techniques, a variety of theoretical criteria is often taken into account The ordinary least-squares estimator in normal regression theory can be shown to have the following characteristics: it is unbiased if its expected value is the unknown parameter, it is consistent in that the distribution of the estimator becomes narrower as the sample size increases, and it is efficient if its variance is less than the variance of any other estimator.

4

Characteristics of Time Series

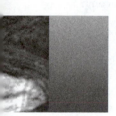

"As to the propriety and justness of representing sums of money, and time, by parts of space, tho' very readily agreed to by most men, yet a few seem to apprehend that there may possibly be some deception in it, of which they are not aware."
WILLIAM PLAYFAIR, *THE COMMERCIAL AND POLITICAL ATLAS,* LONDON, 1786

IN CHAPTER 3, WE HAVE SEEN that data analysis is basic to the forecasting process. In this chapter, we focus on data that are ordered sequentially in time, better known as **time series.** When stock prices, sales volumes, inventory counts, and such are measured or observed as a time series, the data may contain important components that we can effectively analyze, quantify, and create models for. Time series analysis is a useful tool for

- Identifying essential components in historical data so that we can select a good starting model

- Comparing a number of traditional and innovative analytical tools to increase our understanding of data that are typical or representative of the problem being studied (such data are accurate in terms of reporting accuracy and also have been adjusted, if necessary, to eliminate nonrepresentative or extreme values)

This chapter will help you prepare historical data in a variety of graphical ways that are useful in allowing you to see that:

- Not only do data summaries help to explain historical patterns, but the requirements of an appropriate modeling strategy can also be visualized.

117

- Analyses of deseasonalized, detrended, smoothed, transformed data (e.g., logarithms and square roots), fitted values, and residuals can all be most effectively presented graphically.
- Nonstationarity in time series can be analyzed by plotting correlograms of the original and differenced series.

4.1 Visualizing Components in a Time Series

The first known time series using economic data was published in Playfair's 1786 book (Tufte, 1983). Playfair (1759–1823), an English political economist, preferred graphics to tabular displays because he could better show the shape of the data in a comparative perspective. In one example, he plotted three parallel time series—prices, wages, and the reigns of British kings and queens—and noted (quoted in Tufte, 1983, p. 34)

> You have before you, my Lords and Gentlemen, a chart of the prices of wheat for 250 years, made from official returns; on the same plate I have traced a line representing, as nearly as I can, the wages of good mechanics, such as smiths, masons, and carpenters, in order to compare the proportion between them and the price of wheat at every different period. . . . [pages 29–31]

Business forecasters commonly assume that variation in a time series can be expressed in terms of several basic components: a long-term trend plus an economic cycle, a seasonal factor, and an irregular or random term. For a given time series, it may not be possible to observe a particular component directly due to the existence of other components that are more dominant. If appropriate, it is also desirable to correct, adjust, and transform data before creating forecasting models (see Chapter 10).

The purpose of analyzing time series data is to expose and summarize its components as a prelude to a model-building process.

When considering forecasting techniques, such as exponential smoothing models (see Chapter 8), it is useful to classify the trend and seasonal components into an array and associate the components with the appropriate forecasting techniques. Each technique will have a forecast profile, which is the pattern of forecasts that are produced by the technique. For trend and seasonal models, these can be classified in a display known as a Pegels diagram (Pegels, 1969) (Exhibit 4.1).

Each forecasting technique produces a unique pattern in the forecasts. This pattern is called a forecast profile.

EXHIBIT 4.1 Pegels's
Classification of
Trend and Seasonal
Forecast Profiles

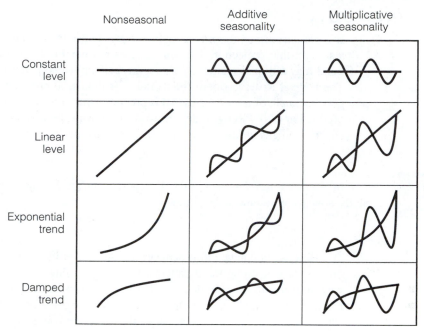

(Source: Gardner, 1985)

After a preliminary examination of the data from a time plot, we may be able to determine which of the dozen profiles seems most appropriate. There are four types of trend profiles (and their meaning) to choose from:

Constant level	Level or no trend at all
Linear trend	Constant amount of change per period
Exponential trend	Constant rate of change per period
Damped trend	Growth pattern that gradually decays to no growth

The same choices may be applied to downwardly trending time series. In terms of seasonality, there is:

Nonseasonal	Time series do not exhibit seasonal variation
Additive seasonality	The amplitude of seasonal fluctuations does not grow (or decline) systematically over time
Multiplicative seasonality	Seasonal fluctuations widen as the average annual level of the data increases

For a downwardly trending time series, multiplicative seasonality appears as steadily diminishing swings about a trend. For level data, the constant level, multiplicative seasonality, and additive seasonality techniques give the same forecast profile.

Trends and Cycles

It is not uncommon for practicing forecasters to use the term trend when referring to a straight-line projection. But a trend does not need to be a straight-line pattern; a trend may fall or rise and can have a more complicated pattern than a straight line. The Federal Reserve Board (FRB) index of industrial production (Exhibit 4.2) is a good example of a time series that is predominantly upward trending. The industrial production (INDPROD) index measures output in the manufacturing, mining, and electric and gas utilities industries; the reference period for the index is 1992.

 In a time series, a trend is seen as the tendency for the same pattern to be predominantly upward or downward over time.

How do we know a time series is trending? The inspection of a time series plot often indicates strong trend patterns. Fitting trend lines is a simple and convenient way of exposing detail in data (see Chapter 3). A useful way of presenting the FRB index is to compare it to some trend line, such as an exponential or straight trend line. This type of analysis brings out sharply the cyclical movements of the FRB index, and it also shows how the current level of output compares with the level that would have been achieved had the industrial sector followed its historical growth rate.

 Trending data are often modeled with exponential smoothing methods (see Chapter 8) and linear regression models (see Chapters 11 and 12).

Although this may not be the best or final trend line for the data, the straight-line trend is a simple summary tool. In order to assess the value of this simple procedure,

EXHIBIT 4.2 Time Plot of FRB Index of Industrial Production (INDPROD)

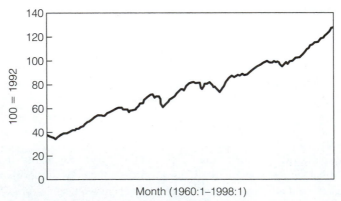

Month (1960:1–1998:1)

(*Source*: *http://www.federalreserve.gov/release/617/download.*)

EXHIBIT 4.3 Time
Plot of the Devia-
tions of the FRB
INDPROD Index
from a Straight-Line
Trend (Exhibit 4.2)

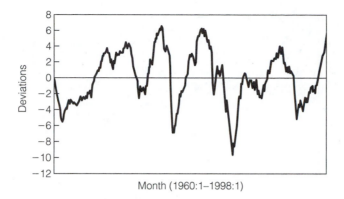

Month (1960:1–1998:1)

the deviations of the data from this trend line (known as **residuals**) of the FRB data are depicted in Exhibit 4.3. It is evident that elimination of trend in the data now reveals a cyclical component that appears to correspond to economic expansions and contractions.

Some financial data can also be useful for analyzing and predicting economic cycles. Exhibit 4.4a shows time plots of U.S. Treasury rates (a plot of the interest rates paid on U.S. securities ranging from 3-month bills to 30-year bonds), which have been used by investors and market analysts to decide which Treasury bond or note offers the best interest rate. Typically, 10-year notes yield between one and two percentage points more than 3-month bills and the yield curve bends up. However, if long-term rates fall below the short-term rates, the curve inverts and arcs down-ward. Economist Frederic Mishkin of the Federal Reserve Bank of New York has discovered that every time the yield curve has inverted, a recession followed a year or so later. A way of looking at this is to plot the difference between the 3-month and 10-year Treasury rates. As shown in Exhibit 4.4b, every time the difference has sunk below zero, a recession followed roughly 12 months later. Business cycle analysis, leading indicators, and a method of cycle forecasting are taken up in Chapter 7.

The definition of a **cycle** in forecasting is somewhat specialized in that the duration and amplitude of the cycle are not constant. This characteristic is what makes cycle forecasting so difficult. Although a business cycle is evident in so many economic series, its quantification is one of the most elusive in time series analysis.

 In practice, trend and cycle are sometimes considered as a single component, known as the trend-cycle.

Seasonality

Certain sales data show strong peaks and troughs within the years, corresponding to a seasonal pattern. When seasonality is removed from these data, secondary patterns

EXHIBIT 4.4 (a) Time Plots of the 3-Month and 10-Year U.S. Treasury Rates; (b) Time Plot of the Differences between 3-Month and 10-Year U.S. Treasury Rates

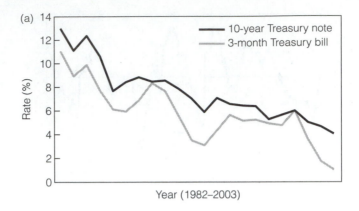

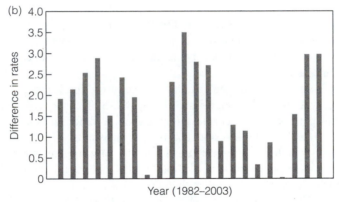

(*Source*: http://www.federalreserve.gov/releases/H15/data/a/tcm10y.txt)

reveal themselves that may still be important. Exhibit 4.5 depicts a time series strongly dominated by seasonal effects, namely, monthly changes in telephone connections and disconnections. The seasonality results from the installation of telephone access lines coincident with school openings and removal of telephones coincident with school closings each year. Thus, the seasonal peaks and troughs appear with regularity each year. The positive trend in telephone access lines is related to the growth in households and the increasing use of telephones by former nonusers. Thus, both trend and seasonal components are superimposed on one another, as well as some residual effects, which are not readily discernible from the raw data. Although seasonality is the dominant pattern, there is also an increasing variability in the highs and lows over time. The seasonality also appears to be additive in the sense that the seasonal deviations from the trend appear relatively constant.

Most commonly, seasonality refers to regular periodic fluctuations that recur every year with about the same timing and intensity. Seasonality can also occur as fluctuations that recur during months, weeks, or days.

EXHIBIT 4.5 Time Plot of Monthly Fluctuations in Telephone Line Access over a 9-Year Period

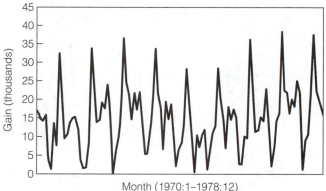

Month (1970:1–1978:12)

(*Source:* Levenbach and Cleary, 1984)

Contrast this seasonal pattern with that of the monthly housing starts shown in Exhibit 4.6. This economic time series is the result of survey data reported by contractors and builders for use by government and private industry. Using a visual comparison with Pegels's diagram (Exhibit 4.1), the seasonality appears to be additive. Exhibit 4.7 is a plot of monthly sale of formal-wear rentals, in which a strong seasonal component is superimposed on an upward trend. In comparison to the patterns in telephone-line access and housing starts, the sale of formal-wear rentals has a pronounced multiplicative seasonal component because seasonality tends to increase with the increase in the level of the data. In Chapter 6, we contrast housing starts and formal-wear rentals using a test to show the difference between additive and multiplicative seasonality.

Many economic series show seasonal variation. For example, income from a farm in the United States may rise steadily each year from early spring until fall and then drop sharply. In this case, the main use of a seasonal adjustment procedure is to remove such fluctuations to expose an underlying trend-cycle. Many industries also have to deal with similar seasonal fluctuations. To make decisions about price and inventory policy and about the commitment of capital expenditures, the business

EXHIBIT 4.6 Time Plot of Monthly Housing Starts over a 40-Year Period

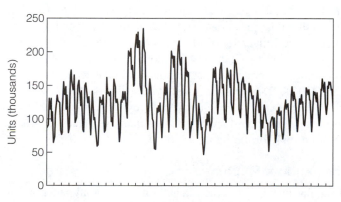

Month (1960:12–1999:11)

(*Source*: http://www.census.gov/const/C20/startsua.xls)

EXHIBIT 4.7 Time
Plot of Monthly
Sales of Formal-
Wear Rentals over an
8-Year Period

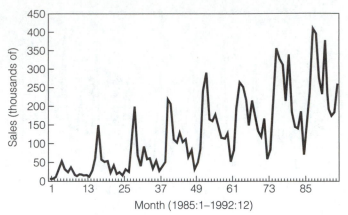

Month (1985:1–1992:12)

(*Source*: Hanke and Reitsch, 1998)

community wants to know whether changes in business activity over a given period of time were larger or smaller than normal seasonal changes. It is important to know whether a recession has reached bottom, for example, or whether there is any pattern in the duration, amplitude, or slope of business cycle expansions or contractions.

The semiconductor business, for instance, plays a crucial role in the size, growth, and importance of information technology. The industry is subject to a number of forces that influence its growth and cyclical behavior. Competitive forces, economic climate, pricing, and political events have a way of making demand forecasting a very complex process. In addition, seasonal patterns are frequently present in the monthly sales demand and unit-shipment figures because of the need for chips in consumer electronics products and the fast-growing market for personal computers. A closely watched indicator of the strength and weakness of this market is the book-to-bill ratio, which compares orders with shipments of semiconductors—a number above 1 indicates positive growth and a ratio of 1.22 indicates that, for every $100 of chips shipped during the month, $122 of chips were ordered. Clearly, to make sense of this indicator we require a seasonal adjustment of the ratio or a seasonal adjustment of the components of the ratio. Seasonal adjustment procedures are treated in Chapter 6.

Other time series data are not strongly dominated by seasonal and trend effects, such as the University of Michigan Survey Research Center's (SRC) index of consumer confidence (Exhibit 4.8). In this case, the dominant pattern is a cycle corresponding to contractions and expansions in the economy. (Compare Exhibit 4.8 with the INDPROD index and monthly housing starts in Exhibits 4.2 and 4.6, respectively.) Of course, a large irregular component is present in this series because there are many unknown factors that significantly affect the behavior of consumers and their outlook for the future.

Irregular or Random Fluctuations

Irregular fluctuation is the catchall category for all patterns that cannot be associated with trend-cycle, or seasonality. Except for some cyclical variation, the plot of the consumer confidence index (Exhibit 4.8) does not suggest any systematic variation

EXHIBIT 4.8 Time
Plot of the SRC
Index of Consumer
Confidence
(monthly) from
1978:1 to 1992:8

(*Source*: University of Michigan Survey Research Center)

that can be readily identified. Irregular fluctuations most often create the greatest difficulty for the forecaster because they are generally unexplainable. A thorough understanding of the source and accuracy of the data is required to recognize the true importance of irregularity.

 The irregular component consists of atypical observations, which may be caused by unusual or rare events, errors of transcription, administrative decisions, and random variation.

An example of an irregular fluctuation, an unusual or rare event arising in a time series, is depicted in Exhibit 4.9, which shows a monthly record of telephone installations in Montreal over a 10-year period. Although dominated by trend and seasonality, the unusually low September 1967 figure is greatly reduced because of the

EXHIBIT 4.9 Impact
of Expo '67 on
Monthly Installations
of Telephones in
Montreal, Canada

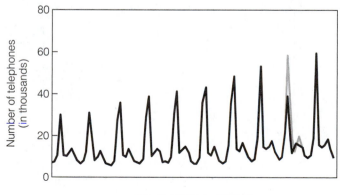

(*Source:* Levenbach and Cleary, 1984)

influence of the 1967 World's Fair held in that city. At the time, residential telephone installations normally accompanied a turnover of apartment leases during September; however, a large number of apartments were held for visitors to the World's Fair that year. The dotted line depicts what might have happened under normal conditions (in the absence of this unusual event).

Exhibit 4.10 shows how an administrative decision can influence a time series. The series represents the number of access lines (telephones for which separate numbers are issued) in service in a specific telephone exchange. The saturation (or filling-up) of a neighboring exchange for a period necessitated a transfer of new service requests from that exchange to the one depicted in Exhibit 4.10, distorting the natural growth pattern. Any modeling effort based on these data must be preceded by an adjustment to account for this unusual event.

Many time series do not exhibit any seasonality, whereas others, such as weather data, are not affected by the business cycle. In practice, we may need to describe additional patterns in data, such as sales cycles, promotions, and other calendar variation.

The components of a time series need not occur simultaneously and with equal strength.

Weekly Patterns

Weekly promotions are frequently used to increase sales of consumer goods, such as canned beverages. Exhibit 4.11 shows a time plot of weekly shipments of a canned beverage product. The sharp peaks can be attributed to periods during which the price of the product was sharply reduced, thereby increasing demand. (These data are reexamined in Chapters 7 and 12 when we quantify the impact of price changes on quantities demanded with elasticities.) When viewed on a monthly basis, such weekly shipment patterns can appear to have a seasonal pattern because of the periodic nature of many marketing promotions.

EXHIBIT 4.10 Time Plot of Access Lines in Service in a Telephone Exchange

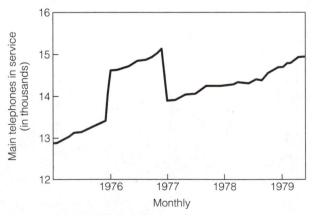

(*Source:* Levenbach and Cleary, 1984)

EXHIBIT 4.11 Time Plot of Weekly Shipments of a Canned Beverage

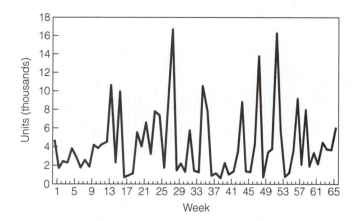

It is important for forecasting demand to clearly differentiate between the seasonality associated with general consumer behavior and the induced seasonal pattern arising from scheduled promotions.

Trading-Day Patterns

Exhibit 4.12 depicts the daily volume of bank transactions for 220 consecutive working days in a year. The task is to forecast future volumes for each business day in the following year. We can see in Exhibit 4.13 that daily volumes experience a yearly pattern that closely follows the average monthly retail seasonality. Displaying box plots of the daily volumes for the working days of a month allows us to note a downward trend in pattern, in that the first part of the month is much heavier than the last. The fifth working day of each month has the greatest amount of variability and the tenth working day shows a higher overall volume than its neighboring days. Tuesdays (marked with asterisks) can also experience some unusually high volumes, perhaps

EXHIBIT 4.12 Time Plot of the Daily Volume of Bank Transactions

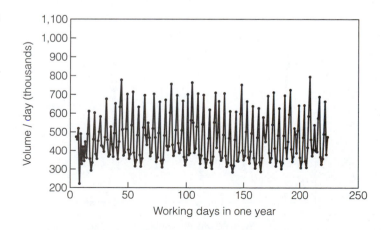

EXHIBIT 4.13 Box Plot of the Daily Volume of Bank Transactions by Working Day of Month

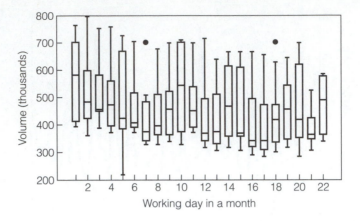

following a holiday on a Monday. Further, if we look at a scatter diagram of the daily volumes by day of week (Exhibit 4.14), we note that the pattern is somewhat parabolic in shape, indicating that the heaviest volumes are on Mondays, the lightest volumes occur midweek, and volume increases again by end of week. These displays are the basis of our analysis of the daily volumes for forecast modeling in later chapters.

4.2 A First Look at Trend and Seasonality

Now that we can visualize trends and seasonality graphically, how can we create a decomposition of a time series in terms of trend and seasonal variations? From the analysis in this section, we will gain insight into the relative strength of these components as well as a measure of the difficulties likely to be encountered in modeling and forecasting. This measure, called the percent residual effect, may turn out to be high because of the presence of outliers or other unknown variation in the data, not explained by trend and seasonal variation.

EXHIBIT 4.14 Scatter Diagram of the Daily Volume of Bank Transactions by Working Day of Week

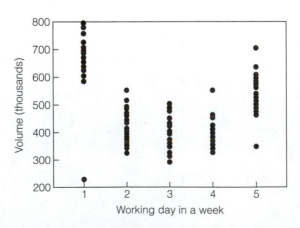

Exploring Components of Variation

Consider a display of 8 years of monthly formal-wear rentals (Exhibit 4.7) in a two-way table, or spreadsheet, as shown in Exhibit 4.15. In this display, the rows represent months and the columns years. Each row shows the trending pattern, if any, for a given month. If a seasonal component is present, this is exhibited by a regularity of the pattern in each column. It is also informative to display the monthly averages per year and the averages per month over 8 years as an additional row and column, respectively. By looking at the yearly totals it is possible to detect a trend component if the annual sums are steadily increasing or decreasing. The high and low seasonal months can be readily determined by comparing the values to the monthly averages.

Consider various graphical displays of the information in Exhibit 4.15. In Exhibit 4.16, we show a plot of each column in the table in a multiple-tier chart. Note how a similar pattern (the seasonal pattern within the year) is shown when the tiers stack on top of another. We can further summarize these by creating a box plot of the values for each month (rows) (Exhibit 4.17a). For a given month, the variation is generally smaller than the variation between months, suggesting an almost constant effect for each month. This is known as the seasonal component. By plotting the values of each month (row) for the formal-wear data in Exhibit 4.17a, we note a seasonality peaking in the fourth and fifth months and also much variability within months. On the other hand, the box plots by years (Exhibit 4.17b) (columns) clearly depict an upward trend, the unusually high months of May in Years 2 and 3, and increasing variability with increasing years. Hence, we note a relationship occurring (1) between the values for successive months in a particular year and (2) between the values for the same month in successive years.

EXHIBIT 4.15 A Month-by-Year Display for Monthly Sales (in dollars) of Formal-Wear Rentals over an 8-Year Period

NUMBER OF YEARS OF DATA = 8
GRAND MEAN OF DATA = 125,872.188

| | Year | | | | | | | | |
| | 1 | 2 | 3 | 4 | 5 | 6 | 7 | 8 | |
Month	1985	1986	1987	1988	1989	1990	1991	1992	AVG
1	6,028	16,850	15,395	27,773	31,416	51,604	58,843	71,043	34,869
2	5,927	12,753	30,826	36,653	48,341	80,366	82,386	152,930	56,273
3	10,515	26,901	25,589	51,157	85,651	208,938	224,803	250,559	110,514
4	32,267	61,494	103,184	217,509	242,673	263,830	354,301	409,567	210,603
5	51,920	147,862	197,608	206,229	289,554	252,216	328,263	394,747	233,550
6	31,294	57,990	68,600	110,081	164,373	219,566	313,647	272,874	154,803
7	23,573	51,318	39,909	102,893	160,608	149,082	214,561	230,303	121,531
8	36,465	53,599	91,368	128,857	176,096	213,888	337,192	375,402	176,608
9	18,959	23,038	58,781	104,776	142,363	178,947	183,482	195,409	113,219
10	13,918	41,396	59,679	111,036	114,907	133,650	144,618	173,518	99,090
11	17,987	19,330	33,443	63,701	113,552	116,946	139,750	181,702	85,801
12	15,294	22,707	53,719	82,657	127,042	164,154	184,546	258,713	113,604
Monthly average per year	22,012	44,603	64,842	103,610	141,381	169,432	213,866	247,231	

(*Source:* Exhibit 4.7)

EXHIBIT 4.16
Monthly Tier Plots
for Formal-Wear
Rentals Data

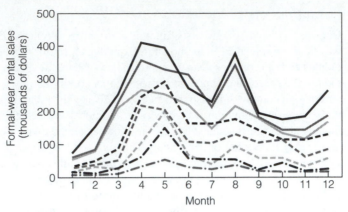

(*Source:* Exhibit 4.15)

This observation will be useful when we consider exponential smoothing models (see Chapter 8) and multiplicative versus additive seasonal modeling for seasonal ARIMA (see Chapter 14).

The method by which a two-way table is constructed is not complex, but the notation can become somewhat cumbersome. A two-way table is a spreadsheet in

EXHIBIT 4.17 Box
Plots of Formal-
Wear Data Showing
(a) Row and (b)
Column Variation

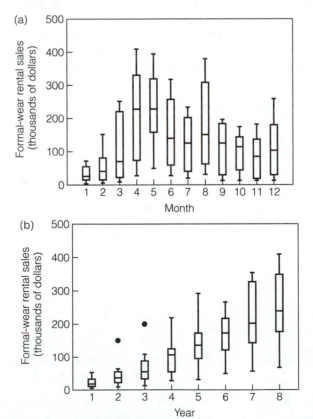

(*Source:* Exhibit 4.15)

which observations Y_{ij} (i = month, j = year) are displayed in a rectangular array, where $i = 1, 2, \ldots, I$ ($I = 12$ for monthly data), and $j = 1, 2, \ldots, J$ (J = number of years). This representation can also be used for quarterly data, and then $i = 1, 2, 3, 4$. Appendix 4A contains a mathematical derivation of the analysis technique.

 A two-way table is a spreadsheet in which the data are arranged in a rectangular array allowing for the display of trend and seasonal influences.

When the data are arranged in a spreadsheet with I (=12) columns (for the months) and J rows (for the years), an additive model that describes a typical observation in terms of a seasonal component S_i, a trend or yearly component T_j, and a residual error term (irregular component ε_{ij}) is given by

$$Y_{ij} = \mu + S_i + T_j + \varepsilon_{ij}$$

where μ denotes a mean effect or typical value. Because row ($\bar{Y}_{i\cdot}$) and column ($\bar{Y}_{\cdot j}$) means in the table represent average seasonal and trend effects, respectively, variations of these averages from the grand mean $\bar{Y}_{\cdot\cdot}$ reflect the respective variation due to an average seasonal pattern and trend. Thus,

$$SS_{Seas} = \sum (\hat{Y}_{i\cdot} - \bar{Y}_{\cdot\cdot})^2 / 11$$

and

$$SS_{Trend} = \sum (\hat{Y}_{\cdot j} - \bar{Y}_{\cdot\cdot})^2 / (J - 1)$$

represent variation in row (monthly) means and column (yearly) means, respectively. It is also possible to quantify the proportion of variability due to seasonal effects, S, by the ratio

$$R^2_{Seas} = J \sum (\hat{Y}_{i\cdot} - \bar{Y}_{\cdot\cdot})^2 / \sum\sum (Y_{ij} - \bar{Y}_{\cdot\cdot})^2$$

and the proportion of variability due to trend effects, T, by

$$R^2_{Trend} = 12 \sum (\hat{Y}_{\cdot j} - \bar{Y}_{\cdot\cdot})^2 / \sum\sum (Y_{ij} - \bar{Y}_{\cdot\cdot})^2$$

Ideally, or for an extreme case, if there is no trend in the data, the column means equals the grand mean and the proportion of variability that is due to trend equals zero. The foregoing computations can be readily made using any statistical package or spreadsheet software that offers a two-way ANOVA capability.

In Exhibit 4.18, we display the layout for performing an ANOVA on the formal-wear sales data.

In Exhibit 4.19, we display an ANOVA decomposition of the data into the seasonal and trend components. Jointly, they appear to explain almost 90% of the total variation. On the other hand, in Exhibit 4.20, the ANOVA decomposition yields a seasonal component in quarterly automobile sales that is quite dominant by itself. Hence, in this case, we look for techniques dealing primarily with seasonality features (see Chapter 8).

For the seasonal time series depicted in Exhibits 4.5–4.7, we find the results in Exhibit 4.21.

EXHIBIT 4.18 Layout for Monthly ANOVA of Sales of Formal-Wear Rentals

	1985	1986	1987	1988	1989	1990	1991	1992
Jan	6028	16850	15395	27773	31416	51604	58843	71043
Feb	5927	12753	30826	36653	48341	80366	82386	152930
Mar	10515	26901	25589	51157	85651	208938	224803	250559
Apr	32267	61494	103184	217509	242673	263830	354301	409567
May	51920	147862	197608	206229	289554	252216	328263	394747
Jun	31294	57990	68600	110081	164373	219566	313647	272874
Jul	23573	51318	39909	102893	160608	149082	214561	230303
Aug	36465	53599	91368	128857	176096	213888	337192	375402
Sep	18959	23038	58781	104776	142363	178947	183482	195409
Oct	13918	41396	59679	111036	114907	133650	144618	173518
Nov	17987	19330	33443	63701	113552	116946	139750	181702
Dec	15294	22707	53719	82657	127042	164154	184546	258713

EXHIBIT 4.19 Results of ANOVA decomposition of Sales of Formal-Wear

Source of Variance	Percent
Trend	56.9%
Seasonality	31.3%
Noise	11.8%
Total	100.0%

EXHIBIT 4.20
Layout for Quarterly
ANOVA of
Automobile Sales in
Quebec City

ANALYSIS OF VARIANCE: QUARTERLY DATA
TITLE1:   AUTOMOBILE SALES IN QUEBEC CITY
TITLE2:   1960-1964
X-AXIS: QUARTER
Y-AXIS: THOUSANDS

Year	Qtr	Per	Data
1960	1	1	27.304
	2	2	42.773
	3	3	24.798
	4	4	27.365
1961	1	5	28.448
	2	6	43.531
	3	7	26.728
	4	8	31.590
1962	1	9	36.824
	2	10	54.115
	3	11	26.708
	4	12	34.313
1963	1	13	36.232
	2	14	58.323
	3	15	28.872
	4	16	42.496
1964	1	17	43.681
	2	18	61.855
	3	19	36.273
	4	20	40.244
1965	1	21	
	2	22	
	3	23	
	4	24	
1966	1	25	

SOURCE OF VARIATION	PERCENT
TREND	27.20%
SEASONALITY	68.20%
IRREGULAR	4.60%
TOTAL	100.00%

We observe that the irregular components are comparable and small enough that reasonable models can be expected for dealing with trend and seasonality. The strong trend component for formal wear suggests that taking a first difference of the data might be beneficial. As randomness increases, forecasting becomes more difficult. Because of possible outliers, data with an irregular component exceeding 15–20% need to be scrutinized more carefully. When the irregular component is quite high, it might be reduced through an adjustment of outliers prior to creating a decomposition or model. As a rule of thumb, experience suggests that monthly and

EXHIBIT 4.21
Analyses of
Telephone Access
Lines, Housing
Starts, and Sales of
Formal-Wear Rentals

Components	Access Lines (4.5)	Housing Starts (4.6)	Formal Wear (4.7)
Trend	7.3	9.8	56.9
Seasonality	81.3	80.9	31.3
Irregular	11.4	9.3	11.8

(*Source:* Exhibits 4.5–4.7)

quarterly time series that show more than 20–40% in the irregular component should not be seasonally adjusted or forecasted with a complex model. The best alternative is to handle such data with simpler models such as exponential smoothing models.

The results of an ANOVA decomposition can provide an important clue to which data patterns can be handled as part of the criteria for technique selection.

The ANOVA model is a simple mechanism for quantifying row and column effects in terms of their relative contributions to the total variation in a table. When used to learn about time series, the ANOVA model is a descriptive tool and is not a practical extrapolative technique. This type of analysis might be useful for characterizing (roughly) the dominant components in the data, identifying potential modeling problems if the irregular component (residual percent) is relatively large, correlating data with similar structures, and suggesting an appropriate transformation to check for the additivity of trend and seasonality.

When the data depart systematically from this additive structure, it may be desirable to reexpress the Y_{ij} using a transformation (see Chapter 10). Later, we introduce the **diagnostic plot** for suggesting a transformation for additivity. Understanding the nature of additive patterns also helps in constructing the appropriate linear regression and ARIMA forecasting models to be discussed in later chapters.

Exhibit 4.19 identifies the reasons for change or variance in a set of data. Variance among the average monthly values between years is classified as trend. Variance among the average monthly variances within the year is classified as seasonality. Any remaining variance in the data cannot be accounted for and is classified as irregular. As mentioned, when the irregular component increases, forecasting becomes more difficult. In cases in which the irregular component dominates the other two, it is better to stick with the simpler models and avoid more complex modeling approaches.

The calculations are described in detail in Appendix 4A. First, the grand mean (average of all the data values) is computed. Then the squared difference between each data value and the grand mean, called the total variation, is determined. This measures the overall variation due to trend, seasonal, and irregular components. The variation due to trend alone is found as well as the variation due to seasonality. The proportion of variation due to trend (for sales of formal-wear rentals, $= 56.9\%$) is found as a percentage of the total variation. Similarly, the proportion of variation due to seasonality ($= 31.3\%$) is determined. The irregular component is $100\% - (56.9\% + 31.3\%) = 11.8\%$, or roughly 12% of the total variation (Exhibit 4.22).

For this analysis to work, we must use complete years of data in the spreadsheet. The data must begin in January of the first year and end in December of the last year. A minimum of 3 years is required and the spreadsheet is preset to handle up to 120 months (10 years). The results are depicted graphically in Exhibit 4.22 for the formal-wear rental data.

EXHIBIT 4.22
ANOVA for Sales
of Formal-Wear
Rentals—Trend/
Seasonal Contribution

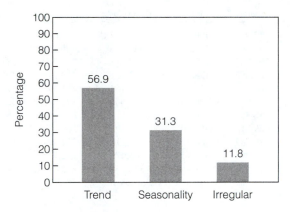

Exhibit 4.20 is similar to Exhibit 4.19 except that it deals with quarterly data. The purpose of this ANOVA decomposition is to determine the reasons for change or variance in this set of quarterly data. The variance among the average quarterly values between years is classified as trend. Variance among the average quarterly values within the year is classified as seasonality. Any remaining variance in the data cannot be accounted for and is classified as irregular (**noise**). The results are depicted graphically in Exhibits 4.22 and 4.23.

Day-of-the-Week Effect

The ANOVA model has so far been used to describe the month and year effects in monthly data. The same technique can be applied to identify a day-of-the-week effect in daily data. The data in Exhibit 4.14 suggest that the daily volume of bank transactions vary by what day of the week it is. In the two-way-table analysis of these data (Exhibit 4.24), certain days (those following major holidays) have been removed because of their unusually high volume. Each cell contains the average volume for that day of week in the particular month. For example, the Jan/Monday cell contains the average of all Monday volumes in the month of January. For forecasting

EXHIBIT 4.23
ANOVA of
Automobile Sales in
Quebec City—Trend
and Seasonal
Contribution

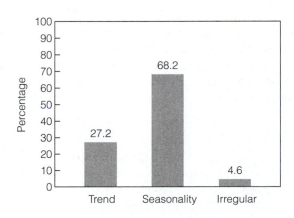

EXHIBIT 4.24		Monday	Tuesday	Wednesday	Thursday	Friday	Row Average
Month of Year by	Jan	477,917	409,518	379,047	393,664	452,449	418,000
Day of Week Table	Feb	636,680	432,902	394,166	408,315	498,308	464,222
for Daily Volumes of	Mar	706,768	420,468	340,188	369,720	538,977	478,260
Bank Transactions	Apr	682,889	438,742	362,521	413,258	497,603	476,059
	May	705,511	411,623	372,164	391,595	577,121	478,848
	Jun	695,630	404,631	332,660	375,995	521,897	466,163
	Jul	658,762	369,258	357,572	338,208	469,491	445,289
	Aug	669,210	364,905	317,892	369,743	528,029	449,859
	Sep	680,885	380,473	360,015	371,994	509,277	464,742
	Oct	669,066	382,439	361,079	392,102	522,753	447,327
	Nov	714,527	401,480	392,144	385,208	545,422	485,524
	Dec	666,161	400,921	361,410	384,294	518,274	461,053

Results:

Day of week	91%
Time of year	3%
Irregular	6%

(Exhibit 4.14)

purposes, it may be easier to forecast the daily volumes first and add a percentage for days following the major holidays.

The data in Exhibit 4.24 are derived from Exhibit 4.14 and have been arranged in a two-way table so that rows = month of the year and columns = day of the week. At first glance, it appears that the time of year is fairly constant in terms of average volumes (Row Average column).

Evidently, the fact that Monday is the big volume day, followed by a midweek low, plays a strong role is the description of this time series and should be reflected in any forecasting model. If we used only the time plot of the data (Exhibit 4.12) or raw data display, we might not have recognized that the day of the week was a key pattern.

4.3 WHAT IS STATIONARITY?

So far, we have seen that a time series can display time dependency in terms of trend and seasonal patterns. However, it is also useful to consider characteristics of time series that are not time dependent. This gives rise to the notion of **stationarity**. In reality, most time series are not stationary and so, the nonstationarity condition in a time series needs to be acknowledged before we can begin to apply the sophisticated autoregressive–moving average (ARMA) theory of stationary noise models discussed in Chapters 13 and 14.

For our immediate purposes, a stationary time series is one in which the mean level and the variability are not dependent on time.

It may be easier to describe what stationarity is not. Naturally trending and seasonal time series are known as nonstationary time series. Generally, many kinds of nonstationarity conditions are present in time series data. The simplest is known as nonstationarity in the level of the mean and occurs when the level of the mean changes or drifts over different segments of the data. A trending time series is a good example. As we have seen, nonstationarity can also occur as a result of a seasonal pattern in the data. A seasonal series is considered nonstationary because the variation in the data is a function of the time of the year. Other kinds of nonstationarity are a result of:

- Increasing or decreasing variability of the data with time. Sales data may have a nonlinear trend or show increasing variability in the seasonal peaks and troughs over time.

- Drastic change in the level of some series. Examples of time series that change drastically are price and unemployment data. An unemployment rate is a series that tends to stay at a high or a low level, depending on economic conditions: The change is usually abrupt and time dependent.

A combination of several kinds of nonstationarity is usually present in most data encountered in forecasting demand.

Detecting Nonstationarity with Correlograms

In Chapter 3, we introduce the correlogram as a device for analyzing autocorrelations in time series. For stationary data, a correlogram will decay rapidly to zero, so any other pattern suggests nonstationarity. The most important forms of nonstationarity result in different patterns in the correlogram due to changes in level, trend, and seasonality.

The correlogram is a useful tool for visually detecting nonstationarity in a time series.

Significant autocorrelations (i.e., large spikes) are detected in a correlogram if individual values fall outside a horizontal band at $\pm 2 / \sqrt{n}$, where n is the number of observations in the time series. This is only an approximate test, but in most practical situations you will find spikes well outside this band. A typical seasonal series shows a correlogram with spikes at the seasonal periods (Exhibit 4.25). A typical trending series, such as the seasonally adjusted housing starts series, will show a correlogram with a gradual decaying pattern (Exhibit 4.26). Because housing starts exhibits trend and seasonality, the correlogram results in a pattern that exhibits

EXHIBIT 4.25
Correlogram of
Housing Starts

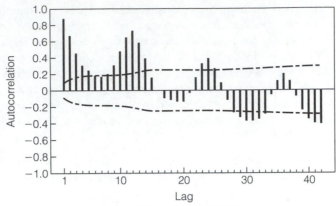

(*Source*: www.census.gov/const/C20/startsua.xls) (Exhibit 4.6)

EXHIBIT 4.26
Correlogram of
Seasonally Adjusted
Housing Starts
(1940:1–1999:11)

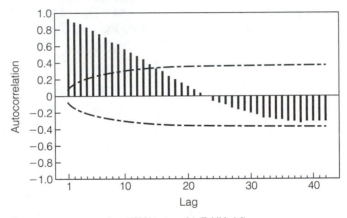

(*Source*: www.census.gov/const/C20/startsua.xls) (Exhibit 4.6)

features of both seasonality and trend. Once nonstationary behavior has been detected, there are ways to remove it through differencing.

A decaying pattern in a correlogram does not necessarily imply that the data are trending. In the case of the seasonally adjusted housing starts, there are many upward and downward trends over the 40-year period. Over this long period, the data appear to fluctuate around 1,500,000 units. On the other hand, when we rely on only the latest 9 years, the seasonally adjusted data show a steady upward trend. The correlogram of the housing starts over this 9-year period is shown in Exhibit 4.27. When we compare this to Exhibit 4.26, we are not able to discern any notable differences other than the rate of decline in the patterns.

 Exercise caution in interpreting correlograms; focus on the pattern, more than the individual spikes in the pattern.

EXHIBIT 4.27
Correlogram of
Seasonally Adjusted
Housing Starts over
the Last 9-Year Period
(1991:1–1999:11)

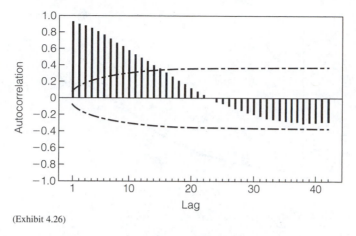

(Exhibit 4.26)

Removing Nonstationarity Using Differencing

A nonstationary series is termed homogeneous if it can be made stationary through differencing. Fortunately, many of the time series that arise in economics and business appear to be of this kind.

Recall that a first difference is the change between Y_t and Y_{t-1}. For example, if Y_t is a straight line given by $Y_t = a + bt$, then the first difference of Y_t results in a new time series $Z_t = [a + bt] - [a + b(t-1)]$. The result is equal to the slope b, which means that the first differences Z_t of a straight line trend are constant and not dependent on time t. Thus, a first difference is defined by

$$Z_t = Y_t - Y_{t-1}$$

To illustrate the calculations, consider a weekly time series of nose-clip shipments from an eyewear manufacturer. $Y_t = \{101, 104, 109, 164, 125, 136, 139, 164, 181, 200\}$. The data are trending up. This trend pattern is removed by taking a first difference of the data. The resulting first differences are given by $Z_t = \{3, 5, 55, -39, 11, 3, 25, 17, 19\}$.

Differencing is useful in ARIMA models (Chapters 14 and 15), especially for forecasting macroeconomic time series. Consider the seasonally adjusted money-supply series shown in Exhibit 4.28, which is a smoothly trending series. This is the U.S. Federal Reserve M2 series, which is a broad measure of the economy's money supply. The M2 money supply consists of private checking deposits and cash, as well as most types of personal savings including money market funds and money market deposit accounts. It excludes some investments such as large certificates of deposit and money market funds sold to institutions. In addition to being affected by seasonal movements between cash and checking deposits, the money supply is impacted by the timing of government payments (e.g., Social Security) and by institutional changes. The M2 series clearly shows nonstationary behavior in its level. This trend pattern is removable by taking a first difference of the data.

Exhibit 4.29 shows the differenced series; it shows increases and decreases in level and has an increasing variability with time. The pattern changes dramatically

EXHIBIT 4.28 A Time Plot of the Seasonally Adjusted Money Supply (the U.S. Treasury M2 series) over a 40-Year Period

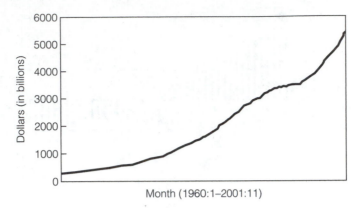

(*Source*: *http://www.federalreserve.gov/H6/*)

during the first few months of 1983 (peaks in the data). This suggests that the data following that period may be more representative of current (and, it is hoped, future) behavior in the M2 series. A forecaster would be advised to investigate these anomalies if the data are to be used in a forecast modeling application.

  In practice, many time series can be made approximately stationary through differencing.

Now let us consider the second difference, the first difference of a first-differenced series:

$$Z_t^{(2)} = Z_t - Z_{t-1} = (Y_t - Y_{t-1}) - (Y_{t-1} - Y_{t-2}) = Y_t - 2Y_{t-1} + Y_{t-2}$$

EXHIBIT 4.29 A Time Plot of First Differences of the Money Supply Data

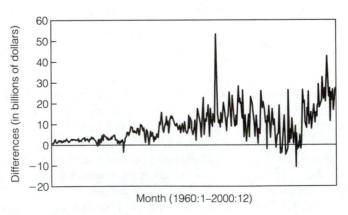

(*Source:* Exhibit 4.28)

For nose-clips shipments, the second differences are obtained by taking the first differences of Z_t. This results in the series $Z_t^{(2)} = \{2, 50, -94, 50, -8, 22, -8, 2\}$. For M2, the result (Exhibit 4.30) gives rise to a series that now appears to have a constant mean but that continues to show uneven variability over time. The corresponding correlograms are shown in Exhibit 4.31. The first differences (Exhibit 4.31a) are not stationary because the pattern does not decay vary rapidly (it takes approximately ten spikes for the pattern to fall within the bounds). In the correlogram of the second differences (Exhibit 4.31b), all spikes, with the exception of the first one, fall within the bounds, suggesting stationary behavior in the underlying data. The negative spike at lag 1 $(= -0.52)$ does not indicate nonstationary behavior in the data; however, it is suggestive of an ARMA model that is discussed in Chapter 14. Overall, it is not necessary to perform any further differencing to achieve stationarity. In fact, differencing more than twice usually leads to overdifferencing, which can be detrimental to modeling the data effectively. Normally, a first or second difference of an economic time series is adequate.

  For most economic data, differences should be taken, at most, twice.

Most economic data are reported on a seasonally adjusted basis. However, with seasonal data, nonstationarity attributed to seasonal variation can be removed through seasonal differencing. Sometimes differencing alone can change a nonstationary series to a reasonably stationary one. More general situations may require a transformation of the Box-Cox type before differencing; this includes taking a square root transformation and a logarithmic transformation (see Chapter 10). The ARIMA class of time series models assumes that stationarity can be achieved fundamentally through differencing and transformations (see Chapter 14).

EXHIBIT 4.30 Time Plot of Second Differences of the Money Supply Data

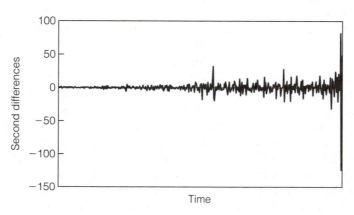

(Exhibit 4.26)

EXHIBIT 4.31 (a)
Correlogram of First
Differences of the
Seasonally Adjusted
Money Supply Data;
(b) Correlogram of
Second Differences
of the Seasonally
Adjusted Money
Supply Data

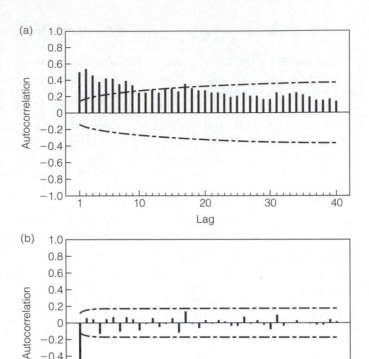

(Exhibit 4.29, Exhibit 4.30)

Logarithmic Transformations with Nonstationary Data

For the money supply data, we have seen (Exhibit 4.28) that it is possible to remove trend through differencing but that the variability of the differenced data is not homogeneous over time. It turns out to be better to make a transformation (such as taking the logarithms) of the data first and then take differences of the log-transformed data. It should be evident that the logarithms of the money supply data are still nonstationary in level. By taking the first differences of the log-transformed data (similar to growth rates), we obtain a time series that appears stationary and displays constancy of variance throughout the time span. When we compare the variability after the spike with the corresponding pattern before the spike in Exhibits 4.29 and 4.32, we note a greater degree of constancy of variability for the transformed data throughout the time span. This is a familiar pattern found in seasonal time series that are trending as well, so taking the logarithms before differencing is a reasonable thing to do. It is also important to do this, not only because of the more sophisticated ARIMA modeling but also because there are now additional models that can be developed for assessing forecasting accuracy.

Another transformation that is sometimes considered before differencing is the square-root transformation. The first differences of the square root of the money

EXHIBIT 4.32 A
Time Plot of First
Differences of the
Log-Transformed
Money Supply Data

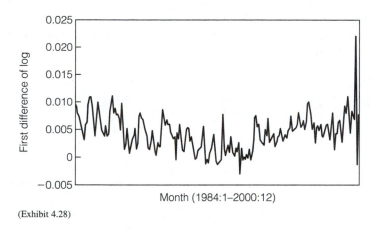

First difference of log

Month (1984:1–2000:12)

(Exhibit 4.28)

supply data do not appear to be very different from the percentage changes of the money supply (not shown). Thus, we can work with the percentage changes (growth rates) of the money supply series as the stationary series for modeling purposes.

 In many forecasting applications, a transformation of the data is required before differencing.

In forecasting applications, there are seasonal time series that appear to grow faster than a straight-line trend over time; if a straight-line trend is fit to such data, a pattern of increasing dispersion in the residuals generally results. The toll revenue and message series in the telecommunications example (in Chapter 3) are two examples in which seasonal fluctuations are increasing with the level of the series. Over the fit period, for a faster-than-straight-line trending time series, the residuals from a straight line fit have a fan-shaped pattern (Exhibit 4.33). After a straight line is fit, the actual values would fall closer to the line in the beginning than in the end part of the data (fitting straight-line trend models is introduced in Chapter 10). The appropriate technique for improving the fit (so that residuals appear to have a random pattern) is to take a logarithmic transformation of the toll revenue data (Exhibit 4.34b). Further, confirmation that taking logarithms is the correct course comes from considering year-by-year percentage changes. If the percentage growth values lie in a relatively small band, near-exponential growth of the data is suggested.

As a final step in examining the effect of transformations and differencing, we review the probability plots for the data used in Exhibits 4.33 and 4.34. If the normal probability plots for residuals look reasonably linear, we are more confident that the data satisfy the assumptions required for normal linear regression models. In comparing the normal probability plots in Exhibits 4.33b and 4.34b, we note that neither pattern exhibits deviations that suggest **nonnormality** in the distribution of the residuals.

Normal probability plotting can be an effective way of looking at the practical value of transformations and differencing operations.

On rare occasions, the residuals from a straight-line regression of a log-transformed series may also show a nonrandom pattern. In this case, it may be necessary to perform another transformation on the transformed series, such as a log-log transformation. Extreme care must be exercised in these situations because projections based on such transformations grow very rapidly and are likely to appear unrealistic.

In the context of regression analysis, we see that transformations also play a key role in achieving normality in the residuals of a model. This becomes important in realizing the consequences of developing linear regression models for non-normal data.

EXHIBIT 4.33
(a) Time Plot of Residuals from a Straight-Line Trend Model for Telephone Toll Revenues (TOLL-REV) over a 10-Year (120-month) Period;
(b) Normal Probability Plot of the Corresponding Residuals

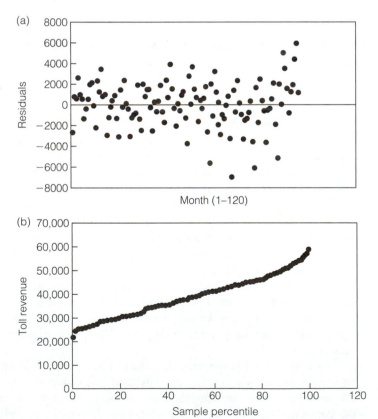

EXHIBIT 4.34 (a)
Time Plot of
Residuals of the
Logarithms of the
Telephone Toll
Revenues with a
Straight-Line Trend
Model; (b) Normal
Probability Plot of
the Corresponding
Residuals

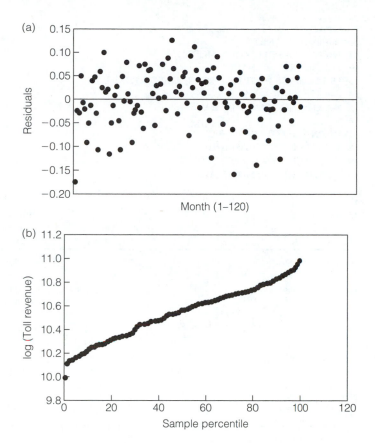

4.4 CLASSIFYING TRENDS

The data displayed in Exhibit 4.35 represent the number of bags (pieces of checked luggage) lost per 100,000 passenger miles by a certain airline over a 12-year period. The ANOVA decomposition results in trend 57%, seasonal 31%, and irregular 12%. To check for trend, we compute the differences between successive data observations. The first-difference values are plotted for Years 2 through 12 along with the original data. If a trend exists, the variance of the difference values will be smaller than the variance of the original data. To understand why this is true, recall that a (sample) variance is a measure of the degree to which individual values in a set of data differ from the average of all values. Because the original data contain a trend, observations near the beginning and end of the data differ radically from their average. By computing the differences between data observations, we eliminate the trend and the difference values cluster around their average.

At this point, we know that a trend exists in the number of bags lost but we do not know the type of trend. We can narrow the possibilities by computing the differences between differences, that is, the second differences.

EXHIBIT 4.35
Layout of the Trend
Analysis for Bags
Lost Per 100,000
Passengers

Filename: DATAANL.XLS
Spreadsheet: TRENDIFF
1 TREND CLASSIFICATION MODEL
2 TITLE1: TREND ANALYSIS
3 TITLE2: BAGS LOST PER 100,000 PASSENGER MILES
4 X-AXIS: YEAR
5 Y-AXIS: THOUSANDS OF BAGS
6
7 MEAN: 0.8375 0.0563 −0.012
8 VARIANCE: 0.0379 0.0040 0.0088
9 VARIANCE INDEX: 100% 10% 23%
10 TYPE OF TREND: MODERATE
11

	A	B	C	D	E	F
12						
13					Differences	Differences
14				Original	between	between
15	Year	Mon	Per	Data	Data	Differences
16	1978	NA	1	0.50	NA	NA
17	1979		2	0.64	0.14	NA
18	1980		3	0.69	0.05	−0.09
19	1981		4	0.79	0.10	0.05
20	1982		5	0.76	−0.03	−0.13
21	1983		6	0.73	−0.03	0.00
22	1984		7	0.80	0.07	0.10
23	1985		8	0.91	0.11	0.04
24	1986		9	0.93	0.02	−0.09
25	1987		10	1.08	0.15	0.13
26	1988		11	1.10	0.02	−0.13
27	1989		12	1.12	0.02	0.00

(Mon, month; NA, not available; Per, period)

If the set of second differences has a smaller variance than the first, we may conclude that there is a strong trend in the data, either linear or exponential. If the set of second differences does not reduce the variance, we conclude that the trend is moderate, probably a damped trend. The Computer Study that follows shows the details of these calculations.

4.5 COMPUTER STUDY: HOW TO DETECT TRENDS

In most business data, two kinds of change must be considered: trends and seasonal cycles. Trends are patterns of growth or decline. For example, the amount of growth each period may be constant. This is called a straight-line or linear trend. The amount of growth each period can be declining (a damped trend) or increasing (e.g., an exponential trend). It is also possible that there is no trend at all. If we guess wrong about the trend pattern in the data, our forecast errors can be embarrassing.

Instead of guessing at trends, we recommend relying on statistical tests; these are quick and easy to implement. The tests help determine the type of trend in the data and also tell us whether a seasonal cycle exists. If the data are collected on an annual basis, we should conduct a trend test before forecasting. If we have quarterly or monthly data, we should test for both a trend and a seasonal cycle.

Get into the habit of testing the data before forecasting. Graphs are often deceptive about trends and seasonal patterns. For example, we suspect that many people would choose a straight-line trend in the graph of the number of bags lost by the airline; however, a test suggests otherwise.

The Basics of Trend Analysis

To explain the rationale for the trend test, we analyze the annual mortgage rate data in Exhibit 2.6. To check for trend, we compute the differences between successive data observations.

If a trend exists, the variance of the differenced values will be smaller than the variance of the original data. To see this, recall that the variance is a measure of the degree to which individual values in a set of data differ from the average of all values (Exhibit 4.36). Because the original data contain a trend, observations near the beginning and end of the data differ drastically from their average. By computing the differences between data observations, we eliminate the trend and the difference values cluster around their average.

This is the case for the data in Exhibit 2.6. The rules in Exhibit 4.37 summarize the analysis of trends.

EXHIBIT 4.36
Computing the
Differences between
Successive Years to
Eliminate Trend

Year	Period	Original Data	Differences between Data	Differences between Differences
1978	1	0.5	NA	NA
1979	2	0.64	0.14	NA
1980	3	0.69	0.05	−0.09
1981	4	0.79	0.1	0.05
1982	5	0.76	−0.03	−0.13
1983	6	0.73	−0.03	0
1984	7	0.8	0.07	0.1
1985	8	0.91	0.11	0.04
1986	9	0.93	0.02	−0.09
1987	10	1.08	0.15	0.13
1988	11	1.1	0.02	−0.13
1989	12	1.12	0.02	0

(bags lost per 100,000 passengers; NA, not available)

EXHIBIT 4.37
Summary of Trend
Analysis (see
Chapter 8)

Minimum Variance Occurs In	Trend Indicated	Recommended Forecasting Technique
Original data	Nonexistent	Simple exponential smoothing
Differences between data observations	Moderate	Damped-exponential trend
Differences between differences	Strong	Linear or exponential trend

EXHIBIT 4.38
Comparing Variances
for Data in Exhibit
4.35

TREND CLASSIFICATION MODEL
TITLE1:   TREND ANALYSIS
TITLE2:   BAGS LOST PER 100,000 PASSENGER MILES
X-AXIS: YEAR
Y-AXIS: THOUSANDS OF BAGS

MEAN:	0.8375	0.056364	−0.012
VARIANCE:	0.03793	0.003965	0.00884
VARIANCE INDEX:	1	0.104548	0.233064
TYPE OF TREND:		MODERATE	

(bags lost per 100,000 passengers)

Building the Trend Analysis Spreadsheet

The trend analysis spreadsheet is shown in Exhibit 4.38. From Exhibit 4.36, we calculate variance of the original data, variance of differences between successive observations, and variances of differences between are shown in the row labeled VARIANCE. The row labeled VARIANCE INDEX converts the variances to indexes to make them easier to interpret. The variance of the original data is taken as a base of 100%. The row labeled TYPE OF TREND is a reminder of the trend indicated when the variance in that column is the minimum. Thus, the difference values display a much smaller variance from their average than the original data. This is the basis of the test for trend. If computing the differences reduces the variance, a trend exists; otherwise, there is no trend.

Exhibit 4.38 compares variances for the original data in Exhibit 4.35 and two sets of differences: between the years (between values in column 3), and between differences (between values in column 4). The smallest variance indicates the type of trend. With a moderate trend, the variance of the differences between years is smallest. With a strong trend, computing the second differences is necessary to achieve the smallest variance.

SUMMARY

Effective graphical displays and simple summaries that help forecasters visualize their data include time plots that provide a visual indication of the predominant characteristics of a time series—trend-cycle, seasonal, and irregular; scatter diagrams that are useful for studying relationships among variables and are widely applied in regression analysis; and ANOVA decompositions that can be used to break down a time series into additive trend and seasonal components by means of an ANOVA in historical data and describe the relative contributions of seasonal and trend effects to the total variation of the data. For monthly data, variance among the average monthly values between years is classified as trend, variance among the average monthly values within the year is classified as seasonality, and any remaining variance in the data is classified as irregular or randomness.

Nonstationarity in time series may be present if the values plotted in the correlogram do not diminish at large lags. It is usually sufficient to look at the correlograms of the original series and of its first and second differences. When the original series or correlogram exhibits nonstationarity, successive differencing is carried out until the correlogram of the differenced series dies out reasonably rapidly.

For most economic data, differences should be taken at most twice. The number of times that the original series must be differenced before a stationary series results is termed the order of homogeneity of the series. As a first step, it is important to identify the minimum amount of differencing required to create a stationary series. Modeling can rarely compensate for overdifferencing.

Modeling techniques can be very versatile. However, all modeling techniques have their limitations, and this should be recognized early. Several techniques may be combined to take into account the patterns discussed in this chapter. Unless examined and accounted for, changes in trend, cycle, seasonal, calendar, promotion, and irregular variations, and possibly their interactions, can cause significant modeling problems. Recognizing these characteristics is a first step to reducing their effect.

REFERENCES

Box, G. E. P. , G. M. Jenkins, and G. Reinsel (1994). *Time Series Analysis, Forecasting and Control.* 3rd ed. Englewood Cliffs, NJ: Prentice Hall.

Cryer, J. D., and R. B. Miller (1994). *Statistics for Business.* Belmont, CA: Wadsworth.

DeLurgio, S. (1998). *Forecasting Principles and Applications.* New York: Irwin/McGraw-Hill.

Dielman, T. E. (1996). *Applied Regression Analysis for Business and Economics.* 2nd ed. Belmont, CA: Wadsworth.

Gardner, E. S. (1985). Exponential smoothing: The State of the Art. J. Forecasting 4, 1–28.

Hanke, J. E., and A. G. Reitsch (1998). *Business Forecasting.* 6th ed. Englewood Cliffs, NJ: Prentice Hall.

Makridakis, S., S. C. Wheelwright, and R. J. Hyndman (1998). *Forecasting Methods and Applications.* 3rd ed. New York: John Wiley & Sons.

Pegels, C. C. (1969). Exponential forecasting: Some new variations. *Manage. Sci.* 12.5, 311–15.

Tryfos, P. (1998). *Methods for Business Analysis and Forecasting: Text and Cases.* New York: John Wiley & Sons.

Tufte, E. (1983). *The Visual Display of Quantitative Information.* Cheshire, CT: Graphics Press.

Wilson, J. H., and B. Keating (1998). *Business Forecasting.* New York: Irwin McGraw-Hill.

PROBLEMS

4.1 As a measure of output from the manufacturing sector of the economy, the FRB INDPROD is expected to be strongly negatively correlated with Unemployment (UEMP), based on economic theory.

Year	INDPROD	UEMP
1973	130	4.8
1974	129	5.5
1975	118	8.3
1976	131	7.6
1977	138	6.9
1978	146	6.0
1979	153	5.8
1980	147	7.0
1981	151	7.5
1982	139	9.5

Source: U.S. Bureau of the Census (1983), *Statistical Abstract of the United States,* 104th ed., Washington, DC: U.S. Government Printing Office.

From the accompanying set of data for INDPROD and UEMP:

a. Display and interpret the scatter diagram. Do the variables appear to be positively or negatively associated?

b. Label the points on the graph sequentially: Let $A = 1973$, $B = 1974$, and so on, and connect the points in this alphabetical order. Interpret the diagram now in light of the theoretical assumptions about the relationship between INDPROD and UEMP.

c. Display a scatter diagram of the annual changes (first differences) in the INDPROD and UEMP. What can you infer from this pattern?

4.2   For the sales of a weight-control product and monthly advertising expenditures data shown in Case 4D:

a. Display the time plots for sales and advertising. Compare the patterns and indicate the nature of the relationship that may be reflected in the data.

b. To further investigate the relationship, plot a scatter diagram and comment on the degree of linear association you expect to see. Which variable should you plot on the vertical axis?

c. Calculate the sample autocorrelation functions and display the correlograms. Contrast the month-to-month correlations (r_1) for the two variables.

d. Market researchers have postulated that the effects of advertising may carry over beyond the month in which the ad is viewed or heard. Display a scatter diagram of sales versus advertising lagged 1 month. Contrast this with the scatter diagram in (b). In which does the association appear stronger?

4.3   A second difference can be viewed as a weighted sum with weights $(1, -2, 1)$. If we construct a polynomial in X in the form $1 - 2X + 1X^2$ with these weights, we see that this is the product of $(1-X)(1-X)$. In turn, this product corresponds to a weighted sum with weights $(1, -1)$ followed by another weighted sum with weights $(1, -1)$.

a. Interpret the weighted sum with weights $(1, -1)$ in terms of the differencing on the original data.

b. Interpret the product of the two individual weighted sums in terms of the differencing operation.

c. Contrast the weighted sum with weights $(1, 0, -1)$ with the second difference.

d. Describe a fourth difference in terms of the weights. How do you interpret a weighted sum with weights $(1, 0, 0, -1)$?

4.4   For the monthly toll revenue data (TOLLREV) and toll messages (REV) on the CD:

a. Plot the time series and interpret the dominant pattern(s) you see in the data.

b. Perform a two-way analysis to determine the relative strength of the seasonal, trend, and irregular components. What is the nature of the seasonality?

c. If seasonality appears predominant, what do you think it could be attributed to?

d. If irregular component appears predominant, what effects do you think contribute to it?

e. Based on your findings, what type of projection technique do you recommend for the generation of future values of the data?

4.5   Repeat Problem 4.4 for one or more of the following monthly time series on the CD:

a. Unemployment rate data (UNEMP, Dielman, 1996, Table 4.7)

b. Silver prices data (SILVER, Dielman, 1996, Table 4.8)

c. Prime rate data (PRIME4, Dielman, 1996, Table 4.14)

d. Retail sales for U.S. retail sales stores (RETAIL, Hanke and Reitsch, 1998)

e. Wheat exports (SHIPMENT, Dielman, 1996, Table 4.9)

f. Shipments of spirits (DSHIP, Tryfos, 1998, Table 6.23)

g. Retail sales in a region (RSALES, Tryfos, 1998, Table 6.25)

h. Australian beer production (AUSBEER, Makridakis et al., 1998, Table 2.2)

i. International airline passenger miles (AIRLINE, Box, Jenkins, and Reinsel, 1994, Series G)

j. Australian electricity production (ELECTRIC, Makridakis et al., 1998, Figure 7-9)

k. French industry sales for paper products (PAPER, Makridakis et al., 1998, Figure 7-20)

l. Shipments of French pollution equipment (POLLUTE, Makridakis et al., 1998, Table 7-5)

m. Sales of recreational vehicles (WINNEBAG, Cryer and Miller, 1994, p. 742)

n. Sales of lumber (LUMBER, DeLurgio, 1998, p. 34)

o. Sales of consumer electronics (ELECT, DeLurgio, 1998, p. 34)

4.6 Repeat Problem 4.4 for one or more of the following quarterly time series on the CD:

a. Sales for Outboard Marine (OMC, Hanke and Reitsch, 1998, Table 4.5)

b. Shoe store sales (SHOES, Tryfos, 1998, Table 6.5)

c. Domestic car sales (DCS, Wilson and Keating, 1998, Table 1-5)

d. Sales of The GAP (GAP, Wilson and Keating, 1998, p.30)

e. Air passengers on domestic flights (PASSAIR-Source DeLurgio, 1998, p. 33)

4.7 Create quarterly time series from the data in Problem 4.4 and repeat the questions for Problem 4.4. If your conclusions change substantially (they might) from your findings in Problem 4.5, what could be a contributing factor?

4.8 Calculate the first three autocorrelation coefficients for INDPROD and UNEMP in Problem 4.1 and interpret the results.

4.9 For the data for the quarterly ABX Company sales on the CD (ABXCO, Dielman, 1996, Table 3.11):

a. Produce a time plot of the ABX Company sales data. Interpret the dominant pattern(s) in the data.

b. Analyze the trend using TRENDIFF worksheet (on the CD). Is there a strong or moderate trend?

c. Contrast your result with a similar analysis using examples from Problem 4.6.

4.10 For the monthly unemployment rate data on the CD (UNEMP, Dielman, 1996, Table 4.7):

a. Produce a time plot and interpret the dominant pattern(s) in the data.

b. Analyze the trend using TRENDIFF worksheet. Is there a strong or moderate trend?

What do you see as the dominant pattern(s) in the data now?

c. Contrast your result with a similar analysis using examples from Problem 4.5.

4.11 For the data in Exhibit 4.12, make a multiple plot of the daily volumes for each month. For a given weekday, contrast its within variation to the variation between weekdays. Does it make sense to state that roughly two-thirds of the total variation in the data may be attributable to a day-of-week pattern?

4.12 The international airline passenger series is a very famous time series analyzed in many time series books (Box et al., 1994, Series G). It depicts 144 monthly totals (in thousands of passengers) from 1949:1 to 1960:12.

a. Plot the time series and give a visual characterization of its pattern.

b. Take logarithms of the data (Box, Jenkins, and Reinsel, 1994, Table 9.1) and plot the logarithmic values. Contrast your visual characterization of part (a) to a similar analysis of part (b). What has happened to the nature of the seasonal fluctuations?

c. Make a multiple time plot of (b) showing a time plot for each of the 12 years. How does the seasonal pattern stack up?

d. Perform a two-way ANOVA on the log-transformed airline passenger data. What is your preliminary view of the relative contribution of trend, seasonal, and irregular components in the data?

4.13 Repeat Problem 4.12 for the monthly data in Problem 4.5.

4.14 For the housing starts and mortgage rates series discussed in this chapter and in Chapter 2:

a. Plot a scatter diagram of Y_{t+1} versus Y_t for each series.

b. Plot a scatter diagram of Y_{t+2} versus Y_t for each series.

c. Do the series appear to be autocorrelated?

d. Plot the sample autocorrelations for each example. What characteristic of the time series is reflected in the pattern of autocorrelations?

4.15 Repeat Problem 4.14 for the monthly data in Problem 4.5.

4.16 For the daily volume of bank transactions by day of week in Exhibit 4.12:

a. Plot a time plot of the series.

b. Plot Y_{t+7} versus Y_t for the series.

c. Plot the first 14 sample autocorrelations. Interpret your results.

4.17 In the following table, determine the month and trend contribution to the total variation in the repair-parts demand. How distinct are the identifiable components of the data (trend and seasonal) from the irregular? Comment on how this might impact your ability to interpret the results in terms of how hard (or easy) it might be to find an adequate model for these data.

1–ANALYSIS OF VARIANCE: MONTHLY DATA
2–TITLE1: Repair part demand
3–TITLE2: 1985-1986
4–X-AXIS: Month
5–Y-AXIS: Units
6–ANOVA-MONTHLY DATA

7	Year	Month	Period	Data
8	1985	1	1	43
9		2	2	55
10		3	3	60
11		4	4	44
12		5	5	39
13		6	6	55
14		7	7	52
15		8	8	61
16		9	9	62
17		10	10	66
18		11	11	39
19		12	12	67
20	1986	1	13	66
21		2	14	51
22		3	15	53
23		4	16	52
24		5	17	40
25		6	18	53
26		7	19	75
27		8	20	43
28		9	21	51
29		10	22	36
30		11	23	42
31		12	24	59
	1987	1	25	67

SOURCE OF VARIANCE	PERCENT
TREND	— %
SEASONALITY	— %
NOISE	— %
TOTAL	100.0%

4.18 Using the formal-wear rental data in Exhibit 4.19 (Hanke and Reitsch, 1998), perform the following preliminary data analyses. Each step of the analysis is intended to provide some insights into aspects of modeling complexities that are dealt with in upcoming chapters.

a. Create a time plot of the logarithms of the time series. Contrast this pattern with the one shown in Exhibit 4.19 in terms of the stability and variability in the seasonal effect.

b. Use a two-way decomposition analysis to determine the relative contributions of the trend, seasonal, and irregular components in the log-transformed data. Contrast this with the results in Exhibit 4.19, and explain the enhanced importance of the seasonal effect.

c. Create a time plot of the first differences of the data in Exhibit 4.19. Contrast this plot with the one shown in Exhibit 4.19 in terms of the level and variability in the seasonal pattern.

d. Use a two-way decomposition analysis to determine the relative contributions of the trend, seasonal, and irregular components in the differenced data. Contrast this with the results in Exhibit 4.19 and (b). Suggest what happened to the relative importance of trend and seasonal components in the transformations.

e. Create a time plot of the first differences of the log-transformed data in (a). Use a two-way decomposition analysis to determine the relative contributions of the trend, seasonal, and irregular components in the differenced log-transformed data. Contrast this with the results in (d). Suggest what happened to the relative importance of trend and seasonal components in the transformations.

f. On the basis of the results in (e), how might you piece together this sequence of

modeling steps to reconstruct an overall
modeling flowchart for these data? Do not
use mathematics.

4.19 Repeat Problem 4.18 for the monthly data in
Problem 4.5.

4.20 For each of the following, comment on the
nature of the nonstationarity found after each
reexpression (transformation) of the original
data. In each case, plot a sample autocorrelation
function and contrast the pattern with the pattern
in the data it is based on.

a. Original data in Exhibit 4.19.

b. Log of the data in Exhibit 4.19; contrast
with (a).

c. First differences of the data in Exhibit
4.19; contrast with (a).

CASE 4A PRODUCTION OF ICE CREAM (CASES 1A AND 3A CONT.)

The entrepreneur who has designed an ice cream plant to be built in an urban region of the country has enlisted your assistance in making better sense of the shipments data.

(1) Examine the degree of nonstationarity in the data (in Case 3A) by constructing time plots and correlograms for (a) first differences and (b) second differences. What is the nature of the nonstationarity in the data?

(2) Create a two-way decomposition table and calculate the percentage variation attributable to the trend, seasonal, and irregular components. Interpret your answer in terms of what you uncovered in (1).

(3) Can you find a couple of reasons why the dominant pattern exists in the data?

(4) Repeat (1) and (2) using the log-transformed data. In qualitative terms, do you need to change your mind about the characteristics of the ice cream data?

CASE 4B DEMAND FOR AIR TRAVEL

The VP of Operations of an airline wants to examine the passenger traffic on a trans-Atlantic route to help understand and improve business opportunities and cost inefficiencies. The VP is aware that the travelers on this route are mostly business-people, so a number of factors affecting demand are worth considering. Among these, you might consider the GNP of the two countries, the exchange rate, and the ticket price. However, the VP has been advised that a preliminary study to gain insight into the nature of the problem is highly recommended and has requested that you give some help using your newly learned forecasting skills. The quarterly data for 4 years is shown in the table.

Air Travel	GNP (Country A)	GNP (Country B)	Passenger Miles (in millions)	Currency Exchange Rate	Ticket Price (U.S. dollars)
YR01: 3	3712.4	407.1	98248	2.08	1306
YR01: 4	3733.6	407.3	88053	2.00	1298
YR02: 1	3783.0	404.5	80505	1.84	1293
YR02: 2	3823.5	407.7	124684	1.81	1274
YR02: 3	3872.8	413.1	146001	1.84	1276
YR02: 4	3935.6	416.6	111542	1.70	1282
YR03: 1	3974.8	423	96979	1.68	1283
YR03: 2	4010.7	422.4	120376	1.70	1284
YR03: 3	4042.7	427.2	144691	1.87	1292
YR03: 4	4069.4	429.2	105751	1.78	1307
YR04: 1	4095.7	440.9	77858	1.85	1374
YR04: 2	4112.2	441.7	113532	1.92	1308
YR04: 3	4129.7	439.8	129591	1.92	1335
YR04: 4	4133.2	444.2	111740	1.81	1363
YR05: 3	4150.6	460.4	99034	1.69	1388
YR05: 4	4155.1	456.3	137179	1.68	1343

(1) Construct a time plot of the data with a four-period moving average to depict a smoother pattern for an overall direction in trend.

(2) The airline industry is highly seasonal, so you can start with a decomposition of the quarterly data to classify the trend, seasonal, and irregular variation in the data. Use the NOISEQTR and TRENDIFF spreadsheets (on the CD) and interpret your findings.

(3) Describe the nature of the nonstationarity in the data using a correlogram analysis for the variables.

CASE 4C DOMESTIC AUTOMOBILE PRODUCTION (CASES 1D AND 3D CONT.)

Your assignment as an analyst for an industry association is to identify the time series characteristics in the data (in Case 3D) in preparation of the modeling steps to follow.

(1) Create an ANOVA decomposition of the monthly data and interpret the results.

(2) Create and interpret a time plot and correlogram of the first-differenced data. Repeat for the second-differenced data. Repeat for the twelfth-differenced data. Look for evidence of nonstationarity in patterns that you may be able to describe in terms of a trend.

(3) For which set of differenced data do you think you are closest to a set of random numbers? For this series, create an ANOVA decomposition. Is there much evidence for any residual trend or seasonality in the data?

(4) Repeat (1)–(3) for the quarterly series. Are your conclusions any different? And if so, how?

(5) Would transforming the data be helpful? Try to make that determination from a visual inspection of the data and the results of the two-way decomposition.

(6) Gather comparable data for a more current period (also containing a recessionary period) and repeat (1)–(3). How much value do you place on the earlier data? Contrast some of the differences and similarities between the two sets of data.

CASE 4D SALES AND ADVERTISING OF A WEIGHT-CONTROL PRODUCT

The VP of Sales has called you into her office to help plan for an upcoming advertising campaign. To date, much of the planning has used a seat-of-the-pants approach and is not entirely satisfactory. To help improve the situation, you recommend investigating some quantitative approaches to the problem. You want to gain some familiarity with some methodologies you find in the literature, and you embark on a preliminary analysis of an existing data set. It is hoped that this investigation will lead to some insights that will help you tackle your company's data. For the following sales of a weight-control product and monthly advertising expenditures data (F. M. Bass and D. G. Clarke (1972), Testing distributed lag models of advertising effect, *J. Marketing Res.* 9.3, 298–308).

Sales Year 1	12.0	20.5	21.0	15.5	15.3	23.5	24.5	21.3	23.5	28.0	24.0	15.5
Advertising	15	16	18	27	21	49	21	22	28	36	40	3
Sales Year 2	17.3	25.3	25.0	36.5	36.5	29.6	30.5	28.0	26.0	21.5	19.7	19.0
Advertising	21	29	62	65	46	44	33	62	22	12	24	3
Sales Year 3	16.0	20.7	26.5	30.6	32.3	29.5	28.3	31.3	32.2	26.4	23.4	16.4
Advertising	5	14	36	40	49	7	52	65	17	5	17	1

(1) Create an ANOVA decomposition of the monthly data and interpret the results. Contrast the decompositions you find for the sales versus the advertising data.

(2) Would transforming the data be helpful? Try to make that determination from a visual inspection of the data and the results of the two-way decomposition.

(3) Use the TRENDIFF spreadsheet with the data and interpret your findings.

(4) Calculate a lagged version of the advertising data and create the following two scatter diagrams: (1) Sales vs. advertising and (2) Sales vs. lagged advertising. How do the scatter diagrams differ, and what is your interpretation of the differences (if any)?

(5) Calculate the product moment and the robust correlation coefficients for the data in the previous question. What is the significance of your preliminary findings up to this point?

APPENDIX 4A A TWO-WAY TABLE DECOMPOSITION

In this chapter, we introduced the year-by-month ANOVA table as a means of constructing the trend and seasonal components of variation in a time series. It is a simple example of a deterministic model but of limited value in creating projections. Nevertheless, it provides an intuitive way of illustrating components of variation while introducing the useful ideas of additivity and residual analysis. Although most statistical software packages and Excel will allow you to perform this analysis using a menu, it is instructive to work this out on a spreadsheet.

Recall that a basic premise underlying a decomposition approach is that a time series forecast can be represented by a number of distinctly interpretable, albeit unobservable, components. Two-way table decompositions can give us a preliminary feel for the relative contributions and additivity of seasonal, trend, cyclical, and irregular variations to the total variation in a time series. This helps the forecaster identify the most important sources of variability and thereby helps in the selection of a suitable forecasting technique.

The method by which a two-way table is constructed is not complex, but the notation is somewhat cumbersome. By working carefully through the analysis, you should be able to understand it completely and create the analysis on a spreadsheet. Assuming monthly data, let Y_{ij} (i = month, j = year) be a representation of a typical observation in the table, where $i = 1, 2 \ldots, 12$, and $j = 1, 2 \ldots, J$ (the number of years). (This representation can also be used for quarterly data, but then $i = 1, 2, 3, 4$). A model that describes a typical observation in terms of a seasonal effect S_i, a trend or yearly effect T_j, and a residual error term ε_{ij} is given by

$$Y_{ij} = \mu + S_i + T_j + \varepsilon_{ij}$$

where μ denotes a mean effect or typical value. Then, the following symbols can be used to summarize certain totals and averages of interest in the analysis:

$$Y_{\cdot j} = \sum Y_{ij}$$

is the total of year j, which has been summed over 12 rows, and

$$\overline{Y}_{\cdot j} = Y_{\cdot j} / 12$$

represents the average per month for year j. Thus $\{Y_{\cdot j}; j = 1, \ldots, J\}$ are yearly totals, which are used to summarize the trend over J years. Summing across columns corresponding to the number of years gives

$$Y_{i\cdot} = \sum Y_{ij}$$

which is the total of month i (sum over J years). Then,

$$\overline{Y}_{i\cdot} = Y_{i\cdot} / J$$

is the average per month of month I. Thus, $\{\overline{Y}_{i\cdot}; i = 1, \ldots, J\}$ are monthly averages, which are used to summarize the average seasonal pattern over J years. To get overall totals, define $Y_{\cdot\cdot}$ by

$$Y_{\cdot\cdot} = \sum \sum Y_{ij}$$
$$= \sum Y_{\cdot j}$$
$$= \sum Y_{i\cdot}$$

The quantity $Y_{\cdot\cdot}$ is known as the grand total, and $\overline{Y} = Y_{\cdot\cdot} / 12J$ is the average per month over J years, or the grand mean.

It is now possible to describe the various contributions to the total variation in the data. The total variation (as measured from the grand mean) is given by

$$\mathrm{SST} = \left[\sum \sum (Y_{ij} - \overline{Y}_{\cdot\cdot})^2 \right] / (12J - 1)$$

This is a measure of the overall variation in the data, which is due to trend, seasonal, and irregular patterns. Specific entries within each row (month) can be measured against the row mean to give a measure of variation for the given month. This variation may be due to trends and changes in the seasonal patterns as well as irregularity. Thus,

$$\mathrm{SS}_{\mathrm{Row}} = \sum (Y_{ij} - \overline{Y}_{i\cdot})^2 / (J - 1)$$

Similarly,

$$SS_{Col} = \sum (Y_{ij} - \bar{Y}_{.j})^2 / 11$$

is a measure of the variation within column J (year) measured from the column mean. This variation may be due to seasonal patterns and changes in trend as well as irregularity. Because the row and column means represent average seasonal and trend effects, variations of these averages reflect the respective variation due to average seasonal pattern and trend. Thus,

$$SS_{Seas} = \sum (\bar{Y}_{i.} - \bar{Y}_{..})^2 / 11$$

and

$$SS_{Trend} = \sum (\bar{Y}_{.j} - \bar{Y}_{..})^2 / (J - 1)$$

represent variation in row (monthly) means and column (yearly) means, respectively.

Contribution of Trend and Seasonal Effects

We can also describe the proportion of variability due to seasonal effects, S, by the ratio $R^2_{Seas} = SS_{Seas}/SST$ and the proportion of variability due to trend effects, T by $R^2_{Trend} = SS_{Trend}/SST$. Ideally, or for an extreme case, if there is no trend in the data, the column means equals the grand mean and the proportion of variability due to trend effects equals zero.

Interpreting the Residual Table

The dominant components usually found in forecasting a time series are the trend, cycle, and seasonal components. At times, it is impractical to distinguish trend and cycle as distinct components. So, in many economic forecasting applications, we speak of trend-cycle as a single source of variation in which the two factors are inseparable. In the discussion here it is instructive to consider trend and cycle as separate components. By removing or subtracting both trend and seasonal effects from a time series, it is possible to examine the residual variation for more subtle patterns; that is:

$$\text{Residuals} = \text{Data} - \text{Fit}$$
$$= \text{Data} - (\text{Trend} + \text{Seasonal effects})$$

The (i, j)th residual represents the deviation from the grand mean less the ith month's seasonal effect and the jth year's trend effect. Thus, the (i, j)th residual is the (i, j)th observation corrected for seasonal and trend effect in the sense of a two-way ANOVA decomposition used in this chapter.

For example, we can examine the residual variation to show the effects of cyclic downturn consistent with empirical data or economic theory. If so, a trend-seasonal decomposition would best describe the dominant patterns in the data. The analysis also shows whether the seasonal pattern is stable from year to year. Have school openings and closings perhaps been shifting, thereby shifting the seasonality of the data? Have pre-Christmas buying patterns of shoppers shifted, thus affecting sales data? Do nontypical values destroy the underlying trend and/or seasonal pattern? All these questions may be reasonably answered with a preliminary ANOVA.

Other useful quantities can be derived from the residual two-way table. The residual variance is the total variation corrected for trend and seasonal effects. This provides a measure of variation for all the residuals. The residual column variance, $S_j^2(C)$, provides a variance measure for each column and the residual row variance, $S_i^2(R)$, provides a variance measure for each row. Large differences among the residual column variances may indicate some outliers or special events in a year that have a large deviation. Large differences among the residual row variances for various months may suggest the relative difficulty of forecasting months with large variability and may also indicate the presence of outliers in certain months.

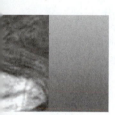

5

Assessing the Accuracy of Forecasts

"Television won't be able to hold on to any market it captures after the first six months. People will soon get tired of staring at a plywood box every night."
A WELL-KNOWN MOVIE MOGUL IN THE EARLY DAYS OF TV

ONE GOAL OF AN EFFECTIVE DEMAND FORECASTING STRATEGY is to improve forecast accuracy through the measurement of forecasting performance of forecasting techniques and modeling. Forecasters use accuracy measures to evaluate the relative performance of forecasting models and assess alternative forecast scenarios. Accuracy measurement provides management with a tool to establish credibility for the way the forecast was derived and to provide incentives to enhance professionalism in the organization.

We recommend the use of relative error measures to see how much of an improvement is made in choosing a particular forecasting model over some benchmark or "naive" forecasting technique. Relative error measures also provide a reliable way of ranking forecasting models on the basis of their usual accuracy over sets of time series.

5.1 THE NEED TO MEASURE FORECAST ACCURACY

It is generally recognized in most demand forecasting organizations that accurate forecasts are essential in achieving significant improvements in their operations. Developers of forecasting models use accuracy measures, and accuracy at all levels is relevant to the users of forecasts in a company. Will the model prove reliable for forecasting units or revenues over a planned forecast horizon, such as item-level product demand for the next 12 weeks or aggregate revenue demand for the next 4 quarters? Will it have a significant impact on marketing and production activities? Will it have an effect on inventory investment or customer service? In a nutshell, inaccurate forecasts can have a direct effect on setting inadequate safety stocks, ongoing capacity problems, massive rescheduling of manufacturing plans, chronic late shipments to customers, and adding expensive manufacturing flexibility resources.

Whether a method that has provided a good fit to historical data will also yield accurate forecasts for future time periods is an unsettled issue. Intuition suggests that this may not necessarily be the case. There is no guarantee that past patterns will persist in future periods. For a forecasting technique to be useful, we must demonstrate that it can forecast reliably and with consistent accuracy. It is not sufficient to simply produce a model that performs well only in a historical (within-sample) fit period. At the same time, the users of a forecast need forecasting results in a timely fashion. Using a forecasting technique and waiting one or more periods for history to unfold in future periods is not practical because our advice as forecasters will not be effective.

5.2 ANALYZING FORECAST ERRORS

Two important aspects of forecast accuracy measurement are bias and precision. **Bias** is a problem of direction: Forecasts are typically too low (downward bias) or typically too high (upward bias). **Precision** is an issue of magnitudes: Forecast errors can be too large (in either direction) using a particular forecasting technique. Consider first a simple situation—forecasting a single product or item. The attributes that should satisfy most forecasters include lack of serious bias, acceptable precision, and superiority over naive models.

Lack of Bias

The Institutional Broker's Estimate System (I/B/E/S), a service that tracks financial analysts' estimates, reported that forecasts of corporate yearly earnings that are made early in the year are persistently optimistic, that is to say, upwardly biased. In all but 2 of the 12 years studied (1979–1991), analysts revised their earnings forecasts downward by the end of the year. In a *New York Times* article (January 10, 1992), Jonathan Fuerbringers writes that this pattern of revisions was so clear that the I/B/E/S urged stock market investors to take early-in-the year earnings forecasts with a grain of salt.

If forecasts are typically too low, we say that they are downwardly biased; if too high, they are upwardly biased. If overforecasts and underforecasts tend to cancel one another out (i.e., if an average of the forecast errors is approximately zero), we say that the forecasts are unbiased.

Bias refers to the tendency of a forecast to be predominantly toward one side of the truth.

EXHIBIT 5.1
(a) Biased Forecasts;
(b) Unbiased
Forecasts

(a)

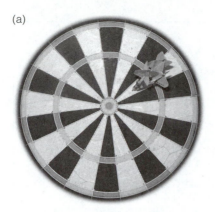

(b)

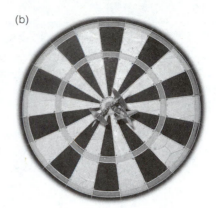

Bias is a problem of direction. If we think of forecasting as aiming at a target, then a bias implies that the aim is off-center, so that the darts land repeatedly toward the same side of the target (Exhibit 5.1a). In contrast, the forecasts in Exhibit 5.1b are unbiased, that is, evenly distributed around the target.

What Is an Acceptable Precision?

Imprecision is a problem if the forecast errors tend to be too large. Exhibit 5.2 shows two patterns that differ in terms of precision. The upper two forecasts are less precise—as a group, they are farther from the target. If bias is thought of as bad aim, then imprecision is a lack of constancy or steadiness. The precise forecast is generally right around the target.

Precision refers to the distance between the forecasts as a result of using a particular forecasting technique and the corresponding actual values.

EXHIBIT 5.2
Precisions in
Forecasts

EXHIBIT 5.3
(a) Bar Chart and
(b) Table Showing
Actuals (A) and
Forecasts for Three
Forecasting
Techniques

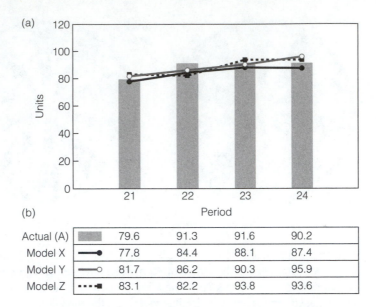

Actual (A)		79.6	91.3	91.6	90.2
Model X	—●—	77.8	84.4	88.1	87.4
Model Y	—○—	81.7	86.2	90.3	95.9
Model Z	···■	83.1	82.2	93.8	93.6

Exhibit 5.3 illustrates the measurement of bias and precision for three hypothetical forecasting techniques. In each case, the fit period is periods 1–20. Shown in the top row of Exhibit 5.3b are actual values for the last four periods (21–24). The other three rows contain forecasts using forecasting techniques X, Y, and Z. These are graphed in Exhibit 5.3a.

Exhibit 5.4 records the deviations between the actuals and their forecasts. Each deviation represents a forecast error (or forecast miss) for the associated period:

$$\text{Forecast error } (E) = \text{Actual } (A) - \text{Forecast } (F)$$

EXHIBIT 5.4
(a) Bar Chart and
(b) Table of the
Forecast Error (E)

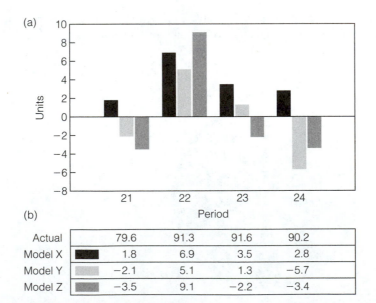

Actual		79.6	91.3	91.6	90.2
Model X	■	1.8	6.9	3.5	2.8
Model Y	▢	−2.1	5.1	1.3	−5.7
Model Z	▨	−3.5	9.1	−2.2	−3.4

Contrast this with a fitting error (or residual) of a model over a fit period, which is:

$$\text{Fitting error} = \text{Actual } (A) - \text{Model fit}$$

Forecast error (or miss) is a measure of forecast accuracy. Fitting error (or residual) is a measure of model adequacy.

In Exhibit 5.4, the forecast error shown for technique X is 1.8 in period 21. This represents the deviation between the actual value in forecast period 21 ($= 79.6$) and the forecast using technique X ($= 77.8$). In forecast period 22, the forecast using technique X was lower than actual value for that period, resulting in a forecast error of 6.9. The period 24 forecast using technique Z was higher than that period's actual value; hence, the forecast error is negative (-3.4). When we *over*forecast, we must make a *negative* adjustment to reach the actual value. Note that if the forecast is less than the actual value, the miss is a positive number; if the forecast is more than the actual value, the miss is a negative number.

To identify patterns of upward- and downward-biased forecasts, we start by comparing the number of positive and negative misses. As Exhibit 5.4 shows, technique X underforecasts in all four periods, indicative of a persistent (downward) bias. Technique Y underforecasts and overforecasts with equal frequency; therefore, it exhibits no evidence of bias in either direction. Technique Z is biased slightly toward overforecasting. As one measure of forecast accuracy, we calculate a percentage error (Exhibit 5.5): $\text{PE} = 100\% \times (A - F)/A$.

EXHIBIT 5.5
(a) Bar Chart and
(b) Table of the
Forecast Error as
Percentage Error
(PE) between Actuals
and Forecasts for
Three Techniques

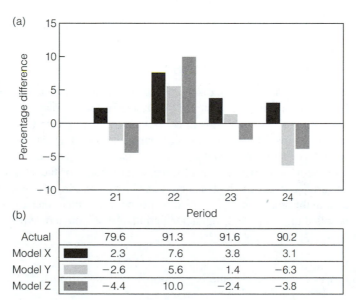

(b)

		21	22	23	24
Actual		79.6	91.3	91.6	90.2
Model X	■	2.3	7.6	3.8	3.1
Model Y	▨	−2.6	5.6	1.4	−6.3
Model Z	▨	−4.4	10.0	−2.4	−3.8

To reduce bias in a forecasting technique, we can either (1) reject any technique that projects with serious bias in favor of a less-biased alternative (after we have first compared the precision and complexity of the methods under consideration) or (2) investigate the pattern of bias in the hope of devising a bias adjustment; for example, we might take the forecasts from method X and adjust them upward to try to offset the tendency of this method to underforecast.

5.3 WAYS TO EVALUATE ACCURACY

A number of forecasting competitions have been held to assess the effectiveness of statistical forecasting techniques and determine which techniques are among the best. Starting with the original M competition, Makridakis et al. (1982) compared the accuracy of about 20 forecasting techniques across a sample of 111 time series. A subset of the methods was tested on 1001 time series. The last 12 months of each series were held out and the remaining data were used for model fitting. Using a range of measures on a holdout sample, the *International Journal of Forecasting* (*IJF*) conducted a competition in 1997 comparing a range of forecasting techniques across a sample of 3003 time series. Known as the M3 competition, these data and results can be found at the website www.maths.monash.edu.au/~hyndman/forecasting/. A number of *IJF* papers have been written summarizing the results of these competitions (Ord et al., 2000). These competitions have become the basis for how we measure forecast accuracy in practice.

Forecast accuracy measurements are performed in order to assess the accuracy of a forecasting technique.

The Fit Period versus the Test Period

In measuring forecast accuracy, a portion of the historical data is withheld and reserved for evaluating forecast accuracy. Thus, the historical data are first divided into two parts: an initial segment (the fit period) and a later segment (the holdout sample). The fit period is used to develop a forecasting model. Sometimes, the fit period is called the within-sample training, initialization, or calibration period. Next, using a particular model for the fit period, model forecasts are made for the later segment. Finally, the accuracy of these forecasts is determined by comparing the projected values with the data in the holdout sample. The time period over which forecast accuracy is evaluated is called the test period, validation period, or holdout-sample period.

A forecast accuracy test should be performed with data from a holdout sample.

To minimize confusion, we introduce some common notation for the concepts used in these calculations. $Y_t(1)$ as the one-period-ahead forecast of Y_t, and $Y_t(m)$ as the m-period-ahead forecast. A forecast for $t=25$ that is the one-period-ahead forecast made at $t=24$ is denoted by $Y_{24}(1)$ and the two-period-ahead forecast for the same time, $t=25$, made at $t=23$ is denoted by $Y_{23}(2)$. Generally, business forecasters are interested in multiperiod-ahead forecasts because lead times longer than one period are required for businesses to act on a forecast. We use the following conventions for a time series:

$\{Y_t, t=1, 2, \ldots, T\}$	the historical data set up to and including period $t=T$
$\{\hat{Y}_t, t=1, 2, \ldots, T\}$	the data set of fitted values that result from fitting a model to historical data
Y_{T+m}	the future value of Y_t, m periods after $t=T$
$\{Y_T(1), Y_T(2), \ldots, Y_T(m)\}$	the one- to m-period-ahead forecasts made from $t=T$

Consequently, we see that forecast errors and fit errors (residuals) refer to:

$Y_1 - \hat{Y}_1, Y_2 - \hat{Y}_2, \ldots, Y_T - \hat{Y}_T$	the fit errors or residuals from a fit
$Y_{T+1} - Y_T(1), Y_{T+2} - Y_T(2),$ $\ldots, Y_{T+m} - Y_T(m)$	the forecast errors (*not* to be confused with the residuals from a fit)

In dealing with forecast accuracy, it is important to distinguish between forecast errors and fitting errors.

Goodness-of-Fit versus Forecast Accuracy

We need to assess forecast accuracy rather than just calculate overall goodness-of-fit statistics for the following reasons:

- Goodness-of-fit statistics may appear to give better results than a forecasting-based calculation, but goodness-of-fit statistics measure model adequacy over a fitting period that may not be representative of the forecasting period.

- When a model is fit, it is designed to reproduce the historical patterns as closely as possible. This may create complexities in the model that capture insignificant patterns in the historical data, which may lead to overfitting.

- By adding complexity, we may not realize that insignificant patterns in the past are unlikely to persist into the future. More important, the subtle patterns of the future are unlikely to have revealed themselves in the past.

- Exponential smoothing models are based on updating procedures in which each forecast is made from smoothed values in the immediate past. For these models, goodness-of-fit is measured from forecast errors made in estimating the next time period ahead from the current time period. These are called one-period-ahead forecast errors (also called one-step-ahead forecast errors). Because it is reasonable to expect that errors in forecasting the more distant future will be larger than those made in forecasting the next period into the future, we should avoid accuracy assessments based exclusively on one-period-ahead errors.

When assessing forecast accuracy, we may want to know about likely errors in forecasting more than one period ahead.

A model's accuracy in fitting the past is likely to be an overly optimistic guide to the accuracy with which the model will forecast the future. Some research has suggested that there is a low correlation between how forecasting techniques rank in reproducing historical data and how they rank in forecasting the future (Makridakis and Winkler, 1985).

Item-Level versus Aggregate Performance

Forecast evaluations are also useful in multiseries comparisons (see Chapter 9). Production and inventory managers typically need demand or shipment forecasts for hundreds to tens of thousands of items (SKUs) based on historical data for each item. Financial forecasters need to issue forecasts for dozens of budget categories in a strategic plan on the basis of past values of each source of revenue. In a multiseries comparison, the forecaster should appraise the method based not only on its performance for the individual item but also on the basis of its overall accuracy when tested over various summaries of the data. In the next section, we discuss various measures of forecast accuracy.

Absolute Errors versus Squared Errors

Both the perspective based on absolute errors and that based on squared errors are useful. There is a good argument for consistency in the sense that a model's forecasting accuracy should be evaluated on the same basis used to develop (fit) the model. The standard basis for model fit is the least-squares criterion, that is, minimizing the MSE between the actual and fitted values. To be consistent, we should evaluate forecast accuracy based on squared error measures, such as the root mean squared error (RMSE). It is useful to test how well different methods do in forecasting using a variety of accuracy measures. Forecasts are put to a wide variety of uses in any organization, and no single forecaster can dictate on how they will be used and interpreted.

Sometimes costs or losses due to forecast errors are in direct proportion to the size of the error—double the error leads to double the cost. For example, when a soft drink distributor realized that the costs of shipping its product between distribution centers was becoming prohibitive, it made a study of the relationship between underforecasting (not enough of the right product at the right place, thus requiring a transshipment, or backhaul, from another distribution center) and the cost of those backhauls. As shown in Exhibit 5.6, overforecasts of 25% or higher appeared strongly related to an increase in the backhaul of pallets of product. In this case, the measures based on absolute errors are more appropriate. In other cases, small forecast errors do not cause much harm and large errors may be devastating; then, we would want to stress the importance of (avoidance of) large errors, which is what squared-error measures accomplish.

EXHIBIT 5.6
Backhauls in Pallets
versus Forecast Error

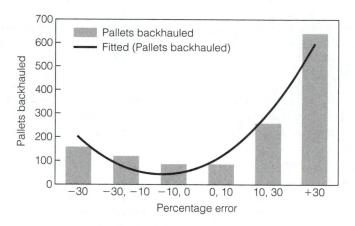

5.4 MEASURES OF FORECAST ACCURACY

Measures of Bias

There are two common measures of bias. The mean error (ME) is the sum of the forecast errors divided by the number of periods in the forecast horizon (h) for which forecasts were made:

$$\text{ME} = \left[\sum_{t=T+1}^{T+h} \left(A_t - F_t \right) \right] / h$$

$$= \left[\sum_{i=1}^{h} \left(A_i - F_i \right) \right] / h$$

The mean percentage error (MPE) is:

$$\text{MPE} = 100 \cdot \left[\sum_{t=T+1}^{T+h} \left(A_t - F_t \right) / A_t \right] / h$$

$$= 100 \cdot \left[\sum_{i=1}^{h} \left(A_i - F_i \right) / A_i \right] / h$$

The MPE and ME are useful supplements to a count of the frequency of under- and overforecasts. The ME gives the average of the forecast errors expressed in the units of measurement of the data; the MPE gives the average of the forecast errors in terms of percentage and is unit-free.

If we examine the hypothetical example in Exhibits 5.4 and 5.5, we find that for technique X ME = 3.8 and MPE = 4.2%. This means that, using technique X, we underforecast by an average of 3.8 per period. Restated in terms of the MPE, this means that the forecasts from technique X were below the actual values by an average of 4.2%. A positive value for the ME or MPE signals downward bias in the forecasts using technique X.

The two most common measures of bias are the mean error (ME) and the mean percentage error (MPE).

Compared to technique X, the ME and MPE for technique Y are much smaller. These results are not surprising. Because technique Y resulted in an equal number of underforecasts and overforecasts, we expect to find that the average error is close to zero.

On the other hand, the low ME and MPE values for technique Z are surprising. We have previously seen that technique Z overforecasts in three of the four periods. A closer look at the errors from technique Z shows that there was a very large underforecast in one period and relatively small overforecasts in the other three periods. Thus, the one large underforecast offset the three small overforecasts, yielding a mean error close to 0. The lesson here is that averages can be distorted by one unusually large error. We should always use error measures as a supplement to the analysis of the individual errors (Armstrong and Collopy, 1992; Makridakis and Winkler, 1983).

Measures of Precision

Certain indicators of precision are based on the absolute values of the forecast errors. By taking an absolute value, we eliminate the possibility that underforecasts and overforecasts negate one another. Therefore, an average of the absolute forecast errors reveals simply how far apart the forecasts are from the actual values. It does not tell us if the forecasts are biased. The absolute errors are shown in Exhibit 5.7, and the absolute values of the percentage errors are shown in Exhibit 5.8.

The most common averages of absolute values are mean absolute error (MAE), mean absolute percentage error (MAPE), and median absolute percentage error (MdAPE).

$$MAE = \left[\sum_{t=T+1}^{T+h} \left| \left(A_t - F_t \right) \right| \right] / h$$

EXHIBIT 5.7
(a) Bar Chart and
(b) Table Showing
Forecast Error as the
Absolute Difference
|E| between Actuals
and Forecasts for
Three Techniques

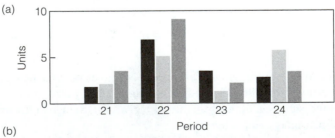

(a)

(b)

		21	22	23	24
Actual		79.6	91.3	91.6	90.2
Model X		1.8	6.9	3.5	2.8
Model Y		2.1	5.1	1.3	5.7
Model Z		3.5	9.1	2.2	3.4

EXHIBIT 5.8
(a) Bar Chart and
(b) Table Showing
Forecast Error as
Absolute Percentage
Difference |PE (%)|
between Actuals and
Forecasts for Three
Techniques. |(PE)| =
100·|(A − F)/A|

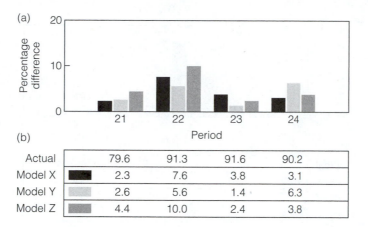

(a)

(b)

Actual		79.6	91.3	91.6	90.2
Model X		2.3	7.6	3.8	3.1
Model Y		2.6	5.6	1.4	6.3
Model Z		4.4	10.0	2.4	3.8

$$= \left[\sum_{i=1}^{h} \left| \left(A_i - F_i \right) \right| \right] / h$$

$$\text{MAPE} = 100 \cdot \left[\sum_{t=T+1}^{T+h} \left| \left(A_t - F_t \right) / A_t \right| \right] / h$$

$$= 100 \cdot \left[\sum_{i=1}^{h} \left| \left(A_i - F_i \right) / A_i \right| \right] / h$$

$$\text{MdAPE} = \text{Median value of} \left\{ \left| \left(A_i - F_i \right) / A_i \right|, i = 1, \ldots h \right\}$$

Interpretations of the averages of absolute error measures are straightforward. In Exhibit 5.7, we calculate that the MAE is 4.6 for technique Z, from which we can conclude that the forecast errors from this technique average 4.6 per period. The MAPE, from Exhibit 5.8, is 5.2%, which tells us that the period forecast errors average 5.2%.

The MdAPE for technique Z, from Exhibit 5.8, is approximately 4.1% (the average of the two middle values: 4.4 and 3.8). Thus, half the time the forecast errors exceeded 4.1% and half the time they were smaller than 4.1%. When there is a serious outlier among the forecast errors, as with technique Z, it is useful to know the MdAPE in addition to the MAPE because medians are less sensitive than mean values to distortion from outliers. This is why the MdAPE is a full percentage point below the MAPE for technique Z. Sometimes, as with technique Y, the MdAPE and the MAPE are virtually identical. In this case, we can report the MAPE because it is the far more common measure.

In addition to the indicators of precision calculated from the absolute errors, certain measures are commonly used that are based on the squared values of the forecast errors (Exhibit 5.9), such as the MSE and RMSE. (RMSE is the square root of MSE.) A notable variant on RMSE is the root mean squared percentage error (RMSPE) which is based on the squared percentage errors (Exhibit 5.10).

EXHIBIT 5.9
(a) Bar Chart and
(b) Table Showing
Forecast Error as
Squared Difference
(E^2) between Actuals
and Forecasts for
Three Techniques

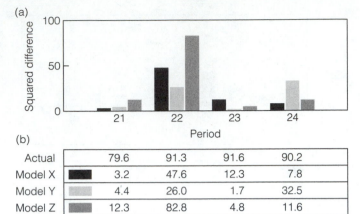

Actual		79.6	91.3	91.6	90.2
Model X	■	3.2	47.6	12.3	7.8
Model Y	▨	4.4	26.0	1.7	32.5
Model Z	▨	12.3	82.8	4.8	11.6

EXHIBIT 5.10
(a) Bar Chart and
(b) Table Showing
Forecast Error as
Squared Percentage
Difference $(PE (\%))^2$
between Actuals and
Forecasts for Three
Techniques

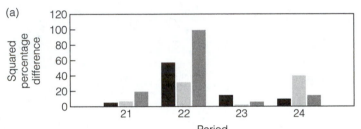

Actual		79.6	91.3	91.6	90.2
Model X	■	5.1	57.1	14.6	9.6
Model Y	▨	6.7	31.2	2.0	39.9
Model Z	▨	19.3	99.3	5.8	14.2

$$MSE = \left[\sum_{t=T+1}^{T+h} \left(A_t - F_t \right)^2 \right] / h$$

$$= \left[\sum_{i=1}^{h} \left(A_i - F_i \right)^2 \right] / h$$

$$RMSPE = \sqrt{ 100 \cdot \left[\sum_{t=T+1}^{T+h} \left\{ \left(A_t - F_t \right) / A_t \right\}^2 \right] / h }$$

$$= \sqrt{ 100 \cdot \left[\sum_{i=1}^{h} \left\{ \left(A_i - F_i \right) / A_i \right\}^2 \right] / h }$$

To calculate RMSPE, we square each percentage error in Exhibit 5.10. The squares are then averaged, and the square root is taken of the average.

The MSE is expressed in the square of the units of the data. This may make it difficult to interpret when referring to dollar volumes, for example. By taking the square root of the MSE, we return to the units of measurement of the original data.

We can simplify the interpretation of RMSE and present it as if it were the mean absolute error (MAE); the forecasts of technique Y are in error by an average of 4. There is little real harm in doing this; just keep in mind that the RMSE is generally somewhat larger than the MAE (compare Exhibits 5.8 and 5.10).

Some forecasters present the RMSE within the context of a prediction interval. Here is how it might sound: "Assuming the forecast errors are normally distributed, we can be approximately 68% confident that technique Y's forecasts will be accurate to within 4.0 on the average." However, it is best to not make these statements too precise because it requires us to assume a normal distribution of errors.

 The RMSE can be interpreted as a standard error of the forecasts.

The RMSPE is the percentage version of the RMSE (just as the MAPE is the percentage version of the MAE). Because the RMPSE for technique Y equals 4.5%, we may suggest that "the forecasts of technique Y have a standard or average error of approximately 4.5%."

If we desire a squared error measure in percentage terms, the following shortcut is sometimes taken. Divide the RMSE by the mean value of the data over the forecast horizon. The result can be interpreted as the standard error as a percentage of the mean of the data and is called the *coefficient of variation* (because it is a ratio of a standard deviation to a mean). Based on technique Y, average actual values over periods 21–24 were 88.2 (Exhibit 5.1), and the RMSE for technique Y was found to be 4.0 (Exhibit 5.10). Hence, the coefficient of variation is 4/88.2 = 4.5%, a result that is virtually identical to the RMSPE.

5.5 COMPARING WITH NAÏVE TECHNIQUES

After reviewing the evidence of bias and precision, we may want to select technique Y as the best of the three candidates. Based on the evaluation, technique Y's forecast for periods 21–24 shows no indication of bias—two underforecasts and two overforecasts—and proves slightly more precise than the other two techniques, no matter which measure of precision is used. The MAE for technique Y reveals that the forecast errors over periods 21–24 average approximately 3.5 per period, and the MAPE indicates that these errors come to just under 4% on average.

To provide additional perspective on technique Y, we might contrast technique Y's forecasting record with that of a very simplistic procedure, one requiring little or no thought or effort. Such procedures are often called naïve techniques. The continued time and effort investment in technique Y might not be worth the cost if it can be shown to perform no better than a naïve technique. Sometimes the contrast with a naïve technique is sobering. The forecaster thinks an excellent model has been developed and finds that it barely outperforms a naïve technique.

If the forecasting performance of technique Y is no better or is worse than that of a naïve technique, it suggests that in developing technique Y we have not accomplished very much.

NAÏVE_1 Technique

A standard naïve forecasting technique has emerged for data that are yearly or otherwise lack seasonality; it is called a NAÏVE_1 or Naïve Forecast1 (NF1). We see later that there are also naïve forecasting techniques for seasonal data. A NAÏVE_1 forecast of the following time period is simply the value in effect during this time period. Alternatively stated, the forecast for any one time period is the value observed during the previous time period.

A NAÏVE_1 is called a no change forecasting technique, because its forecasted value is unchanged from the previously observed value.

Exhibit 5.11 shows the forecast for our example (Exhibit 5.3). Note that one-period-ahead forecasts for periods 22–24 are the respective actuals for periods 21–23. The forecast for period 21 (=73.5) is the actual value for period 20. When we appraise NAÏVE_1's performance, we note it is biased. It underestimated periods 21–23 and overestimated period 24. More generally, by always projecting no change, a NAÏVE_1 underforecasts whenever the data are increasing and overforecasts whenever the data are declining. We do not expect a NAÏVE_1 to work very well when the data are steadily trending in one direction.

EXHIBIT 5.11
Actuals and NAÏVE_1
(one-period-ahead)
Forecasts for
Technique Y

Period	Actual	NAÏVE_1
21	79.6	73.5
22	91.3	79.6
23	91.6	91.3
24	90.2	91.6
MAE	3.6	4.9
MAPE	4	5.60%
RMSE	4	6.6
RMSPE	4.5	7.50%

(*Source*: Exhibit 5.3)

The error measures for NAÏVE_1 in Exhibit 5.11 are all higher than the corresponding error measures for technique Y, indicating that technique Y was more precise than NAÏVE_1. By looking at the relative error measures, we can be more specific about the advantage of technique Y over NAÏVE_1.

Because a naïve technique has no prescription for a random error term, the NAÏVE_1 technique can be viewed as a no-change, no-chance model (in the sense of our forecasting objective in Chapter 1) to predict change in the presence of uncertainty (chance). Hence, its value is as a benchmark forecasting model to beat.

Relative Error Measures

A relative error measure is a ratio; the numerator is a measure of the precision of a forecasting method; the denominator is the analogous measure for a naïve technique.

Exhibit 5.12 provides a compendium of relative error measures that compare technique Y with the NAÏVE_1. The first entry ($=0.73$) is the ratio of the MAE for technique Y to the MAE for the NAÏVE_1 ($3.55/4.87 = 0.73$). It is called the relative absolute error (RAE). The RAE shows that the average forecast error using technique Y is 73% of the average error using NAÏVE_1. This implies a 27% improvement of technique Y over NAÏVE_1.

Expressed as an improvement score, the relative error measure is sometimes called a forecast coefficient (FC), as listed in the last column of Exhibit 5.12. When $FC = 0$, this indicates that the technique in question was no better than NAÏVE_1 technique. When $FC < 0$, the technique was in fact less precise than the NAÏVE_1, whereas when $FC > 0$ the forecasting technique was more precise than NAÏVE_1. When $FC = 1$, the forecasting technique was perfect.

The second row of Exhibit 5.12 shows the relative error measures based on the MAPE. The results almost exactly mirror those based on the MAE, and the interpretations are identical. The relative error measure based on the RMSE (which is the RMSE of technique Y divided by RMSE of the NAÏVE_1) was originally proposed by Theil (1966) and it is known as Theil's U or Theil's U2 statistic. As with the RAE, a score less than 1 suggests an improvement over the naïve technique.

The relative errors based on the squared-error measures, RMSE and RMSPE, suggest that technique Y in Exhibit 5.12 was about 40% more precise than the NAÏVE_1. The relative errors based on MAE and MAPE put the improvement at approximately 25–30%. Why the difference? Error measures based on absolute values, such as the MAE and MAPE, give a weight (emphasis) to each time period in proportion to the size of the forecast error. The RMSE and RMSPE, in contrast, are measures based on squared errors and squaring gives more weight to large errors. A closer look at Exhibit 5.11 shows that the NAÏVE_1 commits a very large error in forecasting period 22.

EXHIBIT 5.12		RAE	FC
Relative Error	RAE	0.73	0.27
Measures for	MAPE	0.71	0.29
Technique Y Relative	RMSE (Theil's U2)	0.61	0.39
to NAÏVE_1	RMSPE	0.60	0.40

(*Source*: Exhibit 5.3)

5.6 TRACKING TOOLS

Tracking tools provide a perspective on how serious a forecasting bias may be. If the bias exceeds a designated threshold, a "red flag" is raised concerning the forecasting approach. The forecaster should then reexamine the data to identify changes in the trend, seasonal, or other business patterns, which in turn suggest some adjustment to the forecasting approach.

Ladder Charts

By plotting the current year's monthly forecasts on a ladder chart, a forecaster can determine whether the seasonal pattern in the forecast looks reasonable. The ladder chart in Exhibit 5.13 consists of six items of information for each month of the year: average over the past 5 years, the past year's performance, the 5-year low, the 5-year high, the current year-to-date, and the monthly forecasts for the remainder of the year. The 5-year average line usually provides the best indication of the seasonal pattern, assuming this pattern is not changing significantly over time. In fact, this is a good check for reasonableness that can be done before submitting the forecast for approval.

A ladder chart is a simple yet powerful tool for monitoring forecast results.

The level of the forecast can be checked for reasonableness relative to the prior year (dashed line in Exhibit 5.13). The forecaster can determine whether or not the actuals are consistently overrunning or underrunning the forecast. In this example, the residuals are positive for 3 months and negative for 3 months. The greatest differences between the actual and forecast values occur in March and April, but here the deviations are of opposite signs; additional research should be done to uncover

EXHIBIT 5.13 Ladder Chart

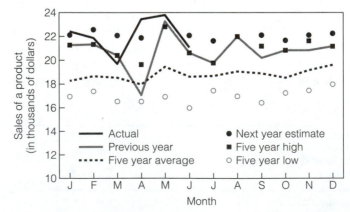

the cause of the unusual March-April pattern. The forecasts for the remainder of the current year look reasonable, although some minor adjustments might be made. The ladder chart is one of the best visual tools for quickly identifying the need for major changes in forecasts.

Prediction-Realization Diagram

Another useful visual approach to monitoring forecast accuracy is the prediction-realization diagram (Theil, 1958). If the predicted values are indicated on the vertical axis and the actual values on the horizontal axis, a straight line with a $45°$ slope will represent perfect forecasts. This is called the line of perfect forecasts (Exhibit 5.14). In practice, the prediction-realization diagram is sometimes rotated so that the line of perfect forecasts is horizontal.

The prediction-realization diagram indicates how well a model or forecaster has predicted turning points and also how well the magnitude of change has been predicted given that the proper direction of change has been forecast.

The diagram has six sections. Points falling in sections II and V are a result of turning-point errors. In Section V, a positive change was predicted, but the actual change was negative. In Section II, a negative change was predicted, but positive change occurred. The remaining sections involve predictions that were correct in

EXHIBIT 5.14
Prediction-
Realization Diagram

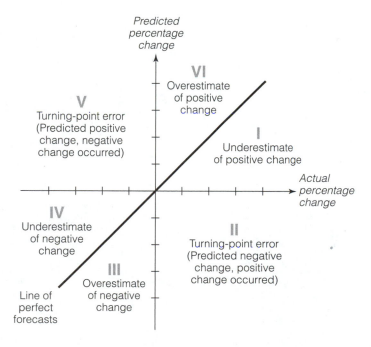

sign but wrong in magnitude. Points above the line of perfect forecasts reflect actual changes that were less than predicted. Points below the line of perfect forecasts represent actual changes that were greater than predicted.

The prediction-realization diagram can be used to record forecast results on an ongoing basis. Persistent overruns or underruns indicate the need to adjust the forecasts or to reestimate the model. In this case, a simple error pattern is evident and we can raise or lower the forecast based on the pattern and magnitude of the errors.

More important, the diagram indicates turning-point errors that may be due to misspecification or missing variables in the model. The forecaster may well be at a loss to decide how to modify the model forecasts. An analysis of other factors that occurred when the turning-point error was realized may result in inclusion of a variable in the model that was missing from the initial specification.

Exhibit 5.15 illustrates a prediction-realization diagram for a model of quarterly telephone access-line gain as a function of housing starts. The model is a simple linear regression of the year-over-year changes (from the same quarter the prior year) in quarterly access-line gain described in terms of annual changes in a quarterly housing starts series.

Prediction Intervals for Time Series Models

When no error distribution is assumed explicitly in a model, as is generally the case with moving averages and exponential smoothing, we can construct empirical prediction intervals (Goodman and Williams, 1971). To construct empirical prediction intervals, we go through the ordered data, making forecasts at each

EXHIBIT 5.15
Prediction-
Realization Diagram
for the Quarterly
Access-Line Gain
versus Housing
Starts Model

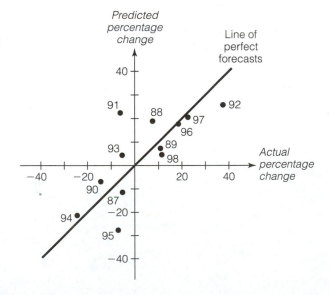

time. Then, the comparison of these forecasts with the actuals (that are known) yield an empirical distribution of forecast errors. If the future errors are assumed to be distributed like the empirical errors, then the empirical distribution of these observed errors can be used to set prediction intervals for subsequent forecasts. In practice, the theoretical size and the empirical size of the intervals have been found to agree closely.

Prediction intervals are used as a way of expressing the uncertainty in the future values that are derived from models, and they play a key role in tracking forecasts from these models.

In regression models, it is generally assumed that random errors are additive to the model and have a normal distribution with zero mean and constant variance. The variance of the errors can be estimated from the time series data. The estimated standard deviation (square root of the variance) is calculated for the forecast period and is used to develop the desired prediction interval. Although 95% prediction intervals are frequently used, the range of forecast values for volatile series may be so great that it might also be useful to show the limits for a lower level of probability, say 75%. This would narrow the interval. It is common to express prediction intervals about the forecast; but, in the tracking process, it is more useful to deal with prediction intervals for the forecast errors (Error = Actual − Forecast) on a period-by-period or cumulative basis.

A forecaster often deals with aggregated data and would like to know the prediction intervals about an annual forecast based on a model for monthly or quarterly data. The annual forecast is created by summing twelve monthly (or four quarterly) predictions. Developing the appropriate probability limits requires that the variance for the sum of the prediction errors be calculated. This can be determined from the variance formula

$$\text{Var}\left(\sum e_i\right) = \sum \text{Var}(e_i) + 2\sum \text{Cov}(e_i, e_j)$$

If the forecast errors have zero covariance (Cov)—in particular, if they are independent of one another—the variance of the sum equals the sum of the prediction variances.

The most common form of correlation in the forecast errors for time series models is positive autocorrelation. In this case, the covariance term is positive and the prediction intervals derived would be too small. In the unusual case of negative covariance, the prediction intervals would be too wide. Rather than deal with this complexity, most software programs assume the covariance is small and inconsequential.

In typical forecasting situations, prediction intervals are probably too conservative or narrow.

Prediction Interval as a Percent Miss

Some users of forecasts find the expression of prediction intervals obtuse. In such cases, the forecaster may find more acceptance of this technique if the prediction intervals for forecasts are expressed in terms of percentages. For example, 95% probability limits for a forecast might be interpreted that we are 95% sure that the forecast will be within ± 15% (say) of the actual numerical value. The percent miss associated with any prediction interval can be calculated from the formula:

$$\text{Percent miss} = [(\text{Estimated standard error of the forecast})\ (t) \times 100\%] / (\text{Predicted value})$$

where t is the tabulated value of the Student t distribution for the appropriate **degrees of freedom** and confidence level. This calculation provides the percent miss for any particular period. Values can be calculated for all periods (e.g., for each month of the year), resulting in a band of prediction intervals spanning the year.

It could also be useful to determine a prediction interval for a cumulative miss. Under the assumption that forecast errors are independent and random, the cumulative percent miss will be smaller than the percent miss for an individual period because the positive and negative errors will cancel to some extent. As an approximate rule, the average period (say monthly or quarterly) percent miss is divided by the square root of the number of periods of the forecast to determine a cumulative percent miss. For example, the annual percent miss is calculated by dividing the average monthly percent miss by $\sqrt{12}$ or the average quarterly percent miss by $\sqrt{4}(=2)$.

We treat confidence intervals for regression parameters in Chapters 11 and 12. Prediction intervals for ARIMA models are discussed in Chapter 14.

Prediction Intervals as Early Warning Signals

One of the simplest tracking signals is based on the ratio of the sum of the forecast errors to the mean absolute error. It is called the cumulative sum tracking signal (CUSUM). For certain CUSUM measures, a threshold or upper limit of $|0.5|$ suggests that the forecast errors are no longer randomly distributed about 0 but rather are congregating too much on one side (i.e., forming a biased pattern).

More sophisticated tracking signals involve the taking of weighted or smoothed averages of forecast errors. The best-known tracking signal, due to Trigg and Leach (Trigg, 1964), is described later in this chapter. Tracking signals are especially useful when forecasting a large number of products at a time, as is the case in inventory management systems.

A tracking signal is a ratio of a measure of bias to a companion measure of precision.

Warning signals can be visualized by plotting the forecast errors over time together with the associated prediction intervals and seeing whether the forecast errors continually lie above or below the zero line. Even though the individual forecast errors may well lie within the appropriate prediction interval for the period (say monthly), a plot of the cumulative sum of the errors may indicate that their sum lies outside its prediction interval.

 An early warning signal is a succession of overruns and underruns.

A type of warning signal is evident in Exhibits 5.16 and 5.17. It can be seen that the monthly errors lie well within the 95% prediction interval for 9 of the 12 months, with two of the three exceptions occurring in November and December. Exhibit 5.16 suggests that the individual forecast errors lie within their respective prediction intervals. However, it is apparent that none of the misses are negative—certainly, they do not form a random pattern about zero. Hence, there appears to be a bias. To determine whether the bias in the forecast is significant, we review the prediction intervals for cumulative sums of forecast errors.

The cumulative prediction interval (Exhibit 5.17) confirms the problem with the forecast bias. The cumulative forecast errors fall on the outside of the prediction interval for all twelve periods. This model is clearly underforecasting. Either the model has failed to capture a strong cyclical pattern during the year or the data are growing rapidly and the forecaster has failed to make a proper specification in the model.

Using these two plots, the forecaster would probably be inclined to make upward revisions in the forecast after several months. It certainly would not be necessary to wait until November to recognize the problem.

Another kind of warning signal occurs when too many forecast errors fall outside the prediction intervals. For example, with a 90% prediction interval, we expect only 10% (approximately 1 month per year) of the forecast errors to fall outside the prediction interval. Exhibit 5.18 shows a plot of the monthly forecast errors for a time series model. In this case, five of the twelve errors lie outside the 95% interval. Clearly, this model is unacceptable as a predictor of monthly values. However, the

EXHIBIT 5.16
Monthly Forecast
Errors and Associated
Prediction Intervals
for a Time Series
Model

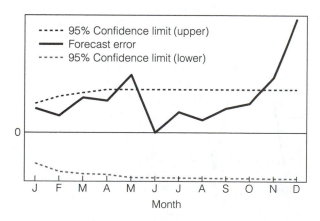

EXHIBIT 5.17
Cumulative Forecast
Errors and Associated
Cumulative Prediction
Intervals for a Time
Series Model

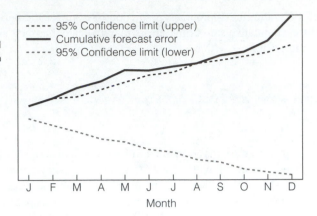

monthly error pattern may appear random and the annual forecast (cumulative sum of 12 months) might be acceptable. Exhibit 5.19 shows the cumulative forecast errors and the cumulative prediction intervals. It appears that the annual forecast lies within the 95% prediction interval and is acceptable.

The conclusion that may be reached from monitoring the two sets of forecast errors is that neither model is wholly acceptable, and that neither can be rejected. In one case, the monthly forecasts were good but the annual forecast was not. In the other case, the monthly forecasts were not good, but the annual forecast turned out to be acceptable. Whether the forecaster retains either model depends on the purpose for which the model was constructed. It is important to note that, by monitoring cumulative forecast errors, the forecaster was able to determine the need for a forecast revision more quickly than if the forecaster were only monitoring monthly forecasts. This is the kind of early warning that management should expect from the forecaster.

EXHIBIT 5.18
Monthly Forecast
Errors and Prediction
Intervals for a Time
Series Model

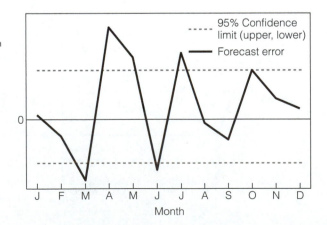

EXHIBIT 5.19
Cumulative Forecast
Errors and Cumula-
tive Prediction Inter-
vals for a Time Series
Model

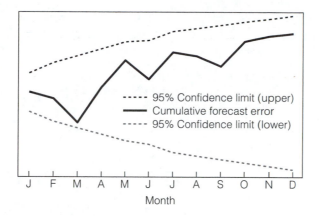

Trigg Tracking Signal

The tracking signal proposed by Trigg (1964) indicates the presence of nonrandom errors; it is a ratio of two smoothed errors E_t and M_t. The numerator E_t is a simple exponential smooth of the errors e_t, and the denominator M_t is a simple exponential smooth of the absolute values of the errors. Thus,

$$T_T = E_T / M_T$$

$$E_T = \alpha e_T + (1-\alpha)E_{T-1}$$

$$M_T = \alpha|e_T| + (1-\alpha)M_{T-1}$$

where $e_T = Y_t - F_t$, the difference between the observed value Y_t and the forecast F_t.

Trigg shows that when T_t exceeds 0.51 for $\alpha = 0.1$ or 0.74 for $\alpha = 0.2$, the errors are nonrandom at the 95% probability level. Exhibit 5.20 shows a sample calculation of the Trigg tracking signal for an adaptive smoothing model of seasonally adjusted airline data. The tracking signal correctly provides a warning at period 15 after five consecutive periods in which the actual exceeded the forecast. Period 11 has the largest error, but no warning is provided because the sign of the error became reversed. It is apparent that the model errors can increase substantially above prior experience without a warning being signaled as long as the errors change sign. Once a pattern of over- or underforecasting is evident, a warning is issued.

5.7 COMPUTER STUDY: HOW TO MONITOR FORECASTS

In this book we discuss a variety of forecasting methods, from simple exponential smoothing to multiple linear regression. Once we implement one of these methods, we cannot expect it to supply good forecasts indefinitely. The pattern of our data may change, and we will have to adjust our forecasting method or perhaps select a new model better suited to the data. For example, suppose that we are using simple exponential smoothing, a model that assumes that the data fluctuate about a constant

		Smoothed	Smoothed	Tracking
Time	Error	Error	Absolute Error	Signal
1	−1.58			
2	2.54	−1.17	1.68	−0.70*
3	5.24	−0.53	2.04	−0.26
4	−0.51	−0.53	1.89	−0.28
5	0.59	−0.42	1.76	−0.24
6	2.26	−0.15	1.81	−0.08
7	1.49	0.01	1.78	0.01
8	1.31	−0.14	1.73	0.08
9	0.43	0.17	1.6	0.11
10	−7.73	−0.62	2.21	−0.28
11	11.57	0.6	3.15	0.19
12	8.98	1.44	3.73	0.39
13	3.82	1.68	3.74	0.45
14	4.17	1.93	3.78	0.51
15	1.06	1.84	3.51	0.53[†]

EXHIBIT 5.20
Trigg's Tracking Signal ($\alpha = 0.1$) for an Adaptive Smoothing Model of Seasonally Adjusted Airline Data

* Starting value—ignore.
† Exceeds 0.51—warning!

level. Now assume that a trend develops. We have a problem because our forecasts will not keep up with the trend. How do we detect this problem?

One answer is to periodically test the data for trends and other patterns. Another is to examine a graph of the forecasts each time that we add new data to our model. Both methods quickly become cumbersome, especially when we forecast a large number of product SKUs. Visually checking all of these data every forecast period is out of the question!

Quick and Dirty Control

The tracking signal, a simple quality control model, is just the tool for the job. When a forecast for a given time period is too large, the forecast error has a negative sign; when a forecast is too small, the forecast error is positive. Ideally, the forecasts should vary around the actual data and the sum of the forecast errors should be near zero. But if the sum of the forecast errors departs from zero in either direction, the forecasting model may be out of control.

Just how large should the sum of the forecast errors grow before we act? To answer this, we need a basis for measuring the relative dispersion, or scatter, of the forecast errors. If forecast errors are widely scattered, a relatively large sum of errors is not unusual. A standard measure of variability such as the standard deviation could be used as a measure of dispersion, but more commonly the mean absolute error is used to define the tracking signal:

$$\text{Tracking signal} = (\text{Sum of errors}) / (\text{MAE})$$

Typically the control limit on the tracking signal is set at ± 4.0. If the signal goes outside this range we should investigate. Otherwise, we can let our forecasting model run unattended. There is a very small probability, less than 1%, that the signal will go outside the range ±4.0 due to chance.

The Tracking Signal in Action

Exhibit 5.21 is a graph of actual sales and the forecasts produced by simple exponential smoothing. For the first 6 months, the forecast errors are relatively small. Then, a strong trend begins in month 7. Fortunately, as shown in Exhibit 5.22, the tracking signal sounds an alarm immediately.

Column F in Exhibit 5.23 contains the mean absolute error. This is updated each month to keep pace with changes in the errors. Column G contains the running sum of the errors, and column H holds the value of the tracking signal ratio. A formula in column I displays the label ALARM if the signal is outside the range ±4.0. We can now take a step-by-step approach to building the spreadsheet. For those unfamiliar with simple exponential smoothing (see Chapter 10), the forecasting formulas are explained as they are entered.

EXHIBIT 5.21
Monthly Forecasts
and Sales for a Time
Series Model

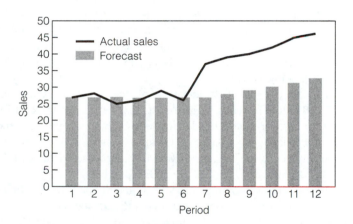

EXHIBIT 5.22
Monthly Forecast
Errors and Tracking
Signal

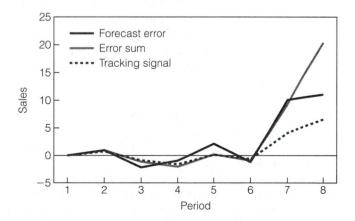

EXHIBIT 5.23

Actuals, Forecasts, and Forecast Tracking Signal

	A	B	C	D	E	F	G	H	I	J	K	L
1	Exponential Smoothing with Tracking Signal											
2	Weight =	0.1		Control Limit =		4.0						
3												
4		Actual	Fcst.	Actual	Abs.		Error	Tracking				
5	Month	Sales	Sales	Error	Error	MAE	Sum	Signal				
6						1.4	0.0	0.0				
7	1	27	27.0	0.0	0.0	1.3	0.0	0.0				
8	2	28	27.0	1.0	1.0	1.3	1.0	0.8				
9	3	25	27.1	-2.1	2.1	1.4	-1.1	-0.8				
10	4	26	26.9	0.9	0.9	1.3	-2.0	-1.5				
11	5	29	26.8	2.2	2.2	1.4	0.2	0.1				
12	6	26	27.0	-1.0	1.0	1.4	-0.8	-0.6				
13	7	37	26.9	10.1	10.1	2.2	9.3	4.2	ALARM	ALARM IN MONTH 7 -->		
14	8	39	27.9	11.1	11.1	3.1	20.3	6.5	ALARM			
15	9	40	29.0	11.0	11.0	3.9	31.3	8.0	ALARM			
16	10	42	30.1	11.9	11.9	4.7	43.2	9.2	ALARM			
17	11	45	31.3	13.7	13.7	5.6	56.9	10.2	ALARM			FORECAST
18	12	46	32.7	13.3	13.3	6.4	70.2	11.0	ALARM	ACTUAL SALES		
19	13		34.0									
20												

Step 1: Set Up the Spreadsheet

Enter the labels in rows 1–5. In cell B2, enter a smoothing weight of 0.1. In cell F2, enter a control limit of 4.0. Number the months in column A and enter the actual sales in column B.

Step 2: Compute the Monthly Forecasts and Errors

To get the smoothing process started, we must specify the first forecast in cell C7. A good rule of thumb is to set the first forecast equal to the first actual data value. Therefore, enter +B7 in cell C7. In cell D7, enter the formula +B7-C7, which calculates the forecast error.

In cell C8, the first updating of the forecast occurs. Enter the formula = C7*B2*D7 in cell C7. This formula states that the forecast for month 2 is equal to the forecast for month 1 added to the smoothing weight times the error in month 1. Copy the updating formula in cell C8 to the range C9 to C19. Copy the error-calculating formula in cell D7 to range D8 to D18. Because the actual sales figure is not yet available for month 13, cells B19 and D19 are blank.

Exponential smoothing works by adjusting the forecasts in a direction opposite to that of the error, working much like a thermostat, an automatic pilot, or a cruise control on an automobile. Examine the way the forecasts change with the errors. The month 1 error is zero, so the month 2 forecast is unchanged. The error is positive in month 2, so the month 3 forecast goes up. In month 3 the error is negative, so the month 4 forecast goes down.

The amount of adjustment each month is controlled by the weight in cell B2. A small weight avoids the mistake of overreacting to purely random changes in the data. But a small weight also means that the forecasts are slow to react to true changes. A large weight has just the opposite effect—the forecasts react quickly to both randomness and true changes. Some experimentation is always necessary to find the best weight for the data.

Step 3: Compute the Mean Absolute Error

Column E converts the actual errors to absolute values. In cell E7, enter the absolute value of D7, and then copy this cell to range E8 to E18. The mean absolute error in column F is updated via exponential smoothing, much like the forecasts. In cell F6, a starting value for the MAE must be specified. This must be representative of the errors or the tracking signal can give spurious results. A good rule of thumb is to use the average of all available errors to start up the MAE. But here we will do an evaluation of how the tracking signal performs without advance knowledge of the surge in sales in month 7. We use the average of the absolute errors for months 2–6 only as the initial MAE (month 1 should be excluded because the error is zero by design). In cell F6, enter the formula for the average of E8 to E12.

To update the MAE, in cell F7 enter B2*E2+(1-B2)*F6. Copy this formula to the range F8 to F12. This formula updates the MAE using the same weight as the forecast. The result is really a weighted average of the MAE, which adapts to changes in the scatter of the errors. The new MAE each month is computed as 0.1 times the current absolute error plus 0.9 times the last MAE.

Step 4: Sum the Errors

This simple operation is considered a separate step because a mistake would be disastrous here. We must sum the actual errors, *not* the absolute errors. The only logical starting value for the sum is zero, which should be entered in cell G6. In cell G7, enter +D7+G6 and copy this to range G8 to G18.

Step 5: Compute the Tracking Signal

The tracking signal value in column H is the ratio of the error sum to the MAE. In cell H6, enter +G6/F6 and copy this to range H7 to H18.

Step 6: Enter the ALARM Formulas

In cell I6, enter an IF statement for ABS (H6)>F3,"ALARM","") and copy it to range I7 to I19. This formula takes the absolute value of the tracking signal ratio. If that value is greater than the control limit, the label ALARM is displayed; otherwise, the cell is blank.

Adapting the Tracking Signal to Other Spreadsheets

The formulas in columns E through I in Exhibit 5.23 are not specific to simple exponential smoothing. They can be used in any other forecasting spreadsheet as long as the actual period-by-period forecast errors are listed in a column. If we use a forecasting method that does not include weights, such as linear regression, we will also have to assign a weight for updating the MAE—we simply use a weight of 0.1 in such applications. Concerning the weights used to update forecasts, experience suggests that there is not much to be gained by experimenting with the weights in the MAE.

In large forecasting spreadsheets, macros can be used to operate the quality control system. For example, we could write a macro that checks the last cell in the tracking signal column of each data set. The macro could sound a beep and pause if a signal is out of range. Or we could write a macro that print out a report showing the location of each offending signal for later investigation.

In Exhibit 5.23, the tracking signal gives an immediate warning when our forecasts are out of control. Do not expect this to happen every time. The tracking signal is by no means a perfect quality control device, and only drastic changes in the data will cause immediate warnings. More subtle changes may take some time to detect.

Statistical theory tells us that the exact probability that the tracking signal will go outside the range ±4.0 depends on many factors, such as the type of data, the forecasting model in use, and the parameters of the forecasting model such as the smoothing weights. Control limits of ±4.0 are a good starting point for tracking most business data. If experiments show that these limits miss important changes in the data, reduce them to 3.5 or 3.0. If the signal gives false alarms, cases in which the signal is outside the limits but there is no apparent problem, increase the limits to 4.5 or 5.0. We recommend adding a tracking signal to every forecasting model. They save a great deal of work by letting us manage by exception.

SUMMARY

To test a forecasting method with a holdout sample, we split the historical data into two parts using the first segment as the fit period. Then, the forecasting method is used to make forecasts for a number of additional periods (the forecast horizon). Because there are actual values withheld in the holdout sample, we can assess forecast accuracy by comparing the forecasts against the known figures. Not only do we see how well the forecasting method has fit the more distant past (fit period) but also how well it would have forecast the test period.

Tracking signals are useful when large numbers of items must be monitored for accuracy. This is often the case in inventory systems. When the tracking signal for an item exceeds the threshold level, the forecaster's attention should be promptly drawn to the problem. The tracking of results is essential to ensure the continuing relevance of the forecast. By properly tracking forecasts and assumptions, the forecaster can inform management when a forecast revision is required. It should not be

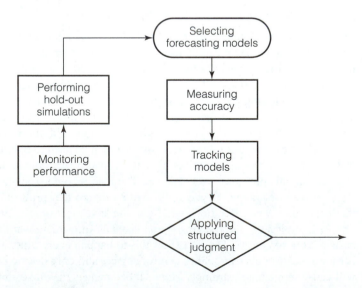

necessary for management to inform the forecaster that something is wrong with the forecast. The techniques for tracking include ladder charts, prediction-realization diagrams, prediction intervals, and tracking signals that identify nonrandom error patterns. Through tracking, we can better understand the models, their capabilities, and the uncertainty associated with the forecasts derived from them.

REFERENCES

Armstrong, J. S., and F. Collopy (1992). Error measures for generalizing about forecasting methods: Empirical comparisons (with discussion). *Int. J. Forecasting* 8, 69–80.

Box, G. E. P., G. M. Jenkins, and G. Reinsel (1994). *Time Series Analysis, Forecasting and Control*. 3rd ed. Englewood Cliffs, NJ: Prentice Hall.

Cryer, J. D., and R. B. Miller (1994). *Statistics for Business*. Belmont, CA: Wadsworth.

DeLurgio, S. (1998). *Forecasting Principles and Applications*. New York: Irwin/McGraw-Hill.

Dielman, T. E. (1996). *Applied Regression Analysis for Business and Economics*. 2nd ed. Belmont, CA: Wadsworth.

Goodman, M. L., and W. H. Williams (1971). A simple method for the construction of empirical confidence limits for economic forecasts. J. Am. Statist. Assoc. 66, 752–54.

Hanke, J. E., and A. G. Reitsch (1998). *Business Forecasting*. 6th ed. Englewood Cliffs, NJ: Prentice Hall.

Makridakis, S., A. Anderson, R. Carbone, R. Fildes, M. Hibon, R. Lewandowski, J. Newton, E. Parzen, and R. Winkler (1982). The accuracy of extrapolation (time series) methods: Results of a forecasting competition. *J. Forecasting* 1, 111–53.

Makridakis, S., S. C. Wheelwright, and R. J. Hyndman (1998). *Forecasting Methods and Applications*. 3rd ed. New York: John Wiley & Sons.

Makridakis, S., and R. Winkler (1983). Averages of forecasts: Some empirical results. *Manage. Sci.* 29.9, 987–96.

Ord, K., M. Hibon, and S. Makridakis (2000). The M3 competition. *Int. J. Forecasting* 16, 433–36.

Tryfos, P. (1998). *Methods for Business Analysis and Forecasting: Text and Cases*. New York: John Wiley & Sons.

Wilson, J. H., and B. Keating (1998). *Business Forecasting*. New York: Irwin McGraw-Hill.

Theil, H. (1958). *Economic Forecasts and Policy*. Amsterdam: North-Holland.

Trigg, D. W. (1964). Monitoring a forecasting system. *Operational Res. Q.* 15, 272–74.

PROBLEMS

5.1 For the annual housing starts in Exhibit 2.4 create a four-period-ahead forecast, using a six-period moving average 6-MAVG$_t$ from $T = 25$ (i.e., determine $F_{25}(1)$, $F_{25}(2)$, $F_{25}(3)$, and $F_{25}(4)$).

a. Measure and interpret the bias by calculating the ME and MPE.

b. Measure and interpret the precision by calculating the MAE, MAPE, and MdAPE.

c. Contrast your findings in (b) with those determined by calculating the RMSE and RMSPE.

5.2 For the mortgage rates data in Exhibit 2.6, create a five-period-ahead forecast, using a five-period moving average 5-MAVG$_t$ from $T = 24$ (i.e., determine $F_{24}(1)$, $F_{24}(2)$, . . . , $F_{24}(5)$).

a. Measure and interpret the bias by calculating the ME and MPE.

b. Measure and interpret the precision by calculating the MAE, MAPE, and MdAPE.

c. Contrast your findings in (b) with those determined by calculating the RMSE and RMSPE.

5.3 For the monthly formal-wear sales data in Exhibit 4.19:

a. Create a 1-year (12 months) holdout sample from the end of the data.

b. Build a 12-period moving average forecasting model 12-MAVG$_t$ and project 12 months.

c. Evaluate the forecast performance over the holdout sample with several measures developed in this chapter.

5.4 Repeat Problem 5.3 for one or more of the following monthly time series on the CD:

a. Unemployment rate data (UNEMP, Dielman, 1996, Table 4.7)

b. Silver prices data (SILVER, Dielman, 1996, Table 4.8)

c. Prime rate data (PRIME4, Dielman, 1996, Table 4.14)

d. Retail sales for U.S. retail sales stores (RETAIL, Hanke and Reitsch, 1998)

e. Wheat exports (SHIPMENT, Dielman, 1996, Table 4.9)

f. Shipments of spirits (DSHIP, Tryfos, 1998, Table 6.23)

g. Retail sales in a region (RSALES, Tryfos, 1998, Table 6.25)

h. Australian beer production (AUSBEER, Makridakis et al., 1998, Table 2.2)

i. International airline passenger miles (AIRLINE, Box et al., 1994, Series G)

j. Australian electricity production (ELECTRIC, Makridakis et al., 1998, Fig. 7-9)

k. French industry sales for paper products (PAPER, Makridakis et al., 1998, Fig. 7-20)

l. Shipments of French pollution equipment (POLLUTE, Makridakis et al., 1998, Table 7-5)

m. Sales of recreational vehicles (WINNEBAG, Cryer and Miller, 1994, p. 742)

n. Sales of lumber (LUMBER, DeLurgio, 1998, p. 34)

o. Sales of consumer electronics (ELECT, DeLurgio, 1998, p. 34)

5.5 For the quarterly automobile sales data in Exhibit 4.21:

a. Create a 1-year (4-quarter) holdout sample from the end of the data.

b. Build a four-period moving average forecasting model 4-MAVG$_t$ and project four quarters.

c. Evaluate forecast performance over the hold-out sample with several measures developed in this chapter.

5.6 Repeat Problem 5.5 for one or more of the following quarterly time series on the CD:

a. Sales for Outboard Marine (OMC, Hanke and Reitsch, 1998, Table 4.5)

b. Shoe store sales (SHOES, Tryfos, 1998, Table 6.5)

c. Domestic car sales (DCS, Wilson and Keating, 1998, Table 1-5)

d. Sales of The GAP (GAP, Wilson and Keating, 1998, p. 30)

e. Air passengers on domestic flights (PASSAIR, DeLurgio, 1998, p. 33)

5.7 a. Determine the Trigg signal with $\alpha = 0.1$ for the annual housing starts forecasts in Problem 5.1.

b. Determine the Trigg signal with $\alpha = 0.1$ for the mortgage rates forecasts in Problem 5.2.

c. Compare the performance of the two models with a time plot.

d. Make a graphical assessment about whether you see a forecasting problem.

5.8 a. Using a three-period moving average, determine the following ten one-period-ahead forecasts for the annual housing starts data in Exhibit 2.4: $F_{19}(1)$, $F_{20}(1)$, . . . , $F_{28}(1)$, and find the corresponding forecast errors $e_{19}(1)$, $e_{20}(1)$, . . . , $e_{28}(1)$.

b. Contrast the adequacy of these forecasts with the NAÏVE_1 technique by calculating a bias measure from the ten forecasting errors in (a).

c. Contrast the adequacy of these forecasts with the NAÏVE_1 technique by calculating a precision measure from the ten forecasting errors in (a).

d. Contrast the adequacy of these forecasts with the NAÏVE_1 technique by calculating a tracking signal based on the ten forecasting errors in (a).

5.9 Repeat Problem 5.8 for the annual mortgage rates in Exhibit 2.6.

5.10 Create a prediction-realization diagram with the results from:

a. Problem 5.8.

b. Problem 5.9.

CASE 5A PRODUCTION OF ICE CREAM (CASES 1A, 3A, AND 4A CONT.)

The entrepreneur has shown you some previous forecasts (the particular technique used is not important at this stage) and would like to get an assessment of how you would measure forecast accuracy. Based on the first 4 years (48 observations), a previous consultant created forecasts for Year 05 and came up with the following 12 values: 57033, 63286, 73347, 74238, 78631, 87388, 87398, 81467, 69630, 61640, 53046, and 54283. The actual values for Year 05 are listed in Case 3A.

(1) Calculate the ME, MPE, MAPE, MAE and the RMSE. Contrast these measures of accuracy.

(2) Create a Naïve_12 (12-period-ahead) forecast for Year 05, by starting with January 05 forecast = January 04 actual, February 05 forecast = February 04 actual, and so on. Construct a relative error measure, such as RAE, comparing the forecasts with the Naïve_12 forecasts. Did the consultant do a better job than a Naïve_12 forecast?

(3) On a month-to-month basis, these accuracy measures might not be meaningful to management (too large, perhaps), so you suggest looking at the results on a quarterly basis. Aggregate the original data and forecasts into quarterly periods (or buckets). Repeat (1) and (2) using a Naïve_4 as the forecast to beat.

CASE 5B DEMAND FOR AIR TRAVEL (CASE 4B CONT.)

The VP of Operations of an airline wants to examine the passenger traffic on a transAtlantic route to help understand and improve business opportunities and cost inefficiencies. The VP is aware that the travel on this route is mostly businesspeople, so a number of factors affecting demand are worth considering. Among these are the GDP of the two countries, the exchange rate, and the ticket price. However, the VP has been advised that a preliminary study to gain an insight into the nature of the problem is highly recommended and has requested that you use your newly learned forecasting skills to help. The quarterly data for 4 years is shown in Case 4B. Based on the first 12 historical values, a previously developed forecasting model gave the following results for periods 13–16.

Period	Forecast	Lower	Upper
13	147473	126790	167845
14	102619	87926	117120
15	84488	72178	96778
16	118980	101766	135986

Forecasts and prediction intervals: 95%

(1) Calculate the ME, MPE, MAPE, MAE, and RMSE. Contrast these measures of accuracy.

(2) Create a Naïve_4 (four-period-ahead) forecast for periods 13–16 by starting with Period 13 forecast = Period 9 actual, Period 14 forecast = Period 10 actual, and so on. Construct a relative error measure, such as RAE, comparing the forecasts with the Naïve_4 forecasts. Do the Naïve_4 forecasts fall within the 95% prediction intervals?

Based on all the historical data, another model gave the following forecasts for periods 17–20.

Period	Forecast	Lower	Upper
17	141229	120506	161471
18	109998	93798	125707
19	91310	77538	105129
20	125497	105911	143706

Forecasts and prediction intervals: 95%

(3) Create a Naïve_4 (four-period-ahead) forecast for periods 17–20 by starting with Period 17 forecast = Period 13 actual, Period 18 forecast = Period 14 actual, and so on. Construct a relative error measure, such as RAE, comparing the forecasts with the Naïve_4 forecasts. Do the Naïve_4 forecasts fall within the 95% prediction intervals? On the basis of what you learned in (2), are there any judgmental adjustments that you would make to these forecasts?

CASE 5C DEMAND FOR BICYCLES

Bicycles have been around for a long time. Many industries are strongly dependent on the health of the economy, but it seems that, no matter how bad the times, parents usually find a way to buy bicycles for their children. Nevertheless, the demand for bicycles is probably affected by a number of socioeconomic and demographic factors.

A local bicycle manufacturer is interested in expanding its business by introducing new products and branching out into new geographical areas of the country. The CEO has hired you, and as your first assignment you are to measure the accuracy of some models for bicycle demand that were developed prior to your arrival on the scene. For the most recent 19 years, the manufacturer reported the following annual unit sales figures: 370, 402, 428, 482, 540, 569, 609, 701, 711, 634, 795, 1182, 1414, 1420, 978, 853, 903, 929, 1027. Based on the first 16 historical values, a previously developed forecasting model gave the following results for periods 17–19:

Period	Forecast	Lower	Upper
17	1257	740	1790
18	1312	782	1851
19	1366	796	1949

Forecasts and prediction intervals: 95%

(1) Calculate the ME, MPE, MAPE, MAE, and the RMSE. Contrast these measures of accuracy.

(2) Create a Naïve_1 (one-period-ahead) forecast for periods 17–19 by starting with Period 17 forecast = Period 16 actual + (Period 16 actual – Period 15 actual), Period 18 forecast = Period 16 actual +2 (Period 16 actual – Period 15 actual), and so on. Construct a relative error measure, such as RAE, comparing the forecasts with the Naïve_1 forecasts. Do the Naïve_1 forecasts fall within the 95% prediction intervals?

Based on all the historical data, another model gave the following forecasts for periods 20–22.

Period	Forecast	Lower	Upper
20	1109	502	1724
21	1121	497	1742
22	1132	503	1784

Forecasts and prediction intervals: 95%

(3) Create a Naïve_1 (one-period-ahead) forecast for periods 20–22 as in (2). Construct a relative error measure, such as RAE, comparing the forecasts with the Naïve_1 forecasts. Do the Naïve_1 forecasts fall within the 95% prediction intervals? On the basis what you learned in (2), are there any judgmental adjustments that you would make to these forecasts?

Forecasting the Aggregate

6

Dealing with Seasonal Fluctuations

"Nature has established patterns originating in the return of events, but only for the most part."
VON LEIBNIZ, 1703

MANY BUSINESS TIME SERIES SHOW seasonal fluctuations. Time series data have been adjusted for seasonal fluctuations in business and financial applications for over 80 years. Businesses may need to know when a change in a time series is due to more than the typical seasonal variation. Government agencies adjust statistical indicators for seasonality before publication and distribution to the public. The X-12 program from the U.S. Census Bureau and the widely applied X-11-ARIMA/88 program from Statistics Canada are accepted standards for the large-scale analysis of publicly reported seasonal adjustments of monthly and quarterly data. These data-driven seasonal-adjustment procedures involve smoothing data to eliminate unwanted irregular variation from the patterns that are meaningful to the analyst.

There are generally three distinct uses of seasonal adjustment: the historical adjustment of available past data, the current adjustment of each new observation, and the predicted seasonal factors for future adjustment. This chapter describes how seasonal effects can be removed and adjusted for in historical data. We examine how the centered moving average and related smoothers play a central role in a number of widely used seasonal-adjustment procedures.

6.1 SEASONAL INFLUENCES

One of the first people to study periodicity in economic time series was the British astronomer William Herschel, who tried to find a relationship between sunspots and wheat prices. Another was the banker James W. Gilbart, who discovered that the Bank of England notes were in high demand in January, April, July, and October due to the payment of dividends in these months. He used this information to argue against attempts by smaller country banks to issue their notes during these periods. Today we see many examples of seasonality in economic time series. Anyone who shops at the end of the calendar year realizes that retail, toy, and card stores have a surge in demand at year end. Some businesses have 25% or more of all their yearly sales in December. Even Peruvian anchovy production shows 7-year repeating patterns caused by recurring changes in ocean currents.

In Chapter 4, we define seasonality as periodic fluctuations that recur every year with about the same timing and intensity. For example, farm income from all farms in the United States may rise steadily each year from early spring until fall and then drop sharply. For agricultural economists, it is important to determine whether a recession has reached bottom or whether there is a predictable pattern in the duration, amplitude, or slope of the business cycle expansions and contractions. In this case, the main use of methods for seasonally adjusting farm income data is to remove any seasonal fluctuations in order to expose an underlying trend-cycle pattern.

 Seasonality is described by periodic fluctuations that recur every year with about the same timing and intensity.

Many business time series are recorded over calendar months, which create a seasonal movement because the number of working days varies from month to month. The timing of certain public holidays (e.g., Christmas and Easter), school openings and closings, dividend payments by corporations, and fiscal tax years all contribute seasonal effects because these events tend to occur at similar times each year.

Often companies have to deal with seasonal fluctuations when making decisions about price and inventory policy or the commitment of capital expenditures. In these situations, the business analyst wants to know whether changes in business activity over a given period were larger or smaller than normal.

Forecasters are not in general agreement on how best to deal with seasonality in forecasting. Some advocate using data-driven methodologies for seasonally adjusting data before forecasting them; others advocate using model-driven approaches* for making seasonality explicit in a model for the data. We follow the data-driven approach in this chapter.

There may be times when seasonally adjusted data are the only data readily available. For instance, computerized data banks containing a wide variety of seasonally adjusted economic data, both national and regional, are available on the Internet. It often makes sense to use commercially available sources rather than adjusting many of these series ourselves, even if the unadjusted data are available.

---

*Model-driven approaches to seasonal adjustments are beyond the scope of this book. Interested readers can find more details in Hillmer and Tiao (1982); Gomez and Maravall (1997).

Removing Seasonality by Differencing

In Chapter 4, we introduce differencing as a means of removing trends from data. For example, a first difference (also called a regular difference of period 1) of the time series Y_t results in a new time series Z_t of successive changes defined by $Z_t = Y_t - Y_{t-1}$. Successive values are separated by one period.

As well as removing trends, differencing can also be used to remove seasonal influences from data.

In the retail industry, for example, it is common to see a high fourth quarter because it is the most important selling season. In order to detect patterns in the data underneath the quarterly seasonal movements, we can compare successive changes between values separated by four time periods. In this fashion, $Z_t = Y_t - Y_{t-4}$ is called a seasonal difference of period 4. Here, Z_t is the difference between a value of the time series and the value of the series four periods earlier. For quarterly data, this represents a year-over-year change and so should be free of any quarterly pattern.

In Exhibit 4.23, we show that the seasonal contribution in a quarterly series of automobile sales in Quebec City was 68.2%. With the differenced data, the seasonal contribution to the total variability is reduced to 4.3%, for trend $= 11.5\%$ and irregular $= 84.2\%$. The calculation and time plot for the seasonal difference of period 4 for these data are in Exhibit 6.1.

The differencing operation needed to remove seasonal fluctuations in monthly data is the seasonal difference of period 12:

$$Z_t = Y_t - Y_{t-12}$$

For the following data on the sales of a weight-control product and expenditures on monthly advertising, an exploratory two-way table analysis (described in Chapter 4) resulted in the following interpretation for the trend and seasonal contributions:

Sales:	trend $= 20\%$	seasonality $= 47\%$	irregular $= 33\%$
Advertising:	trend $= 7\%$	seasonality $= 51\%$	irregular $= 42\%$

When developing formal forecasting models with these variables, a relatively high noise component suggests that much of the variation is not explainable by trend and seasonal influences alone. The time plots in Exhibit 6.2 clearly show the seasonal patterns in the two series, as well as the greater volatility in the advertising expenditures. In addition, peaks and troughs appear to correlate reasonably well. These are important considerations when modeling the data.

With the seasonal differenced data, the seasonal contribution to the total variability in the differenced sales is only approximately 17% (irregular $= 64\%$; trend $= 19\%$). The results for the differenced advertising variable now shows: trend $= 23\%$, seasonal $= 36\%$, and irregular $= 41\%$. Although the seasonal contribution to sales and advertising has been reduced by differencing, this analysis has its weaknesses. The relative scarcity of data (only 24 values of differenced data) does not support

EXHIBIT 6.1
Quarterly Auto-
mobile Sales:
(a) Seasonal
Differences of Period
4 (b) Time Plot of
Seasonal Differences
of Period 4

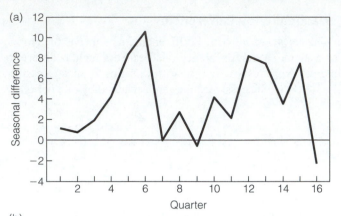

(b)

Sales	Seasonal difference of period 4
27.304	*
42.773	*
24.798	*
27.365	*
28.441	1.137
43.531	0.758
26.728	1.93
31.59	4.225
36.824	8.383
54.115	10.584
26.708	−0.02
34.313	2.723
36.232	−0.592
58.323	4.208
28.872	2.164
42.496	8.183
43.681	7.449
61.855	3.532
36.273	7.401
40.244	−2.252

*No data for these periods.
(*Source*: Exhibit 4.20)

drawing strong conclusions about the actual strength of these components in the data; it only helps us to quantify their relative influences.

Exhibit 6.3 depicts time plots and a scatter diagram for the seasonal differences of period 12 for the sales and advertising data. With the seasonal influences removed, an apparent association still exists between the two variables. We are interested in exploring other patterns, such as a lag relationship, in the data. This is dealt with again in Chapter 11, where we develop a linear regression model.

In a forecasting model, the objective is to predict product sales from advertising expenditures. To do this, we can:

- Correlate unadjusted sales and unadjusted advertising expenditures
- Correlate seasonally adjusted sales and seasonally adjusted advertising expenditures

EXHIBIT 6.2
Time Plots of
Monthly Sales and
Advertising Expen-
ditures over 36
Months

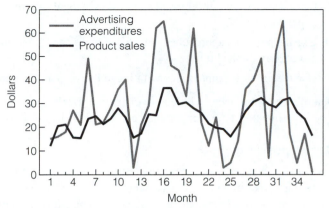

(*Source*: Cryer and Miller, 1994; SALESADS. DAT)

EXHIBIT 6.3
(a) Time Plots and
(b) Scatter Diagram
of Seasonal
Differences for Sales
and Advertising

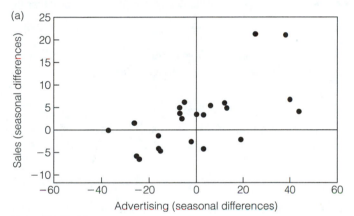

(*Source*: Exhibit 6.1)

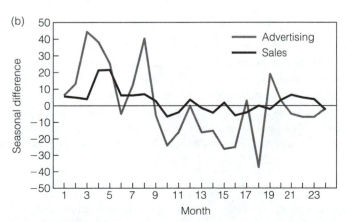

(*Source*: Exhibit 6.1)

- Correlate seasonal differenced sales and seasonal differenced advertising expenditures
- Use the previous approaches with log-transformed sales and log-transformed advertising expenditures
- Use the previous approaches with lagged advertising expenditures

Each of these approaches will lead to different models and hence different forecasts. Only experimentation and experience can uncover the most appropriate model.

Seasonal Decomposition

There are a wide variety of factors that influence economic data, so it is often difficult to determine precisely the way seasonal influences affect changes in a time series. The more commonly used methods of seasonal decomposition for large-scale adjustments are data-driven; they are not based on formal statistical models. These methods use smoothing procedures extensively. However, most methods are based on the assumption that seasonal fluctuations can be measured in terms of a constant set of factors that can be identified apart from underlying trend-cycle and other fluctuations.

The objective of a seasonal decomposition procedure is to measure typical or average seasonal movements in monthly or quarterly data.

If the magnitude of the seasonal increase or decrease is assumed to be essentially constant and independent of the level of the time series, additive decomposition is appropriate:

$$\text{Data} = \text{Trend-cycle} + \text{Seasonal} + \text{Irregular}$$

(recall that irregular is the catch-all word for all unexplained variations including random error). More often, the magnitude of the seasonal change tends to increase or decrease with level, so that seasonality might be assumed to be proportional to the level of the time series. This leads to multiplicative decomposition:

$$\text{Data} = \text{Trend-cycle} \cdot \text{Seasonal factor} \cdot \text{Irregular factor}$$

Even in this circumstance, an additive decomposition could be used if we transform the original time series with logarithms (provided there are no zeros or negatives in the data). This tends to stabilize the magnitude of the seasonal pattern and allows us to use additive decomposition on the transformed series. One major limitation of using the log-transformed model, however, is that the constraint that annual sums of the seasonal factors must be 0 in an additive model does not give the same result as the constraint that the product of seasonal indices must be 1 for the log additive model.

In general, seasonal decomposition is categorized as either additive or multiplicative.

One desired feature of a good seasonal-adjustment procedure is that the seasonal component not change too much over time. The choice between an additive or multiplicative model may be important here. There are also methods that make simultaneous additive and multiplicative adjustments. Because all methods have their limitations, the practitioner needs to be aware of the advantages and disadvantages of seasonal-adjustment procedures in the context of the particular application.

Uses of Seasonal Adjustment

Consider the following simplified example showing how a forecaster uses seasonal factors. Seasonal factors can be used to identify turning points that are not apparent in the raw data, and adjust seasonality out of the data so that forecasting techniques that cannot handle seasonally unadjusted data (e.g., exponential smoothing models found in Chapter 8) can be applied to the seasonally adjusted data.

Exhibit 6.4 shows three rows of numbers. The first row shows the actual demand for a product during a given year. The second row shows seasonal factors that were developed, based on historical data and projected for the same year. The third row shows the seasonally adjusted data under an assumed additive decomposition:

$$\text{Data} - \text{Seasonal factor} = \text{Trend-cycle} + \text{Irregular}$$

In this example, the actual values decline from January through May. The seasonal factors indicate that the first 3 months are generally strong, April has no significant seasonality, and May is generally weak. After adjusting for the seasonal effect, we can see that the adjusted demand grows after February. This might be a result of an economic recovery that is not apparent in the actual values.

Exhibit 6.4 also highlights the importance of assuring ourselves that the seasonal factors are appropriate. Otherwise, inappropriate conclusions can be drawn because of a faulty seasonal adjustment.

In Exhibit 6.5, the same actuals are used, but a different seasonal pattern is assumed. After we adjust for the seasonal effect, the data show a flat demand pattern. In Exhibit 6.6, the same actuals are used, but the seasonal factors have been distorted as a result of severe outliers in the prior year's actuals—that is, the seasonal factors in Exhibit 6.5 are correct, but the method used to derive the seasonal factors in Exhibit 6.6 has incorrectly handled the outliers in the prior year. These distorted

EXHIBIT 6.4 Using Seasonal Factors to Adjust a Data Set	Jan	Feb	Mar	Apr	May
Actual data	2000	1900	1700	1300	1100
Seasonal factors	1000	900	600	0	−400
Seasonally adjusted data	1000	1000	1100	1300	1500

EXHIBIT 6.5 Using Different Seasonal Factors to Adjust the Data Set	Jan	Feb	Mar	Apr	May
Actual data	2000	1900	1700	1300	1100
Seasonal factors	500	400	200	−200	−400
Seasonally adjusted data	1500	1500	1500	1500	1500

EXHIBIT 6.6 Using		Jan	Feb	Mar	Apr	May
Seasonal Factors That						
Have Been Impacted	Actual data	2000	1900	1700	1300	1100
by Outliers in the	Seasonal factors	500	400	200	0	−100
Prior Year's Data to	Seasonally adjusted data	1500	1500	1500	1300	1200
Adjust the Data Set						

factors have then been projected into the current year, altering the April and May seasonal factors. In the seasonally adjusted data, it appears that demand is falling off when it really is not.

6.2 THE RATIO-TO-MOVING-AVERAGE METHOD

In the 1920s and early 1930s, the Federal Reserve Board and the National Bureau of Economic Research were heavily involved in the smoothing of economic time series. Many basic time series such as the monthly manufacturers' shipments, inventories, and orders and retail sales and housing starts needed to be seasonally adjusted and published for use by other government agencies and the public. In 1922, Frederick R. Macauley of the National Bureau of Economic Research developed the ratio-to-moving-average method in a study done at the request of the Federal Reserve Board (Macauley, 1930). Simplicity in the calculations was a necessity in the early days of seasonal-adjustment procedures because of the lack of computing power.

The ratio-to-moving-average method is a simple way to determine seasonal factors in seasonal data.

A general multiplicative decomposition is assumed to take the form:

$$Y_t = TC \times S \times TD \times I$$

where TC = trend-cycle, S = seasonal, TD = trading day, and I = irregular are the components of the original data. (The irregular component includes effects such as strikes, wars, floods, other unusual events, and random errors.)

Step 1: Trading-Day Adjustment

Because trading days are variations that are attributable to the composition of the calendar and that can be measured, the trading-day adjustment TD is usually removed as a component. Typical trading-day factors are obtained by dividing the number of trading days for a given month by the average number of trading days for the same month over a period of the time series. This adjustment ratio is applied by dividing the original monthly value by its adjustment ratio. The seasonal-adjustment procedure of Macauley and those before the Census X-11 variant did not deal with the problem of trading-day variations but simply used the number of working days in a month. This results in an adjustment of the original data, and so the data are assumed to take the form:

$$Y_t = TC \times S \times I$$

Step 2: Calculating a Centered Moving Average

In Chapter 2, we introduce the moving average technique as a means of generating a level projection from an average of the most recent p values in a time series Y_t. For creating a seasonal decomposition, a centered moving average plays a key role. The p-term *centered* moving average P_CMAVG given by:

$$P\_CMAVG_t = \left(Y_{t-m} + Y_{t-m+1} + \cdots + Y_{t-1} + Y_t + Y_{t+1} + \cdots + Y_{t+m}\right)/p$$

where the Y_t values denote the actual (observed) values and p is the number of values included in the average. In this formula, p is an odd integer and $m = (p-1)/2$. In other words, we are using $(p-1)/2$ values on either side of Y_t to produce a smoothed value of Y_t. In particular, the three-term centered moving average 3_CMAVG has the formula:

$$3\_CMAVG_t = \left(Y_{t-1} + Y_t + Y_{t+1}\right)/3$$

The larger the value of p, the smoother the moving average becomes. If p is even, as is the case in smoothing quarterly data ($p=4$) or monthly data ($p=12$), there is no middle position to place the smoothed value. In that case, we place it to the right of the observed value or take another centered two-period moving average.

A centered moving average can be used to smooth unwanted fluctuations in a time series.

To illustrate this calculation, we have extracted the 36 observations of monthly champagne sales from Exhibit 6.7 and displayed the results in Exhibit 6.8. In Exhibit 6.8, the first column displays the data and the second column is a centered 12-term moving average. Note that the first value of the moving average is centered on observation 7 (18.8) of champagne sales. A two-term centered moving average of column 2 results in the values shown in column 3. The first value of this moving average is centered on observation 6 of the original data (26.4). Finally, the time plot in Exhibit 6.9 illustrates the smoothness of the final result.

Step 3: Trend-Cycle and Seasonal Irregular Ratios

The next step in the method is to obtain an estimate of the trend and cyclical factors by use of a p-month centered moving average, where p is the length of the seasonal

EXHIBIT 6.7 Monthly Champagne Sales	Champagne		
	15.0	24.4	31.1
	18.7	24.8	30.1
	23.6	30.3	40.5
	23.2	32.7	35.2
	25.5	37.8	39.4
	26.4	32.3	39.9
	18.8	30.3	32.6
	16.0	17.6	21.1
	25.2	36.0	36.0
	39.0	44.7	52.1
	53.6	68.4	76.1
	67.3	88.6	103.7

(*Source*: The Spreadsheet Forecaster on CD)

EXHIBIT 6.8 Monthly Champagne Sales: Calculation of a Two-Term of a 12-Term Centered Moving Average	Champagne	12_CMAVG	2_12_CMAVG
	15		
	18.7		
	23.6		
	23.2		
	25.5		
	26.4		29.75
	18.8	29.75	29.9115
	16	30.3958	30.3677
	25.2	30.9292	30.9646
	39	31.6042	31.6625
	53.6	32.5125	32.475
	67.3	33.2708	33.2625
	24.4	33.9958	33.951
	24.8	34.5417	34.5344
	30.3	35.0583	35.101
	32.7	35.7458	35.7875
	37.8	36.6	36.7625
	32.3	38.1042	38.0198
	30.3	39.2708	39.1042
	17.6	39.7708	39.8073
	36.0	40.4167	40.3875
	44.7	40.9458	40.8563
	68.4	41.1167	41.1698
	88.6	41.5	41.5073
	31.1	41.9125	41.8698
	30.1	42.1542	42.1302
	40.5	42.3	42.3406
	35.2	42.6083	42.6885
	39.4	43.2375	43.3177
	39.9	44.1875	43.95
	32.6		44.1875
	21.1		
	36.0		
	52.1		
	76.1		
	103.7		

(*Source*: Exhibit 6.7)

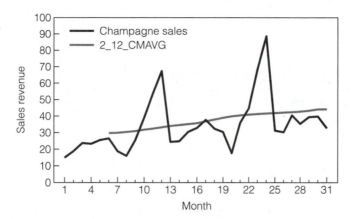

period (Chapter 2). This moving average is divided into the trading-day adjusted data to yield a series of seasonal-irregular ratios:

$$Y_t = S \times I = \left(TC \times S \times I\right)/TC$$

Averaging the SI ratios for a given month over a number of years produces an estimate of the seasonal factor or seasonal index. The irregular factor is assumed to cancel out in the smoothing process.

Step 4: Seasonally Adjusted Data

The final seasonally adjusted data are obtained by dividing each monthly value by the seasonal index for the corresponding month. This corresponds to a multiplicative seasonal-adjustment procedure.

6.3 MULTIPLICATIVE AND ADDITIVE SEASONAL DECOMPOSITIONS

Decomposition of Monthly Data

Recall that a multiplicative seasonal adjustment is appropriate when the range of seasonal fluctuations each year increases as the trend in the data increases. Exhibit 6.10 shows a multiplicative decomposition of monthly champagne sales in a spreadsheet calculation. The data are highly seasonal, which is supported by the two-way table decomposition (Chapter 4): trend = 7%, seasonality = 91%, and irregular = 2%. Exhibit 6.14 (later in the chapter) shows a multiplicative decomposition of quarterly gas grill sales (trend = 14%, seasonality = 75%, irregular = 11%). These are clearly highly seasonal time series.

We also gain some insight from this decomposition into future modeling complexities when building formal forecasting models. The dominant seasonality suggests that particular attention should be paid to accurately representing the seasonal influences in the data.

EXHIBIT 6.10 Monthly Champagne Sales—Multiplicative Decomposition

Filename: SEASADJ.XLS
Worksheet: MULTIMON

1   MULTIPLICATIVE SEASONAL ADJUSTMENT, MONTHLY DATA
2   TITLE1: MULTIPLICATIVE SEASONAL ADJUSTMENT
3   TITLE2: CHAMPAGNE SALES
4   X-AXIS: MONTH
5   Y-AXIS: MILLIONS OF CASES

6	A	B	C	D	E	F	G	H	I	J	K
7				Actual Moving			Sum of	# of	Avg.	Seas.	Adj.
8	Year	Mon	Per	Data	Avg.	Ratio	Ratios	Ratios	Ratio	Index	Data
9	2000	1	1	15.0	0.0	0.00	1.47	2	0.736	0.728	20.60
10		2	2	18.7	0.0	0.00	1.44	2	0.718	0.711	26.32
11		3	3	23.6	0.0	0.00	1.83	2	0.916	0.907	26.02
12		4	4	23.2	0.0	0.00	1.75	2	0.877	0.868	26.74
13		5	5	25.5	0.0	0.00	1.97	2	0.984	0.974	26.18
14		6	6	26.4	0.0	0.00	1.78	2	0.892	0.883	29.90
15		7	7	18.8	29.4	0.64	2.14	3	0.715	0.708	26.57
16		8	8	16.0	30.1	0.53	0.98	2	0.488	0.483	33.13
17		9	9	25.2	30.7	0.82	1.72	2	0.861	0.852	29.57
18		10	10	39.0	31.2	1.25	2.34	2	1.172	1.160	33.62
19		11	11	53.6	32.0	1.68	3.34	2	1.671	1.653	32.42
20		12	12	67.3	33.0	2.04	4.19	2	2.095	2.073	32.46
21	2001	1	13	24.4	33.5	0.73		Sum	12.124	12.000	33.50
22		2	14	24.8	34.5	0.72					34.90
23		3	15	30.3	34.6	0.88					33.40
24		4	16	32.7	35.5	0.92					37.69
25		5	17	37.8	36.0	1.05					38.80
26		6	18	32.3	37.2	0.87					36.59
27		7	19	30.3	39.0	0.78					42.82
28		8	20	17.6	39.6	0.45					36.45
29		9	21	36.0	40.0	0.90					42.24
30		10	22	44.7	40.8	1.09					38.53
31		11	23	68.4	41.1	1.67					41.37
32		12	24	88.6	41.2	2.15					42.74

(*Source*: Exhibit 6.7)

The first step (Exhibit 6.10, column E) is to compute a 12-month centered moving average of the data. The first moving average, covering January to December, is always placed next to month 7. The second moving average, for February to January, is placed opposite month 8 and so on. The result of this procedure is that there is no moving average for the first 6 months and the last 6 months of the data (not shown).

The second step is to use the centered moving averages to compute seasonal indexes. If we divide each data value by its moving average, the result is a preliminary seasonal index. Ratios are computed in column F; each ratio is simply actual sales in column D divided by the moving average in column E. The ratios for the same month in each year vary somewhat, and they are summed in column G and averaged in column I.

The average ratios can be interpreted as follows. Sales in January are predicted to be 73.6% of average monthly sales for the year. Sales in December are predicted to be 209.5% of the average. For this interpretation to make sense, the average ratios must sum to 12 (called normalization) because there are 12 months in the year. The average ratios actually sum to 12.12 because rounding is unavoidable. Thus, the formulas in column J normalize the ratios to sum to 12.

The last step is to adjust the sales data. Each actual value in column D is divided by the seasonal index applicable to that month to obtain the adjusted data in column K. The graph is shown in Exhibit 6.11.

An additive procedure can be developed in a manner analogous to the multiplicative decompostion. Exhibit 6.12 illustrates an additive decomposition of monthly beer production, in which the range of seasonal fluctuation each year is assumed to be relatively constant (Exhibit 6.13).

The first step in column E is to compute a 12-month moving average of the data. The first moving average, covering January to December, is always placed next to month 7. The second moving average, for February to January, is placed opposite month 8 and so on. The result of this procedure is that there is no moving average for the first 6 months and the last 6 months of the data.

The second step is to use the moving averages to compute seasonal indexes. If we subtract each moving average from its actual value, the result is a preliminary seasonal index. The differences are computed in column F (the actual sales in column D minus the moving average in column E). The differences for the same month in each year vary somewhat, so they are summed in column G and averaged in column I.

The average differences can be interpreted as follows. The data in January are predicted to be 8.68 units *less* than the average monthly data for the year. The data in July are predicted to be 22.43 units *more* than the average. For this interpretation to make sense, the average differences must sum to zero. The average differences actually sum to 2.23 because rounding is unavoidable. Thus, the formulas in column J normalize the differences to sum to zero.

The last step is to adjust the data. The final seasonal index is subtracted from each actual data value in column D to obtain the adjusted data in column K.

EXHIBIT 6.11
Time Plot of Monthly
Champagne Sales
and Seasonally
Adjusted Values by
Multiplicative
Decomposition

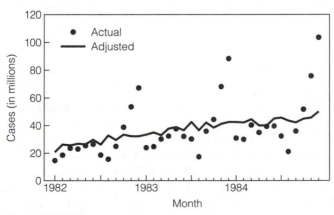

(*Source*: Exhibit 6.10)

EXHIBIT 6.12   Monthly Beer Production—Additive Decomposition

Filename: SEASADJ.XLS
Worksheet: ADDITMON

1	ADDITIVE SEASONAL ADJUSTMENT, MONTHLY DATA										
2	TITLE1: ADDITIVE SEASONAL ADJUSTMENT										
3	TITLE2: BEER PRODUCTION										
4	X-AXIS: MONTH										
5	Y-AXIS: BARRELS										
6	A	B	C	D	E	F	G	H	I	J	K
7				Actual Moving			Sum of	# of	Avg.	Seas.	Adj.
8	Year	Mon	Per	Data	Avg.	Diff.	Diffs.	Diffs.	Diff.	Index	Data
9	1987	1	1	18.7	0.0	0.00	−17.36	2	−8.68	−8.87	27.57
10		2	2	20.2	0.0	0.00	−15.02	2	−7.51	−7.69	27.89
11		3	3	20.5	0.0	0.00	−9.96	2	−4.98	−5.17	25.67
12		4	4	22.1	0.0	0.00	−9.30	2	−4.65	−4.84	26.94
13		5	5	30.1	0.0	0.00	8.83	2	4.42	4.23	25.87
14		6	6	36.8	0.0	0.00	23.82	2	11.91	11.73	25.07
15		7	7	49.2	27.2	22.03	67.30	3	22.43	22.25	26.95
16		8	8	39.6	27.3	12.35	22.86	2	11.43	11.24	28.36
17		9	9	24.2	27.4	−3.16	−8.92	2	−4.46	−4.64	28.84
18		10	10	22.4	27.6	−5.19	−8.28	2	−4.14	−4.33	26.73
19		11	11	21.2	27.8	−6.56	−12.83	2	−6.41	−6.60	27.80
20		12	12	21.1	28.0	−6.94	−14.26	2	−7.13	−7.32	28.42
21	1988	1	13	19.6	28.3	−8.73			Sum 2.23	0.00	28.47
22		2	14	21.5	28.6	−7.13					29.19
23		3	15	23.3	28.7	−5.38					28.47
24		4	16	24.1	28.7	−4.58					28.94
25		5	17	33.5	29.1	4.42					29.27
26		6	18	40.3	29.3	10.96					28.57
27		7	19	52.7	29.5	23.16					30.45
28		8	20	40.3	29.8	10.51					29.06
29		9	21	24.2	30.0	−5.76					28.84
30		10	22	27.2	30.3	−3.09					31.53
31		11	23	24.3	30.6	−6.27					30.90
32		12	24	23.5	30.8	−7.32					30.82

(*Source*: The Spreadsheet Forecaster on CD)

Decomposition of Quarterly Data

The MULTIQTR worksheet in Exhibit 6.14 is similar to the MULTIMON worksheet (Exhibit 6.10) except that it removes a quarterly seasonal pattern from the data instead of a monthly seasonal pattern. The first step in column E is to compute a 4-quarter centered moving average of the data. The first moving average, covering quarters 1–5, is always placed next to quarter 3. The second moving average, for quarters 2–6, is placed opposite quarter 4 and so on. The result of this procedure is that there is no moving average for the first 2 quarters and the last 2 quarters of the data.

The second step is to use the moving averages to compute seasonal indexes. If we divide each value by its moving average, the result is a preliminary seasonal

EXHIBIT 6.13
Time Plot of Monthly
Beer Production and
Seasonally Adjusted
Values by Additive
Decomposition

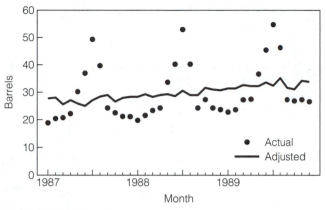

(*Source*: Exhibit 6.12)

index. Ratios are computed in column F; each ratio is simply actual sales in column D divided by the moving average in column E. The ratios for the same month in each year vary somewhat, so they are summed in column G and averaged in column I.

The average ratios can be interpreted as follows. The data in quarter 1 are predicted to be 80.8% of average quarterly data for the year. The data in quarter 3 are predicted to be 138.5% of the average. For this interpretation to make sense, the average ratios must sum to 4 because there are 4 quarters in the year. The average ratios actually sum to 4.008 because rounding is unavoidable. Thus, the formulas in column J normalize the ratios to sum to 4.

The last step is to adjust the sales data. Each actual value in column D is divided by the seasonal index applicable to that month to obtain the adjusted data in column K. A time plot of the quarterly gas grill data and seasonally adjusted values are shown in Exhibit 6.15.

The ADDITQTR worksheet (Exhibit 6.16) deals with additive seasonality, a pattern in which the range of seasonal fluctuations each year is assumed to be relatively constant.

The first step in column E is to compute a 4-quarter centered moving average of the data. The first moving average, covering quarters 1–5, is always placed next to quarter 3. The second moving average, for quarters 2–6, is placed opposite quarter 4 and so on. The result of this procedure is that there is no moving average for the first 2 quarters and the last 2 quarters of the data.

The second step is to use the moving averages to compute seasonal indices. If we subtract each moving average from its actual value, the result is a preliminary seasonal index. The differences are computed in column F (actual sales in column D minus the moving average in column E). The differences for the same month in each year vary somewhat, so they are summed in column G and averaged in column I.

The average differences can be interpreted as follows. The data in quarter 1 are predicted to be 3.725 units *less* than the average quarterly data for the year. The data in quarter 4 are predicted to be 7.438 units *more* than the average. For this interpretation to make sense, the average differences must sum to zero. The average differences

EXHIBIT 6.14　Quarterly Gas Grill Sales—Multiplicative Decomposition

Filename:　SEASADJ.XLS
Worksheet:　MULTQTR

1　MULTIPLICATIVE SEASONAL ADJUSTMENT, QUARTERLY DATA
2　TITLE1: MULTIPLICATIVE SEASONAL ADJUSTMENT
3　TITLE2: GAS GRILL SALES
4　X-AXIS: QUARTER
5　Y-AXIS: UNITS

6	A	B	C	D	E	F	G	H	I	J	K
7				Actual	Moving		Sum of	# of	Avg.	Seas.	Adj.
8	Year	Qtr	Per	data	Avg.	Ratio	Ratios	Ratios	Ratio	Index	Data
9	1978	1	1	201	0.0	0.00	1.62	2	0.808	0.807	249.2
10		2	2	253	0.0	0.00	1.83	2	0.913	0.912	277.5
11		3	3	312	250.8	1.24	4.16	3	1.385	1.383	225.7
12		4	4	237	252.8	0.94	1.80	2	0.901	0.899	263.6
13	1979	1	5	209	254.5	0.82			Sum 4.008	4.000	259.1
14		2	6	260	275.0	0.95					285.2
15		3	7	394	276.3	1.43					285.0
16		4	8	242	280.0	0.86					269.1
17	1980	1	9	224	281.8	0.80					277.7
18		2	10	267	303.0	0.88					292.9
19		3	11	479	322.5	1.49					346.4
20		4	12	320	0.0	0.00					355.8
21	1981	1	13		0.0	0.00					0.0
22		2	14		0.0	0.00					0.0
23		3	15		0.0	0.00					0.0
24		4	16		0.0	0.00					0.0
25	1982	1	17		0.0	0.00					0.0
26		2	18		0.0	0.00					0.0
27		3	19		0.0	0.00					0.0
28		4	20		0.0	0.00					0.0
29	1983	1	21		0.0	0.00					0.0
30		2	22		0.0	0.00					0.0
31		3	23		0.0	0.00					0.0
32		4	24		0.0	0.00					0.0

(*Source*: The Spreadsheet Forecaster on CD)

EXHIBIT 6.15
Time Plot of
Quarterly Gas Grill
Sales and Seasonally
Adjusted Values by
Multiplicative
Decomposition

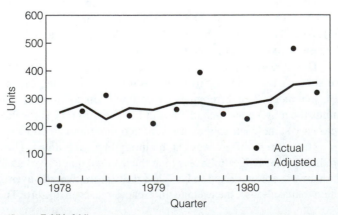

(*Source*: Exhibit 6.14)

EXHIBIT 6.16   Quarterly Calculator Sales—Additive Decomposition

Filename: SEASADJ.XLS
Worksheet: ADDITQTR

1   ADDITIVE SEASONAL ADJUSTMENT, QUARTERLY DATA
2   TITLE1: ADDITIVE SEASONAL ADJUSTMENT
3   TITLE2: CALCULATOR SALES
4   X-AXIS: QUARTER
5   Y-AXIS: THOUSANDS OF UNITS

6	A	B	C	D	E	F	G	H	I	J	K
7				Actual Moving			Sum of	# of	Avg.	Seas.	Adj.
8	Year	Qtr	Per	data	Avg.	Diff.	Diffs.	Diffs.	Diff.	Index	Data
9	1978	1	1	73.9	0.0	0.00	−7.45	2	−3.725	−3.778	77.7
10		2	2	76.8	0.0	0.00	−4.55	2	−2.275	−2.328	79.1
11		3	3	77.7	78.8	−1.10	−3.67	3	−1.225	−1.278	79.0
12		4	4	86.8	79.3	7.50	14.88	2	7.438	7.384	79.4
13	1979	1	5	75.9	79.8	−3.85			Sum 0.213	0.000	79.7
14		2	6	78.6	80.4	−1.78					80.9
15		3	7	80.2	80.8	−0.63					81.5
16		4	8	88.6	81.2	7.38					81.2
17	1980	1	9	77.5	81.1	−3.60					81.3
18		2	10	78.1	80.9	−2.78					80.4
19		3	11	79.3	81.3	−1.95					80.6
20		4	12	90.1	0.0	0.00					82.7
21	1981	1	13		0.0	0.00					0.0
22		2	14		0.0	0.00					0.0
23		3	15		0.0	0.00					0.0
24		4	16		0.0	0.00					0.0
25	1982	1	17		0.0	0.00					0.0
26		2	18		0.0	0.00					0.0
27		3	19		0.0	0.00					0.0
28		4	20		0.0	0.00					0.0
29	1983	1	21		0.0	0.00					0.0
30		2	22		0.0	0.00					0.0
31		3	23		0.0	0.00					0.0
32		4	24		0.0	0.00					0.0

(*Source*: The Spreadsheet Forecaster on CD)

actually sum to 0.213 because rounding is unavoidable. Thus, formulas in column J normalize the differences to sum to zero.

The last step is to adjust the sales data. Each actual data value in column D is subtracted from the seasonal index applicable to that quarter to obtain the adjusted data in column K. A time plot of quarterly calculator sales and seasonally adjusted values by additive adjustment are shown in Exhibit 6.17.

Seasonal Decomposition of Weekly Point-of-Sale Data

With the advent of scanner technology, point-of-sale (POS) data have become the basic data used to improve the forecasting accuracy of detailed product sales by store in the retail industry. POS data are typically summarized in weekly periods for product detail at a store (referred to as "door" in the retail industry) level. We describe here an analy-

EXHIBIT 6.17
Time Plot of Quarterly Calculator Sales and Seasonally Adjusted Values by Additive Decomposition

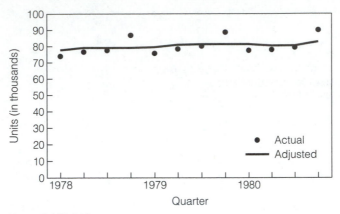

(*Source*: Exhibit 6.16)

sis of weekly cycles and show how weekly forecasts can be effectively combined with monthly seasonal patterns to improve the forecasts of very short-term trends.

The retail industry is faced with increasingly shortened **lead-time demands** due to changing consumer-supplier relationships and overall competitive and profitability pressures. Various industry initiatives, such as **QR** (**Quick Response**), **ECR** (**Efficient Consumer Response**), **CRP** (**Continuous Replenishment Programs**), and **VMI** (**Vendor-Managed Inventory**), have surfaced but share a similar goal: To make every order count. Retail and consumer goods manufacturers, in particular, strive for a seamless flow of product from manufacturers to retailers through tightly integrated information systems.

Typical weekly POS data are characterized by volatile patterns consisting of calendar effects, promotional spikes, and seasonality.

Exhibit 6.18 depicts a sample of POS data over a 112-week period that shows seasonality according to a monthly cycle, promotional peaks for a duration of 1 to 3 weeks, an outlier due to an isolated unexplained event, and minimal trend. The forecasting challenge is to isolate these patterns and determine their relationships to the underlying dynamics of location-specific consumer demand. By structuring patterns according to readily understood factors driving consumer demand at a store level, forecasters can improve the ordering and inventory processes for the large numbers of SKUs required for replenishment and production planning (see Chapter 9).

Practice suggests that promotion patterns consist of a prepromotion dip, a promotional peak, and then a forward-buy dip. Such patterns are difficult to discern in real data because of the inherent volatility of weekly patterns. Although calendars for promotions can be documented, consistent patterns are usually difficult to see in POS data for promoted products. With little trending evident in the disaggregated data, seasonality can be displayed by showing a week as a percentage of the annual

EXHIBIT 6.18
Time Plot of Weekly
Sales of Clips for
Sunglasses

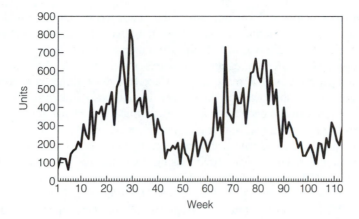

average of weekly sales. Shown over the same period, this can lead to patterns in which the weekly variation can easily overshadow the underlying seasonal cycle in the data. Plotting the weekly sales for 52 weeks for each year in a ladder chart can also show this pattern (Exhibit 6.19).

 Promotions are characterized by a pattern consisting of a prepromotion dip, a promotional peak, and then a forward-buy dip.

However, when we view the same data in monthly periods the seasonal pattern is much clearer, suggesting that seasonal patterns in POS data might be better analyzed at an aggregated level, such as months. For instance, it is a common practice to create monthly aggregates based on a 4-4-5-4-4-5-4-4-5 pattern for the 52 weeks in a year. When weeks are aggregated into months according to a 4-4-5 pattern, it

EXHIBIT 6.19
Ladder Chart of
Seasonal Pattern in
Weekly Sales of Clips
for Sunglasses

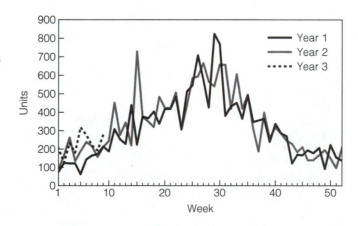

means that month 1 is the sum of the first 4 weeks, month 2 is the sum of the next 4 weeks, and month 3 is the sum of the next 5 weeks. This pattern is repeated for each quarter in the year. Exhibit 6.20 illustrates a year-over-year pattern for monthly aggregates; after aggregating the weeks into months we show the data as a time plot in Exhibit 6.21. The seasonal peaks and troughs within a year appear smoother and are not overshadowed by the weekly fluctuations within months.

To start the decomposition, we can make a seasonal adjustment of the monthly aggregated POS data, either in a multiplicative or additive model. Because POS data tend to have small values and are generally not strongly trending, the additive decomposition might be preferred. Using the ratio-to-moving-average decomposition ADDITMON (as in Exhibit 6.12) on the monthly data results in a set of 12 additive indices or components $\{-629, 9, 168, 216, 1064, 874, 288, 329, -340, -623, -635, -712\}$ for January through December. To complete an additive decomposition, the monthly seasonal components can be applied to the weeks within the months by reversing the operation we used to translate weeks to months and using proportional week/month ratios from the same month that the adjustments are applied to. This results in a set of weekly seasonal components and a seasonally adjusted weekly series.

Once a weekly POS time series has been adjusted using the weekly seasonal components, we can continue to analyze the residual pattern for a secondary seasonal pattern. For example, a pattern may be found that reflects the consumers' monthly buying behavior or an account manager's attempts to meet sales goals. Using a decomposition approach in which weekly cycles are analyzed within quarter seasons (by using 13 periods per quarter as seasonality), it is at times possible to reveal these consumer patterns. If properly interpreted, they can be useful in improving forecasting accuracy by improving our understanding of the buying behavior of consumers and the selling pressures of account managers. For example, such a secondary pattern in the POS data was found for a costume-jewelry manufacturer (see CD for data). The pattern showed a rise in the second week of the month when consumers may have more discretionary income for nonessentials than in the last week of a month. Moreover, there appeared to be a strong increase in sales at the end of the quarter, perhaps due to sales efforts to reach quotas.

EXHIBIT 6.20
Ladder Chart of
Seasonal Pattern in
Monthly Sales of
Clips for Sunglasses

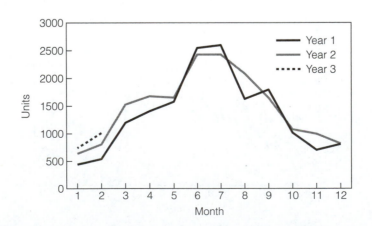

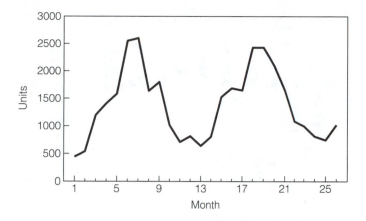

6.4 CENSUS SEASONAL ADJUSTMENT METHOD

Government agencies adjust statistical indicators for seasonality before publication and distribution to the public. In many industries, it is vital to use an accepted standard high-quality approach to seasonal adjustment. In regulated industries, the regulatory bodies may require this kind of reliability for the economic and cost studies submitted as part of a hearing. For example, in the semiconductor industry regional bookings and billings data are of interest to monitor changes in the book-to-bill ratio, a leading indicator for the industry. The book-to-bill ratio, reported on a seasonally adjusted basis, compares orders with shipments: a value above 1 indicates positive growth, and a value of 1.22 indicates that, for every $100 of chips shipped during the month, $122 of chips were ordered. The U.S., European, and Japanese and Asian markets experienced shifts in their quarterly patterns as the industry became more global and the U.S. market share declined during the 1980s.

Seasonal adjustments are very important in the analysis of economic business cycles and short-term trends and the development of large-scale econometric models.

The X-11 and X-12 programs from the Bureau of Labor Statistics and the X-11-ARIMA/88 program from Statistics Canada are the best-known programs for the large-scale analysis and seasonal adjustment of monthly and quarterly data.

Why Use the X-11 and X-12 Programs?

Sometimes unusual variation can appear in ordinary situations in demand forecasting applications. For instance, in the hotel/motel room demand data DCTOUR for the Washington, D.C. metropolitan area (see the CD), a basic seasonal decomposition can lead to a series of seasonal-irregular ratios in which the variation for a particular month (January) appears much larger than the variation for ratios in the

EXHIBIT 6.22
Bar Chart of
Seasonal Factors for
Hotel/Motel Demand
(Long, 1987–1994,
with two inaugural
years; Short,
1990:1–1992:12,
with no inaugural
year)

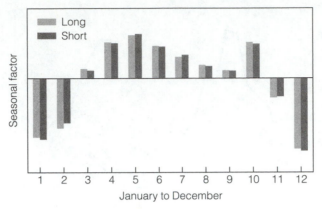

(*Source*: Frechtling, 1996, App. 1)

remaining months. Let us look carefully at the causes that lead to this situation. Exhibit 6.22 depicts the seasonal factors for a ratio-to-moving-average (multiplicative) seasonal-adjustment process for two periods: "Long" is the period 1987–1994, in which two U.S. presidential inaugurations (January 1989 and January 1993) took place and "short" is the period 1990–1992, in which no inaugurals took place. March, August, and September are the typical months for these data; they are right on trend. The winter months November through February run 8–30% below trend on average. The remaining 5 months run 10–20% above trend. Exhibit 6.23 shows the resulting seasonally adjusted data for the two decompositions. Although most seasonal factors remain fairly constant (except for February), the January inauguration following an election year appears to have an influence on the stability of the seasonal pattern. Evidently, there are significant differences in the adjusted values for the months adjacent to each January. To provide a consistent, reliable picture of the seasonal impact on the industry, we may have to rely on more sophisticated approaches, such as X-11 or X-12, than a basic ratio-to-moving-average decomposition method.

EXHIBIT 6.23
Time Plot of
Seasonally Adjusted
Hotel/Motel
Demand (Long,
1987–1994, with
two inaugural years;
Short, 1990:1–
1992:12, with no
inaugural year)

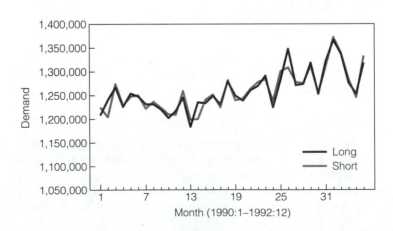

The X-11/ X-12 Programs

In 1954, Julius Shiskin first developed a computerized approach (known as Method I) at the U.S. Census Bureau for decomposing large numbers of time series. An improved Method II, which essentially contained refinements to the ratio-to-moving-average method, followed the first Census program very closely. Subsequent experimental variants of Method II (known as X-1, X-2, and so on) included moving seasonal-adjustment factors and smoother and more flexible trend-cycle curves. Adjustments for variations in the number of working days were included in the last major release of the program (the X-11 variant), which appeared in 1965 (Shiskin et al., 1965). A description of the Census Method II seasonal-adjustment procedure is given in Kallek (1978). Variable holidays (such as Easter) were later included in the X-11-ARIMA/88 and X-12 programs.

The basic goal of the X-11 method is to estimate seasonal factors from seasonal data.

An important development over the past 3 decades combines ARIMA models with the X-11 seasonal-adjustment procedure to produce future seasonal factors (Dagum, 1976). The technique has shown demonstrable improvements over the X-11 method, and the X-11-ARIMA/88 implementation from Statistics Canada is being used widely as a replacement program for the X-11 (Dagum, 1988; www.delphus.com).

There have also been releases of an X-12 version by the U.S. Census Bureau (Findley et al., 1997; ftp://ftp.census.gov/pub/ts/x12a/); a complete regression and time series modeling language has been added as an integral part of the program. However, for the basic seasonal decomposition, the X-11 and X-12 programs produce essentially identical outputs.

The U.S. Census Bureau's X-12-ARIMA offers refinements over Statistics Canada's industry-standard program X-11-ARIMA/88 in that it can handle more complex economic data and provides some new, very useful diagnostics.

The moving averages employed by X-11 require data for up to 3 additional years before symmetric moving averages may be applied. This is because X-11 estimates trends with a 12-month moving average and also smoothes the seasonal component across years with a 3×5 moving average (a 3×5 moving average is a five-term moving average applied to a three-term moving average). The latter represents a seven-term weighted moving average, so symmetry is lost for data within 3 years of the end of the sample. To circumvent the need to apply asymmetric moving averages for the last few observations, Estela Bee Dagum at Statistics Canada developed an important modification to the procedure in 1975—enlarging the original time series by 1 additional year with forecasts from ARIMA models. This enhancement allowed for symmetric moving averages to be applied and resulted in improved estimates of current seasonal factors.

The X-11 and X-12 programs are now very widely used by government agencies, central banks, and corporations, creating a standardized way to obtain deseasonalized data for publishing official statistics and to use in regulatory studies. These programs have seasonally adjusted literally thousands of economic and demographic time series reported by federal agencies for public use. There are separate programs to deal with monthly and quarterly data.

The basic goal of the X-11 method is to estimate seasonal factors from seasonal data. Then we can remove the seasonal component and produce an adjusted series that most clearly shows the trend-cycle and irregular variations. The basic strategy of the program is to remove the influence of extreme values so as to reveal the underlying movement in the data in a better way. The basic tactic is the use of iteration to achieve refinement.

The assumptions underlying the X-11 program are that a time series is composed of seasonal, trend-cycle, trading-day, and irregular components. There are two versions of this program available: additive and multiplicative. In the multiplicative version, the time series Y_t (the subscript t can be suppressed for convenience) is assumed to be a product of a seasonal factor S, trend-cycle TC, trading day (the number of active working or business days) TD, and irregular component I:

$$Y = TC \cdot S \cdot TD \cdot I$$

The alternative additive formulation is that the original time series is a summation of these components:

$$Y = TC + S + TD + I$$

Generally speaking, the multiplicative model produces the best seasonal factors for most series. However, it will not work for series that have negative values and for series that are highly volatile; the additive model is more appropriate for these.

Program Output

The X-11 and X-12 programs make the following sequence of computations: (1) trend-cycle, (2) seasonal-irregular ratios, (3) replacement of extreme irregular ratios, (4) seasonal factors, and (5) seasonally adjusted series. The procedure is iterative. As each factor is isolated and removed, the remaining factors are recomputed. This procedure continues until each factor is isolated.

There are three major computational runs within the X-11 program producing a series of tables labeled B, C, and D. Run 1 produces a series of B-tables, which are considered preliminary; run 2 results in the C-tables, which are semifinal; and run 3 results in the final D-tables and subsequent analytical tables.

Step 1: Trend-Cycle

Because a seasonally adjusted series consists of trend-cycle and irregular components, it is sometimes desirable to remove the irregular component and look at trend-cycle alone. Smoothing operations can do this. Centered moving averages based on an even number of terms, such as a four-term or 12-term moving average, are not centered in the middle of the value range. When the length n is even, the moving average can be centered in two stages:

1. Calculate the moving average and place it halfway between time period $n/2$ and $(n + 1)/2$.

2. Calculate a two-term moving average of the result.

The first estimate of the trend-cycle is computed using a two-term average of a 12-term moving average. The X-11 program creates two different trend-cycle series—the Months for Cyclical Dominance (MCD) series and **Henderson curves.** In the MCD series, the MCD value indicates the minimum period over which the average absolute change can be attributed to cyclical change rather than unexplained fluctuations. It is an unweighted moving average of, at most, 6 months and is determined as follows:

1. The irregular component is divided by the trend-cycle.

2. The number of months that must be added together before that ratio is less than 1 becomes the MCD.

3. If the months for cyclical dominance exceed 6, then 6 months is used as the maximum term in the smoothing.

A Henderson weighted moving average or Henderson curve gives an estimate of the trend-cycle component.

The reason for using the MCD series is to ensure that we have current values. Using a smoothing operation in which more than six terms are needed would introduce a significant lag in the data and many months would be lost at both ends of the data. Clearly, the MCD series is particularly important when the most current data are of interest. The Table F-2 of the X-11 program (produced by a run of the program) contains the MCD series.

A Henderson curve is a 9-, 13-, or 23-term weighted moving average; this is particularly useful for series with strong cyclical patterns. In the Henderson calculations, an attempt is made to overcome the lag introduced by long-term moving-average operations by applying different weights to the varying months. Estimates are also made of what the last $(n - 1)/2$ months would be if future data were available because with any centered moving average the end values are lost. The Table D-12 of the X-11/X-12 programs (produced by a run of the program) contains data for Henderson curves. A 13-term Henderson curve is shown for the hotel/motel demand series in Exhibit 6.24.

The methods of seasonal adjustment in the X-11 and X-12 programs isolate the seasonal and irregular factors, leaving a composite trend and cycle component in the form of a long-term Henderson weighted moving average. An MCD moving average is a short-term alternative for this trend-cycle component.

EXHIBIT 6.24
A Time Plot of
13-Term Henderson
Curve for Hotel/
Motel Demand with
the Historical Data
from 1987:1–1994:12

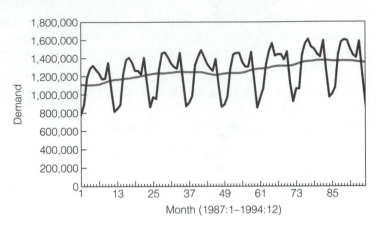

(Exhibit 6.17) (*Source*: ftp://ftp.census.gov/pub/ts/x12a/)

Step 2: Seasonal-Irregular Ratios

After computing the trend-cycle from the B-1 series, the seasonal-irregular (SI) ratios are computed. An estimate of the seasonal factors S comes from a smoothing of SI ratios.

Steps 3–5

The remaining steps, 3–5, result in the determination of seasonal factors and the adjustment of the data using the seasonal factors.

A Forecast Using X-12

To forecast the hotel/motel demand data, there are now several options:

- Create a forecast for the trend-cycle component with a trending technique, such as nonseasonal exponential smoothing (Chapter 9), linear regression (Chapters 11–12), or nonseasonal ARIMA models (Chapters 14–15). Then, seasonalize the projections with the seasonal factors.

- Create a forecast for the original data with a seasonal technique, such as seasonal exponential smoothing (Chapter 9), linear regression (Chapters 11–12), or seasonal ARIMA models (Chapters 14–15).

- Create a forecast for the seasonally adjusted data with a trending technique.

Exhibit 6.25a displays the results of a forecast for monthly hotel/motel demand using the last 12 months of data (1994:1–1994:12) as a holdout sample. Using data through 1993:12, we derived a projection for the 12 months of 1994 using the X-12 program. A comparison of these forecasts and holdout sample is shown along with the percentage errors for the months of 1994 (Exhibit 6.25b). It appears that the projections are mostly higher than the actuals, with the greatest errors in the winter months. This suggests that the impact of the presidential inaugurations has not yet been adequately discounted in the analysis.

EXHIBIT 6.25
(a) Time Plot of the
12-Month Forecast
for Monthly Hotel/
Motel Demand Using
X-12-ARIMA;
(b) Bar Chart of the
Forecast Percentage
Errors over the
Holdout Period
(1994:1–1994:12)

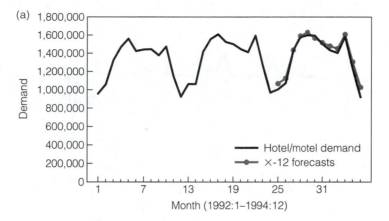

(a)

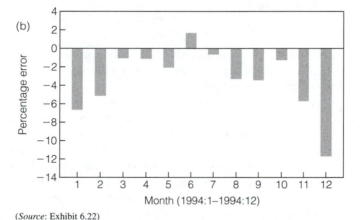

(b)

(*Source*: Exhibit 6.22)

6.5 RESISTANT SMOOTHING

A resistant smoothing procedure is used to estimate a typical value within a range of data as a sliding window across the data. To illustrate, we have used a short section of logarithms of airline data compiled in Box et al. (1994, series G). Exhibit 6.26 shows a plot of the data along with three outliers that were arbitrarily inserted for June 1957, September 1958, and July 1959 (7.0, 8.0, 3.0) to demonstrate how such extreme values can be handled.

 Resistance to outliers is also important in the smoothing of seasonal time series, in which underlying trends should not be unduly distorted by extreme values in the pattern.

Exhibit 6.27 shows the monthly values for 1956–1959. Column 1 contains the basic data and column 2 contains a 12-month moving median of the data. To

EXHIBIT 6.26
Time Plot of the
Logarithms of the
Airline Data,
Showing Three
Inserted Outliers

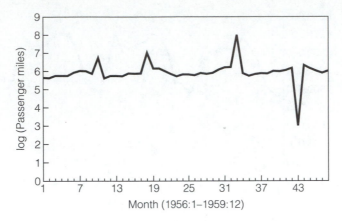

Month (1956:1–1959:12)

(*Source*: Box et al., 1994, series G)

compute the medians, the data have been resequenced (put in order from smallest to largest, 12 months at a time) and the average of the sixth and seventh values has been taken in every 12-month period.

In column 3 of Exhibit 6.27, a 12-month moving average of the median is calculated. The first entry in this column is positioned between the sixth and seventh moving median entries in the December 1956 row. A 3-month moving average of the means (column 4) further smoothes the data and results in a new time series with exactly 1 year of data missing at the beginning and end of the original data.

Next, a tapered moving average (or tapered smooth) is used. This involves calculating weighted moving averages in which the weights diminish with distance from the month for which the calculation is being made. A scheme is used in which the weights follow a bisquare function of the form:

$$B(u) = \begin{cases} (1-u^2)^2 & \text{if } |u| \leq 1 \\ 0 & \text{if } |u| > 1 \end{cases}$$

The user selects the amount of trend smoothing required. For a small, medium, and large amount of smoothing, the recommended smooth window is 7, 15, and 31, respectively. The calculation of the weights for a seven-period window bisquare weighting function is illustrated in Exhibit 6.28.

A tapered smooth, like a bisquare, is a weighted moving average in which weights diminish with distance from the month for which the calculation is being made.

The tapered mean is calculated by (1) multiplying the values by the appropriate weights, (2) summing up the weighted values, and (3) dividing by the sum of the weights. A sample calculation is shown in Exhibit 6.29. Here the months of January through July in 1957 figure into the tapered mean for April 1957. Column 5

EXHIBIT 6.27   Logarithms of the Monthly Airline Data for 1956–1959, with Three Outliers

Year	Month	(1) Data	(2) 12-Mo Moving Median of (1)	(3) 12_MAVG of (2)	(4) 3_MAVG of (3)	(5) Tapered MAVG of (4)
1956	Jan	5.649				
	Feb	5.624				
	Mar	5.759				
	Apr	5.746				
	May	5.762				
	Jun	5.924	5.753			
	Jul	6.023	5.756			
	Aug	6.004	5.756			
	Sep	5.872	5.758			
	Oct	6.724	5.807			
	Nov	5.602	5.862			
	Dec	5.724	5.862	5.822		
1957	Jan	5.753	5.862	5.831	5.831	5.84
	Feb	5.707	5.862	5.84	5.84	5.844
	Mar	5.875	5.862	5.849	5.849	5.848
	Apr	5.852	5.862	5.858	5.856	5.854
	May	5.872	5.862	5.862	5.861	5.858
	Jun	7*	5.862	5.863	5.863	5.862
	Jul	6.146	5.862	5.864	5.864	5.864
	Aug	6.146	5.862	5.865	5.865	5.865
	Sep	6.001	5.862	5.865	5.865	5.866
	Oct	5.849	5.862	5.866	5.866	5.867
	Nov	5.72	5.872	5.868	5.868	5.869
	Dec	5.817	5.872	5.871	5.871	5.871
1958	Jan	5.829	5.872	5.873	5.873	5.873
	Feb	5.762	5.872	5.875	5.875	5.876
	Mar	5.892	5.872	5.877	5.877	5.879
	Apr	5.852	5.888	5.88	5.881	5.883
	May	5.894	5.888	5.886	5.887	5.889
	Jun	6.075	5.888	5.896	5.896	5.896
	Jul	6.196	5.889	5.906	5.904	5.903
	Aug	6.225	5.889	5.911	5.911	5.91
	Sep	8*	5.89	5.917	5.917	5.917
	Oct	5.883	5.938	5.922	5.923	5.924
	Nov	5.737	5.994	5.931	5.931	5.932
	Dec	5.82	5.994	5.939	5.94	5.941
1959	Jan	5.886	5.934	5.949	5.949	5.95
	Feb	5.835	5.934	5.959	5.959	5.959
	Mar	6.006	5.934	5.969	5.969	5.969
	Apr	5.981	5.994	5.98	5.979	5.977
	May	6.04	5.994	5.988	5.987	5.985
	Jun	6.157	6.005	5.992	5.992	5.993
	Jul	3*	6.008	5.995	5.998	6.001
	Aug	6.326	6.008	6.008	6.008	6.01
	Sep	6.138	6.021	6.02	6.02	6.021
	Oct	6.009	6.036	6.033	6.033	6.031
	Nov	5.892	6.036	6.045	6.045	6.039
	Dec	6.004	6.036	6.056	6.056	6.045
				6.067		

*Original values: June 1957, 6.045; September 1958, 6.001; July 1959, 6.306.
(MAVG, moving average)

EXHIBIT 6.28
Calculation of the
Weights for a Seven-
Period Bisquare
Weighting Function

$$W(t) = B\left[t/(T+1)\right] \quad t = -T, \ldots, 0, \ldots, T$$

$$W(-3) = B\left(-\frac{3}{4}\right) = \left[1-\left(-\frac{3}{4}\right)^2\right]^2 = 0.191$$

$$W(-2) = B\left(-\frac{2}{4}\right) = \left[1-\left(-\frac{2}{4}\right)^2\right]^2 = 0.563$$

$$W(-1) = B\left(-\frac{1}{4}\right) = \left[1-\left(-\frac{1}{4}\right)^2\right]^2 = 0.879$$

$$W(0) = B\ (1-0) = \left[1-(-0)^2\right]^2 = 1.0$$

$$W(1) = B\left(\frac{1}{4}\right) = 0.879$$

$$W(2) = B\left(\frac{2}{4}\right) = 0.563$$

$$W(3) = B\left(\frac{3}{4}\right) = 0.191$$

EXHIBIT 6.29
Calculation for a
Tapered Mean of
Log-Transformed
Airline Data for
April 1957

Month	Data*	Weight	Weighted Values	Tapered Moving Average
Jan	5.831	0.191	1.114	
Feb	5.84	0.563	3.288	
Mar	5.849	0.879	5.141	
Apr	5.856	1	5.856	5.854
May	5.861	0.879	5.152	
Jun	5.863	0.563	3.301	
Jul	5.864	0.191	1.12	
Total	4.266	24.972		

Tapered mean = 24.972/4.266 = 5.854

*From column 3, Exhibit 6.27.

of Exhibit 6.27 was compiled in this manner and shows the tapered moving averages of the values given in column 4. The tapered mean is resistant to outliers because it is a mean of medians, which are themselves resistant to the distortion of extreme values.

Exhibit 6.27 can be extended to include additional columns. Exhibit 6.30 shows the continuation of the calculations to develop a more refined trend smooth. Column 6

EXHIBIT 6.30
Continuation of the
Calculation of the
Trend Smooth

Year	Month	(6) (4) − (5)	(7) Tapered MAVG of (6)	(8) Initial Trend Smooth (7) + (5)	(9) Data Trend (1) − (8) (= T + I)
1956	Jan				
	Feb				
	Mar				
	:				
	:				
	Oct				
	Nov				
	Dec				
1957	Jan	−0.009	−0.004	5.836	−0.083
	Feb	−0.004	−0.003	5.84	−0.133
	Mar	0.001	−0.001	5.847	0.026
	Apr	0.002	0	5.854	−0.002
	May	0.003	0.001	5.859	0.013
	Jun	0.001	0.001	5.863	1.137
	Jul	0	0.001	5.865	0.281
	Aug	0	0	5.865	0.281
	Sep	−0.001	−0.001	5.865	0.136
	Oct	−0.001	−0.001	5.866	−0.017
	Nov	−0.001	−0.001	5.868	−0.148
	Dec	0	−0.001	5.87	−0.053

(*Source*: Exhibit 6.27)

is the moving average (column 4 of Exhibit 6.27) minus the tapered moving average (column 5). This result can be viewed as a systematic noise pattern about the trend and is similar to autocorrelated residuals about a regression line. These residual values are then smoothed (column 7) and added to the trend approximation (tapered moving average in column 5) to develop the initial trend smooth of the data that represents trend-cycle (column 8). The seasonal values are computed next. They are then subtracted from the original data to yield trend plus irregular (the notation for this is T + I). The T + I series can then be smoothed to produce a resistant approximation of trend.

6.6 HOW TO DETECT SEASONAL CYCLES—FORMAL-WEAR RENTAL REVENUE

The Basics of Seasonal Analysis

If there is a possibility that our data contain a seasonal cycle, a similar test for trends using differences of the data can be performed. To illustrate, let us use the formal-wear data from Exhibit 4.19. The original data are 8 years of sales rentals of tuxedos. There appears to be a seasonal cycle—notice that the first part of the year is always the low

EXHIBIT 6.31

Computing the
Differences between
the Same Month in
Different Years
Eliminates a
Seasonal Cycle

TREND CLASSIFICATION MODEL

Mean:	43,819	21,415	878	950
Variance:	1,565,532,927	481,510,777	856,756,632	2,390,105,150
Variance:	100%	31%	55%	153%
Minimum:		****		

	Formal-wear rentals	Difference by Month	Differences	Differences of differences
Year 1-1	6028			
Year 1-2	5927			
Year 1-3	10515			
Year 1-4	32267			
Year 1-5	51920			
Year 1-6	31294			
Year 1-7	23573			
Year 1-8	36465			
Year 1-9	18959			
Year 1-10	13918			
Year 1-11	17987			
Year 1-12	15294			
Year 2-1	16850	10822		
Year 2-2	12753	6826	−3,996	
Year 2-3	26901	16386	9,560	13,556
Year 2-4	61494	29227	12,841	3281
Year 2-5	147862	95942	66,715	53,874
Year 2-6	57990	26696	−69,246	−135,961
Year 2-7	51318	27745	1049	70,295
Year 2-8	53599	17134	−10,611	−11,660
Year 2-9	23038	4079	−13,055	−2,444
Year 2-10	41396	27478	23,399	36,454
Year 2-11	19330	1343	−26,135	−49,534
Year 2-12	22707	7413	6,070	32,205
Year 3-1	15395	−1455	−8,868	−14,938
Year 3-2	30826	18073	19,528	28,396
Year 3-3	25589	−1312	−19,385	−38,913
Year 3-4	103184	41690	43,002	62,387
Year 3-5	197608	49746	8,056	−34,946
Year 3-6	68600	10610	−39,136	−47,192
Year 3-7	39909	−11409	−22,019	17,117
Year 3-8	91368	37769	49,178	71,197
Year 3-9	58781	35743	−2,026	−51,204
Year 3-10	59679	18283	−17,460	−15,434
Year 3-11	33443	14113	−4,170	13,290
Year 3-12	53719	31012	16,899	21,069

(*Source:* Hanke and Reitsch, 1998)

point within each year, whereas the last part of the year is always the high point.
However, the seasonal cycle should be confirmed before it is used in forecasting.

To confirm the seasonal cycle, recall that we compute the differences between
data for the same month for 2 years. The first value of the differenced data plotted is

for January of Year 2—the data for the first month of Year 2 minus the data for the first month of Year 1. The next difference value is for February of Year 2—the data for the second month of Year 2 minus the data for the second month of Year 1—and so on.

The next step is to check for trend in the rental sales. Because the seasonal cycle has been removed from the monthly differences, we can analyze them to see if a trend exists. First, we compute the differences between the monthly differences. Then, we compute the differences between the last set of differences.

How much data do we need to carry out the tests? We would not attempt to test for trend in annual data with fewer than six observations. In data with a possibility of a seasonal cycle, we need at least two complete cycles, that is 8 quarters or 24 months of data. Like other statistical procedures, the more data we have, the more reliable the results are likely to be.

If a seasonal cycle exists, the variance of the differences between the same month in different years is smaller than the variance of the original data. By computing differences, we eliminate the fluctuations caused by the seasonal cycle and thereby reduce the variance. In Exhibit 6.31, notice that the deviations for the average are much smaller for the differences than for the original data. If we find the variance reduced, this suggests there is a moderate trend in addition to the seasonal cycle. In our example, the differenced values display a much smaller variance than the original data and indicate that a seasonal cycle exists.

If the variance is reduced again after computing the differences between the previous set of differences, then there is a strong trend in addition to the seasonal cycle. In our example, the variance increases, so there could be minimal trend in addition to the seasonal cycle.

If the data are quarterly, we use the same procedure except that we analyze differences between the same quarter in 2 years.

SUMMARY

Seasonal adjustment is a useful procedure that helps identify turning points in the economy or the trends in the demand for products and services. Knowledge of the seasonal pattern also helps in planning employee workloads and inventory levels. If we can remove seasonality from a time series, we can apply a number of forecasting techniques that otherwise would not handle seasonal data (e.g., some of the exponential smoothing techniques discussed in Chapter 8).

Many business managers find the decomposition of a time series into subcomponents appealing as a forecasting tool. It allows them to explain the variations in the data and to predict the changes in subpatterns in terms that are easy to interpret. Because they are able to relate their knowledge of economic or industry patterns to the forecasting process, managers not only favor the method as a forecasting tool, but also consider it a means of achieving greater management control.

The X-11-ARIMA/88 and X-12 seasonal adjustment programs provide a standardized data-driven procedure for deseasonalizing and creating trend-cycle decompositions for thousands of time series data in government and business. These programs are capable of mass-processing data and producing detailed analyses of seasonal factors and of trend-cycle and irregular variations; they can be run in an additive or multiplicative form for quarterly or monthly data.

The use of resistant smoothing techniques is demonstrated in the calculation of a tapered smooth. Twelve-month moving medians of a time series eliminate the influence of extreme values. Bisquare weights (which diminish in magnitude with distance from the time for the calculation) are calculated and then used to smooth these medians. In later chapters, these robust/resistant techniques are reinforced in the selection of specific models and in the interpretation of the reasonableness of models.

REFERENCES

Box, G. E. P., G. M. Jenkins, and G. M. Reinsel (1994). *Time Series Analysis—Forecasting and Control.* 3rd ed. Englewood Cliffs, NJ: Prentice Hall.

Cryer, J. D., and R. B. Miller (1994). *Statistics for Business.* Belmont, CA: Wadsworth.

Dagum, E. B. (1976). Seasonal factor forecasts from ARIMA models. *Proc. Int. Statist. Inst.* 40, 206–19.

Dagum, E. B. (1978). Modeling, forecasting, and seasonally adjusting economic time series with the X-11-ARIMA method. *Statistician* 27, 203–16.

DeLurgio, S. (1998). *Forecasting Principles and Applications.* New York: Irwin/McGraw-Hill.

Dielman, T. E. (1996). *Applied Regression Analysis for Business and Economics.* 2nd ed. Belmont, CA: Wadsworth.

Findley, D. F., B. C. Monsell, W. R. Bell, M. C. Otto, and B. C. Chen (1997). New capabilities and methods of the X-12-ARIMA seasonal adjustment program. *J. Business Econ. Statist.* 16, 127–52.

Frechtling, D. C. (1996). *Practical Tourism Forecasting.* Oxford: Butterworth-Heinemann.

Hanke, J. E., and A. G. Reitsch (1998). *Business Forecasting.* 6th ed. Englewood Cliffs, NJ: Prentice Hall.

Gomez, V. and A. Maravall (1997). Program TRAMO and SEATS. Bank of Spain (www.bde.es).

Hillmer, S. C. and G. C. Tiao (1982). An ARIMA model–based approach to seasonal adjustment. *J. Amer. Statist.* 77, 63–70.

Kallek, S. (1978). An overview of the objectives and framework of seasonal adjustment. In *Seasonal Analysis of Economic Time Series,* edited by A. Zellner. Washington, DC: U.S. Government Printing Office.

Macauley, E. R. (1930). *The Smoothing of Time Series.* Cambridge, MA: National Bureau of Economic Research.

Makridakis, S., S. C. Wheelwright, and R. J. Hyndman (1998). *Forecasting Methods and Applications.* 3rd ed. New York: John Wiley & Sons.

Shiskin, J., A. H. Young, and J. C. Musgrave (1965). The X-11 variant of Census Method II seasonal adjustment program. Technical Paper No. 15, U.S. Department of Commerce, Bureau of the Census. Washington, DC: U.S. Government Printing Office.

Tryfos, P. (1998). *Methods for Business Analysis and Forecasting: Text and Cases.* New York: John Wiley & Sons.

Wilson, J. H., and B. Keating (1998). *Business Forecasting.* New York: Irwin McGraw-Hill.

PROBLEMS

6.1   Based on past data, your firm's sales show a seasonal pattern. The seasonal index for November is 1.08, for December 1.38, and for January 0.84. Sales for November were $285,167.

   a. Would you ordinarily expect an increase in sales from November to December in a typical year? How do you know?

   b. Find November's sales on a seasonally adjusted basis.

   c. Take the seasonally adjusted November figure in (b) and seasonalize it using the December index to find the expected sales level for December.

   d. Sales for December have just been reported as $430,106. Is this higher or lower than expected, based on November's sales?

   e. Find December's sales on a seasonally adjusted basis.

f. On a seasonally adjusted basis, were sales up or down from November to December? What does this tell you?

6.2 Moving averages that are based on an even number of terms, such as a four-term moving average, are not centered in the middle of the value range. When applying moving averages with quarterly data in Census Method II, the length n is 4. Show that a four-term moving average followed by a two-term moving average results in a five-term weighted moving average with weights {1/8, 1/4, 1/4, 1/4, 1/4, 1/8}. This is known as a 2×4 moving average.

6.3 Moving averages that are based on an even number of terms, such as 12-term moving average, are not centered in the middle of the value range. When applying moving averages with monthly data in Census Method II, the length n is 12. Show that a 12-term moving average followed by a two-term moving average results in a 13-term weighted moving average with weights {1/24, 1/12, 1/12, ..., 1/24}. This is known as a 2×12 moving average.

6.4 A 3×3 moving average (three-term followed by three-term) is used in Census Method II in the "replacement of extreme values" step. It is equivalent to a five-term weighted moving average. Determine the weights.

6.5 A 3×5 moving average (five-term followed by three-term) is equivalent to a seven-term weighted moving average. Determine the weights.

6.6 A Spencer moving average is used in the X-11 program. Show that this $5 \times 5 \times 4 \times 4$ moving average is equivalent to a 15-term weighted moving average. Determine the weights.

6.7. The average number of trading days for the months of January through December over a number of years is listed in the following table.

a. If Year 2 in Exhibit 6.2 represents the *current* year's actual sales, determine the trading-day-adjusted sales figures for Year 1.

b. If Year 3 in Exhibit 6.2 represents *next* year's trade-adjusted forecasts, determine the unadjusted forecasts for next year.

6.8 The causes of seasonality in tourism demand include the following. Provide several examples for each cause.

a. Climate/weather

b. Social customs/holidays

c. Business customs

d. Calendar effects

6.9 Use the ratio-to-moving-average method to calculate seasonal factors in the table on page 230. A multiplicative model is assumed.

a. For the observations in column 1, calculate a centered 12-month moving average. Because there is an even number of months, place the first smoothed value between June and July.

b. Smooth column 2 with a two-term moving average. The first value appears opposite July. This is the trend-cycle component.

c. Because the trend-cycle starts in July of the first year, there are six fewer observations in this series than in the original data. True or false?

d. Calculate seasonal-irregular ratios by dividing the entries in column 1 by the corresponding entries in column 3. *Note:* With multiple years, the SI ratios can be averaged by individual month to obtain a single factor by month (i.e., all January SI ratios averaged, all February SI ratios averaged, etc.).

e. Seasonal factors for a year should sum to 12.0. Ratio the factors in column 4 so that the sum is 12.0.

6.10 For the monthly hotel/motel room demand data (DCTOUR; on the CD) for the Washington, D.C., metropolitan area, 1987–1994, produce a seasonal adjustment using the ratio-to-moving-average method.

a. Over the 6 sample years, for which month do the estimated seasonal factors display the greatest spread (variability)? Which 2 years appear to be unusual in this set of six numbers? Can you give an interpretation in terms of the U.S. presidential election cycle?

Month	Jan	Feb	Mar	Apr	May	Jun	Jul	Aug	Sep	Oct	Nov	Dec
Average	20.8	20.3	22.4	21.3	21.1	21.6	20.9	22.4	20.4	22.0	20.6	21.0

		(1)	(2)	(3)	(4)	(5)
Year	Month	Data	12_MAVG of (1)	2_MAVG of (2) = TC	(1)/(3) = SI ratio	Total = 12 for 1 year
1	Jan	100				
	Feb	90				
	Mar	95				
	Apr	100				
	May	105				
	June	110				
	Jul	105	106.7	107.8		
	Aug	105	108.9			
	Sep	110				
	Oct	115				
	Nov	120				
	Dec	125				
2	Jan	125				
	Feb	110				
	Mar	115				
	Apr	115				
	May	120				
	Jun	125				
	Jul	110				
	Aug	110				
	Sep	120				
	Oct	130				
	Nov	130				
	Dec	135				

b. Estimate a differential seasonal impact for that particular month and year in the election cycle. How would you use that information in forecasting monthly hotel/motel demand?

c. Assume a seasonally adjusted value for December 1994 to be 1,340,000 room-nights sold with a straight line projected increase in trend of 33,000 room-nights per month through 1997. Provide a monthly seasonal forecast through 1997.

6.11 For the data for the quarterly ABX Company Sales on the CD (ABXCO, Dielman, 1996, Table 3.11), perform a ratio-to-moving-average decomposition of the series into identifiable trend and seasonal and irregular patterns. What do you see as the dominant pattern(s) in the data now?

6.12 Repeat Problem 6.11 for one or more of the following quarterly time series on the CD:

a. Sales for Outboard Marine (OMC, Hanke, and Reitsch, 1998, Table 4.5)

b. Shoe store sales (SHOES, Tryfos, 1998, Table 6.5)

c. Domestic car sales (DCS, Wilson, and Keating, 1998, Table 1-5)

d. Sales of The GAP (GAP, Wilson, and Keating, 1998, p. 30)

e. Air passengers on domestic flights (PASSAIR, DeLurgio, 1998, p. 33)

6.13 For the monthly toll revenues (TOLLREV) and toll messages (MSG) on the CD:

a. Plot the time series and interpret the dominant pattern(s) you see in the data.

b. Perform a ratio-to-moving-average decomposition to determine specific patterns in the seasonal, trend, and noise components. What is the nature of the seasonality?

c. What type of projection do you recommend for future values of the data?

6.14 Repeat Problem 6.13 for one or more of the following monthly time series on the CD:

a. Unemployment rate data (UNEMP, Dielman, 1996, Table 4.7)

b. Silver prices data (SILVER, Dielman, 1996, Table 4.8)

c. Prime rate data (PRIME4, Dielman, 1996, Table 4.14)

d. Retail sales for U.S. retail sales stores (RETAIL, Hanke and Reitsch, 1998)

e. Wheat exports (SHIPMENT, Dielman, 1996, Table 4.9)

f. Shipments of spirits (DSHIP, Tryfos, 1998, Table 6.23)

g. Retail sales in a region (RSALES, Tryfos, 1998, Table 6.25)

h. Australian beer production (AUSBEER, Makridakis et al., 1998, Table 2.2)

i. International airline passenger miles (AIRLINE, Box et al., 1994, series G)

j. Australian electricity production (ELEC-TRIC, Makridakis et al., 1998, Fig. 7-9)

k. French industry sales for paper products (PAPER, Makridakis et al., 1998, Fig. 7-20)

l. Shipments of French pollution equipment (POLLUTE, Makridakis et al., 1998, Table 7-5)

m. Sales of recreational vehicles (WINNEBAG, Cryer and Miller, 1994, p. 742)

n. Sales of lumber (LUMBER, DeLurgio, 1998, p. 34)

o. Sales of consumer electronics (ELECT, DeLurgio, 1998, p. 34)

6.15 For each of the following, perform a ratio-to-moving-average decomposition on one or more of the monthly time series in Problem 6.14. In each case, plot a sample autocorrelation function of the irregular component and contrast the pattern with the pattern in the data that it is based on, as indicated.

a. Original data

b. Log of the data; contrast with (a).

c. First differences of the data; contrast with (a).

d. First differences of logarithms of the data; contrast with (a)–(c).

CASE 6A PRODUCTION OF ICE CREAM (CASES 1A, 3A, 4A, AND 5A, CONT.)

Ice cream is a seasonal product. The entrepreneur needs to have the data deseasonalized so that she can look at trends and business cycles.

(1) Create an additive and multiplicative decomposition of the monthly data.

(2) Use the TRENDIFF spreadsheet to assess the degree of trending in the data.

(3) Looking at the residuals (difference between the actual and seasonally adjusted data) in a time plot, decide whether an additive or multiplicative seasonality is the most appropriate.

(4) What is the nature of the business cycle?

CASE 6B DEMAND FOR AIR TRAVEL (CASE 4B CONT.)

From Case 4B, the VP of Operations of the airlines has learned about the importance of seasonality in the data and now instructs you to quantify the seasonality.

(1) Run a multiplicative and additive decomposition of the data and recommend which type of decomposition is most suitable for this data set.

(2) Create a time plot of the original data with the two seasonally adjusted series. Does the seasonality appear to be additive or multiplicative?

(3) Considering the general trend in the data, would one of the adjustments be more appropriate to re-seasonalize a trend projection? Explain which quarters could be most impacted.

CASE 6C DEMAND FOR APARTMENT RENTAL UNITS

Cozy House Apartments is a 112-unit rental complex. The complex is divided into five buildings, each with apartments ranging from efficiencies to two- or three-bedroom units. Only 110 units are available for rent because the superintendent uses a three-bedroom apartment and an efficiency unit is used as the office. The owner wants to improve the efficiency of the apartment complex and has asked you to create predictions of future vacancies. If there is an expectation of high vacancies, advertising expenditures may need to be increased and pricing incentives established. To get a feel for the problem, he recommends that you perform a preliminary analysis of the seasonality in the monthly demand for rental units. The management supplies you with the following data for an 11-year period.

Month	YR01	YR02	YR03	YR04	YR05	YR06	YR07	YR08	YR09	YR10	YR11
Jan	12	10	6	13	6	5	7	5	11	14	12
Feb	15	8	7	14	8	7	8	10	8	10	6
Mar	13	8	8	15	7	5	10	9	10	6	3
Apr	16	7	4	10	7	9	12	10	6	5	3
May	13	7	3	9	9	10	5	5	3	6	2
Jun	12	8	5	7	11	6	3	6	11	11	2
Jul	7	9	9	3	9	7	8	4	3	7	12
Aug	8	12	11	5	10	5	9	7	8	8	6
Sep	7	10	8	4	6	6	7	12	13	10	7
Oct	4	8	6	4	0	7	4	13	18	10	6
Nov	5	7	8	5	2	5	4	8	12	9	10
Dec	10	7	11	10	2	10	5	10	14	9	10
Average	10.2	8.4	7.2	8.3	6.4	6.8	6.8	8.3	9.8	8.8	6.6
SD	3.8	1.5	2.4	4.0	3.3	1.8	2.6	2.8	4.2	2.4	3.5

(1) What insight does the analyis give you about the nature of the data?

(2) Rebuild the table into quarterly periods and repeat your decomposition. Is there any change?

(3) If management offers you rental data by apartment size, how would you refine your analyses?

(4) At this stage of the analysis, what other data could you suggest as being drivers of the demand for rental units?

CASE 6D DOMESTIC AUTOMOBILE PRODUCTION (CASES 1D, 3D, AND 4D CONT.)

Your assignment as an analyst for an automobile industry association is to understand the nature of the seasonal fluctuations and create time series that are seasonally adjusted. The data are found in Case 3D.

(1) Calculate first, second, and twelfth differences for the monthly automobiles sales data. Also calculate first and twelfth differences of the data. Display the corresponding time plots and correlograms. Compare your findings in the correlograms with the visual depiction in the time plots in terms of nonstationarities.

(2) Use the ratio-to-moving-average method (both additive and multiplicative) to find seasonal indexes for the data. Which type of decomposition do you prefer for these data? Provide a rationale for your choice.

(3) What are the peak periods for sales? Create a seasonally adjusted series for the data.

(4) What kind of projection can you make using this technique for the year 1986?

CASE 6E SALE OF TWIN SCREW EXTRUDERS TO THE FOOD INDUSTRY

Food extruders, a high-capital-expense item for a food manufacturer, are used in the production of snack foods, candy, cereal, and animal feed for pets. (See, for example, Riaz, M. N., ed. (2000). *Extruders in Food Applications*. Weimar, TX: C.H.I.P.S.) Your company, a manufacturer in the food industry, has been selling twin screw extruders for the past decade. As a sales manager you are interested in looking for new opportunities in the market place and your CEO has asked you to develop some sales forecasts for the next 8 quarters. The sales data are shown in the table.

Quarter	Extruder Sales	Quarter	Extruder Sales
YR01 : I	0	YR04 : III	1
YR01 : II	1	YR04 : IV	3
YR01 : III	1	YR05 : I	2
YR01 : IV	1	YR05 : II	4
YR02 : I	0	YR05 : III	3
YR02 : II	3	YR05 : IV	7
YR02 : III	1	YR06 : I	1
YR02 : IV	4	YR06 : II	11
YR03 : I	3	YR06 : III	3
YR03 : II	5	YR06 : IV	7
YR03 : III	1	YR07 : I	4
YR03 : IV	3	YR07 : II	6
YR04 : I	2	YR07 : III	2
YR04 : II	5	YR07 : IV	9

You embark on an analysis that requires you to complete the following tasks:

(1) Determine seven key factors you may need to consider that impact the demand for extruders.

(2) Do you expect these factors to have a positive or negative impact on demand?

(3) Provide a measure of the trend and seasonal contribution in the data.

(4) Perform a seasonal decomposition of the data. Which is the appropriate approach in your opinion—multiplicative or additive?

(5) What are the peak periods for sales? Why is this important to know in your business?

(6) Do any sales periods strike you as unusual?

(7) What kind of projection do you recommend for Years 8 and 9 by quarter?

7

Forecasting the Business Environment

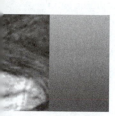

"The only function of economic forecasting is to make astrology look respectable."
—John Kenneth Galbraith

When statistical techniques are combined with economic theory, it makes up what is known as econometrics. Econometric techniques have been widely applied as a way to model the macroeconomy. Macroeconomic demand analysis and econometric methods are used extensively in forecasting, structural analysis, and policy analysis in a wide variety of business planning applications. In this chapter, we offer an overview, without much technical detail, of the uses and pitfalls of econometric analysis in business forecasting. We discuss the method of leading indicators as the most important aspect of any macroeconomic forecasting activity dealing with forecasts of the levels of economic indicators in econometric models; measurement of the impact of expansions and contractions on businesses, government, or public-sector organizations; and policy studies to assess the impact of changing economic and demographic assumptions on business and public programs. We also discuss how price elasticities, both own-price and cross-elasticity, for the major services and products of the business, are essential for intelligent and effective macroeconomic forecasting and planning, where own-price is the price of the item under consideration and cross-elasticity measures the percentage change in the demand for good A as a result of a given percentage change in the price of good B. The determination of elasticities is an important function essential for understanding business growth and for predicting revenue growth as well as quantity growth.

Before creating forecasting models for products and services, forecasters need to identify the important features, uses, and interpretations of the factors that model the macroeconomy so that maximum benefit can be derived from these techniques in practical situations. The goal is to acquire the information needed to select the indicators that will help structure a framework for forecast modeling and judgment from a macro or top-down perspective.

7.1 FORECASTING WITH ECONOMIC INDICATORS

Macroeconomic Demand Analysis

Econometric methods have been widely applied as a way to model the macroeconomy. Macroeconomic demand deals with the aggregates of income, employment, and price levels. The uses of econometric modeling and analysis techniques can be classified by the way outputs are required. The outputs produce three classes of applications for econometric modeling (structural models, policy analysis, and forecasting), of which the most widely used application is that of econometric forecasting.

  Applications of econometrics include structural models policy analysis, and forecasting.

In econometric forecasting, the general focus is the development of a set of equations based on an economic rationale whose parameters are estimated using a statistical methodology. The model is designed to provide the business variable(s) with some explanatory underpinnings, but is also able to generate extrapolative values for future periods. That is, the model offers predictive values for the output variable(s) outside the sample of data actually observed. In practice, the statistical estimation procedures are evaluated from the perspective of forecasting performance through the up- and downturns of business cycles and using leading indicators.

Origin of Leading Indicators

The method of leading indicators dates back to the sharp business recession of 1937–1938. At that time, an effort was initiated by the National Bureau of Economic Research (NBER) to devise a system that would signal the end of a recession (Burns and Mitchell, 1946). Burns and Mitchell first developed a comprehensive description of business-cycle activity in the economy that became the foundation of classical methods of business-cycle analysis.

A considerable amount of data, assembled by the NBER since the 1920s, has been analyzed to gain a better understanding of business cycles. These data, which included monthly, quarterly, and annual series on prices, employment, and production, resulted in a collection of 21 promising economic indicators that were selected on the basis of past performance and future promise as reliable indicators of business revival. Over the years, this effort has been greatly expanded to other public and private agencies (Shiskin and Moore, 1967; Moore and Shiskin, 1972). To this day, the NBER publishes turning points of business cycle peaks and troughs. The longest economic expansion on record ended in March 2001 and gave way to the first recession in a decade and the tenth since World War II. This latest recession lasted 8 months, ending in November 2001. Exhibit 7.1 illustrates business-cycle turning points for historical U.S. data from the mid 1800s to 2001.

A number of time series, such as employment, indexes of consumer and producer prices, and manufacturers' orders, are published in newspapers, business journals, and websites. As indicators of the nation's economic health, professional economists and the business community follow them very closely, especially during periods of rapid change in the pace of business activity.

For convenience of interpretation, economic indicators have been classified into three groups: leading, coincident, and lagging. **Leading indicators** are those that provide advance warning of probable changes in economic activity. Indicators that confirm changes previously indicated are known as lagging indicators. Coincident indicators are those that reflect the current performance of the economy, providing a measure of current economic activity. They are the most familiar and include the GDP industrial production, personal income, retail sales, and employment.

Economic indicators are classified as leading, lagging, and coincident with changes in economic activity.

Exhibit 7.2a shows the quarterly GDP in billions of chained 2000 dollars (real GDP) from quarter I of 1947 to quarter I of 2004. Mostly trending, the economic cycles are nevertheless in evidence. To highlight the quarter-to-quarter changes, Exhibit 7.2b depicts the percentage changes in seasonally adjusted annual rates for the recent period 1996:I to 2003:IV. Exhibit 7.3 shows the time plot of the real disposable income per capita series from January 1999 to December 2001 in chained (meaning adjusted for inflation) 1990 dollars. Exhibit 7.4a depicts the monthly retail sales from January 1992 to December 2001 in chained 1996 dollars; Exhibit 7.4b shows monthly, seasonally adjusted civilian labor force employment in millions from January 1989 to December 2003. These are all coincident indicators of the U.S. economy.

Use of Leading Indicators

It would be very useful to forecasters and planners to have some advance warning of an impending change in the local, national, or world economy. Whereas coincident indicators are used to indicate whether the economy is currently experiencing expansion, recession, or inflation, leading indicators help forecasters assess short-term trends in the coincident indicators. In addition, leading indicators help planners and policy makers anticipate adverse effects on the economy and examine the feasibility of taking corrective steps.

EXHIBIT 7.1 Business-Cycle Turning Points

BUSINESS CYCLE REFERENCE DATES		DURATION IN MONTHS			
				Cycle	
Peak	Trough	Contraction: Peak to Trough	Expansion: Previous trough to this peak	Trough from Previous Trough	Peak from Previous Peak
	December 1854 (IV)	—	—	—	—
June 1857 (II)	December 1858 (IV)	18	30	48	—
October 1860 (III)	June 1861 (III)	8	22	30	40
April 1865 (I)	December 1867 (I)	**32**	**46**	**78**	**54**
June 1869 (II)	December 1870 (IV)	18	18	36	50
October 1873 (III)	March 1879 (I)	65	34	99	52
March 1882 (I)	May 1885 (II)	38	36	74	101
March 1887 (II)	April 1888 (I)	13	22	35	60
July 1890 (III)	May 1891 (II)	10	27	37	40
January 1893 (I)	June 1894 (II)	17	20	37	30
December 1895 (IV)	June 1897 (II)	18	18	36	35
June 1899 (III)	December 1900 (IV)	18	24	42	42
September 1902 (IV)	August 1904 (III)	23	21	44	39
May 1907 (II)	June 1908 (II)	13	33	46	56
January 1910 (I)	January 1912 (IV)	24	19	43	32
January 1913 (I)	December 1914 (IV)	23	12	35	36
August 1918 (III)	March 1919 (I)	**7**	**44**	**51**	**67**
January 1920 (I)	July 1921 (III)	18	10	28	17
May 1923 (II)	July 1924 (III)	14	22	36	40
October 1926 (III)	November 1927 (IV)	13	27	40	41
August 1929 (III)	March 1933 (I)	43	21	64	34
May 1937 (II)	June 1938 (II)	13	50	63	93
February 1945 (I)	October 1945 (IV)	**8**	**80**	**88**	**93**
November 1948 (IV)	October 1949 (IV)	11	37	48	45
July 1953 (II)	May 1954 (II)	**10**	**45**	**55**	**56**
August 1957 (III)	April 1958 (II)	8	39	47	49
April 1960 (II)	February 1961 (I)	10	24	34	32
December 1969 (IV)	November 1970 (IV)	**11**	**106**	**117**	**116**
November 1973 (IV)	March 1975 (I)	16	36	52	47
January 1980 (I)	July 1980 (III)	6	58	64	74
July 1981 (III)	November 1982 (IV)	16	12	28	18
July 1990 (III)	March 1991(I)	8	92	100	108
March 2001 (I)	November 2001 (IV)	8	120	128	128
Average, all cycles:		17	38	55	56*
1854–2001 (32 cycles)		22	27	48	49**
1854–1919 (16 cycles)		18	35	53	53
1919–1945 (6 cycles)		10	57	67	67
1945–2001 (10 cycles)					
Average, peacetime cycles:		18	33	51	52***
1854–2001 (27 cycles)		22	24	46	47****
1854–1919 (14 cycles)		20	26	46	45
1919–1945 (5 cycles)		10	52	63	63
1945–2001 (8 cycles)					

* 31 cycles
** 15 cycles
*** 26 cycles
**** 13 cycles

Figures printed in bold are the wartime expansions (Civil War, World Wars I and II, Korean War, and Vietnam War), the wartime contractions, and the full cycles that include the wartime expansions.

Quarterly dates are in parentheses.

(*Source*: NBER; U.S. Department of Commerce, *Survey of Current Business*, October 1994, Table C-51)

EXHIBIT 7.2(a)
Time Plot of a
Coincident Indicator
of the U.S. Economy,
GDP in Billions of
Chained 1972
Dollars for 1947:II
to 2003:IV; (b)
Percentage Changes
of Seasonally Ad-
justed Annual Rates
for GDP from 1996:I
to 2003:IV

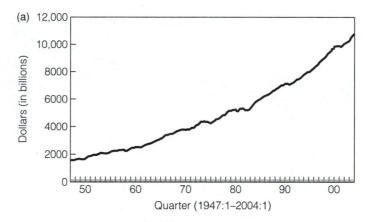

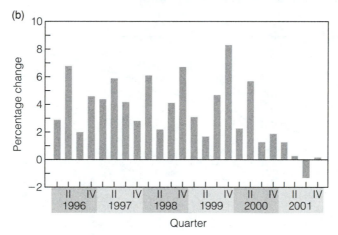

(*Source*: http://www.bea.doc.gov)

EXHIBIT 7.3 Time
Plot of a Coincident
Indicator of the U.S.
Economy, Personal
Income

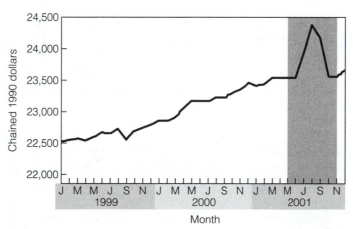

(*Source*: http://www.bea.doc.gov)

EXHIBIT 7.4 Time Plots of Coincident Indicators of the U.S. Economy: (a) Retail Sales and (b) Civilian Labor Force Employment (shaded area represents recession; the break in January 1994 is due to the redesign of the survey)

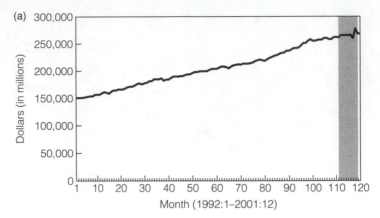

(a)

Y-axis: Dollars (in millions)
X-axis: Month (1992:1–2001:12)

(*Source*: http://www.census.gov/svsd/advretl/view/adv44000.txt)

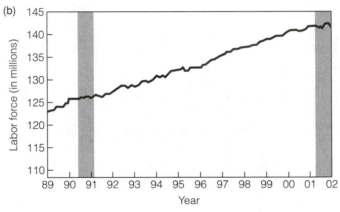

(b)

Y-axis: Labor force (in millions)
X-axis: Year

(*Source*: http://stats.bls.gov)

In individual sectors, such as agriculture, leading indicators have played a major part in short-term production forecasting (Allen, 1994). For example, the estimation of the number of acres planted to spring wheat is a good indication of harvested acreage. Economic indicator analysis has also been used to assist investors in optimizing the rate of return in their asset allocation between stocks and fixed income securities (Moore et al., 1994).

Knowledge of current economic conditions can be found in the duration, rate, and magnitude of recovery or contractions in business cycles.

Among the leading indicators in business forecasting, housing starts, new orders for durable goods, construction contracts, formation of new business enterprises, hiring rates, and average length of workweek are the most commonly quoted. In recent

EXHIBIT 7.5 Time Plot of a 12-Period Moving Average of the Monthly Housing Starts (New Private Housing Units Started), January 1972 to December 1982

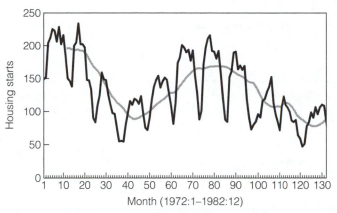

(*Source*: www.census.gov/briefin/esbr/www/esbr020.html)

times, weekly initial employment claims, expressed in terms of a 4-week moving average, are getting a great deal of attention in the media. Housing starts, a key leading indicator plotted in Exhibit 7.5, tend to lead fluctuations in overall economic activity. The data are used in this book as an explanatory variable related to telephone access-line gain and are generally useful in forecasting future activity (both short and long term) in construction, construction-related manufacturing, real estate, finance and communications, and so on.

A useful set of indicators for revealing and explaining the economy's broad cyclical movements includes manufacturers' shipments and orders (Exhibit 7.6). These are comprehensive indicators of industrial activity and are especially important to forecasters because the durable goods sector (plant equipment and durable machinery, automobiles, etc.) is the economy's most volatile component. Exhibit 7.6a displays total manufacturers' shipments and the 3-month moving average in billions of dollars for the time period May 1999 to May 2001, Exhibit 7.6b shows total manufacturers' orders, Exhibit 7.6c shows total inventory, and Exhibit 7.6d displays the ratios of unfilled orders and total inventory to shipments.

Shipments are an indicator of current economic activity, measuring the dollar value of products sold by all manufacturing establishments. Orders, on the other hand, are a valuable leading indicator. They measure the dollar value of new orders and the net order cancellations received by all manufacturers. The two series are distorted by inflation because there is no relevant price index to convert it to real terms. It is the difference between shipments and orders, which shows what is happening to the backlog of unfilled orders, that gives insight into the degree of sustainability of current national output.

The data are widely used by private economists, corporations, trade associations, investment consultants, and researchers for market analysis and economic forecasting; and by the news media in general business coverage and specialized commentary.

An example of a lagging indicator is the unemployment rate (Exhibit 7.7). Although it is frequently quoted in the press, business forecasters should realize that the unemployment rate is not an indicator of future or even current labor market conditions.

EXHIBIT 7.6 Time Plots of (a) Total Manufacturers' Shipments, (b) Total Manufacturers' Orders, (c) Total Inventory, and (d) Ratios of Unfilled Orders and Total Inventory to Shipments in Billions of Seasonally Adjusted Current Dollars for May 1999 to May 2001

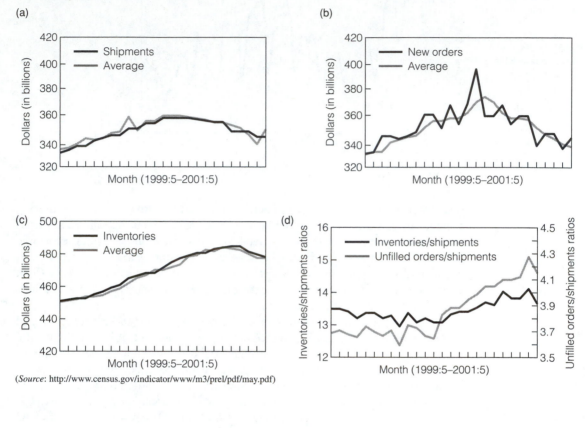

(*Source*: http://www.census.gov/indicator/www/m3/prel/pdf/may.pdf)

EXHIBIT 7.7
Time Plot of U.S.
Unemployment
Rates

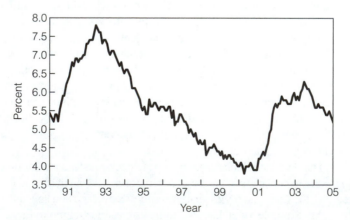

(*Source*: http://www.economagic.com/)

Composite Indicators

Economists have developed composite indicators to reduce the number of series that must be reviewed and at the same time not lose a great deal of information. These series provide single measures of complicated economic activities that experience common fluctuations. The procedure involved includes amplitude adjustment, in which the month-to-month percentage change of each series in the composite is standardized so that all series are expressed in comparable units. The average month-to-month change, without regard to sign, is 1.0. The score it receives from the scoring plan weights each individual series.

A composite indicator provides a single measure of complicated economic activities that experience common fluctuations.

If an index shows an increase of 2.0 in a month, it is rising twice as fast as its average rate of change in the past. If an index increases by 0.5, it is rising only one-half as fast as its historical rate of increase. Composite indicators have been developed for the leading, coincident, and lagging series.

In order to have a more comprehensive coverage of the economy, the Conference Board publishes a composite index of ten economic leading indicators. This index is a weighted combination of individual indicators, for example, of average work-week of production workers in manufacturing, money supply, and an index of stock prices. Exhibit 7.8 shows the components of the Conference Board's index of leading indicators and their net weighted contribution to the index's change from July to August 1997. The two-tenths of 1% advance in the Conference Board's index of leading indicators, covering August, reflected increases in seven of the ten components, with the main impact from an expansion of the money supply and a jump in orders at manufacturers making consumer goods. The main drag on the index came from an increase in the number of people filing first-time claims for unemployment benefits.

EXHIBIT 7.8 Components of the Index of Leading Indicators and Their Net Weighted Contribution to the Index's Monthly Change		
Average workweek of production workers in manufacturing		0.03
Average initial weekly claims for state unemployment insurance[a]		−0.11
New orders for consumer goods and materials, adjusted for inflation		0.1
Vendor performance (companies delivering slower deliveries from suppliers)		0.02
New orders for nonmilitary capital goods, adjusted for inflation		0.01
New building permits issued		−0.01
Index of stock prices		0.01
Money supply (M2, adjusted for inflation)		0.16
Spread between rates on 10-year Treasury bonds and Federal funds		0.01
Index of consumer expectations		−0.03

[a]Series is inverted in computing the index; that is, a decrease in the series is considered upward movement.
(*Source*: Conference Board)

EXHIBIT 7.9 Time
Plot of a Composite
Index of Leading
Indicators

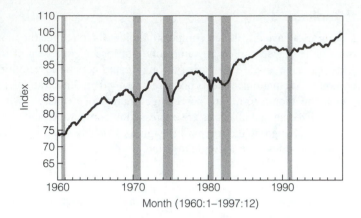

One problem with interpreting an index of leading indicators is that its month-to-month changes can be erratic (Exhibit 7.9); however, comparing movements of the index over a longer span helps to bring out the underlying cyclical movements. For example, Exhibit 7.10 shows the percentage change in the current level of the leading index from the average level of the preceding 12 months. On that basis, the leading indicators have declined (i.e., fallen below zero) before every one of the six recessions since 1970.

Reverse Trend Adjustment of the Leading Indicators

Economists have been concerned about two aspects of the leading indicators: (1) the lead at the business cycle peak is much longer than the lead at the trough and (2) leading indicators do not have the long-term trend that the economy has, as measured by coincident indicators. Because the objective of macroeconomic forecasting is to predict current levels rather than detrended levels, the reverse trend adjustment procedure adds a trend to the leading indicators (rather than removing the trend from the coincident indicators). First, however, it is necessary to eliminate whatever trend

EXHIBIT 7.10 Time
Plot of the Percent-
age Change in the
Current Level of the
Leading Index from
the Average Level of
the Preceding 12
Months

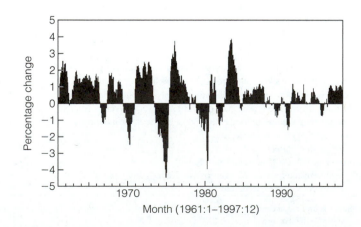

already exists in the leading indicators. Then the trend of the coincident indicators (based on full cycles) is added to the detrended leading indicators.

The effect of reverse trend adjustment is to shorten the lead time at business-cycle peaks and increase the lead time at troughs. It also tends to reduce the number of false signals of recession that are evident when the unadjusted index turns down but a recession does not occur. Because reverse trend adjustment helps to reduce the lead time at peaks and increase the lead time at troughs, it makes the two lead times more equal. This lessens the reaction time at the peak, however. Even with reverse trend adjustment, the lead at the peaks is approximately 1 or 2 months longer, on the average, than the lead at the troughs.

In forecasting with regression models (Chapter 12), the generally different lead times must be reckoned with before regression models are used. Because regression models do not offer us the chance to vary the lead or lag times in the explanatory variables at different periods in the cycle, the indicators tend to average the impact of the lesser lead or lag at either the peak or the trough. It is possible to have one model in which the indicator has as its lead time the appropriate lead for a peak and a second model that has as its lead time the appropriate lead for a trough. Then, either the first or second model is used to generate forecasts, depending on the state of the business cycle.

Sources of Indicators

On a monthly basis, the Conference Board publishes current data for many different indicators. The charts and graphs cover the national income and product accounts series, cyclical indicators, series on anticipations and intentions, analytical measures, and international comparisons. The series are usually seasonally adjusted and the NBER reference dates for recessions and expansions are shown. It is apparent from the plots that business contractions are generally shorter than business expansions. The average peacetime cycle is slightly less than 4 years.

Selecting Indicators

Specified criteria have been applied by the NBER to hundreds of economic series from which a list of indicators could be selected. A score can be given for each of six criteria, and those series with the highest scores can then be retained. The scoring is subjective in many aspects.

The criteria for selecting indicators include economic significance, statistical adequacy, historical conformity to business cycles, consistency of lead or lag, smoothness of the data, and timeliness of the data.

Economic Significance

Some aspects of criterion of significance have already been discussed—that is, the role a given economic process has in theories that purport to explain how business cycles come about or how they may be controlled or modified.

A consideration in indicator selection and scoring is the breadth of coverage. A broad indicator covers all corporate activity, total consumption, or investment; a narrow indicator relates to a single industry or to minor components of the broad series. A broad economic indicator (e.g., nonfarm employment) may continue to perform well even when some components deteriorate because of technological developments, changes in customer tastes, or the rapid growth or decline of single products or industries. Therefore, a broad indicator receives a higher score than a narrow indicator.

Statistical Adequacy

The characteristics we should consider in evaluating the statistical adequacy of a series include a good reporting system and good coverage; that is, the data should cover the entire period they represent, benchmarks should be available, and there should be a full account of survey methods, coverage, and data adjustments.

A good reporting system is based on primary rather than indirect sources or estimates. Some important series, such as the index of industrial production, the index of net business formation, and GDP, are based largely on indirect sources. Employment and retail sales are based on direct reporting from primary sources.

Good coverage means that if sampling is required, it should be a probability sample with stated measurement error regarding sample statistics. Moreover, coverage means, for example, that monthly data should include all days and not be a figure based on one day or week. In addition, the availability of a benchmark is important as a check on the accuracy of data. For example, the U.S. Census provides a benchmark for estimates of population.

Historical Conformity to Business Cycles

The NBER developed an initial index to measure how well the fluctuations in a series conformed to business-cycle variations. A series that rose through every business expansion and declined during every contraction received an index score of 100. This particular index did not include extra cycles, such as occurred in 1966–1967, which are not classified as recession troughs. The index did not indicate whether the lack of conformity occurred early in the data or later, and it did not take into account the amplitude of the cycles. The scoring system subsequently developed by Shiskin and Moore (1967) takes these considerations into account.

Consistency of Timing

A number of considerations govern scoring a series on the basis of consistency of timing. The first is the consistency of lead or lag time relative to cycle peak or trough. The second is the variability about the average lead or lag time. A third consideration is the difference in lead time for a peak compared to the lead time for a trough. A final consideration is whether there has been any recent departure from historical relationships. Leading indicators have a median lead time of 2 or more months; lagging indicators have a median lag of 2 or more months. Coincident indicators have a median timing of -1, 0, or $+1$ month. Occasionally, median leads of $+2$ months are possible when the lead or lag is not constant over many cycles.

Smoothness and Timeliness of Data

The factors that are weighed in arriving at a score for smoothness and timeliness include the prompt availability of data and their smoothness. It is easier to identify changes in direction in a smooth series than in an irregular series. Generally speaking, because of the irregularity of data, comparisons over spans greater than 1 month must usually be made to detect cyclical changes. The smoothing of some irregular series may result in some delay, but may still provide a longer lead time than for other series that are less irregular but have shorter lead times.

Generally speaking, leading indicators are the most erratic; lagging indicators are the smoothest. Coincident indicators have the shortest publication lag and the highest conformity scores. For example, corporate profits after taxes received an average score of 68 in the NBER index. This indicator also received fairly high scores for economic significance, statistical adequacy, conformity, and timing. However, it received a score of 60 for smoothness because it is irregular and only 25 for timeliness because it is a quarterly series subject to slow reporting.

Leading indicator methodology is not without its problems. Some basic limitations of leading indicators are that they are too focused on manufacturing, whose importance is declining; they can produce many false signals for turning points; and they may not be available on a timely basis, thereby reducing their lead advantages.

7.2 TREND-CYCLE FORECASTING WITH TURNING POINTS

This section describes a cycle or turning-point technique using economic indicators to help determine the timing of the turning point and the rate, duration, and magnitude of the cyclical component.

Pressures Analysis

Some observers of economic patterns believe that the business cycle is dead and that recessions are a thing of the past. Others state that the Federal Reserve Board has been able to avoid recessions by bringing the economy in for a "soft landing" during periods of declining growth. During a time of economic uncertainty, analyzing cycles in business data may still prove to be a valuable addition in a business forecaster's toolkit. Information about business cycles is extremely valuable for business planning, especially developing a budget, an inventory plan, or a production schedule. The technique can be applied to any demand-related data collected at monthly time intervals. We can use it to analyze company or industry sales, orders, shipments, inventories, production, and so on. As in all forecasting work, the more data the better, although we can get by with only a couple of years of history.

The aims of pressures analysis are to detect changes in the rate of growth in data and to identify the turning points (peaks and troughs) of cycles.

The hotel/motel industry experienced a slump in the early 1990s in the eastern part of the United States. We illustrate here the use of pressures to identify an approximate turning point. A pressure is a ratio calculation between two points in time. A spreadsheet demonstration of pressures analysis is shown in Exhibit 7.23 (see Section 7.5). The three graphs in Exhibits 7.11 and 7.12 illustrate the 1/12, 3/12, and 12/12 pressures for the Washington, D.C., hotel/motel demand series. How should we interpret the behavior of hotel/motel demand? Will growth pick up again? Has a turning point occurred? To help answer these questions, a ratio is calculated of the demand in each month compared to the same month a year before. If the index for a given month is 100%, the demand for that month is unchanged from a year before; if the index is greater than 100%, demand has grown from a year before. Such index numbers are called 1/12 pressures.

In the 1/12 pressures graph (Exhibit 7.11a), there is little evidence of a turning point. In the 3/12 pressures graph (Exhibit 7.11b), which compares 3-month totals with the same period the previous year, a minimum begins to appear toward the end of 1990. This is further confirmed in the 12/12 pressures graph (Exhibit 7.12). Does the decline in 1994 suggest another downturn, and when will it turn up again? Ask the experts in the tourism industry.

EXHIBIT 7.11
Pressures for
Hotel/Motel
Demand Data for
Washington, D.C.
(DCTOUR): (a) 1/12
Pressures; (b) 3/12
Pressures

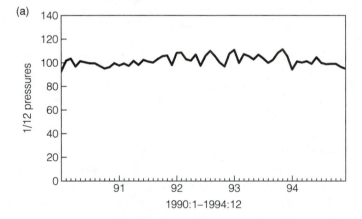

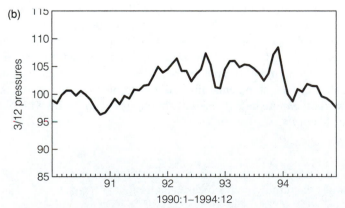

(*Source*: Exhibits 1.4 and 6.17–6.20)

EXHIBIT 7.12
Pressures for
Hotel/Motel
Demand Data for
Washington, D.C.
(DCTOUR): 12/12
Pressures

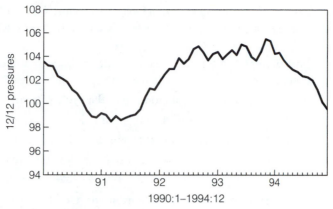

(*Source*: Exhibits 1.4 and 6.17–6.20)

If the timings of the possible turning points are substantially different, we should be more cautious about the future. Pressure analysis is not foolproof and randomness may prevent us from seeing any definite patterns. 1/12, 3/12, and 12/12 are the pressures most used in practice, but we may also want to look at 6/12 and 9/12 pressures when the results are ambiguous. For quarterly data, we would use 1/4, 2/4, and 4/4, and for weekly data use 1/52, 4/52, 13/15, and 52/52 pressures.

Forecasters may want to temper or boost their projections when a turning point appears in the pressures. We should also keep an eye on the trends in pressures; for example, projections of strong growth in sales are less likely when the recent trend in pressures is down.

Ten-Step Procedure for Making a Turning-Point Analysis

The ten-step procedure for making a turning-point analysis (TPA) forecast for cyclical time series is based on and extends the seasonal decomposition methodology presented in Chapter 6. A TPA forecast is partly qualitative in nature and involves some statistical analysis and regression modeling. However, it is primarily intuitive and thus serves as a valuable starting point in establishing future scenarios of a planning forecast.

A turning-point analysis is a systematic approach for establishing forecast scenarios for demand planning.

Here are the steps:

1. Plot the time series.
2. Remove seasonality (seasonal adjustment or differencing).
3. If necessary, remove irregularity with a low-order smoother (e.g., moving average or moving median).
4. Fit a trend line to the series in step 3 and plot deviations from trend. This is a representation of a business cycle. If appropriate, transform the time series

first so that the cycle pattern appears reasonably symmetric about the trend line (i.e., avoid cup-shaped patterns resulting from fitting straight-line trends to highly growing series).

5. Follow steps 1–4 for other national, regional, local, or industry series for a comparison of business-cycle patterns.

6 a. If there is a historic relationship in the cycle patterns:

   - Obtain forecasts of the other variable.

   - Plot these forecasts in terms of deviations from a trend.

   - Forecast the cycle for the series based on the cycle forecast of the economic variable. Take leading or lagging relationships into account.

   b. If there is no historical relationship, develop the cycle forecast based on a pattern analysis that considers: (1) the peak-to-trough or trough-to-peak historical duration (months, quarters), peak-to-peak or trough-to-trough duration; (2) the magnitudes (amplitudes) of peaks or troughs (amount, percent); (3) the slopes of the peaks or troughs (speed of recovery, decline); and (4) anticipated future cyclical patterns based on economic, market, or industry information.

7. Project the trend line of step 4, and add to it the cycle forecast from step 6 to obtain trend-cycle forecasts.

8. To reintroduce seasonality (if desired), add the forecasts of the seasonal factors.

9. If appropriate, retransform the series (e.g., exponentiate, raise it to a power) to the original scale in step 7 or 8 if a transformation was taken in step 4 or 2.

10. Plot the history and forecast together for reasonableness.

Preparing a Cycle Forecast for Revenues

Let us examine some results of the ten-step procedure. Exhibit 7.13 is a plot of a monthly toll revenue series for a 10-year period (step 1); Exhibit 7.13 also shows a plot of the seasonally adjusted series (step 2). Next, a 3-month moving average of

EXHIBIT 7.13
Time Plot of the Original Historical Toll Revenues (TOLL_REV) and of the Seasonally Adjusted Data (the data are seasonally adjusted with X-12-ARIMA)

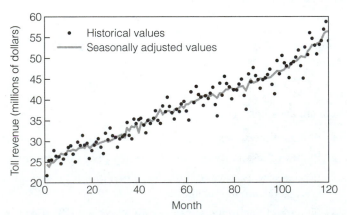

(*Source*: Levenbach and Cleary, 1984)

EXHIBIT 7.14 Time Plot of Residuals from a Straight-Line Fit to the 3-Month Moving Average of the Seasonally Adjusted Toll Revenues; The Deviations from Trend Suggest a Need for a Transformation

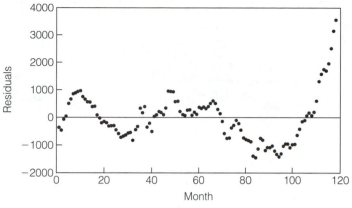

(Exhibit 7.13)

the seasonally adjusted series was calculated (step 3). This smoothes the irregular component and results in a smoother cycle pattern for the later stages of analysis.

Exhibit 7.14 shows the residuals to a trend line fitted to the data from step 3 (step 4). The deviations from trend show a cup-shaped pattern in the later years, indicating the need to transform the series. The peak-to-trough reference dates for the first-year and fifth-year recessions, as determined by the NBER (see Exhibit 7.1), do not correlate well with the two dips below trend, as we might expect. The corresponding regression output, shown in Exhibit 7.15, does not give any insight to this intuitive observation. However, it does substantiate that the statistical fit is satisfactory.

A logarithmic transformation was then taken of the smoothed series, and a straight-line trend was fitted; the deviations from trend are shown as a cycle in Exhibit 7.16. Similar steps were followed for an economic series (nonfarm employment), and the deviations from trend are shown in Exhibit 7.17 (step 5). From the peaks and troughs in Exhibit 7.1 for the nation's economy, it can be seen that employment in the

EXHIBIT 7.15 Summary Regression Output for a Model of a Straight-Line Fit to the 3-Month Moving Average of the Seasonally Adjusted Toll Revenues (df, degrees of freedom; F, F statistic)

Regression Statistics	
Multiple R	0.994
R^2	0.989
Adjusted R^2	0.989
Standard error	875.73
Observations	118

ANOVA

	df	SS	MS	F
Regression	1	7.735E + 09	7.735E + 09	10085.4
Residual	116	88961091	766905.95	
Total	117	7.824E + 09		

	Coefficients	Standard Error	t Statistic
Intercept	24591.152	162.266	151.549
X variable	237.685	2.367	100.426

EXHIBIT 7.16
Deviations from
Trend for
Transformed (with
logarithms),
Smoothed (3-month
moving average),
and Seasonally
Adjusted Toll
Revenues

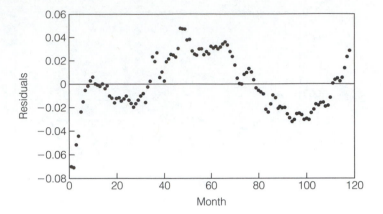

region for which the forecast is being made peaks at approximately the same time as in the rest of the nation but reaches bottom (June 1976) a year or more after the national economy has bottomed (March 1975).

A comparison of the regional employment cycle with the revenue cycle shows similar patterns—especially following the third year. A forecast of the employment series must be obtained (in terms of deviation from trend) to complete the analysis. With this prediction as a starting point, we could then produce three scenarios (optimistic, most likely, and pessimistic) for the revenue cycle that is consistent with the economic outlook reflected in the regional employment variable.

The most likely scenario approximates the relationship that existed between the two series in the past. The optimistic scenario should show a somewhat shallower decline and a more rapid recovery. The pessimistic scenario should project a much sharper decline and more gradual recovery that is more similar to Years 7–9. Because of the steadily worsening economic news in the first quarter of the latest year, the pessimistic scenario is a more likely alternative forecast than the optimistic forecast.

EXHIBIT 7.17
Deviations from a
Straight-Line Trend
Fitted to Nonfarm
Employment Series
(NFRM)

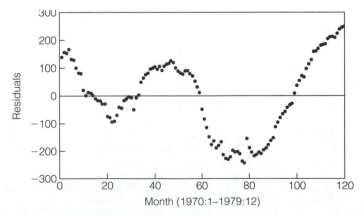

(*Source*: Levenbach and Cleary, 1984)

Finally, to complete the presentation, we show a plot of the history and forecast in terms of a smoothed seasonally adjusted series. The trend line is extrapolated and the cycle forecast is then superimposed on the trend. Because the revenue cycle is created in logarithmic units, this result needs to be exponentiated and plotted. If desired, seasonality could be reintroduced by adding the historical and projected seasonal factors. The irregular component has been smoothed over the historical period and is projected to be zero over the forecast period.

Alternative Approaches to Turning-Point Forecasting

Strong arguments against the use of seasonally adjusted data for econometric regression models have been made (Jenkins, 1979). The objections are related, in part, to the fact that we cannot be quite sure what the statistical properties of the residuals are after we have subjected the series to a seasonal adjustment procedure. This has an impact on the inferences that can be drawn—specifically for establishing probability limits about forecasts. In the cycle forecasting approach, illustrated in the previous sales-revenue example, the subjective nature of the forecast and the intentional omission of prediction intervals recognize that this is a highly subjective approach.

As we see in Chapters 13 and 14, an alternative to modeling seasonally adjusted data is modeling appropriately differenced data. We can also take differences of economic data and correlate the patterns of deviations from trend of the two series, thereby developing cycle forecasts. A final step is to "undifference" the series by adding the predicted differences to the appropriate actual (and later predicted) values of the revenue series.

Despite the subjective nature of the turning-point forecast, it is intuitively appealing to a nontechnical audience, a consideration that cannot be dismissed. Even if not used to establish the final forecast, a turning-point forecast is an effective way of presenting a strategic forecast to higher management. In this instance, the deviations from trend can be modeled instead of subjectively projected.

7.3 USING ELASTICITIES

Two important determinants of a firm's profitability—indeed, its survival—are cost and the demand for its products or services. Demand must exist or be created if the business is to survive. It must also be high enough at least to cover fixed costs. Because of its key role, all business-planning activities require a careful analysis of demand over time. Business forecasters are also concerned with the relationship between the quantity demanded and price (Taylor, 1980).

Forecasters play an important role in helping to make pricing decisions by estimating price elasticities for products and services with their models.

Determinants of Demand

Economists have long attempted to determine what causes people to behave as they do in the marketplace. Over the years, one aspect of this research has evolved into a theory of demand. In theory, demand expresses the inverse relationship between price and quantity; it shows the maximum amount of money consumers are willing and able to pay for each additional unit of some commodity or the maximum amount of the commodity they are willing and able to purchase at a given price. There may not be enough of the commodity available to satisfy the demand. Economists concern themselves not with a single item purchased by members of a group (a market) but rather with a continuous flow of purchases by that group. Therefore, demand is expressed in terms of the amount desired per day, per month, or per year.

There are a number of determinants of demand. The demand for international holiday tourism, for instance, is known to depend on a number of factors, including (1) the origin population (the higher the number of people resident in a country, the greater the number of trips taken abroad), (2) the origin country real income and personal disposable income, (3) the cost of travel to the destination and the cost of living for the tourist in the destination, and (4) a relative price index relating substitution between tourist visits to a foreign destination and domestic tourism (Witt and Witt, 1995).

Demand varies with tastes, total market size, average income, the distribution of income, the price of the good or service, and the prices of competing and complementary goods.

The Price Elasticity

In analyzing the demand for gasoline in a market, the quantity demanded can be measured by gasoline consumption (gross expenditure divided by the price index). Exhibit 7.18 shows the consumption of gasoline in the U.S. market from 1960 to

EXHIBIT 7.18 Time Plot of Gasoline Consumption in the United States, 1960–1995

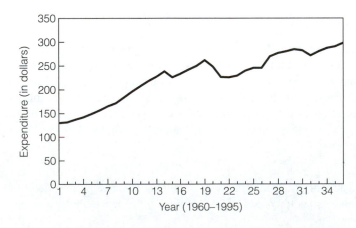

1995. We note a steady rise in consumption through the 1960s as gasoline prices fell in real terms and incomes rose. The dips in years 15, 20–21, and 31 are due to OPEC formation, the Iranian revolution, and the war in Iraq, respectively. Slow recovery in the early 1980s occurred following a severe recession in the U.S. economy. The market appears to have become very unstable since 1974, compared to the 1960s. Some of the variables that may determine the consumption of gasoline include (1) a price index for gasoline, (2) per capita income, (3) a price index for new cars, (4) a price index for used cars, (5) a price index for public transportation, (6) a general price index for consumer durables, (7) a general price index for consumer nondurables, and (8) a general price index for consumer services.

The quantity demanded for a product will increase as the price of the product decreases, all other determinants held constant (known as the *ceteris paribus* condition). As price falls, a product becomes cheaper relative to its substitutes, and thus it becomes easier for the product to compete for the consumer's dollar. This relationship is known as the demand curve.

The forecast of the quantity demanded at various price levels is important to businesses because it permits them to maximize profitability by considering price and cost trade-offs.

In Chapter 4, we display the weekly shipments of a canned beverage (Exhibit 4.11). These shipments are related to the price of a can of the beverage for that week, which shows a general increase in shipments with declining prices. The missing element is elasticity, which explains the responsiveness of changes in demand to changes in price or any of the other variables.

Most often we make note of a price elasticity, defined as the percentage change in the quantity demanded Q as a result of a given percentage change in price P:

$$\frac{\text{Percentage change in } Q}{\text{Percentage change in } P} = \frac{\Delta Q/Q}{\Delta P/P}$$

An important condition in the definition of elasticity is that all factors influencing demand other than own-price (price of the item under consideration) are held constant while own-price is varied (*ceteris paribus* condition). In general, price elasticity is determined by at least four factors:

- Whether or not the good is a necessity
- The number and price of close substitutes
- The proportion of the budget devoted to the item
- The length of time the price change remains in effect

If the product or service is a necessity, its demand will be inelastic; consumers will pay any reasonable price for a necessity. Lack of substitutes for a product will also cause demand to be inelastic. If substitutes are available, consumers will switch their purchases to those substitutes that have not increased in price. If the proportion of income spent for a good is small, price changes may not have too great an impact on the demand for a good. If the proportion of the income is large, price increases will cause postponements in demand or reductions in the quantity demanded.

 If a good is both a necessity and without a substitute, demand will tend to be very inelastic.

The longer a price change remains in effect, the more elastic the demand for a product. Consumers become aware of price changes and adjust their consumption habits to the new circumstances. Elastic as used here is a relative term; demand becomes more elastic as time goes by, but it could still be inelastic (i.e., less than unity). Other factors influencing price elasticity are the frequency of purchases and the presence or absence of complements (e.g., automobiles and gasoline). Frequently purchased and relatively inexpensive products may be more inelastic than infrequently purchased expensive items.

Price Elasticity and Revenue

Because revenue equals unit price multiplied by quantity demanded, a price change can result in an increase, no change, or a decrease in total revenue (Exhibit 7.19). Let us

EXHIBIT 7.19 The Relationship of Price to Revenue: (a) Table, (b) Unitary, (c) Inelastic, (d) Elastic

(a)

Behavior of model	Unitary	Inelastic	Elastic
Price rise	Total revenue remains the same	Total revenue increases	Total revenue decreases
Price decline	Total revenue remains the same	Total revenue decreases	Total revenue increases
Gain versus loss	Gain = Loss	Gain > Loss	Gain < Loss

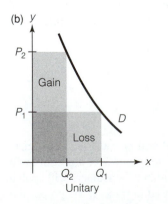

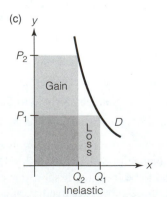

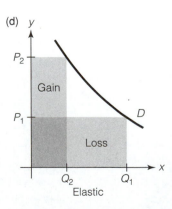

consider first the impact of own-price elasticity on revenue by means of an example, deferring the impact of cross-elasticity until later. Suppose a forecast for a service predicts $1000 in revenues for a particular future year. Assume the own-price elasticity E is -0.2. If P = the existing price, Q = the forecast of quantity demanded at the existing price, and R = the forecast of revenue at the existing price, then $R = P \times Q = \$1000$. What will be the impact of a 10% price increase in the service under consideration, effective at the start of the future year?

The new price is clearly just $1.10P$, but the quantity demanded will be somewhat less than before. With $E = -0.2$, a 10% price increase will result in approximately a 2% loss. Hence, the new quantity demanded will be approximately $0.98Q$, and the revenue R' after the price change will be

$$R' = (1.10P)(0.98Q) = 1.08R$$

Thus, a 10% price increase will increase revenues by 8%, in this example. In general, the own-price elasticity effect will cause an $X\%$ price increase to yield less than an $X\%$ increase in revenues. The reprice value of a rate or price increase is the incremental revenue that would result if there were no demand reaction; it is $100 in the example. The amount by which revenues fall short of the reprice value, $20 here, is called revenue repression. A revenue repression factor may be defined as the ratio of the revenue repression to the reprice value; in this example, it is 0.2. An elasticity of -0.2 implies a minimum revenue repression factor of 0.2—that is, at least $20 for a price increase worth $100 on a reprice basis. The revenue repression factor often goes up with the magnitude of the price change.

 Price elasticities can be used to show how price changes can affect total revenue.

To continue with the $1000 service example, suppose it has been determined that this service must produce $1100 to enable the company to reach its profit objective. A 10% price increase will not be sufficient, owing to the revenue repression of $20 (revenue repression factor 0.2). Evidently, to net $100 in incremental revenues, almost a 12.5% increase in price will be necessary. The reprice value ($1000 $\times$ 1.125) of such an increase is $1125. A repression of $25 (0.2 $\times$ $125) leaves a net of $1100.

Cross-Elasticity

The availability or prices of other goods or services frequently affect a service, in addition to its own price. The cross-elasticity measures the percentage change in the demand for good A as a result of a given percentage change in the price of good B. If a good has a close substitute, a price increase for one will create increased demand for the other. When airfares increase, for instance, travelers may find it more desirable to travel by train, bus, or automobile. When products are complementary (used together), a decrease in the price of one will lead to an increase in demand for both products. For example, in the travel industry, close destinations, such as Australia and New Zealand, are complementary in that package tours frequently package such destinations together at more attractive prices than the destinations individually. The cross-elasticity is positive for substitutes and negative for complementary goods. As before, this is a dimensionless number because it is defined in terms of percentage changes.

A cross-elasticity coefficient is a measure of the interaction effect between the price of a good and the prices of other goods.

Other Demand Elasticities

Our attention so far has been concentrated on price elasticity. However, there is elasticity associated with each independent variable in the demand function. Income is another determinant of demand that also receives attention. For most goods and services, we expect a positive relationship between demand and income. Income elasticity is the percentage change in demand divided by the percentage change in income. However, the elasticity for normal goods is now positive instead of negative. If the demand for a good is income-inelastic, the increase in demand will not be proportional to the percentage increase in income. As national income rises, for instance, a business firm will not experience a proportional growth in revenues, and its share of national income will decline.

Income elasticity is a two-edged sword. If the economy contracts and income declines, the revenues of an income-inelastic firm will shrink less than the revenues for an income-elastic firm. The firms whose goods are income-elastic are more concerned with anticipating business-cycle expansions and contractions.

Estimating Elasticities

In forecasting retail sales of durable goods (plant equipment and durable machinery, automobiles, etc.), we might expect monthly retail sales to be very sensitive to changes in the gross earnings of workers and to the **Consumer Price Index (CPI)** for durable goods. If earnings rise, we expect retail sales to increase. On the other hand, if the CPI rises, we expect retail sales to fall. Because the explanatory variables in a demand function can change simultaneously, a general model for the retail sales of product Y may take the form:

$$Q_Y = Q(P_Y, I, P_X, A, B, \varepsilon)$$

where the quantity demanded Q_Y is a function of the (deflated) price P_Y of the product, income I, the price P_X of a competing (complementary) product, the market potential A, advertising B, any number of other variables, and a random error term ε.

Demand theory does not help us with the specific form for Q. However, several forms appear to give a useful interpretation of the elasticity. In the application of demand models, the additive and multiplicative forms are the most commonly used.

Elasticities can be derived for either model, but the multiplicative form gives a particularly simple result. A multiplicative demand model takes the form:

$$Q_Y = \beta_0(P_Y)^{\beta_1} (I)^{\beta_2} (P_X)^{\beta_3} (A)^{\beta_4} (B)^{\beta_5} e^{\varepsilon}$$

where the β_i are parameters and e^{ε} is a multiplicative error term. The estimation of these parameters from data is performed by regression analysis, which is treated in Chapters 11 and 12. The important point to note here is that the price elasticity (the partial derivative of $\ln Q_Y$ with respect to $\ln P_Y$) is the constant β_1. Hence, it is called the constant elasticity model.

 Elasticities can be estimated from additive and multiplicative demand models.

The theory can be applied to determine the responsiveness of tax revenues (TAX_REV) to changes in either the tax rate (RATE) or the tax base (BASE) in a state's budget. For example, if a (hypothetical) model takes the form:

$$\log \text{TAX\_REV} = -3.8 - 0.6 \log \text{RATE} + 1.4 \log \text{BASE}$$

a suitably qualified statement could be made that says: (1) a 1% increase in the tax rate will lead to a 0.6% decrease in tax revenues, and (2) a 1% increase in the taxable base, retail sales, will increase revenues by 1.4%.

Getting back to the previous consumption of gasoline example, Exhibit 7.20 shows the results of an analysis for the parameters in a constant elasticity model. The coefficients are elasticities. For example, the price elasticity of demand is estimated to be -0.518 and the income elasticity is 1.222. All else being equal, we expect consumption to increase 0.89% per year. Because the coefficients for the price indexes for new automobiles are less than zero, these products are complementary with gasoline. The elasticity for consumer durables as a group are closer to unity than the elasticities for a more narrowly defined category of new automobiles, which is consistent with economic theory.

To distinguish long-term from short-term elasticities, we can construct a constant elasticity model with a lagged term for the **dependent variable** gasoline consumption. Using multiple linear regression analysis, the resultant coefficient

EXHIBIT 7.20
Elasticities of a
Constant Elasticity
Model for the
Consumption of
Gasoline in the
United States,
1960–1995

Constant	−23.34
log Price index for gasoline	−0.518
log Per-capita Income	1.222
Year (time trend)	0.009
log Price index for new cars	0.038
log Price index for used cars	−0.119
log Price index for public transportation	0.125
log General price index for consumer durables	1.058
log General price index for consumer nondurables	1.149
log General price index for consumer services	−1.362

EXHIBIT 7.21	Constant	−23.314
Determining Short-	log Price index for gasoline (X_1)	−0.402
and Long-Term	log Per-capita income (X_2)	0.597
Elasticities for the	Year (time trend)	0.011
Consumption of	log Price index for new cars	−0.205
Gasoline in the	log Price index for used cars	−0.017
United States,	log Price index for public transportation	0.101
1960–1995	log General price index for consumer durables	0.588
	log General price index for consumer nondurables	1.146
	log General price index for consumer services	−0.950
	log Consumption of gasoline lagged one period (X_3)	0.383

estimates are shown in Exhibit 7.21 for this data set. This conforms with the theory that demand becomes more elastic as time goes by; but it could still be inelastic, that is, less than unity. With this formulation, the elasticities are as shown in the next table.

	Price	Income
Short-term	Coeff of $X_1 = -0.402$	Coeff of $X_2 = 0.597$
Long-term	$(1 - \text{Coeff of } X_3)/\text{Coeff of } X_1 = -0.651$	$(1 - \text{Coeff of } X_3)/\text{Coeff of } X_2 = 0.967$

7.4   ECONOMETRICS AND BUSINESS FORECASTING

Widespread use has been made of econometric techniques for forecasting business demand. In many business applications, it may be desirable to achieve certain targets to meet particular objectives. By employing an econometric model, we can simulate various alternative inputs (our controlling variables) to find the input that best achieves the desired target. The degree of accuracy will be directly dependent on the formulated model and the appropriate use of econometric techniques to estimate the model. Consulting firms, government agencies, and academic institutions provide the business community with a plethora of economic forecasts from sophisticated, computerized econometric systems. These forecasts are an important asset to the decision maker in a rapidly changing economic environment.

Although there is a widespread use of forecasts from surveys and econometric systems, there does not appear to be a universal acceptance that econometric forecasting systems have produced consistently reliable and defensibly accurate forecasts over the past 3 decades (see, for example, Armstrong, 1978; Granger and Newbold, 1986; McNees and Ries, 1983). In 1946, the late columnist Joseph Livingston started a survey of professional forecasters, known as the Livingston Survey, which has been conducted by the Federal Reserve Bank of Philadelphia since 1990 (http://www.phil.frb.org/econ/liv/). There are often simpler approaches yielding projections that are more accurate. Nevertheless, a vital role of econometric forecasting is to provide an economic rationale for the process along with the mechanism for producing forecasts.

Uses of Econometric Models

Structural Analysis Models

Structural analysis is the process of deriving information about an underlying economic relationship through the specifications of a mathematical model. Parameters in the model have special meaning to the investigator. For example, in the analysis for the demand of a product or service, a product manager requires estimates of the price elasticities, which explain the responsiveness of changes in demand to changes in prices. For certain demand functions, these price elasticities are found to be functions of the parameters in the model, and these parameters are estimated using relevant historical or cross-sectional data. Generally, attention is focused on the economic and theoretical aspects of the mathematical model to approximate the relationship between demand and prices rather than on the statistical niceties of the model.

Structural models relate underlying economic relationships in terms of a mathematical model.

Models formulated for structural analysis can be very simple or quite complex and typically include many simultaneous equations. In general, it is possible and frequently necessary to segment a complex model so that we can extract the most essential information about the output from a complex maze of input activity.

Policy Analysis Models

Policy analysis models are formulated for structural activity or forecasting and can be used to make scenarios using alternative assumptions about exogenous factors influencing the system. Thus, policy analysis asks "what-if" questions about the future state of a system or about alternative conditions in the system's recent past. The econometric model provides an approximation of the general structure of the system within the sample period of the data.

Policy models establish "what-if" questions about the future state of a system in a mathematical model.

Static versus Dynamic Models

When using macroeconomic models, the estimation procedure may produce a static or dynamic response in the endogenous variables. If the model is static, we assume that the exogenous variables in the system impact the endogenous variables at the same time. This assumption may be adequate when the periods measured span large periods, such as a year. On the other hand, a static model can be expected to be less accurate when data are monthly or quarterly.

In econometrics, variables determined within the system are called endogenous and those determined outside the system are called exogenous.

In general, we may expect the impact of the exogenous variables on the endogenous variables to be dynamic, that is, for the effects to be distributed over time. The appropriate distribution may not be known a priori, and thus we will require the use of econometric distributed-lag estimation procedures. When we need to analyze short-term changes in a system, the inclusion of a dynamic element may be well worth the effort.

Types of Econometric Models

We can also classify econometric models by how they use data, ranging from time series regressions to simultaneous equations. In a time series regression model, an econometric relationship is expressed by predicting a dependent variable from the knowledge of one or more related independent variables. Within this context, the models use time series data with a static or dynamic mathematical specification (Swanson and White, 1997).

Alternatively, there are models that employ cross-sectional data. For example, in modeling the per-capita consumption of electricity for a country, we may consider pooling annual time series available for 20 countries for 5 years. We can build 20 time series regression models, five cross-sectional models, or one pooled time-series cross-sectional model. The pooled model integrates the two types of data by pooling or grouping together the individual time series for each of the cross sections. The sample data from the different cross sections may be at the same or different time periods. What is relevant is the sampling scheme, and the analysis is done without regard to the chronological dimension. By combining each time series with the cross section of other systems, we are effectively pooling information to be used for the estimation process. Generally, we prefer to use as much information as possible for estimation purposes, so pooling the data can satisfy this need.

Econometric models with more than one equation can be as simple as sets of single-equation models with little or no relationship among them; each is explained by exogenous information. Other models are more complex and simultaneous, in that some endogenous variables are jointly determined by other variables within the system, thus requiring more complex estimation procedures. Difficulties in estimating such models can be overcome when the number of equations is kept small or when the system can be grouped into several independent recursive systems.

A Recursive System

The flowchart in Exhibit 7.22 represents a recursive model for telephone access-line gain in a geographic area. Each individual regression model is built sequentially, starting at the bottom of Exhibit 7.22 until all the regressions have been run. The predictions from the previous equations are available as forecasts for the exogenous variables

EXHIBIT 7.22
A Recursive Econo-
metric Model for
Telephone Access-
Line Gain in a Region
(R, regional; N,
national)

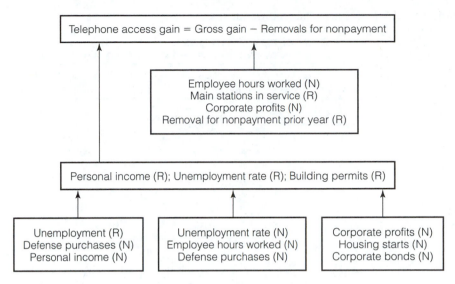

in later equations. In this example, the exogenous variables include regional, national, economic, and telephone data. This is an econometric model in that it contains a set of interrelated equations based on a specification of economic relationships.

The net change in access lines is assumed to be a function of personal income in a region, regional unemployment rate, and building permits issued in the region. Removals for nonpayment are considered to be a function of employee hours worked in the nation, access lines in service (the potential number of lines that could be terminated), national corporate profits, and the number of removals for nonpayment in the previous year (a lagged variable).

Each of the variables that determine the gross gain in access lines is a function of other variables. Personal income is a function of the unemployment rate, national defense purchases, and national personal income. The regional unemployment rate is a function of the national unemployment rate and employee hours worked and defense purchases. Regional building permits are assumed to be a function of national corporate profits in constant dollars, national housing starts, and the interest rate.

The underlying theory is that the national economy drives the country and its various regions; therefore, the economic forecasts for the region are related to the economic forecasts for the whole country. The demand for telephone access lines is then related to these forecasts of economic activity. The regression models establish the linkages between the national economy and the region and, in turn, the regional economy and the demand for access lines.

Some Pitfalls

The success of an econometric modeling effort lies in a number of factors related to quality of data, sensitivity of model specifications, and appropriateness of estimation methodology. Measurement error and representativeness affect the quality

EXHIBIT 7.23
Pressures Analysis of
Business Cycles
(Mon, month; Per,
period)

FILENAME: DATAANAL.XLS
WORKSHEET: CYCLES
ANALYSIS OF BUSINESS CYCLES
TITLE2:   NEW ORDERS FOR METALWORKING MACHINERY

Year	Mon	Per	Actual data	1/12 Ratio	3-mon. Total	3/12 Total	12-mon. Total	12/12 Ratio
1972	1	1	226					
	2	2	284					
	3	3	331		841			
	4	4	292		907			
	5	5	301		924			
	6	6	336		929			
	7	7	315		952			
	8	8	277		928			
	9	9	332		924			
	10	10	314		923			
	11	11	335		981			
	12	12	362		1011		3705	
1973	1	13	370	163.7%	1067		3849	
	2	14	407	143.3%	1139		3972	
	3	15	498	150.5%	1275	151.6%	4139	
	4	16	411	140.8%	1316	145.1%	4258	
	5	17	406	134.9%	1315	142.3%	4363	
	6	18	421	125.3%	1238	133.3%	4448	
	7	19	387	122.9%	1214	127.5%	4520	
	8	20	382	137.9%	1190	128.2%	4625	
	9	21	393	118.4%	1162	125.8%	4686	
	10	22	487	155.1%	1262	136.7%	4859	
	11	23	423	126.3%	1303	132.8%	4947	
	12	24	500	138.1%	1410	139.5%	5085	137.2%
1974	1	25	375	101.4%	1298	121.6%	5090	132.2%
	2	26	517	127.0%	1392	122.2%	5200	130.9%
	3	27	628	126.1%	1520	119.2%	5330	128.8%
	4	28	581	141.4%	1726	131.2%	5500	129.2%
	5	29	573	141.1%	1782	135.5%	5667	129.9%
	6	30	589	139.9%	1743	140.8%	5835	131.2%
	7	31	524	135.4%	1686	138.9%	5972	132.1%
	8	32	519	135.9%	1632	137.1%	6109	132.1%
	9	33	690	175.6%	1733	149.1%	6406	136.7%

of data. Because data can be experimental or historical, the investigator cannot always directly control their quality. In most econometric applications, historical data are gathered at various government or corporate sources where accuracy considerations cannot be effectively controlled. Most data go through various levels of aggregation, stages of revision and possible redefinition, so validity cannot always be assumed.

In the model-building process, the log-linear formulation has special appeal and is very often used for good reasons. As we have seen in Section 7.3, elasticity coefficients are constant and thus are easy to interpret. Empirical analyses do not always support a constant elasticity assumption. At the same time, data limitations can preclude the inclusion of a dynamic description of elasticities. Thus, a forecaster may inappropriately estimate and project constant elasticity effect.

Econometric estimation techniques can be very complex and not readily comprehended without a strong mathematical and statistical background. Any methodology has its limitations, and these methods are also prone to unusual variation in the data and can lead to imprecise and misleading parameter estimates. Even sophisticated methods cannot overcome basic data problems. Users are well advised to complement complex estimation methods with simpler, more widely understood techniques as a hedge against losing credibility. Sometimes a simple analysis can balance the overbearing importance that a sophisticated analysis can have on a management audience.

Last, the needs of business forecasters, their users, and their management can introduce unexpected constraints on the success of a complex project. Each has special interests and constraints, which may work to the detriment of all parties involved. Time constraints, budget cuts, and technical specializations have ways of affecting best intentions and project designs. If econometric modeling objectives and expectations are clearly articulated, the chances of obtaining disappointing results can be minimized.

7.5 USING PRESSURES TO ANALYZE BUSINESS CYCLES

Consider Exhibit 7.23. Turning points are rarely obvious in a time plot of raw data until some time has passed. To illustrate, Exhibit 7.24 shows a time plot of new orders for the metalworking machinery industry in millions of dollars. The orders were logged during the period 1972–1974, a period of great economic uncertainty; the economy peaked in November 1973 and then gradually declined through the

EXHIBIT 7.24 Time Plot of New Orders for Metalworking Machinery, 1972–1974.

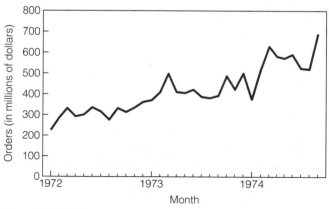

(*Source*: Exhibit 7.23)

official trough month of the recession, February 1975. New orders for metalworking machinery were essentially flat from March 1973 through January 1974. In February and March 1974, orders increased but fell off again from April through August. Suppose we are forecasting in this industry during the first half of 1974. How should we interpret the behavior of new orders? Will growth pick up again? Has a turning point occurred?

1/12 Pressures

Exhibit 7.25 is a graph of index numbers that compare orders for a given month to the same month the previous year. The first point plotted is [(Jan. 1973 orders) / (Jan. 1972 orders)] · 100. The second point plotted is [(Feb. 1973 orders) / (Feb. 1972 orders)] · 100. Multiplying each ratio by 100 forms an index number that is easy to interpret. If the index for a given month

- =100, orders for that month are unchanged from the previous year
- >100, orders have grown from the previous year
- <100, orders have declined from the previous year

Such index numbers are called 1/12 pressures. The number 1 means that the index numbers are based on monthly totals and the number 12 means that the totals are separated by 12 months. The 1/12 pressures decline in an erratic pattern during 1973. In January 1974, a turning point occurred when the 1/12 pressures bottomed out at near 100%. Starting in February, growth picked up again, reaching a pressure level of more than 175 by September.

3/12 Pressures

By including more data in the totals used in the pressure calculations, we can get additional evidence to verify the turning points. In general, the more data used in the totals, the more reliable the results. Exhibit 7.26 shows 3/12 pressures, consecutive 3-month totals compared to the same totals a year before. The first point plotted is

EXHIBIT 7.25 Plot of 1/12 Pressures of New Orders for Metalworking Machinery, 1972–1974

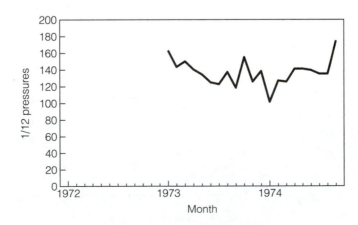

[(Jan.–Mar. 1973 orders) / (Jan.–Mar. 1972 orders)] · 100. The second point plotted is [(Feb.–Apr. 1973 orders) / (Feb.–Apr. 1972 orders)] · 100.

12/12 Pressures

Exhibit 7.27 shows 12/12 pressures, consecutive 12-month totals compared to the same totals a year before. The first point plotted is [(Jan.–Dec. 1973 orders) / (Jan.–Dec. 1972 orders)] · 100. The second point plotted is [(Feb. 1973–Jan. 1974 orders) / (Feb. 1972–Jan. 1973 orders)] · 100.

Exhibits 7.26 and 7.27 show that both the 3/12 and 12/12 pressures mark a turning point in March 1974. Because of the randomness in the data, the 1/12 pressures seem to disagree with the other two about the timing of the turning point. So long as the timings of the turning points based on different pressures are close, we can have some confidence that things will get better (or get worse if we see a turning point at the top of a cycle).

EXHIBIT 7.26 Plot of 3/12 Pressures of New Orders for Metalworking Machinery, 1972–1974

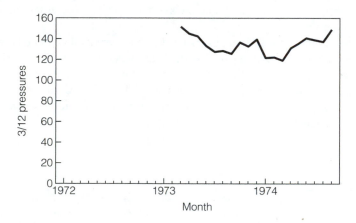

EXHIBIT 7.27 Plot of 12/12 Pressures of New Orders for Metalworking Machinery, 1972–1974

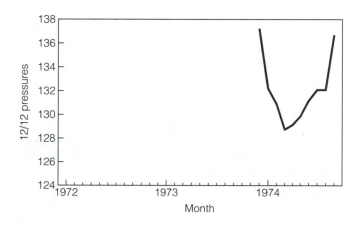

SUMMARY

Simple approaches to obtain cycle forecasts are based on the decomposition of a time series into trend, cyclical, seasonal, and irregular components. Seasonally adjusted data are required to help identify the trend-cycle component. The trend-cycle component is correlated with economic indicators to relate turning points in the economy with the demand for products and services. Forecasted cycles can be recombined in a TPA to develop a forecast for the time series under study.

The underlying assumptions behind an econometric system include that economic behavior can be described by a system of mathematical equations, the representation will capture the essential features of the economic relationships, future values can be obtained for predictive purposes, and, as a strategic tool, varying input variables can develop alternative economic scenarios. The description of economic behavior with systems of mathematical equations has prompted forecasters to take considerable interest in the field of econometrics, has led to the extensive use of econometric systems in forecasting and planning models, and has contributed to the study of how alternative economic assumptions affect business planning.

Forecasters and managers need to be aware of the strengths and weaknesses of econometric modeling so that they can make the soundest decisions possible. This chapter has provided an overview of methods, uses, and pitfalls of macroeconomic analysis and econometrics for managerial uses. With this background, a manager will possess the basic knowledge to offer support and direction to the econometric forecasting effort.

REFERENCES

Allen, P. G. (1994). Economic forecasting in agriculture. *Int. J. Forecasting* 10, 81–135. (Winner of 1994/95 Best Paper award in the *IJF*.)

Armstrong, J. S. (1978). Forecasting with econometric methods: Folklore versus fact. *J. Business* 51, 549–64.

Burns, A. F., and W. C. Mitchell (1946). *Measuring Business Cycles*. New York: National Bureau of Economic Research.

DeLurgio, S. (1998). *Forecasting Principles and Applications*. New York: Irwin/McGraw-Hill.

Dielman, T. E. (1996). *Applied Regression Analysis for Business and Economics*. 2nd ed. Belmont, CA: Wadsworth.

Granger, C. W. J., and P. S. Newbold (1986). *Forecasting Economic Time Series*. 2nd ed. New York: Academic Press.

Hanke, J. E., and A. G. Reitsch (1998). *Business Forecasting*. 6th ed. Englewood Cliffs, NJ: Prentice Hall.

Jenkins, G. M. (1979). *Practical Experience with Modeling and Forecasting Time Series*. Jersey, UK: GJ&P (Overseas) Ltd.

Levenbach, H., and J. P. Cleary (1984). *The Modern Forecaster*. Belmont, CA: Wadsworth Press.

McNees, S. K., and J. Ries (1983). The track record of macroeconomic forecasts. *N. Engl. Econ. Rev.* Sept./Oct., 10.

Moore, G. H., and J. Shiskin (1972). Early warning signals for the economy in statistics. In *Statistics: A Guide to the Unknown*, edited by J. M. Tanur et al. San Francisco: Holden-Day.

Moore, G. H., E. A. Boehm, and A. Banerji (1994). Using economic indicators to reduce risk in stock market investments. *Int. J. Forecasting* 10, 405–17.

Pankratz, A. (1991). *Forecasting with Dynamic Regression Models*. New York: John Wiley & Sons.

Shiskin, J., and G. H. Moore (1967). *Indicators of Business Expansions and Contractions*.

Cambridge, MA: National Bureau of Economic Research.

Swanson, N. R., and H. White (1997). Forecasting economic time series using flexible versus fixed specifications and linear versus nonlinear econometric models. *Int. J. Forecasting* 13, 439–61. (Winner of 1996/97 Best Paper award in the *IJF*.)

Taylor, L. D. (1980). *Telecommunications Demand: A Survey and Critique*. Cambridge, MA: Ballinger Press.

Wilson, J. H., and B. Keating (1998). *Business Forecasting*. New York: Irwin McGraw-Hill.

Witt, S. F., and C. A. Witt (1995). Forecasting tourism demand: A review of empirical research. *Int. J. Forecasting* 11, 447–75.

PROBLEMS

7.1. In a constant elasticity model, what do you expect the signs of the parameters to be for:

   a. Price P

   b. Income I

7.2. For an additive demand model $Q_Y = \beta_0 + \beta_1 P_Y + \beta_2 I + \beta_3 P_X + \beta_4 A + \beta_5 B + \varepsilon$, the point elasticity for price is given by $\beta_1 P_Y / Q_Y$. For a simplified fitted model $Q_Y = b_0 - b_1 P_Y$, derive an expression for the price elasticity of demand from the formula $\{-\Delta Q/Q_Y\}/\{\Delta P/P_Y\}$, where $\Delta Q = Q_t - Q_{t-1}$ and $\Delta P = P_t - P_{t-1}$.

7.3. In a model for revenues $R = QP$, assume a multiplicative demand model for Q and derive the point elasticity of revenue for:

   a. Price P, $\quad E_P = \{-\Delta R/R\}/\{\Delta P/P\}$

   b. Income I, $\quad E_I = \{\Delta R/R\}/\{\Delta I/I\}$

7.4. For the following steel-consuming industries, suggest some economic time series a business analyst would want to consider for forecasting purposes: (a) railroads, (b) oil and gas consumption, (c) automotive, (d) mining, (e) shipbuilding, (f) farm machinery, (g) appliances, and (h) furniture.

7.5. Both automobiles and beer are subject to intense advertising. Discuss the relative impact of advertising on sales volatility and market share in each industry over the course of a business cycle.

7.6. An aircraft engine company with multinational sales is making a forecast for jet-engine sales.

   a. Set up the economic, political, and technological assumptions the company forecaster should consider in order to prepare the study.

   b. List the types of supply and demand data the forecasting group might be interested in using.

   c. Suggest a number of economic indicators that may be relevant in the study for correlational analysis.

   d. Without using formal statistical calculations, how would you effectively display such correlations as graphs?

7.7. A revenue forecaster is preparing a recommendation as to the most productive way to increase tax revenues to balance the budget (Bails, D. G., and L. C. Peppers (1982). *Business Fluctuations*. Englewood Cliffs, NJ: Prentice Hall, Table 7-2). The sales tax revenues in millions of dollars (Y) for a 12-year period is analyzed against the sales tax rate (X_1) and retail sales (X_2) in millions of dollars. The final equation is

$$\log Y = -3.841 - 0.562 \log X_1 + 1.352 \log X_2$$

   a. What would a 1% increase in the tax rate lead to for tax revenues?

   b. What would a 1% decrease in retail sales lead to for tax revenues?

   c. Can the state government rely on sales tax increases as a means of raising taxes based on this model?

7.8. For the U.S. housing permits data (Pankratz, 1991, Series 9), perform a business-cycle analysis using the CYCLES worksheet (see the CD). What are your interpretations of the timing of turning point(s) in the historical period 1947:I–1967:II? Because the data are quarterly, how do you need to modify the CYCLES worksheet?

7.9. Repeat Problem 7.8 for the quarterly time series:

   a. Consumer Confidence Index (Conference Board), shown in Exhibit 1.6

b. Domestic car sales (DCS, Wilson and Keating, 1998, Table 1-5)

c. Sales of The GAP (GAP, Wilson and Keating, 1998, p. 30)

7.10. Repeat Problem 7.8 for the monthly time series:

a. Orders of U.S. capital equipment data (PLTEQP20, DeLurgio, 1998, Table 14-7)

b. Index of industrial production (INDPRO47, DeLurgio, 1998, Table 14-2)

c. Unemployment rate data (UNEMP, Dielman, 1996, Table 4.7)

d. Formal-wear rental sales data (Hanke and Reitsch, 1998, Case Study 2.2)

e. New housing units data, shown in Exhibit 1.8

f. Consumer Confidence Index (SRC, University of Michigan), shown in Exhibit 4.5

CASE 7A PRODUCTION OF ICE CREAM (CASES 1A, 3A, 4A, 5A, AND 6A CONT.)

Through a seasonal decomposition, the entrepreneur was advised of a business cycle in the shipments data. Your suggestion is to analyze the business cycle with the Pressures spreadsheet model DATAANAL.XLS (see CD).

(1) Can you identify any turning points suggestive of past recessionary periods?

(2) In preparation for a more sophisticated modeling exercise, you offer the entrepreneur an intuitive TPA forecast. If the past can be a lesson for the future, what do you suggest might happen in a future business cycle?

An industry report suggests that changes in relative prices and income are the most important factors influencing annual changes in per-capita consumption of ice cream and dairy products.

(3) In addition to income-related variables, which other socioeconomic and demographic variables should be considered in modeling the demand for ice cream?

(4) Are these factors likely to have a short-term or long-term effect?

(5) What should be some good sources for information on the frozen dairy industry?

CASE 7B TOP-DOWN FORECAST OF INTEGRATED CIRCUIT PRODUCTION

In 1947, the vacuum-tube age of electronics technology came to a rapid close with the invention of the transistor at Bell Laboratories. This was the beginning of the integrated circuit (IC) age. The sphere of applications of IC technology has become very broad, including consumer goods such as automobiles, television sets, washing machines, and microwave ovens. Other areas of application include communications equipment, computers, the military, and spacecraft. The large-scale expansion of the IC market in Japan began in the latter half of the 1960s, when the IC began to be used in calculators, watches, television sets, audio equipment, and other goods. However, the IC's range of application extends far beyond these few examples. Consequently, to forecast the demand for ICs is a complex process impacted by changes in the economy, exchange rates, technological innovation, changes in finished goods and their uses, trade friction, and globalization.

To get an overall sense of the market trends, your VP of Strategic Planning requests a top-down view of the trends in IC production. In the table that follows, you are given some data from a much earlier period to analyze (EIAJ, Electrical Industry Association of Japan). The data represent annual IC production (in thousands of Japanese yen) for the electronics market over a 10-year period, 1978–1987.

		1978	1979	1980	1981	1982	1983	1984	1985	1986	1987
Consumer		2185	2289	2932	3668	3506	3834	4719	4912	4435	3949
	Video	919	1013	1354	1882	2001	2335	3069	3210	2887	2561
	Audio	1266	1276	1578	1786	1505	1499	1650	1702	1548	1378
Industrial		2372	2703	3069	3412	3891	4601	6112	6994	7425	8151
	Wire communication	504	571	602	705	801	939	1156	1326	1339	1557
	Radio communication	307	305	380	446	480	537	599	639	663	739
	Computers	1066	1290	1465	1650	1949	2391	3468	4021	4559	5021
	Measurement equipment	224	266	315	359	394	425	514	582	482	481
	Office equipment	271	271	307	252	267	309	375	426	382	353
Components		1874	2098	2677	3333	3555	4329	6064	6027	5918	6147
	Parts	1052	1164	1450	1797	1921	2273	2851	2913	2916	2975
	Active devices	822	934	1227	1502	1591	2013	3169	3062	2938	3086
	LCD	0	0	0	34	43	43	44	52	64	86

(1) From the total IC production, calculate ratios for consumer, industrial, and components in each of the 10 years.

(a) Comment on the stability (regularity) in these ratios.

(b) From the consumer production total, calculate ratios for audio and video segments.

(c) From the industrial production total, calculate the ratios of its segments.

(d) From the component production total, calculate the ratios of its segments.

(2) (a) Using only the first 8 years of the data, calculate the growth rate in each of the three subcomponents of total IC production. A growth rate can be estimated by taking the first difference of the log-transformed data.

(b) Average the growth rates of each subcomponent.

(c) Apply this growth rate to the 1985 production number to project 1986 and 1987.

(d) Compare your projected production with the actual values.

(3) (a) Plot the ratios determined in (1) and smooth the ratios with a simple moving average, arithmetic mean, or some other suitable projection technique and make a projection for the 1986 and 1987 ratios.

(b) Normalize them so that their totals are one within each segment.

(c) Apply the projected ratios to obtain forecasts for each of the segments for 1986 and 1987.

(d) How does this work for the LCD segment, a new-product introduction at that time? How could you handle this situation differently?

(e) Calculate percentage errors and summarize your findings. Did you end up with about the same number of over- and underforecasts?

(4) (a) Repeat the analysis in (1)–(3) for a more recent 10-year period. Contrast the results and speculate as to what might be behind any differences you find.

(b) Create a list of products that use ICs, such as television, video, audio, home electric, home information, fax, PC, computer peripherals, communication equipment, measurement equipment, AI machines, medical equipment, and automobiles.

(c) For each product in (b), speculate on the possible growth and decline factors affecting the product (e.g., for television, high picture quality is a growth factor and limited distribution is a decline factor).

CASE 7C DEMAND FOR BICYCLES (CASE 5C CONT.)

The bicycle company CEO asks you to create some models for forecasting the market, the company's share in the market, inventory, and production runs. The following data on the annual sales of bicycles (BIKE), Unemployment (UNEMPL), and automobile registrations (CARREG) are provided.

YEAR	BIKE	UNEMPL	CARREG (000)
01	370	6.7	593
02	402	5.5	671
03	428	5.7	734
04	482	5.2	795
05	450	4.5	903
06	569	3.8	901
07	609	3.8	852
08	701	3.6	922
09	711	3.5	936
10	634	4.9	859
11	795	5.9	956
12	1182	5.6	1031
13	1414	4.9	1110
14	1420	5.6	918
15	978	8.5	848
16	853	7.7	946
17	903	7.0	1053
18	929	6.0	1083
19	1027	5.8	1049

(1) Create a flowchart outlining the steps needed to create an econometric modeling system for this manufacturer.

(2) Propose five more key factors affecting the demand for bicycles in the market. Create a rationale for your selections.

(3) Create time plots and scatter diagrams for the variables you have selected in your proposed modeling plan. Are recessionary periods evident in the data?

Do you detect any other unusual periods for which you may want to find an explanation?

(4) Search the web or literature for sources of data related to the estimation of bicycle demand.

(5) Search the web or literature for examples of existing demand modeling techniques.

(6) What uses might a community or county make of forecasts of bicycle demand?

CASE 7D SALES OF DOMESTIC AUTOMOBILES (CASES 1D, 3D, 4D, AND 6D CONT.)

In order to create a forecast of the domestic automobile market, an economic consulting firm needs you to develop an industry model for the domestic automobile market. For a given set of macroeconomic and demographic factors, you attempt to calculate how many automobiles are likely to be in use in the next 5 years.

(1) Determine approximately seven top factors that will affect the total demand for automobiles in the United States (or your own country).

(2) For each of these factors, write a couple of paragraphs describing its impact on the demand for domestic automobiles.

(3) Collect sufficient (quarterly) data to include at least one business cycle (one recessionary period), and perform a pressures analysis to analyze the business cycle.

(4) Create a flowchart of how you would proceed to make a short-term and a long-term forecast.

(5) What use would the production manager make of such a forecast? The materials manager? The strategic planner?

(6) Postulate the structure of an econometric model for the long term (10 years) in terms of the factors identified in (1).

(7) In addition to the formal models, are there other considerations management should take into account, perhaps subjectively, to arrive at a consistent forecast that the corporation will be able to use for long-range planning?

CASE 7E EXPORT OF JAPANESE AUTOMOBILES

The export of Japanese automobiles to the United States and other countries is vital to the health of the Japanese economy. Japanese automobile manufacturers are interested in the trends in this industry on a worldwide basis. As an analyst for an economic research firm, you are charged to understand the export market and develop some projections. Naturally, you are inspired to investigate the econometric modeling approach for this study.

(1) Determine approximately seven top factors that will affect the export of Japanese automobiles in Japan.

(2) For each of these factors, write a couple of paragraphs describing its impact on the export of automobiles from Japan to the United States (or your own country).

(3) Create a flowchart of how you would proceed to make a short-term and a long-term forecast.

(4) Collect adequate (quarterly) data to include in an econometric model.

(5) What use would the production manager of a Japanese firm make of such a forecast? The materials manager? The strategic planner?

(6) Postulate the structure of an econometric model for the long term (10 years) in terms of factors identified in (1).

(7) In addition to the formal models, are there other considerations management should take into account, perhaps subjectively, to arrive at a consistent forecast that the corporation will be able to use for long-range planning?

Applying Bottom-Up Techniques

8

The Exponential Smoothing Method

"The most striking outcome of the M2-Competition is the good and robust performance of exponential smoothing methods."
[M2-COMPETITION, 1993]

EXPONENTIAL SMOOTHING METHODS provide a viable framework for forecasting disaggregate demand patterns. For short-term planning and control systems, these techniques are extremely valuable and have a more than adequate track record in forecast accuracy. Each technique employs weights that give more emphasis to data from the recent past than to the older data. As a result of weighting of the past data, a fitted model appears smoother (less volatile) than the original data.

This chapter deals with the description and evaluation of techniques that:

- Are widely used in the areas of sales, inventory, and production management as well as in quality control, process control, financial planning, and marketing planning

- Are based on the mathematical extrapolation of past patterns into the future, accomplished by using forecasting equations that are simple to update and require a relatively small number of calculations

- Capture level (a starting point for the forecasts), trend (a factor for growth or decline), and seasonal factors (an adjustment for seasonal variation) in data patterns

- Can be described in terms of a state-space modeling framework that provides prediction intervals and procedures for model selection

- Are especially suitable for large-scale, automated forecasting applications, because they require little forecaster intervention, thereby releasing the time of the forecaster to concentrate on the few problem cases

8.1   WHAT IS EXPONENTIAL SMOOTHING?

In Chapter 2, we introduce forecasting with simple and weighted moving averages as a smoothing technique for the short-term forecasting of level data. With exponential smoothing, we can create short-term forecasts with a wider variety of data having trends and seasonal patterns. As a methodology, exponential smoothing provides a unified approach to weighting past historical data for smoothing and extrapolation purposes. This exponentially declining weighting scheme contrasts with the equal weighting scheme that underlies the simple moving average technique.

Exponential smoothing was invented during World War II by Robert G. Brown (1963), who was involved in the design of a tracking system for fire-control information on the location of enemy submarines. Later on, the principles of exponential smoothing were applied to business data, especially in the analysis of the demand for service parts in inventory systems (Brown, 1982).

Exponential smoothing is a method of forecasting that extrapolates historical patterns such as trends and seasonal cycles into the future.

There are many types of exponential smoothing techniques, each appropriate for a particular pattern. As a forecasting tool, exponential smoothing is very widely accepted and a proven tool for a wide variety of short-term forecasting applications. Most inventory planning and production control systems rely on exponential smoothing to some degree.

We will see that the process for assigning smoothing weights is simple in concept and versatile for dealing with diverse types of data. Other advantages of exponential smoothing are that the methodology takes account of trend and seasonal patterns in a time series; embodies a weighting scheme that gives more weight to the recent past than to the distant past; is easily automated, making it especially useful for large-scale forecasting applications; and can be described in a modeling framework needed for deriving useful statistical prediction limits and confidence intervals.

For forecasting, the disadvantages are that exponential smoothing techniques do not easily allow for the inclusion of explanatory variables into a forecasting model and cannot handle business cycles. Hence, when forecasting economic variables, such techniques are not expected to perform well on business data that exhibit cyclical turning points.

In applications for which the techniques are designed, such limitations are not of critical concern. Exponential smoothing provides an essential simplicity and ease of understanding for the practitioner, and has been found to have a good track record for accuracy in many business applications.

8.2 SMOOTHING WEIGHTS

To understand how exponential smoothing works, we need first to understand the concept of exponentially decaying weights. Consider a time series of production rates (number of completed assemblies per week) for a 4-week period in the following table.

Week	Production
Three periods ago $(T-3)$	266
Two periods ago $(T-2)$	411
Previous $(T-1)$	376
Current $t = T$	425

In order to predict next period's $(T+1)$ production rate without having knowledge of or information about future demand, we assume that the following week will have to be an average week for production. A reasonable projection for the following week can be based on taking an average of the production rates during past weeks. However, what kind of average should we propose?

Equally Weighted Average

The simplest option, described in Chapter 2, is to select an equally weighted average, which is obtained by given equal weight to each of the weeks of available data:

$$(425 + 376 + 411 + 266) / 4 = 370$$

This equally weighted average is simply the arithmetic mean of the data. The forecast of next week's production rate is 370 assemblies. Implicitly, we are assuming that events of 2 and 3 weeks prior (e.g., the more distant past) are as relevant to what may happen next week as were events of the most current and prior week.

In Exhibit 8.1, a weight is denoted by w_i, where the subscript i represents the number of weeks into the past. For an equally weighted average, the weight given to each of the terms is $1/n$, where n is the number of time periods. With $n = 4$, each weight in column 3 is equal to 1/4.

If we consider only the latest week, we have another option, shown in column 4 of Exhibit 8.1, which is the Naïve_1 forecast; it places all weight on the most recent data value. Thus, the forecast for next week's production rate is 425, the same as the current week's production. This forecast makes sense if only the current week's events are relevant in projecting the following week. Whatever happened before this week is ignored.

EXHIBIT 8.1
Various Weighting Schemes

Week	Weights	Equal	Naïve_1	Linear Decay	Exponential Decay	Adjusted Weights
$T-3$	w_4	0.25	0	0.1	0.0625	0.0667
$T-2$	w_3	0.25	0	0.2	0.125	0.1333
$T-1$	w_2	0.25	0	0.3	0.25	0.2667
T	w_1	0.25	1	0.4	0.5	0.5333
Sum		1.00	1.00	1.00	0.9375	1.0000

Exponentially Decaying Weights

Most business forecasters find a middle ground more appealing than either of the two extremes, equally weighted or Naïve_1. In between lie weighting schemes in which the weights decay as we move from the current period to the distant past.

$$w_1 > w_2 > w_3 > w_4 > \cdots$$

The largest weight, w_1 , is given to the most recent data value. This means that to forecast next week's production rate, this week's figure is most important, last week's is less important, and so forth.

Many other patterns are possible with decaying weight schemes. As illustrated by column 5 of Exhibit 8.1, the weight starts at 40% for the most recent week and declines steadily to 10% for week $T - 3$. Our forecast for week $t = T + 1$ is the weighted average with decaying weights:

$$425 \times 0.4 + 376 \times 0.3 + 411 \times 0.2 + 266 \times 0.1 = 392$$

This weighted average gives a production rate forecast that is more than that of the equally weighted average and less than that of the Naïve_1.

 An exponentially weighted average refers to a weighted average of the data in which the weights decay exponentially.

The most useful example of decaying weights is that of exponentially decaying weights, in which each weight is a constant fraction of its predecessor. A fraction of 0.50 implies a decay rate of 50%, as shown in column 6 of Exhibit 8.1. In forecasting next period's value, the current period's value is weighted 0.5, the prior week half of that at 0.25, and so forth with each new weight 50% of the one before. (These weights must be adjusted to sum to unity as in column 7.) From Exhibit 8.1, we can see that the adjusted weights are obtained by dividing the exponential decay weights by 0.9375.

Exhibit 8.2 illustrates the weighted average of all past data, with recent data receiving more weight than older data. The most recent data is at the bottom of the spreadsheet. The weight on each data value is shown in Exhibit 8.3. The weights decline exponentially with time, a feature that gives exponential smoothing its name.

Simple Exponential Smoothing

All exponential smoothing techniques incorporate an exponential-decay weighting system, hence the term exponential. Smoothing refers to the averaging that takes place when we calculate a weighted average of the past data. To determine a one-period-ahead forecast of historical data, the projection formula is given by

$$Y_t(1) = \alpha Y_t + (1 - \alpha)Y_{t-1}(1)$$

EXHIBIT 8.2
Computation of
Weights on
Historical Data

ANALYSIS OF WEIGHTS ON PAST DATA

TITLE1:	WEIGHTS ON PAST DATA			SMOOTHING WEIGHT =	0.64
TITLE2:	SIMPLE EXPONENTIAL SMOOTHING			NBR. OF PERIODS =	12
X-AXIS:	PERIOD				
Y-AXIS:	WEIGHT				

Year	Month	Per	Data	Weight	Data*Weight
		0	14	0.000005	0.000066
1989	1	1	14	0.000008	0.000118
	2	2	11	0.000023	0.000257
	3	3	12	0.000065	0.000780
	4	4	14	0.000181	0.002528
	5	5	12	0.000502	0.006018
	6	6	14	0.001393	0.019504
	7	7	18	0.003870	0.069657
	8	8	17	0.010750	0.182742
	9	9	19	0.029860	0.567337
	10	10	16	0.082944	1.327104
	11	11	18	0.230400	4.147200
	12	12	16	0.640000	10.240000
	SUM			1.000000	16.563312

(*Source*: SMOOTH.XLS, The Spreadsheet Forecaster on the CD)

EXHIBIT 8.3
Exponentially
Decaying Weights

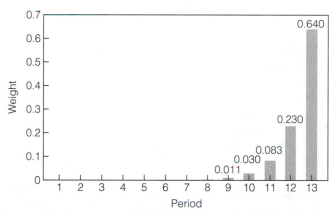

where $Y_t(1)$ is the smoothed value at time t, based on weighting the most recent observation Y_t with a weight α (α is a smoothing parameter) and the current period's forecast (or previous smoothed value) with a weight $(1 - \alpha)$. By rearranging the right-hand side, we can rewrite the equation as

$$Y_t(1) = Y_{t-1}(1) + \alpha[Y_t - Y_{t-1}(1)]$$

which can be interpreted as the current period's forecast $Y_{t-1}(1)$ adjusted by a proportion α of the current period's forecast error $[Y_t - Y_{t-1}(1)]$.

The simple exponential smoothing technique produces forecasts that are a level line for any period in the future, but it is not appropriate for projecting trending data or patterns that are more complex.

We can now show that the one-step-ahead forecast $Y_t(1)$ is a weighted moving average of all past observations with the weights decreasing exponentially. If we substitute for $Y_{t-1}(1)$ into the first smoothing equation, we find that:

$$Y_t(1) = \alpha Y_t + (1-\alpha)[\alpha Y_{t-1} + (1-\alpha)Y_{t-2}(1)]$$
$$= \alpha Y_t + \alpha(1-\alpha)Y_{t-1} + (1-\alpha)^2 Y_{t-2}(1)$$

If we next substitute for $Y_{t-2}(1)$, then for $Y_{t-3}(1)$, and so, we obtain the result

$$Y_t(1) = \alpha Y_t + \alpha(1-\alpha)Y_{t-1} + \alpha(1-\alpha)^2 Y_{t-2} + \alpha(1-\alpha)^3 Y_{t-3} + \alpha(1-\alpha)^4 Y_{t-4}$$
$$+ \cdots + \alpha(1-\alpha)^{t-1} Y_1 + (1-\alpha)^t Y_0(1)$$

The one-step-ahead forecast $Y_T(1)$ represents a weighted average of all past observations. For three selected values of the parameter α, the weights that are assigned to the past observations are shown in the following table.

Weight Assigned to:	$\alpha = 0.1$	$\alpha = 0.3$	$\alpha = 0.5$	$\alpha = 0.9$
Y_T	0.1	0.3	0.5	0.9
Y_{T-1}	0.09	0.21	0.25	0.09
Y_{T-2}	0.081	0.147	0.125	0.009
Y_{T-3}	0.0729	0.1029	0.0625	0.0009
Y_{T-4}	0.0656	0.0720	0.0313	0.00009

In Exhibit 8.4, we calculate a forecast of the production data, assuming that $\alpha = 0.5$. (The production data are repeated in Exhibit 8.4, in the Actual column.) To use the formula, we need a starting value for the smoothing operation—a value that represents the smoothed average at the earliest week of our time series, here $t = T - 3$. The simplest choice for the starting value is the earliest data point. In our example, the starting value for the exponentially weighted average is the production rate for week $t = T - 3$, which was given as 266. The final result, $Y_T(1) = 391$ (rounded) for week $t = T$, is called the current level. It is a weighted average of 4 weeks of data, where the weights decline at a rate of 50% per week.

We defined a one-period-ahead forecast made at time $t = T$ to be $Y_T(1)$. Likewise, the m-period-ahead forecast is given by $Y_T(m) = Y_T(1)$, for $m = 2, 3, \ldots$ For a time series with a relatively constant level, this is a good forecasting technique. We called this simple smoothing in Chapter 2, but it is generally known as simple exponential smoothing.

EXHIBIT 8.4
Updating an
Exponentially
Weighted Average

Week	Y	Actual	$Y_t(1)$	Formula	Error $Y_t - Y_{t-1}(1)$
$T-3$	Y_{T-3}	266	266	$Y_{(t-3)}(1) = Y_{(t-3)}$	
$T-2$	Y_{T-2}	411	339	$Y_{(t-2)}(1) = 0.5 \times Y_{(t-2)} + 0.5 \times Y_{(t-3)}(1)$	72
$T-1$	Y_{T-1}	376	357	$Y_{(t-1)}(1) = 0.5 \times Y_{(t-1)} + 0.5 \times Y_{(t-2)}(1)$	19
T	Y_T	425	391	$Y_t(1) = 0.5 \times Y_t + 0.5 \times Y_{(t-1)}(1)$	34

The forecast profile of the simple exponential smoothing technique is a flat or level line.

Simple smoothing works much like an automatic pilot or a thermostat. At each time period, the forecasts are adjusted according to the sign of the forecast error (actual data minus forecast.) If the current forecast error is positive, the next forecast is increased; if the error is negative, the forecast is reduced.

To get the smoothing process started (Exhibit 8.5), we set the first forecast (cell E8) equal to the first data observation (cell D8). We can also use the average of the first few data observations. Thereafter, the forecasts are updated as follows: In column F, each error is equal to actual data minus forecast. In column E, each forecast is equal to the previous forecast plus a fraction of the previous error. This fraction is called the smoothing weight (cell I2).

But how do we select the smoothing weight? The smoothing weight is usually chosen to minimize the MSE, a statistical measure of fit introduced in Chapter 5. This smoothing weight is called optimal because it is our best estimate based on a prescribed criterion (MSE). Forecasts, errors, and squared errors are shown in columns E, F, and G. The one-step-ahead forecast ($=16.60$ in cell E20) extends one period into the future. The travel expense data, smoothed values, and the one-period-ahead forecast are shown graphically in Exhibit 8.6.

EXHIBIT 8.5
Forecasting with
Simple Exponential
Smoothing:
Company Travel
Expenses

	1	SIMPLE EXPONENTIAL SMOOTHING					
	2	TITLE1: SIMPLE EXPONENTIAL SMOOTHING				WEIGHT: 0.64	
	3	TITLE2: COMPANY TRAVEL EXPENSES				MSE: 4.7056	
	4	X-AXIS: MONTH					
	5	Y-AXIS: THOUSANDS OF $					
	6	A	B	C	D	E	F	G
	7	Year	Mon	Per	Data	Fcst	Error	Error^2
	8	1989	1	1	14	14.0	0.0	0.0
	9		2	2	11	14.0	−3.0	9.0
	10		3	3	12	12.1	−0.1	0.0
	11		4	4	14	12.0	2.0	3.9
	12		5	5	12	13.3	−1.3	1.7
	13		6	6	14	12.5	1.5	2.4
	14		7	7	18	13.4	4.6	20.7
	15		8	8	17	16.4	0.6	0.4
	16		9	9	19	16.8	2.2	5.0
	17		10	10	16	18.2	−2.2	4.8
	18		11	11	18	16.8	1.2	1.5
	19		12	12	16	17.6	−1.6	2.4
	20	1990	1	13		16.6		

(*Source*: SMOOTH.XLS, The Spreadsheet Forecaster on the CD)

EXHIBIT 8.6
Simple Exponential
Smoothing:
Company Travel
Expenses

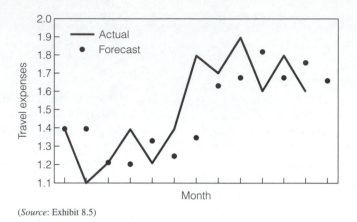

(*Source*: Exhibit 8.5)

8.3 TYPES OF SMOOTHING TECHNIQUES

A system of exponential smoothing techniques can be classified by the type of trend and/or seasonal pattern generated as the forecast profile. The most appropriate technique to use for any forecasting should match the profile expected or desired in an application. Exhibit 8.7 shows the extended Pegels classification for 12 forecasting profiles for exponential smoothing (Pegels, 1969; Gardner, 1985).

A Pegels classification of exponential smoothing techniques gives rise to 12 forecast profiles for trend and seasonal patterns.

After a preliminary examination of the data from a time plot, we may be able to determine which of the dozen techniques seems most suitable. There are four types of trend profiles to choose from:

Constant level	Level or no trend at all
Linear trend	Constant amount of change per period
Exponential trend	Constant rate of change per period
Damped trend	Growth pattern that gradually decays to no growth

The same choices may be applied to downwardly trending time series as well. There are three types of seasonality:

Nonseasonal	Time series that do not exhibit seasonal variation
Additive seasonality	Seasonal fluctuations whose amplitude does not grow (or decline) systematically over time
Multiplicative seasonality	Seasonal fluctuations that widen as the average annual level of the data increases

Each profile can be directly associated with a specific exponential smoothing procedure (Exhibit 8.8), as described in the next section (some of which are referred

For a downwardly trending time series, multiplicative seasonality appears as steadily diminishing swings about a trend. For level data, the constant-level multiplicative and additive seasonality techniques give the same forecast profile.

EXHIBIT 8.7
Pegels Classification of Exponential Smoothing Techniques Extended to Include the Damped Trend Technique

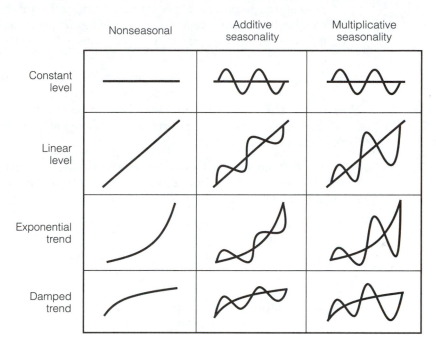

EXHIBIT 8.8
Commonly Used Exponential Smoothing Techniques

Model type	Trend profile	Seasonal profile
Simple (single)	None	None
Holt	Linear	None
Holt-Winters	Linear	Additive or multiplicative

to by a common name attributed to their authors). We now explain how each procedure works to generate forecasts; that is, we describe how each procedure produces the appropriate forecasting profile.

8.4 SMOOTHING LEVELS AND CONSTANT CHANGE

An exponential smoothing method comprises one or more of the following components: the current level, the current trend, and the current seasonal index.

- The *current level* serves as the starting point for the forecast. It is calculated to represent an exponentially weighted average of the time series at the end of the fit period. We can regard the current level as the value the time series would now have if there were nothing at all unusual going on at present. An alternative is to use the last observation as the starting point, but doing so might set the forecasts off at the wrong level if the most recent data is abnormal.

- The *current trend* represents the amount by which we expect the time series to grow or decline per time period into the future. It is often calculated as an exponentially weighted average of past period-to-period changes in the level of the series. In this way, recent growth or decline in the time series is given more weight than changes farther back in time.

- The *current seasonal index* is interpreted the same way as a conventional seasonal index—as the amount or degree by which the season's value tends to exceed or fall short of the norm. Recall that in the **classical decomposition** of a time series, a (multiplicative) seasonal index measured the norm as a moving average (see Chapter 6).

The determination of the current seasonal index, a key part of Holt-Winters' method (Holt et al., 1960; Holt, 2004; Winters, 1960), differs from the conventional indexes in two respects:

1. **Difference in weighting years.** In a ratio-to-moving-average method, the data used over the years to determine a monthly or quarterly index are weighted equally. In contrast, the Holt-Winters method takes an exponentially weighted average of the ratios to level, thus giving more weight to recent years than to those of the past.

2. **Representing the norm.** The Holt-Winters method uses the current level of the series instead of a moving average of four quarters or 12 months.

The formulae to describe the current level, trend, and seasonal indexes are presented in Appendix 8A. The interpretation of the parameters in the smoothing algorithms is a complex matter and we will not deal with the estimation issues. Optimal or near-optimal parameter settings are readily derived using prescribed criteria for optimal selection found in modern software tools. In addition, combinations of multiple parameter values severely limit an intuitive feel for their impact on the forecasts. Moreover, for inventory-replenishment planning purposes, a forecaster needs to rely on automatic forecasting features of modern software because of the very large volume of data involved.

We now illustrate the use of the equations in Appendix 8A for (1) calculating the current level, trend, and seasonal indexes and (2) combining these values into the forecasting formula. Consider Exhibit 8.9 for weekly shipments of a canned beverage. Exhibit 8.10 is a time plot of weekly shipments of a canned beverage for 65 weeks, shown previously in Chapter 4. The series is highly variable but not trending. The average level is approximately 4000–5000 units per week with a standard deviation of 3639. Some of the high peaks may be attributed to holidays or promotions, but we will choose to forecast it using simple exponential smoothing. This technique is appropriate for data lacking trend and seasonality.

EXHIBIT 8.9 Weekly Shipments of a Canned Beverage—No Trend or Seasonality

Filename: SMOOTH.XLS
Worksheet: SIMSMOOT.XLS
SIMPLE EXPONENTIAL SMOOTHING
TITLE1: SINGLE EXPONENTIAL SMOOTHING WEIGHT: 0.62
TITLE2: SHIPMENT OF CANNED BEVERAGE MSE: 1668955
X-AXIS: WEEK
Y-AXIS: UNITS

Year	Mon	Per	Data	Fcst	Error	Error^2
		1	4671	4671	0	0
		2	1670	4671	−3001	9006001
		3	2431	2810.38	−379.38	143929.2
		4	2322	2575.164	−253.164	64092.21
		5	3809	2418.202	1390.798	1934318
		6	2908	3280.497	−372.497	138754
		7	1765	3049.549	−1284.55	1650066
		8	2528	2253.129	274.8714	75554.31
		9	1850	2423.549	−573.549	328958.3
		10	4160	2067.949	2092.051	4376679
		11	3843	3365.02	477.9795	228464.4
		12	4303	3661.368	641.6322	411691.9
		13	4523	4059.18	463.8202	215129.2
		14	10547	4346.748	6200.252	38443121
		15	2316	8190.904	−5874.9	34514501
		16	9920	4548.464	5371.536	28853403
		17	718	7878.816	−7160.82	51277288
		18	918	3439.11	−2521.11	6355996
		19	1147	1876.022	−729.022	531472.9
		20	5554	1424.028	4129.972	17056666
		21	4004	3984.611	19.38924	375.9428
		22	6556	3996.632	2559.368	6550364
		23	3242	5583.44	−2341.44	5482342
		24	7805	4131.747	3673.253	13492786
		25	7367	6409.164	957.836	917449.9
		26	1728	7003.022	−5275.02	27825860
		27	7989	3732.508	4256.492	18117720
		28	16695	6371.533	10323.47	1.07E+08
		29	1446	12772.08	−11326.1	1.28E+08
		30	2182	5749.911	−3567.91	12729992
		31	1284	3537.806	−2253.81	5079643
		32	5694	2140.446	3553.554	12627743
		33	1407	4343.65	−2936.65	8623911
		34	1277	2522.927	−1245.93	1552334
		35	10469	1750.452	8718.548	76013076
		36	7708	7155.952	552.0482	304757.2
		37	884	7498.222	−6614.22	43747929
		38	1125	3397.404	−2272.4	5163821
		39	656	1988.514	−1332.51	1775593
		40	2210	1162.355	1047.645	1097560
		41	1032	1811.895	−779.895	608236.2
		42	1316	1328.36	−12.3601	152.7718
		43	3410	1320.697	2089.303	4365188
		44	8788	2616.065	6171.935	38092784
		45	1304	6442.665	−5138.66	26405874
		46	1293	3256.693	−1963.69	3856088
		47	4338	2039.203	2298.797	5284467
		48	13735	3464.457	10270.54	1.05E+08
		49	743	9832.194	−9089.19	82613443
		50	3309	4196.894	−887.894	788355.1
		51	3790	3646.4	143.6004	20621.08
		52	16171	3735.432	12435.57	1.55E+08
		53	5847	11445.48	−5598.48	31343024
		54	736	7974.424	−7238.42	52394781
		55	1152	3486.601	−2334.6	5450362
		56	3509	2039.148	1469.852	2160464
		57	9188	2950.456	6237.544	38906950
		58	2026	6817.733	−4791.73	22960709
		59	7966	3846.859	4119.141	16967325
		60	1876	6400.726	−4524.73	20473148
		61	3325	3595.396	−270.396	73114
		62	2079	3427.75	−1348.75	1819128
		63	4461	2591.525	1869.475	3494936
		64	3692	3750.6	−58.5996	3433.91
		65	3637	3714.268	−77.2678	5970.319
		66	6025	3666.362	2358.638	5563174
				5128.717		
				5128.717		
				5128.717		
				5128.717		

EXHIBIT 8.10
Weekly Shipments of
Canned Beverage

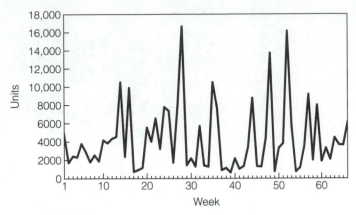

(*Source*: Exhibit 4.11)

EXHIBIT 8.11
Forecast Model
for Canned Beverage
Shipments: Simple
Exponential
Smoothing: No
Trend, No
Seasonality

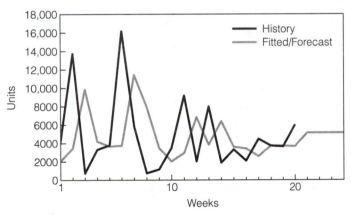

(*Source*: Exhibit 8.10)

The simple exponential smoothing model is fit to all 65 weeks and forecast for 4 weeks. The optimal estimate (based on the MSE criterion) of the smoothing parameter is 0.62 with MSE = 1,668,955. The multiperiod forecasts are a constant level (=5129). Thus, it represents a "typical" level. Exhibit 8.11 displays the most recent 20 weeks of historical shipments, 20 weeks of fitted values, and the four forecasts.

Because simple exponential smoothing views the future of the time series as lacking both trend and seasonality, the forecasting equation does not contain these terms, leaving the current level as the sole component.

The current level L_t is calculated by an equation for an exponentially smoothed average of the past data: $Y_T(m) = L_t$.

Consider Exhibit 8.12 for annual car registrations (data shown in Case 7C). Exhibit 8.13 is a time plot of the data. Because the data are annual, the time series is necessarily nonseasonal. The global trend appears to be linear, although there are a number of

EXHIBIT 8.12   Annual Car Registrations—Linear trend (Case 7C)

Filename: CARREG.XLS
Worksheet: LINSMOOT.XLS
LINEAR-TREND SMOOTHING
TITLE1: LINEAR-TREND SMOOTHING   LEVEL WEIGHT: 0.30
TITLE2: CAR REGISTRATIONS   TREND WEIGHT: 0.20
X-AXIS: YEAR   MSE: 5594.010
Y-AXIS: NUMBER

Year	Mon	Per	Data	Fcst	Error	Error^2	Level	Trend
		0					525.67	67.33
78	N/A	1	593	593.00	0.00	0.00	593.00	67.33
79		2	671	660.33	10.67	113.78	663.53	69.47
80		3	734	733.00	1.00	1.00	733.30	69.67
81		4	795	802.97	−7.97	63.47	800.58	68.07
82		5	903	868.65	34.35	1179.92	878.96	74.94
83		6	901	953.90	−52.90	2798.23	938.03	64.36
84		7	852	1002.39	−150.39	22617.90	957.27	34.29
85		8	922	991.56	−69.56	4838.58	970.69	20.37
86		9	936	991.07	−55.07	3032.17	974.55	9.36
87		10	859	983.91	−124.91	15601.45	946.43	−15.62
88		11	956	930.81	25.19	634.38	938.37	−10.58
89		12	1031	927.79	103.21	10653.23	958.75	10.06
90		13	1110	968.81	141.19	19934.85	1011.17	38.30
91		14	918	1049.46	−131.46	17282.75	1010.02	12.00
92		15	848	1022.03	−174.03	30286.23	969.82	−22.80
		16	946	947.02	−1.02	1.04	946.71	−23.01
		17	1053	923.71	129.29	16716.29	962.50	2.85
		18	1083	965.35	117.65	13841.72	1000.64	26.38
		19	1049	1027.03	21.97	482.78	1033.62	30.78
		20	1064.40				1064.40	30.78
		21	1095.18				1095.18	30.78
		22	1125.95				1125.95	30.78
		23	1156.73					

(*Source*: SMOOTH.XLS, The Spreadsheet Forecaster on the CD)

local variations on the trend. Exhibit 8.14 is the graph of the output from the Holt method—linear trend, no seasonality (Holt et al., 1960; Holt, 2004). We see that from the vantage point of Year 19, the current level of car registrations is estimated to be 1034 and the current trend is estimated to be an increase of 31 registrations per year.

EXHIBIT 8.13
Time Plot of Annual
Car Registrations for
a 19-Year Period

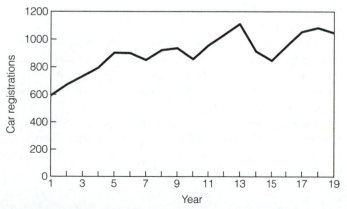

(*Source*: Exhibit 8.12)

EXHIBIT 8.14
Time Plot of the
Forecast Model for
Car Registrations:
Holt Method—Linear
Trend, No Seasonality

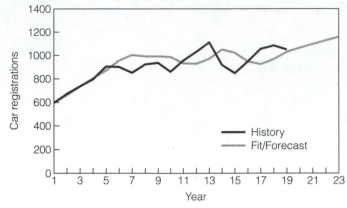

(*Source*: Exhibit 8.12)

For nonseasonal data, the seasonal index terms are not present in the forecasting equations. What remains in the Holt method, however, is the forecasting equation for a linear trend:

$$Y_T(m) = [L_t + m \times T_t]$$

The forecast for year 22 is 1126, based on calculating a 3-year-ahead projection from the base year $T = 19$:

$$Y_{19}(3) = [1033.62 + 3 \times 30.78] = 1126$$

8.5 DAMPED AND EXPONENTIAL TRENDS

In the seasonal methods, the trend component, if one is present, is assumed to be linear. Trends can also be nonlinear. For example, damped (upward) trend assumes that the series will continue to grow but that the growth gradually dampens out. An exponential growth trend assumes that the series will grow by a progressively larger amount. Exponential growth is equivalent to a constant percentage rate of growth. As the base grows over time, the constant percentage increase on the base translates into larger and larger increments in volume, in the manner of compound-interest growth on an investment.

In the case of a downward trend, the damped and exponential patterns are similar. During a phase out or decline under adverse market conditions, a forecast profile will be decaying without becoming negative. We may refer to the pattern of shipments of a cosmetic product (Exhibit 8.15) either as a downwardly damped trend or as exponential decay.

Like a no-trend or linear trend, the exponential trend may be used in conjunction with multiplicative, additive, or no seasonality patterns. We use the nonseasonal case here to illustrate damped and exponential trends. (When seasonality is included, it is called the Holt-Winters procedure, discussed later.)

EXHIBIT 8.15
Time Plot of Weekly
Shipments of a
Cosmetic Product—
Damped Exponential
Trend Model

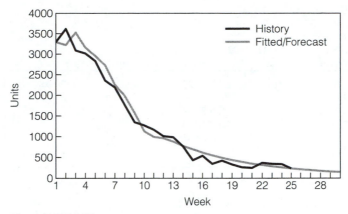

(*Source*: Exhibit 8.12)

Both damped and exponential trends can be represented in a single forecasting equation, given by

$$Y_T(m) = L_t + \sum_i^m \phi^i \times T_t$$

where m is the length of the forecast horizon. The symbol ϕ is called the trend-modification parameter. Depending on the value of ϕ, the forecast profile can be an exponential trend, linear trend, damped trend, or constant level. Here are the cases.

If $\phi > 1$ the trend is *exponential*.

If $\phi = 1$ the trend is *linear*.

If $\phi < 1$ the trend is *damped*.

If $\phi = 0$ there is no trend.

Exhibit 8.16 shows the historical, fitted, and forecast values for a damped trend model of the annual car registration series. The growth in the forecasts, the change from the prior year's forecast, has dampened. With $\phi = 0.83$, the forecasted trend is slowing by $1 - 0.83 = 0.17$ or 17% per period. The estimates of level and trend weights are 0.6 and 0.2, respectively.

We can illustrate how the forecast was obtained for Year 20 ($= T + 1$); the fitted value for year 19 is 1078.55 , $L_t = 1060.86$, and $T_t = 18.75$. Setting $m = 1$,

$$Y_{19}(1) = L_t + \phi \times T_t$$

$$= 1060.86 + 0.83^1 \times (18.75)$$

$$= 1076.42$$

The Year 21 forecasts are calculated as follows:

$$Y_{19}(2) = L_t + (\phi^1 + \phi^2) \times T_t$$

$$= 1060.86 + (0.83^1 + 0.83^2) \times (18.75)$$

$$= 1089.34$$

EXHIBIT 8.16
Exponential Trend
Forecast Model for
Car Registrations
(dampening factor =
1.0)

	Level weight:	0.6				
	Trend weight:	0.2				
	Trend modifier:	1				
	MSE:	4238.633				
Period	Data	Forecast	Error	Error^2	Level	Trend
19	1049	1083.323	−34.32299	1178.068	1062.729	22.80642
20		1085.536			1085.536	22.80642
21		1108.342			1108.342	22.80642
22		1131.148			1131.148	22.80642
23		1153.955			1153.955	22.80642
24		1176.761			1176.761	22.80642

(*Source*: Exhibit 8.12)

EXHIBIT 8.17
Damped Trend
Forecast Model for
Car Registrations
(dampening factor =
0.83)

	Level weight:	0.6				
	Trend weight:	0.2				
	Trend modifier:	0.83				
	MSE:	3568.477				
Period	Data	Forecast	Error	Error^2	Level	Trend
19	1049	1078.654	−29.65415	879.3687	1060.862	18.74755
20		1076.422			1076.422	15.56047
21		1089.337			1089.337	12.91519
22		1100.057			1100.057	10.71961
23		1108.954			1108.954	8.897274
24		1116.339			1116.339	7.384737

(*Source*: Exhibit 8.12)

Alternatively, we can obtain the forecast for $T + 2$ by calculating

$$Y_T(2) = Y_T(1) + \phi^2 \times T_t$$

Exhibits 8.16 and 8.17 show a comparison of the forecasts from the exponential trend and the damped trend for car registrations data. Based on the MSE criterion, the MSE of the damped trend model is approximately 16% smaller than the MSE of the linear trend model. The five-period-ahead forecast (fifth-year projection) for the damped trend model is approximately 5% below the linear trend projection. Whether the difference in the 5-year projections is significant depends on the context in which these forecasts are used. A comparison of the two techniques is shown in Exhibit 8.18.

Consider the trend shown in Exhibit 8.19 for the sales of a cosmetic product. The sales for this cosmetic product, for the period 1978–2002, show a declining trend. The forecasts decline exponentially, modeled by a nonlinear trend exponential smoothing technique. Exhibit 8.20 presents a comparison of a linear trend and damped exponential trend model for the cosmetic product. Note that minimizing MSE may not be the only criterion for selecting a model; the eventual profile should also be considered. In this case, the linear trend model with the lowest MSE also yields much lower forecasts over the forecast period than the damped trend model. It may require some judgment on the part of the forecaster to determine the most appropriate profile for the data at hand.

EXHIBIT 8.18 Time Plot of Historical Car Registration Data with Two Exponential Smoothing Forecasts

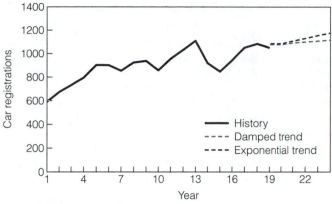

(*Source*: Exhibit 8.12)

EXHIBIT 8.19 Sales of a Cosmetic Product—Damped Exponential Trend

Filename:
Worksheet: EXPSMOOT.XLS
EXPONENTIAL SMOOTHING: EXPONENTIAL AND DAMPED-EXPONENTIAL TRENDS
TITLE1: DAMPED-EXPONENTIAL TREND
TITLE2: COSMETIC PRODUCT
X-AXIS: YEAR
Y-AXIS: COSMETIC SALES

LEVEL WEIGHT: 0.64
TREND WEIGHT: 0.34
TREND MODIFIER: 0.88
MSE: 62312.86

Year	Mon	Per	Data	Fcst	Error	Error^2	Level	Trend
		0					3386.333	−92.3333
1978	N/A	1	3294	3294	0	0	3294	−81.2533
1979		2	3616	3222.50	393.50	154844.56	3474.34	62.29
1980		3	3089	3529.15	−440.15	193734.17	3247.45	−94.84
1981		4	3017	3164.00	−147.00	21608.16	3069.92	−133.44
1982		5	2831	2952.49	−121.49	14760.94	2874.74	−158.73
1983		6	2359	2735.05	−376.05	141416.19	2494.38	−267.54
1984		7	2189	2258.94	−69.94	4891.83	2214.18	−259.22
1985		8	1761	1986.07	−225.07	50655.29	1842.02	−304.63
1986		9	1350	1573.95	−223.95	50151.74	1430.62	−344.22
1987		10	1272	1127.71	144.29	20820.49	1220.05	−253.85
1988		11	1170	996.66	173.34	30045.71	1107.60	−164.46
1989		12	1013	962.88	50.12	2512.35	994.96	−127.68
1990		13	985	882.60	102.40	10486.36	948.13	−77.54
1991		14	773	879.90	−106.90	11427.27	811.48	−104.58
1992		15	435	719.45	−284.45	80912.53	537.40	−188.75
1993		16	538	371.31	166.69	27786.78	477.99	−109.42
1994		17	346	381.70	−35.70	1274.52	358.85	−108.43
1995		18	423	263.44	159.56	25460.82	365.56	−41.16
1996		19	333	329.33	3.67	13.46	331.68	−34.98
1997		20	265	300.90	−35.90	1288.74	277.92	−42.99
1998		21	248	240.10	7.90	62.48	245.15	−35.14
1999		22	369	214.23	154.77	23953.45	313.28	21.70
2000		23	343	332.38	10.62	112.84	339.18	22.71
2001		24	339	359.16	−20.16	406.31	346.26	13.13
2002		25	236	357.81	−121.81	14837.43	279.85	−29.86
2003		26		253.57			253.57	−26.28
2004		27		230.45			230.45	−23.13
2005		28		210.10			210.10	−20.35
2006		29		192.19				

(*Source*: SMOOTH.XLS, The Spreadsheet Forecaster on the CD)

EXHIBIT 8.20		Linear Trend		Damped Exponential Trend		Linear Trend	
Forecast Models for a Cosmetic Product: Damped and Linear Trend Exponential Smoothing	Smoothing Component	Weight	Final Value (2002)	Weight	Final Value (2002)	Weight	Final Value (2002)
	Level	0.64	281.98	0.64	279.85	0.81	
	Trend	0.34	−26.92	0.34	−29.86	0.18	
	Dampen	1.0		0.88		1.0	
	MSE		62,780		62,313		61,597

(*Source*: Exhibit 8.19)

To illustrate how the damped exponential trend forecast was obtained for the year 2005 ($=T + 3$), we set $m = 3$ and $T = 2002$:

$$Y_{2005}(3) = L_t + (\phi + \phi^2 + \phi^3) \times T_t$$

$$= 279.85 + (0.88 + 0.88^2 + 0.88^3) \times (-29.86)$$

$$= 210.10$$

The difference in the forecast profiles for the linear trend and damped trend arises from the value of the trend modification parameter, which is below unity ($\phi = 0.88$) for the damped trend model. This value of ϕ leads to a decrease over prior year's forecast and is characteristic of an exponential decline. On the other hand, the damped exponential trend model, for which the trend modification parameter is constrained to $\phi = 1$, gives rise to a linear trend forecast profile. A comparison of the two techniques yields a statistical summary (Exhibit 8.21) and forecast profiles for the cosmetics sales data (Exhibit 8.22) that indicate the difference in the rounded results.

Exponential trends should be applied with caution, and careful consideration should be made of the underlying business environment of the data. Blind acceptance of optimal parameter estimates and best MSE values should be avoided.

Now consider Exhibit 8.23. By taking a transformation of the data, exponential growth patterns can also be modeled by applying a linear trend (Holt model) to the natural logarithm of the time series. The results are shown in Exhibit 8.24. The forecasts in Exhibit 8.24 are calculated first in terms of the logarithm of the series and

EXHIBIT 8.21 A		Linear Trend	Damped Trend	Linear Trend
Four-Period Forecast Comparison for Sales of a Cosmetic: Linear and Damped Trend Exponential Smoothing	Period	LT (0.64, 0.34, 1.0)	DT (0.64, 0.34, 0.88)	LT (0.81, 0.18, 1.0)
	2002 actual	236	236	236
	2003	255	253	214
	2004	228	230	180
	2005	201	210	144
	2006	174	192	109

(*Source*: Exhibit 8.19)

EXHIBIT 8.22 Time Plot of History and Forecast Profiles of Sales of a Cosmetic: Damped Exponential Trend ($\phi = 0.88$) Compared to Linear Trend ($\phi = 1.0$)

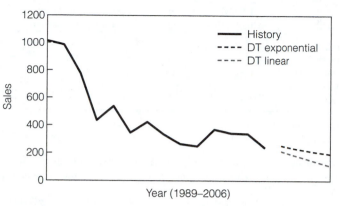

(*Source*: Exhibit 8.19)

EXHIBIT 8.23 Sale of a Cosmetic Product: Log Transform, Linear Trend

Worksheet: LINSMOOT.XLS
EXPONENTIAL SMOOTHING: EXPONENTIAL AND DAMPED-EXPONENTIAL TRENDS
TITLE1: DAMPED-EXPONENTIAL TREND
TITLE2: COSMETIC PRODUCT
X-AXIS: YEAR
Y-AXIS: SALES

LEVEL WEIGHT: 0.64
TREND WEIGHT: 0.34
TREND MODIFIER: 1.00
MSE: 0.002

Year	Mon	Per	Data	Fcst	Error	Error^2	Level	Trend
		0					3.53	20.01
1978	N/A	1	3.5177236	3.52	0.00	0.00	3.52	−0.01
1979		2	3.5582284	3.51	0.05	0.00	3.54	0.01
1980		3	3.4898179	3.54	−0.05	0.00	3.51	−0.01
1981		4	3.4795753	3.50	−0.02	0.00	3.49	−0.02
1982		5	3.4519399	3.47	−0.01	0.00	3.46	−0.02
1983		6	3.3727279	3.43	−0.06	0.00	3.39	−0.04
1984		7	3.3402458	3.35	−0.01	0.00	3.34	−0.05
1985		8	3.2457594	3.30	−0.05	0.00	3.26	−0.06
1986		9	3.1303338	3.20	−0.07	0.00	3.16	−0.09
1987		10	3.1044871	3.07	0.04	0.00	3.09	−0.08
1988		11	3.0681859	3.02	0.05	0.00	3.05	−0.06
1989		12	3.0056094	2.99	0.01	0.00	3.00	−0.05
1990		13	2.9934362	2.95	0.05	0.00	2.98	−0.04
1991		14	2.8881795	2.94	−0.05	0.00	2.91	−0.05
1992		15	2.6384893	2.85	−0.21	0.05	2.72	−0.13
1993		16	2.7307823	2.59	0.14	0.02	2.68	−0.08
1994		17	2.5390761	2.60	−0.06	0.00	2.56	−0.10
1995		18	2.6263404	2.46	0.16	0.03	2.57	−0.04
1996		19	2.5224442	2.52	0.00	0.00	2.52	−0.04
1997		20	2.4232459	2.48	−0.06	0.00	2.44	−0.06
1998		21	2.3944517	2.38	0.01	0.00	2.39	−0.06
1999		22	2.5670264	2.33	0.24	0.06	2.48	0.02
2000		23	2.5352941	2.50	0.03	0.00	2.52	0.03
2001		24	2.5301997	2.56	−0.03	0.00	2.54	0.02
2002		25	2.372912	2.56	−0.19	0.04	2.44	−0.04
2003		26		2.40			2.40	−0.04
2004		27		2.36			2.36	−0.04
2005		28		2.32			2.32	−0.04
2006		29		2.28			2.28	−0.04

(*Source*: SMOOTH.XLS, The Spreadsheet Forecaster on the CD)

	LT (0.64, 0.38, 1.0)	Linear Trend model	DT (0.64, 0.38, 0.8)
Period	(log-transformed data)	(untransformed column 1)	(from Exhibit 8.21)
2002 Actual	2.373	236	236
2003	2.40	251	253
2004	2.36	229	230
2005	2.32	209	210
2006	2.28	190	192

EXHIBIT 8.24
Forecast Model for Sales of a Cosmetic Product (logarithmic transformation): Linear Trend

(*Source*: Exhibit 8.23)

then transformed into the original data by exponentiation. To illustrate how the forecast was obtained for the year 2006 ($t = T + 4$), we set $m = 4$ and $T = 2002$:

$$\text{Forecast for } \log_{10}(T + 4) = 2.44 + 4 \times (-0.04) = 2.28$$

$$\text{Transform back to original data} = 10^{2.28} = 190$$

The results can be compared with the corresponding forecasts from the damped trend exponential model (DT) in Exhibit 8.21. The log-transform approach yields forecasts that are slightly lower than those shown in Exhibit 8.21; however, both depict a forecast profile with a negative exponential trend. Depending on the context in which these projections are used in practice, the differences could become substantial. This illustrates the limitation of using these types of techniques for extrapolating highly trending annual time series for more than a couple of periods. More important, when we calculate prediction limits on the forecasts, these two approaches also give different interpretations.

Because of the transformation, the log-transformed model will result in asymmetric (right-skewed) prediction limits, whereas the original model will give symmetric limits.

Prediction Limits for Exponential Smoothing Models

Prediction limits on forecasts have not generally been available for exponential smoothing models. To derive probability models, we need a modeling framework in which uncertainty (random-error assumption) is made explicit for the forecasting techniques. Up to this point, we have limited ourselves to the deterministic portions of the technique.

Earlier work on establishing prediction limits on forecasts for exponential smoothing appeared in Chatfield and Yar (1991), Yar and Chatfield (1990), and Koehler et al. (2001). A state-space framework for forecasting techniques using exponential smoothing has been developed in recent years by Hyndman et al. (2002). The state-space models that underlie the exponential smoothing techniques come in two forms: a model with additive errors and a model with multiplicative errors. Although the forecast profiles for the two models are identical, there are important differences in the prediction limits produced by the two models. With this distinction in error structure, the state-space framework effectively doubles the number of models in the Pegels classification from 12 to 24.

There are differences in their use as well. The **multiplicative error** models are not well defined if there are zeros or negative values in the data. Similarly, additive error models should not be used with **multiplicative trend** or multiplicative seasonality if any value is zero.

The estimation and model selection steps are beyond the scope of this book; the reader is referred to Hyndman et al. (2002). The software for exponential smoothing provided with this book uses the state-space formulations.

A Pegels classification of exponential smoothing techniques in a state-space modeling framework gives rise to 24 forecast profiles with prediction limits for trend and seasonal patterns.

Linear Trend

Consider Exhibit 8.25a. The initial values for level and trend are in the top cells of the columns Level and Trend, respectively. Thereafter, adding a fraction of the error to each smooths the level and trend. This is the same self-correcting idea used in the simple smoothing technique. The cells for LEVEL WEIGHT and TREND WEIGHT show that we used individual weights to smooth the level and trend. Each forecast is just the sum of the latest estimate of level and trend. For example, the linear trend forecast for 1990 is 48.02. This is the sum of the level and trend at the end of 1989

Exhibit 8.25 Company Sales: Comparison of Exponential Smoothing of (a) Linear and (b) Exponential Trends

(a)
LINEAR-TREND SMOOTHING
TITLE1: LINEAR-TREND SMOOTHING
TITLE2: COMPANY SALES
X-AXIS: YEAR
Y-AXIS: SALES

LEVEL WEIGHT: 0.80
TREND WEIGHT: 0.17
MSE: 4.410

Year	Mon	Per	Data	Fcst	Error	Error^2	Level	Trend
		0					17.73	3.07
78	NA	1	20.8	20.80	0.00	0.00	20.80	3.07
79		2	23.1	23.87	−0.77	0.59	23.25	2.94
80		3	27.2	26.19	1.01	1.02	27.00	3.11
81		4	30.0	30.11	−0.11	0.01	30.02	3.09
82		5	36.0	33.11	2.89	8.34	35.42	3.58
83		6	37.6	39.00	−1.40	1.97	37.88	3.34
84		7	38.0	41.22	−3.22	10.39	38.64	2.79
85		8	39.4	41.44	−2.04	4.16	39.81	2.45
86		9	41.6	42.26	−0.66	0.43	41.73	2.34
87		10	43.8	44.07	−0.27	0.07	43.85	2.29
88		11	42.0	46.14	−4.14	17.18	42.83	1.59
89		12	46.5	44.42	2.08	4.35	46.08	1.94
90		13		48.02				
91		14		49.96				
92		15		51.91				

(b)

EXPONENTIAL SMOOTHING: EXPONENTIAL AND DAMPED-EXPONENTIAL TRENDS

TITLE1: DAMPED-EXPONENTIAL TREND

TITLE2: COMPANY SALES

X-AXIS: YEAR

Y-AXIS: SALES

LEVEL WEIGHT: 0.80

TREND WEIGHT: 0.17

TREND MODIFIER: 0.80

MSE: 5.172

Year	Mon	Per	Data	Fcst	Error	Error^2	Level	Trend
		0					17.73	3.07
78	NA	1	20.8	20.80	0.00	0.00	20.80	2.45
79		2	23.1	22.76	0.34	0.11	23.03	2.02
80		3	27.2	24.65	2.55	6.51	26.69	2.05
81		4	30.0	28.33	1.67	2.79	29.67	1.92
82		5	36.0	31.20	4.80	22.99	35.04	2.35
83		6	37.6	36.92	0.68	0.46	37.46	2.00
84		7	38.0	39.06	−1.06	1.13	38.21	1.42
85		8	39.4	39.35	0.05	0.00	39.39	1.14
86		9	41.6	40.30	1.30	1.68	41.34	1.13
87		10	43.8	42.25	1.55	2.41	43.49	1.17
88		11	42.0	44.43	−2.43	5.89	42.49	0.52
89		12	46.5	42.91	3.59	12.92	45.78	1.03
90		13	46.61					
91		14	47.27					
92		15	47.79					
93		16	48.22					
94		17	48.55					
95		18	48.82					
96		19	49.04					
97		20	49.21					
98		21	49.35					
99		22	49.46					

(*Source*: SMOOTH.XLS, The Spreadsheet Forecaster on the CD)

(46.08 + 1.94). At the end of the data, we can forecast as many years ahead as we like. The forecast for 2 years ahead (1991) is 49.96, the 1990 forecast plus another increment of trend (48.02 + 1.94).

For most business data, the magnitudes of the level and trend can differ greatly. In this example, the numbers in the level column range from 7 to 25 times larger than the numbers in the trend column. Thus, the level weight has to be larger than the trend weight to give reasonable forecasts. The weights shown minimize the MSE. (See Exhibit 8.26 for the time plot.)

Damped Trend

Exhibit 8.25b is identical to Exhibit 8.25a, except that a new parameter has been added to modify the trend. Each forecast is the latest estimate of the level plus the trend times a trend modifier (shown in the cell for TREND MODIFIER). For example, the forecast for 1990 is 46.61. The formula is:

$$\text{Forecast} = (\text{Level at the end of 1989}) + (\text{Trend modifier})$$
$$\times (\text{Trend at the end of 1989})$$

$$= 45.78 + 0.80 \times 1.03 = 46.61$$

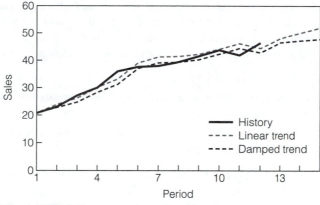

Exhibit 8.26 Time Plot of Company Sales: Linear and Damped Trend Smoothing

(*Source*: Exhibit 8.25)

To generate forecasts for more than one period into the future, the formula is

$$\text{Forecast} = (\text{Last forecast}) + (\text{Trend modifier})^{(\text{Number of periods into the future})} \times (\text{Final trend estimate})$$

The forecast for 1991 is thus

$$46.61 + 0.80^2 \times 1.03 = 47.27$$

Note that the trend modifier in Exhibit 8.25 is a fraction. Raising a fraction to a power produces smaller numbers as we move farther into the future. The result is called a damped trend because the amount of trend added to each new forecast declines.

By changing the trend modifier, we can produce different kinds of trend. A modifier equal to 1.0 yields a linear trend, exactly the same result as the linear trend worksheet. A modifier greater than 1.0 yields an exponential trend, one in which the amount of growth gets larger each time period. A modifier of 0 produces no trend at all and the results are the same as simple exponential smoothing. The weights shown minimize the MSE, given that the trend modifier has been set. (See Exhibit 8.26 for the time plot.)

In general, the damped trend techniques appear better suited for smoothing short-term patterns for operational forecasting in inventory and production planning than for smoothing long-term forecasting.

8.6 SEASONAL MODELS

The forecasting equations for additive and multiplicative seasonality are:

$$\text{Additive: } Y_T(m) = L_t + m \times T_t + \text{Seasonal index}$$

$$\text{Multiplicative: } Y_T(m) = [L_t + m \times T_t] \times \text{Seasonal index}$$

where $Y_T(m)$ denotes a forecast made at time T, the final season in the fit period, for m periods into the future. This technique is known as the Holt-Winters procedure.

Unlike the nonseasonal exponential smoothing techniques, seasonality brings in a complexity that makes interpreting the smoothing equations directly less intuitive. Fortunately, these algorithms are widely available in software, so we show here only that the forecast for m seasons ahead can be compiled using the steps:

1. Start at the current level, L_t.
2. Add the product of current trend, T_t, and the number of periods m ahead that we are projecting trend.
3. Adjust the resulting sum of level and trend for seasonality through a multiplicative or additive seasonal index.

Exhibit 8.27 is a time plot of a quarterly automobile sales time series that we analyzed in a preliminary two-way table decomposition in Chapter 4. Approximately 68% of the total variation was attributed to seasonality. The trend appears to be linear with seasonal peaks occurring in the second quarter of each year. The key results are summarized in Exhibit 8.28. (Recall that the various error measures—MAPE, MAE, MAD, and RMSE—are defined and interpreted in Chapter 5). Each measure for the model is compared with an appropriate naïve technique, such as Naïve_3 for seasonal data. The fit coefficient is $1 - $ [MSE of your model)/(MSE of Naïve_3)]. Naïve_3 is computed as follows:

1. The series is deseasonalized so that the seasonal pattern is removed.
2. The preliminary forecast for each period is taken as the seasonally adjusted value from the previous period.
3. The final forecasts are computed by reseasonalizing the preliminary forecasts.

Errors are computed by subtracting final forecasts from actual data.

The Naïve_3 technique assumes no trend, but has valuable information about seasonality. In a seasonal series with little trend, we cannot expect to do much better than Naïve_3.

The prediction limits are shown in Exhibit 8.27. These 95% prediction limits indicate that we are 95% sure that the true (in the sense that the model is correct)

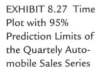

EXHIBIT 8.27 Time Plot with 95% Prediction Limits of the Quartely Automobile Sales Series

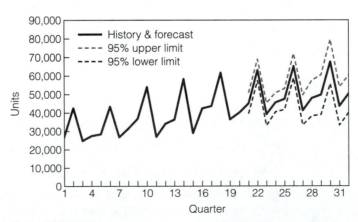

(*Source*: Exhibit 4.20)

EXHIBIT 8.28
Model-Fitting
Summary for
Quarterly
Automobile Sales—
Linear Trend,
Additive Seasonality

(a)

	Additive		
Level = 0.02	Trend = 0.115		Seasonality = 0
	Your Model		Naïve_3
MAPE	7.10%		8.60%
ME	120		742
MAE	2335		2825
RMSE	2815		3739
Fit coefficient =	0.43		

(b)

	Multiplicative		
Level = 0.34	Trend = 0.065		Seasonality = 0
	Your Model		Naïve_3
MAPE	6.10%		6.60%
ME	558		818
MAE	2190		2426
RMSE	2486		3011
Fit coefficient =	0.32		

(*Source*: Exhibit 4.20)

forecasts will lie within these limits. The forecasts produced by the fitted model (not shown) fall halfway between these limits.

The smoothing weights (level = 0.02, trend = 0.115, and seasonal = 0.00) in Exhibit 8.29(a) reveal the relative emphasis given to the data from the recent and more distant past in the calculation of the current level, trend, and seasonal indexes. The values for the current level, trend, and seasonal indexes are called final values. Note that period T is the second quarter (spring) of Year 6. To forecast from this time origin, the current level is 46,841, the current trend is 561.8 (units per quarter), and each season has its own seasonal index. The summer index (Q3 = -2225.8) indicates that automobile sales during the summer tend to be approximately 2226 units below the norm.

Starting from spring Q2 of Year 6, the automobile sales forecast for Q3 is (setting $m = 1$):

$$Y_T(m) = [L_t + m \times T_t] + \text{Seasonal index}$$

$$Y_T(1) = [46,841 + 561.8] + (-2225.8)$$

$$= 45,177 \text{ units}$$

EXHIBIT 8.29
Forecast Model for
Automobile Sales:
Additive Winters—
Linear Trend,
Additive Seasonality

Smoothing Component	Weight	Final Value (Year 6:II)
Level	0.02	46,841
Trend	0.115	562
Seasonal	0	$-3388, -2225, 15,216, -9602$

First four forecasts: 45,178; 63,181; 38,925; 45,701

(*Source*: Exhibit 4.20)

To forecast three periods ahead to winter Q1 of Year 7, we set $m = 3$ and use the seasonal index for Q1:

$$Y_T(3) = [L_t + 3 \times T_t] + \text{Seasonal index}$$

$$= [46{,}841 + 3 \times 561.8] + (-9602.5)$$

$$= 38{,}924 \text{ units}$$

The forecast for the winter of Year 7 is lower than that for the previous summer for two reasons. First, the trend is growing only by approximately 562 units per quarter. Second, and more substantially, the seasonal index for winter is 7377 units lower than in summer.

In Exhibit 8.30, the multiplicative seasonal model produces the following forecast for the winter quarter of Year 7:

$$Y_T(3) = [L_t + 3 \times T_t] \times \text{Seasonal index}$$

$$= [46{,}961.5 + 3 \times 957.1] \times 0.755$$

$$= 37{,}624 \text{ units}$$

In this example, the two versions of Holt-Winters procedure give very similar values for the current level and trend. The seasonal indexes are in a different form and result in different projections. The additive index for the summer season tells us that summer sales tend to be approximately 15,000 units above the norm; this will be a constant amount for all future years. In contrast, the multiplicative index for the summer season is estimated to be approximately 39% above the norm, which represents an increasing amount as long as the data are trending up.

Which model is preferable? On a strictly statistical basis, the multiplicative model has the better summary results. However, this may not necessarily mean that forecast performance will be better as well. In Exhibit 8.31, we provide some forecasting performance comparisons between an additive and multiplicative model for a monthly time series of hotel/motel demand (Exhibit 8.32).

In the examples so far, we have been concerned with how well a technique fits a time series. However, fit may not be a good indication of postsample forecast accuracy. The reason is that the same data are used to estimate the parameters and evaluate its suitability as a forecasting tool.

EXHIBIT 8.30
Forecast Model for
Automobile Sales:
Multiplicative
Winters—Linear
Trend, Multiplicative
Seasonality

Final Value (Year 6:II)

46,961.5
957.1
0.94, 1.39, 0.76, 0.91
First four forecasts (rounded):
45,043; 67,937; 37,873; 46,219;

(*Source*: Exhibit 8.20)

EXHIBIT 8.31
Model-Fitting
Summary for Monthly
Hotel/Motel Demand
(DCTOUR)—Linear
Trend, Additive
Seasonality and
Linear Trend,
Multiplicative
Seasonality

	Additive	Multiplicative
Level	1.00	1.00
Trend	0.205	0.275
Season	0.00	0.00
MAPE	3.4%	3.1%
ME	−7436	−6419
MAE	38,511	35,839
RMSE	58,665	51,245
Fit coefficient	−2.1	−1.4

(*Source*: Exhibit 8.32)

The preferred way to simulate forecast accuracy is by creating a holdout period and distinguishing a fit period (to estimate parameters) from a forecast period (to determine forecast accuracy).

Evaluation procedures are taken up in detail in Chapter 15. However, here we display and discuss some results for exponential smoothing techniques that can be readily created on spreadsheets.

The first analysis is to compare the summary statistics over the fit period for two competing seasonal models for the monthly hotel/motel demand data (DCTOUR): additive and multiplicative seasonality with linear trend, also known as the Holt-Winters technique (Exhibit 8.28). The historical period consists of 96 monthly values. The data are clearly seasonal (Exhibit 8.32).

The smoothing parameters come from minimizing MSE over the fit period. It is noteworthy, and perhaps somewhat unsettling, that the fit coefficients are negative, which suggests that the Naïve_3 model is better than these Holt-Winters models for the data (Naïve_3 has smaller MSE over the fit period). On the basis of the error measures, the multiplicative version appears preferable to the additive version. Hence, we will proceed with the multiplicative Holt-Winters technique.

EXHIBIT 8.32 Time
Plot of Monthly
Hotel/Motel
Demand (DCTOUR)

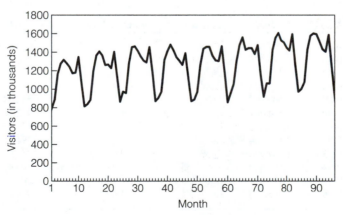

(*Source*: D. C. Frechtling, 1996, App. 1)

Because our primary interest is in forecasting performance, we started with a holdout period of the last 12 months and created forecasts based on the first 84 monthly values. The summary statistics led to the same conclusions we made for the longer time series. Level and seasonal parameter estimates were unchanged and the trend parameter estimates changed slightly to 0.215 (additive) and 0.285 (multiplicative). The stability of these model parameter estimates may be relatively unimportant because frequently parameter estimates can vary widely with little impact on forecast accuracy.

The evaluation in Exhibit 8.33 is a static test in the sense that all forecasts were made from a single time point ($T = 1984$) for a fixed horizon ($m = 12$). In a dynamic evaluation, both the starting point and the horizon change. In Exhibit 8.34, we summarize the forecasting evaluations using the MAD, MAPE, and RMSE performance statistics. For one-period-ahead forecasts (lead = 1), there are 12 possible forecasts that could be generated by moving the starting point through the 12-month horizon. The average of the 12 one-period-ahead forecast percentage errors is 2.4%. This suggests that one-period-ahead forecasts can be accurate within 3% or so. For longer horizons, the accuracy decreases, but is still within 10%. The lead 12 accuracy is based on only one forecast, so the improved accuracy is an anomaly.

When we repeat this dynamic evaluation for the additive seasonal version of Holt-Winters model (not shown), the range of MAPEs is between 2.5% and 14%. Hence, the multiplicative model is clearly a better choice for these data.

The one-step-ahead forecast errors play a special role in the analysis of forecast performance because prediction limits are based on them. Exhibit 8.35 summarizes the 12 one-period-ahead forecasts made with the damped trend, multiplicative seasonal model. These forecasts are compared with the actuals in the holdout sample and forecast errors are calculated. The final level and trend components are shown, along with the seasonal index for that month. Evidently, the peak month is May (index = 1.187), meaning that May is almost 19% above norm. This may be attributable to the attractions of the spring season in this location. As expected for these data, the low months are in the winter.

EXHIBIT 8.33	Period	Actual	Forecast	Percentage Error	Lower Probability	Upper Limits
Monthly Hotel/Motel Demand Series— Linear Trend, Multiplicative Seasonality; Forecast Errors over 12-Month Horizon	1985	1,001,666	1,068,239	−6.65	1,004,435	1,132,043
	1986	1,073,196	1,128,479	−5.15	1,056,364	1,200,594
	1987	1,421,423	1,436,573	−1.07	1,360,753	1,512,392
	1988	1,577,321	1,594,802	−1.11	1,515,704	1,673,901
	1989	1,600,991	1,634,021	−2.06	1,552,044	1,715,997
	1990	1,594,481	1,568,101	1.65	1,482,740	1,653,463
	1991	1,510,052	1,519,918	−0.65	1,431,133	1,608,703
	1992	1,436,164	1,483,799	−3.32	1,391,555	1,576,042
	1993	1,404,978	1,453,293	−3.44	1,357,560	1,549,027
	1994	1,585,409	1,605,685	−1.28	1,506,434	1,704,937
	1995	1,234,848	1,305,234	−5.70	1,202,439	1,408,029
	1996	923,115	1,031,071	−11.69	924,812	1,137,329

(*Source*: Exhibit 6.20)

	Lead	Number of Forecasts	MAD	MAPE	RMSE
EXHIBIT 8.34 Performance Evaluation of the Hotel/Motel Demand Series—Damped Trend, Multiplicative Seasonality Evaluation (Dynamic)	1	12	30,871	2.40%	37,779
	2	11	43,041	3.09%	55,160
	3	10	48,202	3.43%	60,343
	4	9	52,679	3.75%	68,680
	5	8	70,180	5.19%	98,091
	6	7	72,493	5.56%	99,852
	7	6	72,736	5.12%	99,632
	8	5	93,286	7.23%	119,237
	9	4	116,838	9.08%	149,592
	10	3	93,531	8.31%	122,273
	11	2	88,349	9.11%	103,814
	12	1	44,061	4.77%	44,061

(*Source*: Exhibit 8.32)

	Date	Actual	Forecast	Error	Level	Trend	Index
EXHIBIT 8.35 One-Period-Ahead Forecasts over the Holdout Period for the Hotel/Motel Demand Series—Damped Trend, Multiplicative Seasonality	1994—1	1,001,666	1,063,949	−62,283	1,326,148	−2660	0.77
	1994—2	1,073,196	1,045,975	27,221	1,345,926	−623	0.79
	1994—3	1,421,423	1,398,118	23,306	1,359,538	337	1.04
	1994—4	1,577,321	1,578,977	−1656	1,358,871	178	1.16
	1994—5	1,600,991	1,614,232	−13,241	1,352,017	−256	1.19
	1994—6	1,594,481	1,537,137	57,344	1,383,392	1525	1.14
	1994—7	1,510,052	1,502,676	7376	1,388,664	1250	1.09
	1994—8	1,436,164	1,456,457	−20,293	1,377,391	186	1.05
	1994—9	1,404,978	1,412,361	−7383	1,373,012	−117	1.02
	1994—10	1,585,409	1,593,154	−7745	1,368,758	−303	1.16
	1994—11	1,234,848	1,261,615	−26,767	1,350,392	−1179	0.92
	1994—12	923,115	951,879	−28,764	1,324,091	−2162	0.70

(*Source*: Exhibit 8.32)

8.7 HANDLING SPECIAL EVENTS WITH SMOOTHING MODELS

Special events arise whenever periodic actions of the organization, such as special promotions, scheduled disruptions for maintenance, unusual weather, and holiday effects, cannot be modeled as seasonality because they do not fall within the same period (week) each year. A special event adjustment can be combined with an exponential smoothing technique to improve the accuracy of a forecast.

A special event refers to a sudden change in the level of the time series that is expected to recur.

When dealing with an outlier, it is generally not possible to simply delete it because doing so would leave a gap or missing value event between adjacent time periods. One procedure to remedy missing values is to interpolate, or replace the value by averaging the two adjacent values. For example, a missing value for June can be replaced by the average for May and July. An alternative to interpolating missing values is to use dummy variable events. We first identify the specific time period(s) when the outlier occurs and then create dummy variable events—a time series of 0s and 1s with 1 for each outlier time period and 0 for each normal time period. Once this dummy time series is incorporated into a forecasting technique, the outlying values are effectively removed from the calculations of the coefficients of the forecasting formula. The resulting forecasts are generated as if the unusual time periods never occurred.

Dummy variable events can be used with other forecasting technique as well. We consider regression and ARIMA applications in later chapters. For exponential smoothing, event adjustments can be applied both to outliers and special events.

Event Adjustments for Outliers

First, the dummy variable event is created as an index in the exponential smoothing forecasting formula. For example, in the Winters method, the forecasting formula is an extension of the exponential smoothing technique for multiplicative and additive forms:

Additive: $Y_T(m) = L_t + m \times T_t + \text{Seasonal index} + \text{Event index}$

Multiplicative: $Y_T(m) = [L_t + m \times T_t] \times \text{Seasonal index} \times \text{Event index}$

A multiplicative event index is interpreted as a multiple of the level of the series. For example, if the current level is 100 and the event index is 0.75, the result indicates that the average effect of the outlier value(s) was to reduce the level of the series by 25%. The same result in the additive model becomes an event index of -25. Note that the effect of the event index in the procedure is to remove the outlier values from the calculation of the trend component and seasonal indexes. Then, the event index is set ($=0$ for additive, $=1$ for multiplicative) to generate forecasts on the assumption that no outliers will occur over the forecasting horizon.

Event Adjustments for Promotions and Other Special Events

When data are available on the timing and magnitude of product promotions (price changes), we can effectively use this information in a **causal forecasting** approach, such as a regression model, in which the promotion variable enters as one or more explanatory (independent) variable.

Although exponential smoothing does not explicitly incorporate explanatory variables, it can be extended to incorporate dummy variables. In this manner, a dummy (0,1) coding can be used to distinguish those periods in which a promotion occurred from those free of promotion. The introduction of such a dummy event series effectively models the timing of the promotions, without specifying their magnitude.

Exhibit 8.36 shows a summary from an analysis of the monthly sales of a consumer product during a 24-month period. There were two peaks in the data associated with promotions during December of Year 1 and a promotion in August of Year 2. Three versions of a Holt-Winters (multiplicative) procedure were fit.

Version 1 is a direct application of a multiplicative Winters seasonal exponential smoothing method. The MAPE is 16.3% for the 24-month fit period. The strong effect of the twin promotions is viewed as seasonality, giving the months December,

Model	Promotions Variable	MAPE	Promotions Index	Peak Seasonal Indexes
Version 1	None	16.30%	n.a.	Jan, Aug, Sep, Dec 1.24, 1.22, 1.07, 1.12
Version 2	(0,1)	11.20%	1.57	Jan, Feb, Sep, Dec 1.56, 1.17, 1.14, 1.04
Version 3	(0,1,2)	6.90%	Event 1: 1.71 Event 2: 1.75	Jan, Feb, Nov, Dec 1.38, 1.23, 1.14, 1.19

EXHIBIT 8.36
Forecast Performance
with Three Models
(n.a., not available)

January, August, and September the high seasonal indexes of the year. However, promotions will appear as the seasonal indexes only if they are scheduled for the same months every year. This was not the case for this product in Year 3.

Version 2 employs an event adjustment model in which the occurrence of promotions is represented by a (0,1) dummy variable. For the historical period, the dummy variable is set equal to 1 in December of Year 1 and August of Year 2 and set equal to zero otherwise. The Winters procedure results in a substantial reduction in the MAPE from 16.3% to 11.2%, along with a much altered seasonal pattern. Now the high seasonal indexes are attributed to January, February, and September because December of Year 1 and August of Year 2 are identified as promotion months. The event index is reported as 1.57, a result that implies that, on average, sales during a promotions month will be approximately 57% higher than sales in a normal month.

Although this is an improvement over Version 1, the seasonal pattern in Version 2 is still problematic. The high indexes for January and September are felt to be, at least in part, a response to the promotions occurring in December of Year 1 and August of Year 2 and to be unrepresentative of the plans for Year 3.

Version 3 replaces the (0,1) coding for promotions by a (0,1,2) coding. In general, a (0,1,2) coding indicates that there are two types of special events being dealt with: Event 1 and Event 2. In this example, the Event 2 was used to represent the delayed response of sales to an earlier promotion. Hence, the promotion time series was set equal to 1 for December of Year 1 and August of Year 2, 2 during January of Year 2 and September of Year 2, and 0 for all other months during these 24 months.

Version 3 very substantially improves the fit to the historical data, with the MAPE falling to 6.9%. The event indexes imply that when a promotion occurs, sales can be expected to rise by 71% in the same month (Event 1) and 75% in the following month (Event 2). The January seasonal index has fallen from Version 2 because part of the January strength is now attributed to the December of Year 1 promotion.

Extensions of Event Modeling

In Version 3, we have illustrated the use of a (0,1,2) coding to distinguish the immediate (Event 1) from the delayed effect (Event 2) of a promotion. In the same manner, the (0,1,2) coding can be used to try to discriminate between strong and weak promotions; or between two types of promotions; between a promotion and an outlier; or between two special events, neither of which is necessarily related to a promotion. Further extensions are feasible, for example to three events, by using a (0,1,2,3) coding.

SUMMARY

This chapter provides an introduction to a family of exponential smoothing techniques useful for forecasting trending and seasonal data. The components can describe a current level, trend, and seasonal index. The current level is the starting point, the trend is the growth or decline factor, and seasonal index is the adjustment for seasonality. All three components are exponentially weighted averages, rather than equally weighted averages, of the historical data. In calculating the current level, an exponentially weighted average is taken of the past data. The current trend is an exponentially weighted average of the past changes in the level and each seasonal index is an exponentially weighted average of the past ratios of data to level.

The estimation of parameters in the exponential smoothing algorithms is not emphasized because in practical situations estimates can vary widely without significantly affecting the forecast profile of the algorithm. Optimal or near-optimal parameter settings are readily derived with modern software tools. In addition, combinations of multiple parameter values severely limit an intuitive feel of their impact on the forecasts. Moreover, for inventory-replenishment planning purposes, a forecaster needs to rely on the automatic forecasting features of modern software because of the very large volume of data involved.

We have described the key characteristics of exponential smoothing as a projection technique using (exponentially decaying) weights that give more emphasis to more recent periods in the data. Several examples are used to illustrate the nature of exponentially decaying weights for different smoothing techniques. The techniques can be classified in terms of forecast profiles generated by the techniques, which helps the forecaster select the most appropriate technique for handling trend and seasonal patterns. We have shown how the forecasting formula works for Holt and Winters methods and also for nonlinear trend procedures such as those with damped and exponential trends.

REFERENCES

Box, G. B. P., G. M. Jenkins, and G. Reinsel (1994). *Time Series Analysis, Forecasting and Control*. 3rd ed. Upper Saddle River, NJ: Prentice Hall.

Brown, R. G. (1963). *Smoothing, Forecasting, and Prediction of Discrete Time Series*. Englewood Cliffs, NJ: Prentice Hall.

Brown, R. G. (1982). *Advanced Service Parts Inventory Control*. 2nd ed. Norwich, VT: Materials Management Systems.

Chatfield, C., and M. Yar (1991). Prediction intervals for the multiplicative Holt-Winters. *Int. J. Forecasting* 7(1), 31–38.

Cryer, J. D., and R. B. Miller (1994). *Statistics for Business*. Belmont, CA: Wadsworth.

DeLurgio, S. (1998). *Forecasting Principles and Applications*. New York: Irwin/McGraw-Hill.

Dielman, T. E. (1996). *Applied Regression Analysis for Business and Economics*. 2nd ed. Belmont, CA: Wadsworth.

Frechtling, D. C. (1996). *Practical Tourism Forecasting*. Oxford: Butterworth-Heinemann.

Gardner, E. S. Jr. (1985). Exponential smoothing: The state of the art. *J. Forecasting* 4, 1–28.

Hanke, J. E., and A. G. Reitsch (1998). *Business Forecasting*. 6th ed. Englewood Cliffs, NJ: Prentice Hall.

Holt, C. C. (2004). Forecasting seasonals and trends by exponentially weighted moving averages. *Int. J. Forecasting* 20, 5–10.

Holt, C. C., F. Modigliani, J. F. Muth, and H. A. Simon (1960). *Planning Production, Inventories, and Work Force*. Englewood Cliffs, NJ: Prentice Hall.

Hyndman, R. J., A. B. Koehler, R. D. Snyder, and
R. D. Grose (2002). A state space framework
for automatic forecasting using exponential
smoothing methods. *Int. J. Forecasting* 18,
439–54.

Koehler, A. B., R. D. Snyder, and J. K. Ord (2001).
Forecasting models and prediction intervals for
the multiplicative Holt-Winters method. *Int.
J. Forecasting* 17, 269–86.

Makridakis, S., S. C. Wheelwright, and R. J. Hyndman
(1998). *Forecasting Methods and Applications*.
3rd ed. New York: John Wiley & Sons.

Pegels, C. C. (1969). Exponential forecasting: Some
new variations. *Manage. Sci.* 12, 311–15.

Tryfos, P. (1998). *Methods for Business Analysis and
Forecasting: Text and Cases*. New York: John
Wiley & Sons.

Wilson, J. H., and B. Keating (1998). *Business Forecasting*. New York: Irwin McGraw-Hill.

Winters, P. R. (1960). Forecasting sales by exponentially
weighted moving averages *Manage. Sci.* 6, 324–42.

Yar, M., and C. Chatfield (1990). Prediction intervals
for the Holt-Winters forecasting procedure. *Int.
J. Forecasting* 6, 127–37.

PROBLEMS

8.1 By setting $F_{t+1} = S_t$ in the smoothing equation
$S_t = \alpha Y_t + (1 - \alpha) S_{t-1}$, the basic exponential
smoothing forecasting model becomes $F_{t+1} =$
$\alpha Y_t + (1 - \alpha) F_{t-1}$. Using expressions for F_{t-1},
F_{t-2}, F_{t-3}, and so on and making successive
substitutions, derive an expression for F_{t+1} in
terms of past values of Y_t.

    a. What are the weights?

    b. Show that the weights sum to unity.

8.2 Consider a time series {200, 135, 195, 198, 310,
175, 155, 130, 220, 278, 235}.

    a. Compute the smoothed (fitted) values with
values for $\alpha = 0.1$, 0.5, and 0.9.

    b. Plot the results on a single graph, and inter-
pret the effect that the value of α has on the
amount of smoothing done.

    c. Calculate the squared differences of the
smoothed values from the corresponding values of the time series, and compare the sum
of squared differences for $\alpha = 0.1$, 0.5, and
0.9. How might you interpret the rankings of
these three sums of squared errors?

8.3 Show that the N-term moving average can be
written in the form $S_t = \alpha Y_t + (1 - \alpha) S_{t-1}$.
What simplifying assumption do you have to
make? What is the value of α in terms of N?

8.4 To develop the comparison of exponential
smoothing and a moving average further, it may
be useful to calculate the average age of the data
in the two methods. For a p-period moving average, the average age is

$$P = (0 + 1 + 2 + \cdots + p - 1)/p = (p - 1)/2$$

For simple exponential smoothing, the weight
given to data p periods ago is $\alpha(1 - \alpha)^p$.

    a. Show that the average age is $(1 - \alpha)/\alpha$.

    b. Show that an exponential smoothing model
that has the same average age as the p-period
moving average has a smoothing constant
given by $\alpha = 2/(p + 1)$.

    c. Show that the same result can be obtained by
equating the variances of S_t and M_t.

8.5. For the annual mortgage rate series depicted in
Exhibit 2.5:

    a. Create a time plot and a 4-year holdout sam-
ple from the end of the data.

    b. Build a single exponential smoothing model
with the reduced data set.

    c. Evaluate forecast performance over the hold-
out sample using at least two accuracy measures developed in Chapter 5.

    d. Comment on the appropriateness of the tech-
nique for the particular time series.

8.6. Repeat Problem 8.5 using Holt-Winters (linear
trend) for several of the following monthly time
series, choosing between the additive or multiplicative seasonal version based on inspection of
the time plot (e.g., sales revenue data tend to be
multiplicative and unit shipment data tend to be
additive):

    a. Unemployment rate data (UNEMP, Dielman,
1996, Table 4.7)

    b. Silver prices data (SILVER, Dielman, 1996,
Table 4.8)

c. Prime rate data (PRIME4, Dielman, 1996, Table 4.14)

d. Retail sales for U.S. retail sales stores (RETAIL, Hanke and Reitsch, 1998)

e. Wheat exports (SHIPMENT, Dielman, 1996, Table 4.9)

f. Shipments of spirits (DSHIP, Tryfos, 1998, Table 6.23)

g. Retail sales in a region (RSALES, Tryfos, 1998, Table 6.25)

h. Australian beer production (AUSBEER, Makridakis et al., 1998, Table 2.2)

i. International airline passenger miles (AIRLINE, Box et al., 1994, Series G)

j. Australian electricity production (ELECTRIC, Makridakis et al., 1998, Fig. 7-9)

k. French industry sales for paper products (PAPER, Makridakis et al., 1998, Fig. 7-20)

l. Shipments of French pollution equipment (POLLUTE, Makridakis et al., 1998, Table 7-5)

m. Sales of recreational vehicles (WINNEBAG, Cryer and Miller, 1994, p. 742)

n. Sales of lumber (LUMBER, DeLurgio, 1998, p. 34)

o. Sales of consumer electronics (ELECT, DeLurgio, 1998, p. 34)

8.7. Repeat Problem 8.5 using the Holt-Winters (linear trend) model for several of the following quarterly time series, choosing between additive or multiplicative seasonal version based on inspection of the time plot (e.g., sales revenue data tend to be multiplicative and unit shipment data tend to be additive):

a. Sales for Outboard Marine (OMC, Hanke and Reitsch, 1998, Table 4.5)

b. Shoe store sales (SHOES, Tryfos, 1998, Table 6.5)

c. Domestic car sales (DCS, Wilson and Keating, 1998, Table 1-5)

d. Sales of The GAP (GAP, Wilson and Keating, 1998, p. 30)

e. Air passengers on domestic flights (PASSAIR, DeLurgio, 1998, p. 33)

8.8 Repeat Problem 8.6 using the damped trend model, choosing between additive or multiplicative seasonal version based on inspection of the time plot.

8.9 For the time series shown in Problem 8.6, determine by visual inspection whether a multiplicative seasonal version is preferred over an additive seasonal version. For a multiplicative seasonal choice:

a. Repeat steps in Problem 8.5 with the damped trend, multiplicative seasonal exponential smoothing technique.

b. Take logarithms of the data and repeat Problem 8.5 with the damped trend, additive seasonal smoothing technique.

c. Exponentiate the projections determined in (b).

d. Create forecast errors (residuals over the holdout period) from (c) and determine at least two measures of accuracy.

e. Compare and comment on your results in (a) with your findings in (c).

CASE 8A PRODUCTION OF ICE CREAM (CASES 1A, 3A–7A CONT.)

Having performed a thorough preliminary data analysis looking for the important patterns, the entrepreneur feels prepared to build some models on the monthly shipments data.

(1) Reviewing the Pegels diagram, select three of the most appropriate model formulations for the ice cream data.

(2) Run the models, using the first 5 years (60 months) and project the remaining 8 months.

(3) Summarize and interpret the performance measures over the fit period.

(4) Plot a correlogram of the residuals. Is there any suggestion of any nonstationarity?

(5) Use at least two forecast accuracy measures and evaluate the performance of the models over the 8-month holdout period.

(6) Summarize your results and make a recommendation as to what model(s) should be retained for forecasting in the future.

(7) With your best model(s), create forecasts for the 16 months following the last data period.

(8) Create a time plot of the history, forecasts, and upper and lower prediction limits for presentation to the entrepreneur.

CASE 8B DEMAND FOR APARTMENT RENTAL UNITS (CASE 6C CONT.)

Following your preliminary analysis of Cozy House Apartments in Case 6C, you want to examine forecasts from exponential smoothing models.

(1) On the basis of the preliminary results, which models in the Pegels diagram can you exclude? Which ones are worth considering?

(2) By holding out YR11 from the analysis, create your best forecasting model for the data.

(3) Accumulate the data into quarterly periods, and repeat (2).

(4) Using your results from (2) and (3), what are your suggested forecasts for the first quarter of the holdout period? For the whole year?

CASE 8C ENERGY FORECASTING (CASE 1C CONT.)

Before embarking on an econometric modeling project, you want to gain further insight into the data by running some exponential smoothing models. Because the data are nonseasonal, you will only be looking at the nonseasonal models.

The table that follows contains 42 years of annual data (1947–1988) for U.S. energy consumption and U.S. energy production (in quadrillion BTUs) (1989).

Year	Energy Consumption	Energy Production	Year	Energy Consumption	Energy Production
1947	33.0	35.0	1968	61.7	56.6
1948	33.9	35.9	1969	65.0	58.7
1949	31.5	30.6	1970	67.1	62.5
1950	34.0	34.4	1971	68.3	61.7
1951	36.8	37.6	1972	71.6	62.8
1952	36.5	36.7	1973	74.6	62.1
1953	37.6	37.0	1974	72.7	60.8
1954	36.3	35.3	1975	70.5	59.9
1955	39.7	39.1	1976	74.3	59.9
1956	41.7	41.8	1977	76.2	60.2
1957	41.7	42.1	1978	78.0	61.1
1958	41.7	39.2	1979	78.8	63.8
1959	43.1	40.7	1980	75.9	64.8
1960	44.6	41.6	1981	73.9	64.4
1961	45.3	42.3	1982	70.8	63.9
1962	47.4	43.9	1983	70.5	61.2
1963	49.3	46.0	1984	74.0	65.8
1964	51.2	47.6	1985	73.9	64.8
1965	53.3	49.1	1986	74.2	64.2
1966	56.4	51.9	1987	76.8	64.8
1967	58.3	54.8	1988	79.9	65.8

First, consider trend models.

(1) Create time plots for the energy consumption and energy production variables. Create a scatter diagram between the two variables and interpret your findings.

(2) If trend is the predominant component in the data, use the linear, exponential, and damped trend exponential smoothing models as follows:

   (a) Fit the models through the first 22 years and project the remaining 20 years.

   (b) Summarize forecast accuracy for the 1- to 20-year forecast horizon with at least two accuracy measures.

   (c) Is there a tendency to over- or underforecast?

   (d) What does that suggest about the overall pattern in the data?

(3) Repeat (2a) and (2b) by fitting through the first 27 years, the first 32 years, and the first 37 years.

(4) Is any one model always best? If not, review the 1-year-ahead and 2-year-ahead forecasts for the models.

(5) What degree of confidence do you have in the one- and two-period-ahead forecasts?

Next, consider growth rate models.

(6) Take first differences of the logarithms of the data. These are the approximate relative growth rates.

(7) Repeat (1) through (5).

(8) Because the growth rates could be relatively constant, use the Naive_1 model as a benchmark for comparison.

Now consider univariate forecasts for 1988–1998.

(9) From your preferred trend and growth rate models, create a new 10-year forecast, using the performance results to adjust up or down based on your best judgment. (Note: We do not suggest there is a "best" model).

(10) If possible, compare your results with an updated data source for U.S. total gross consumption and production of energy sources.

CASE 8D A PERSPECTIVE OF THE AUTOMOTIVE INDUSTRY—JAPANESE AUTOMOBILE PRODUCTION (CASES 1D, 3D, 4D, 6D, AND 7E CONT.)

In Case 1D, you investigated the changes in the factors affecting the demand of automobiles in the most recent decade and proposed a modeling strategy for forecasting the next decade. Your assignment now is to create some models for the historical data used in the 1988 study and update them with data from a recent decade. The following data on Japanese production of passenger cars has been made available to you.

Japanese Production of Passenger Cars, 1956–1987

Year	Number of Cars	Year	Number of Cars
1956	32,056	1972	4,022,289
1957	47,121	1973	4,470,550
1958	50,643	1974	3,931,842
1959	78,598	1975	4,567,854
1960	165,094	1976	5,027,792
1961	249,508	1977	5,431,045
1962	268,784	1978	5,975,698
1963	407,830	1979	6,175,771
1964	579,660	1980	7,038,108
1965	696,176	1981	6,974,131
1966	877,656	1982	6,881,586
1967	1,375,755	1983	7,151,888
1968	2,055,821	1984	7,073,173
1969	2,611,499	1985	7,646,816
1970	3,178,708	1986	7,809,809
1971	3,717,858	1987	7,891,087

First, consider trend models.

(1) Create a time plot for Japanese production of passenger cars (1956–1987). Find and plot comparable data for (1988–present). Comment on the periods when the trend appears interrupted.

(2) If trend is the predominant component in the data, use the linear, exponential, and damped trend exponential smoothing models as follows:

    (a) Fit models through the first 22 years (1956–1977) and project the remaining 10 years.

    (b) Repeat (a) for 1957–1978, and project the remaining 9 years.

    (c) Continue in this way, shifting 1 year and using a fixed length of 22 years for fitting and forecasting the remaining data.

    (d) Summarize forecast accuracy for the 1- to 10-year forecast horizon with at least two accuracy measures.

    (e) Is there a tendency for some models to over- or underforecast?

    (f) Create a box plot of the forecast errors by year for each of the forecasted years 1978, 1979, . . . , 1987 using all your models

    (g) What does (f) suggest about the overall behavior of the models for this data set?

(3) Repeat (2a) and (2b) by fitting through the first 23 years, the first 24 years, and so on.

(4) Is any one model always best? If not, review the 1-year-ahead and 2-year-ahead forecasts for the models.

(5) What degree of confidence do you place in the one- and two-period-ahead forecasts?

Next, consider growth rate models.

(6) Take first differences of the logarithm of the data. These are approximate relative growth rates

(7) Repeat (1) through (5).

(8) Because the growth rates could be relatively constant, use the Naive_1 model as a benchmark for comparison.

Now consider univariate forecasts for 1988–1998.

(9) From your preferred trend and growth rate models, create a new 10-year forecast, using the performance results to adjust up or down based on your best judgment. (Note: We do not suggest there is a "best" model).

(10) If possible, compare your results with an updated data source for Japanese production of passenger cars.

CASE 8E THE DEMAND FOR CHICKEN

Chicken is one of the major food commodities in our daily life. The accompanying table shows the per-capita consumption of chicken (in pounds) in the United States. (Economic Research Service, U.S. Department of Agriculture). A research firm specializing in agricultural economics has recruited you. Your first assignment is to assist one of the principals in the firm with a study for a chicken franchise on some long-term trend forecasts for various meat products.

Year	Chicken	Year	Chicken	Year	Chicken
1965	33.7	1977	42.8	1989	59.3
1966	35.6	1978	44.9	1990	61.5
1967	35.6	1979	48.3	1991	64.0
1968	37.1	1980	49.0	1992	67.8
1969	38.5	1981	49.4	1993	70.3
1970	40.3	1982	49.6	1994	71.1
1971	40.2	1983	49.8	1995	70.4
1972	41.7	1984	51.6	1996	71.3
1973	39.9	1985	53.1	1997	72.4
1974	39.7	1986	54.3	1998	72.9
1975	39.0	1987	57.4	1999	77.5
1976	39.1	1988	57.5	2000	77.9
				2001	77.6

(1) Create a time plot of the data and of the logarithms of the data. Comment on the nature of the nonstationarity.

(2) Divide the data into three groups. Use the damped trend model on each group, and create forecasts for the next 12 periods. Evaluate the forecasts with the actuals provided in the following group. How would you use the performance results obtained from the first two groups to tune the forecasts for the third group?

(3) Divide the data into two groups and repeat (2). Compare the estimated model coefficients. Are they similar? Contrast the forecasts in terms of patterns of over- and underforecasting. Are there any big differences to note?

(4) Repeat (2) with log-transformed data. What are the qualitative differences in the two analyses? Do you recommend transforming the data in this case?

CASE 8F SALE OF TWIN SCREW EXTRUDERS TO THE FOOD INDUSTRY (CASE 6E CONT.)

In Case 6D, you created a very simplistic projection of YR08 and YR09 by quarter using a seasonal decomposition approach.

(1) Create a time plot of the data in Case 6D, and decide whether you need an additive or multiplicative model for the seasonality.

(2) Create a modeling flowchart to map out a strategy that is both cost-effective and not overly time consuming.

(3) Run several exponential smoothing models with different forecast profiles using the first 6 years of the historical data (24 observations). Forecast the next four periods and measure the accuracy of the forecasts against the actuals for YR07.

(4) Repeat (2) using the full history (28 observations) and make a projection for YR08 and YR09. Based on what you learned in (2), use your judgment and the forecast prediction limits to tune your forecasts up or down, if required.

(5) What do you recommend for a next step in the modeling process?

APPENDIX 8A

This appendix contains the formulae for the exponential method (Gardner, 1985).

Model Formulations

Some techniques are known by special names. The linear trend, no seasonal technique is known as a Holt two-parameter exponential smoothing method. The linear trend, multiplicative seasonal technique is also known as the Holt-Winters model. The nonlinear-trend formulations contain both damped and exponential trends. In the damped trend, ϕ is constrained to lie between 0 and 1. In the exponential trend, ϕ is constrained to lie between 1 and 2. The parameter search may yield an optimal ϕ value of 1 for either the damped or exponential trend techniques. If ϕ is 1, both techniques reduce to a linear trend. Optimal parameters are always listed on the model-fitting summary reports under the name of the technique.

Notation

Y_t = observed value of the series in period t

$\hat{Y}_t(m)$ = forecast made at the end of t for m steps ahead

e_t = forecast error in t

L_t = level (mean) of the series at the end of t

T_t = trend at the end of t

I_t = seasonal index for t

h_1 = smoothing parameter for the level of the series

h_2 = smoothing parameter for trend

h_3 = smoothing parameter for the seasonal index

ϕ = trend modification parameter

p = number of periods in one season

Nonseasonal Techniques

Constant level

$$L_t = L_{t-1} + h_1 e_t$$
$$\hat{Y}_t(m) = L_t$$

Linear trend

$$L_t = L_{t-1} + T_{t-1} + h_1 e_t$$
$$T_t = T_{t-1} + h_2 e_t$$
$$\hat{Y}_t(m) = L_t + m\,T_t$$

Nonlinear trend

$$L_t = L_{t-1} + \phi\,T_{t-1} + h_1 e_t$$
$$T_t = \phi\,T_{t-1} + h_2 e_t$$

$$\hat{Y}_t(m) = L_t + \sum_{j=1}^{m} \phi_i T_t$$

Additive-Seasonal Techniques

Constant level

$$L_t = L_{t-1} + h_1 e_t$$
$$I_t = I_{t-p} + h_3 e_t$$
$$\hat{Y}_t(m) = L_t + I_{t-p+m}$$

Linear trend

$$L_t = L_{t-1} + T_{t-1} + h_1 e_t$$
$$T_t = T_{t-1} + h_2 e_t$$
$$I_t = I_{t-p} + h_3 e_t$$
$$\hat{Y}_t(m) = L_t + m\,T_t + I_{t-p+m}$$

Nonlinear trend

$$L_t = L_{t-1} + \phi\,T_{t-1} + h_1 e_t$$
$$T_t = \phi T_{t-1} + h_2 e_t$$
$$I_t = I_{t-p} + h_3 e_t$$

$$\hat{Y}_t(m) = L_t + \sum_{j=1}^{m} \phi_i T_t + I_{t-p+m}$$

Multiplicative-Seasonal Techniques

Constant level

$$L_t = L_{t-1} + h_1 e_t / I_{t-p}$$
$$I_t = I_{t-p} + h_3 e_t / L_t$$
$$\hat{Y}_t(m) = L_t \times I_{t-p+m}$$

Linear trend

$$L_t = L_{t-1} + T_{t-1} + h_1 e_t / I_{t-p}$$
$$T_t = T_{t-1} + h_2 e_t / I_{t-p}$$
$$I_t = I_{t-p} + h_3 e_t / L_t$$
$$\hat{Y}_t(m) = (L_t + m\,T_t) I_{t-p+m}$$

Nonlinear trend

$$L_t = L_{t-1} + \phi T_{t-1} + h_1 e_t / I_{t-p}$$
$$T_t = \phi\,T_{t-1} + h_2 e_t / I_{t-p}$$
$$I_t = I_{t-p} + h_3 e_t / L_t$$

$$\hat{Y}_t(m) = \left(L_t + \sum_{j=1}^{m} \phi_i T_t \right) I_{t-p+m}$$

9

Disaggregate Product-Demand Forecasting

"Not everything that counts can be counted, and not everything that can be counted counts."
ALBERT EINSTEIN

DEMAND FORECASTERS ARE IN THE BUSINESS of making statements about future demand for products and services in the face of uncertainty. Demand planning and forecasting is the process that drives inventory levels to improve a company's ability to replenish or fulfill its product to meet customer needs in a timely and cost-effective way. If forecasting does not have a good link to drive inventory stocks, improving it will not necessarily improve customer-**service levels** or reduce costs. A forecast is not just a number, outcome, or task. It is part of an ongoing process affecting sales, marketing, inventory, production, and all other aspects of the supply chain.

Supply Chain Management (SCM) makes use of computerized intelligence to synchronize and optimize the essential elements of manufacture and distribution. Demand planners manage item-level (disaggregated) data from a number of sources to create a clear view of what product demand is likely to be and then link inventory and replenishment processes to that future view. A bottom-up forecast incorporates a logical and coherent series of steps that, if performed in an organized management-supported fashion, can improve forecasting effectiveness, reliability, and accuracy throughout the supply chain.

In this chapter you will learn:

- The role of demand forecasting in the supply chain
- How to design an effective demand forecasting system
- How to implement a demand planning process
- How to manage demand forecasts for the supply chain

9.1 FORECASTING FOR THE SUPPLY CHAIN

In a traditional supply chain, product flows sequentially through a system from one level to another (Exhibit 9.1) in a linear fashion. Driven by manufacturers, the traditional supply chain is the furthest away from the ultimate consumer or end user. Traditionally, each operation tended to maintain its own information systems and communication flows that occurred between individual departments. Nowadays, in a world dominated by global economics, the term supply chain has taken on a much broader meaning. The Council of Supply Chain Management (http://cscmp.org) defines a supply chain as the "material and informational interchanges in the logistical process stretching from acquisition of raw materials to delivery of finished products to the end user. All vendors, service providers and customers are links in the supply chain." This definition is still a mouthful, but it underlies the recognition that competition is no longer limited to individual companies vying against one another.

In an integrated supply chain (which arose during the past decade), information in the form of orders also flows back in the opposite direction, so that all operations have complete visibility to the whole supply process (Exhibit 9.2). Instead of being driven or supplied by the manufacturer, consumers are the drivers of demand, demanding cheaper, faster, and higher-quality products. A firm's success is a combination of an integrated supply chain, a sound infrastructure, and a focus on consumers.

A traditional supply chain is any sequential set of business operations leading from raw material through conversion processes, storage, distribution, and delivery to an end customer. In the integrated supply chain, information flows in the reverse direction as well.

EXHIBIT 9.1
Traditional Supply
Chain

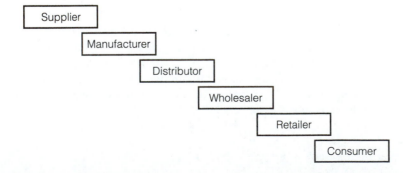

EXHIBIT 9.2
Integrated Supply
Chain

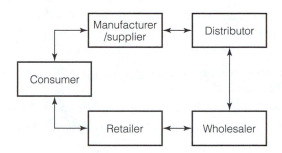

Material flow is from suppliers to the manufacturer through the distribution channel to the consumer. Information flow is in the reverse direction, from the customer to the suppliers. **Quick Response** (QR), **Efficient Consumer Response** (ECR; Fisher et al., 1994), and **Vendor Managed Inventory** (VMI) are all terms used in the trade for strategies for making manufacturers responsible for keeping the retailer in stock. These acronyms represent industry initiatives to facilitate the flow of good information in a timely manner. By implementing these management strategies, companies have reduced costs, increased sales, gained competitive advantage, and taken market share away from laggards.

Although the analogy of a chain is useful in visualizing this process, it is far too simplistic to describe what really happens. Within the supplier and manufacturer, the supply chain includes the possibility of multiple sources of supply at every stage. In the distribution channel, multiple centers can supply multiple factories and provide service to multiple retail outlets. The supply chain model includes a number of highly integrated processes for sourcing/suppliers (production, scheduling, and supply sourcing), distribution (channel management, transportation, and warehouse operations), and customer interface/point-of-sale (demand management, order management, inventory management, and store operations).

In today's competitive environment, companies must achieve both in-stock levels and high **inventory turns.** In addition to competitive pressures, many companies have found it necessary to share information and forecasts with their business partners. Retailers, in particular, frequently share forecasting information with their supply chain partners. Manufacturers have also recognized the importance of historical-based forecasting and top-down planning along with joint collaborations in forecasting with suppliers and customers. Because of the high volume of items involved, statistical forecasting is being adopted widely.

Supply chain management (SCM) refers to getting the right amount of the right product to where it is needed while managing unproductive inventory levels to achieve maximum return on assets.

Systems for the Supply Chain Pipeline

The manufacturing/distribution pipeline starts with raw materials and purchased parts required by the manufacturing plant. At the manufacturing level, we can add the fabricated components, subassemblies, and assemblies used to produce the finished-goods inventory. At the distribution level, we generally have finished goods. Analytical systems have similar underlying logic, but different factors/parameters affect the inventory plan at each point in this pipeline (Exhibit 9.2):

- **Manufacturing resource planning** (MRP) plans the raw materials, purchased parts, and components.
- Master production scheduling (MPS) plans the finished goods.
- **Distribution resource planning** (DRP) plans the finished goods at the distribution centers.

The material flowing through a supply chain can be viewed from any one of three perspectives: the customer view, the distribution view, or the supplier/manufacturer view (SKUs). The customer view defines how the end customer uses product descriptions, product numbers (SKUs), and product options to uniquely identify a complete product configuration. The distribution view defines the individual SKU, its contents including documentation and accessories, and its packaging and labeling. A complete customer configuration may require the shipment of several different SKUs. The supplier/manufacturer view, like an engineering-parts list, tends to consider a product or assembly to be complete without regard for the packaging, documentation, software, or accessories that will make it a SKU.

Depending on the industry and business model, companies use forecasting systems in a variety of ways. For instance, distribution-oriented companies are likely to use systems to help organize the replenishment and flow of goods into distribution centers (Exhibit 9.3). These companies are also likely to send the output of forecasting systems to transportation management or other order-fulfillment systems.

On the other hand, manufacturing companies generally use forecasting systems to help synchronize production schedules and finished-goods inventory with actual customer demand. Therefore, they are more likely to feed forecast information to the **MRP** module of an **ERP** system or even to an advanced planning system (APS). In addition, forecast data are becoming part of the sales and operations planning (S&OP) process, which brings people from different functional areas together to agree on a single forecast that drives the activities of the entire enterprise.

The sales and operations planning (S&OP) process brings people from different functional areas in the organization together to agree on a single forecast.

Each industry has its own production and distribution needs. Systems designed to manage the supply chain are focused on vertical markets in process manufacturing or discrete/repetitive/to-order manufacturing. Process manufacturers, which are predominantly batch-processing operations, include companies in the energy/petrochemical, chemical, and pharmaceutical industries. Electronics, fabricated metals, and automotive supplies are examples of discrete manufacturing markets.

EXHIBIT 9.3
Forecasting Is a
Crucial Link in the
Supply Chain

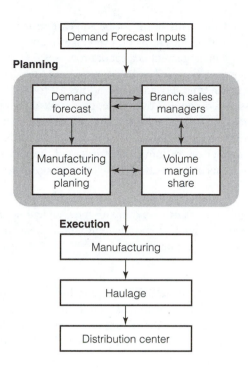

Operating Lead Times

Lead time influences inventory at different levels—the time it takes to get raw material, to manufacture, to ship product, and to process data all influence cumulative inventory and lead times. Although reducing **cycle time** is a major objective of SCM, the variation in cycle times experienced by manufacturers and retailers can be quite substantial. Typical cycle-time variations for product development can be between 5 days and 15 weeks, for production 5 days to 6 weeks, for inbound transportation from supplier to distribution center 1 day to 1.5 weeks, and for outbound transportation from distribution center to final destination 2 days to 1.5 weeks. Accurate forecasts are essential to the success of SCM in order to yield shorter lead times and, hence, higher turns and lower costs.

 The amount of time it takes for information and goods to flow through a supply chain pipeline is known as cycle time; it is also called lead time.

What Is Demand Management?

Demand planners frequently discuss dependent and **independent demand** forecasts. Independent demand, which must be forecasted, comes from the customer and includes the demand for finished goods as well as service parts. In contrast, dependent demand applies to raw materials and other components that are used in production.

The dependent demand for items need not be forecasted; it is calculated from the schedules of the item required for production and distribution.

At its core, demand management is a set of processes that produces plans or sets of time-phased numbers (forecasted orders) representing the best estimate of what demand will be at a given time. For instance, a forecast for an item at a distribution center shows the estimated demand over time, by the week or by the month, going forward. An order needs to be placed with the manufacturer or supplier against these requirements so that the requested item can arrive at the distribution center in time for shipment to the retailer or consumer. The timing of these orders is a function of the lead times of the items and the **safety stock** that assures adequate supply.

Demand management is the process of managing all independent demands for a company's product line and effectively communicating these demands to the master planner and top management production function.

The safety stock for that same SKU is a plan for that component of inventory each week into the future. A replenishment plan for the same SKU shows the quantity of product arriving weekly at the distribution center. A shipment plan for the same SKU shows the quantity of product that should be shipped to the distribution center weekly.

To fully balance the supply chain, an increased awareness and exchange of information must be established between the demand creation side of sales and marketing and the supply side of manufacturing and distribution. This should include both short-term communications about promotional programs that will affect demand and long-term communication for capacity planning. It is essential to keep the goal of balancing supply with demand in mind and communicate it across all functional groups at the outset of each new sales initiative.

Distributional Resource Planning: A Time-Phased Planned Order Forecast

Distribution managers face considerable complexities in managing inventory at various distribution points. Many variables, such as changing customer demand, transportation time, and shifting production schedules, make it difficult to ensure correct inventory levels at the proper locations, at the proper time. DRP systems plan and manage the many variables that cause distribution problems (Brown, 1982; Martin, 1995; DeLurgio and Bhame, 1991).

A DRP system uses forecasts of independent demand—the demand of the end user—instead of the dependent demand of the distribution center (DC) on the supplier/manufacturer. DRP starts with a forecast of end-user demand and calculates how long it will take to manufacture and move products through a distribution network to the customer.

DRP is part of the demand management function that creates long-term schedules designed to meet customer needs without holding excess inventory demand.

EXHIBIT 9.4 A Basic
DRP Calculation

	Period 1	Period 2	Period 3	Period 4	Period 5	Period 6	Period 7
Forecast	100	100	100	100	100	100	100
Receipts		200	100	100	100	100	100
Order quantity	200	100	100	100	100	100	100
Inventory	0	100	100	100	100	100	100
Lead time	1 period						
Safety stock	1 period						
On-hand	100						

Exhibit 9.4 shows a simplified but typical DRP allocation for a single product, in which the requirements we need are tied to the **order quantity** in a one-for-one relationship (i.e., we need one; we get one). The forecasts are assumed to be 100 units per period (typically a month or a week). With an on-hand inventory of 100 units, the ending inventory in period 1 is 0, which is also the beginning inventory for period 2. In order to keep a one-period supply of safety stock, we need to order 200 units in period 1, which will be received in period 2 (because lead time = 1). This same logic is used for the future periods.

In Exhibit 9.5, some more logic is added in that the requirements needed are tied to the order quantity based on a minimum requirement (the lowest quantity that must be ordered). If the requirement is only one unit, the order must be 200 because the minimum is set to 200.

Exhibit 9.6 shows a DRP allocation in which the requirements are tied to the order quantity based on a minimum requirement and multiple (quantities above and beyond the minimum amount). For example, if the total need is 22, the minimum is 20, and the multiple is 5, then the total order should be 25.

EXHIBIT 9.5 Basic
DRP Calculation
with a Minimum
Order Quantity of
200 Units

	Period 1	Period 2	Period 3	Period 4	Period 5	Period 6	Period 7
Forecast	100	100	100	100	100	100	100
Receipts		200		200		200	
Order quantity	200		200		200		200
Inventory	0	100	100	100	100	100	100
Lead time	1 period						
Safety stock	1 period						
On-hand	100						
Minimum	200						

EXHIBIT 9.6 Basic
DRP Calculation
with a Minimum
Order Quantity of
200 and Multiple of
Five Units

	Period 1	Period 2	Period 3	Period 4	Period 5	Period 6	Period 7
Forecast	10	20	32	47	51	68	73
Receipts		20	35	45	50	70	75
Order quantity	20	35	45	50	70	75	
Inventory	0	0	3	2	1	2	2
Lead time	1 period						
On-hand	15						
Minimum	20						
Multiple	5						

A manufacturing and distribution schedule, usually covering several weeks or months, is created to meet that order forecast. For example, a manufacturer of service parts may ship parts to several DCs that service dealers worldwide. If each DC tracks its own inventory and places orders independently, it will create a demand on the supplier/manufacturer that varies unpredictably. By using DRP, the supplier/manufacturer obtains a greater visibility of upcoming orders. With accurate information about demand and inventory in each area, the DRP system can calculate a long-term plan for when each part should be produced and in what quantity, thus ensuring that each DC has the product it needs.

Benefits from a DRP system include reduced transportation costs; higher customer-service levels; fewer stock outs; improved communication among sales, distribution, and production; and having the right product at the right place at the right time.

9.2 A Framework for an Integrated Demand Forecasting System

One of the biggest paybacks of an effective demand management process is the creation of a one-number forecasting system (Fildes and Beard, 1992). Sales, marketing, operations, and financial planners can then view their own forecasts in their own units of measure with the knowledge that these forecasts will be reconciled and communicated companywide. This sharing of forecasts with other organizations, and more recently with business partners, results in a streamlined process, allowing for reduced costs and increased sales and profits.

Typically in many firms across most industries, there is a lack of integrated planning at the operational planning and business execution levels. Unexpected demands, which must be met at any location, wreak havoc on operations. At the higher levels of planning, such as strategic and tactical levels, forecasters have more options and more time to effect change. At the operational levels, however, forecasters are more strapped for quality data, modeling options, and effective software systems.

 The single forecast of customer demand, at the item-location level, provides the unifying perspective from which to integrate all forecasting activities in the supply chain.

Achieving the benefits of a one-number forecast requires the integration of data and information across the whole firm, or enterprise as it is called these days. The planning systems to support these activities are called forecast support systems (FSS). Forecasters need to build responsiveness into their operations and the implementation of planning systems that integrate the different levels of planning across the enterprise.

A forecasting support system (FSS) is a computer-based set of procedures that supports the forecasting process. It allows the practitioner to readily prepare data, execute models, evaluate performance, and reconcile a variety of forecast-related information for the purpose of driving a firm's objectives for a future period. It should also have the functionality to incorporate human judgment and track forecast accuracy.

A System Architecture

In an integrated demand management system, projected demand plans are visible in a common database accessed by those along the supply chain. The source of the corporate (internal) data is a legacy database residing on a secure corporate server that can be accessed by its users through an intranet or network server. This enables the firm to guarantee the integrity and accuracy of the essential input information. To collaborate with forecast users outside the firm, the forecasting system needs to be able to communicate with vendors' planning systems through an extranet connection. In addition, when external information on economic, demographic, and competitive factors is required, forecasters can access many of these data sources through the Internet.

The demand forecasting system needed to support a supply chain has a number of components for linking to data sources.

Once the data sources are in place, a forecaster can interact with the data through a presentation component of the forecasting system. This client-centric part of the system allows for flexible data input, data conversions, prices, graphs, note pad, scheduled receipts and on-hand inventory, data displays (year-to-date, percentages of annual totals), and communication.

To adequately perform the planning function, the forecaster needs to have ready access to multiple modeling methods. This modeling component of the system should allow for quantitative assessments (promotion analysis, statistical techniques, and causal modeling), qualitative assessments (event management, field sales, and new product introductions), **batch forecasting** engine, best-fit evaluation capability, user overrides on models, integrated forecasting and planning, exception handling, and user feedback.

Information needs to be distributed to forecast users through a data-reporting component. This provides output in terms of standard reports, user-defined reports, data export to SCM and ERP systems, and electronic linkages with customers and suppliers.

Dimensions of Demand

Effective demand planning requires that forecasters incorporate data into their forecasts, whatever their aggregation, to adequately support their clients and the sales, marketing, and financial and operational planners in the firm. Hence, demand forecasting is done at different levels of detail, involving

period (time) granularity (annually, quarterly, monthly, weekly, daily, shifts, or hours)

product hierarchy (business operating units, category, brand, product flavors, sizes, or special packs)

place (geographical/customer/location) segments (global, national, market zone, channel, sales regions, warehouses, plants, zip codes, stores, and customers)

The forecast dimensions of period, product, and customer/location can be represented by a triangle, in which each side depicts a multilevel forecasting hierarchy (Exhibit 9.7).

Demand forecasting is performed at different levels of detail incorporating dimensions of period, product, and customer/location.

Period

Depending on the forecasting environment, a forecasting system needs to be able to support a calendar. For typical SCM applications, the weekly and monthly time buckets are most frequently used for forecasting. Because weeks do not roll up neatly into months, there are different weekly patterns in use. For example, a 4-4-5 pattern means that the first month of a quarter is made up 4 weeks, the next month is 4 weeks, and

EXHIBIT 9.7
Multilevel Forecasting
Hierarchy for Product, Customer/
Location (place),
and Period

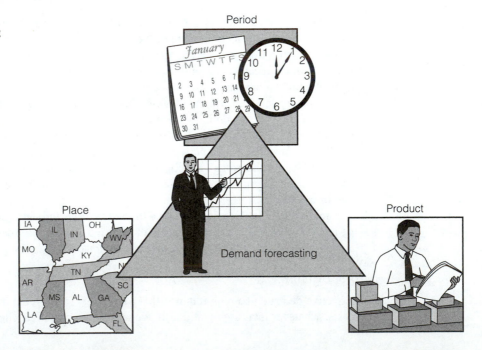

the last month is 5 weeks. There are also firms that operate on a 13-period year. Quarterly and yearly figures do not need to be stored because they accumulate naturally from months. In the energy utility industry, for example, a calendar is disaggregated even further into hours and days.

In terms of period granularities,

- The *forecast cycle* describes how often we should forecast. For many companies this is a monthly process, but with the higher service levels required to satisfy customers these days, it is not unusual to see a forecast cycle every week.

- The *forecast horizon* tells how far out we should forecast. Many companies use a 12- to 18-month rolling forecast horizon. Today, with e-business, companies are shortening the horizon to 1–3 months to ensure that their businesses can respond to the increased volatility driven by new market dynamics.

- The *forecast granularity* tells how detailed we should make the forecast. Commonly used quarterly granularity often does not provide the level of detail needed to address customer-service level requirements, efficient supply planning, or the need of marketing to respond to critical issues. Thus, demand and supply forecasting may need to implement a weekly forecast granularity for forecasts within critical **manufacturing lead times.**

Product

To support a forecasting function, the demand planning system needs to maintain a variety of product-specific detail. The lowest level item identification is the SKU. For each SKU, the system needs to maintain fields on unit price, unit costs (labor, material, etc., so as to be able to calculate margins), unit shipments (carton and pallet quantity, unit weight, and unit cube), lead time, and other attributes (description, cut-in and cut-out dates, and summary category identifiers).

Place

Location/customer-specific information starts with a designation of a lowest level location code. Typically, this is a customer account code that other segments can be mapped into. Additional fields might include description, discount rates, codes to map customer/locations into regions, channels, warehouses, and field sales accounts. In addition, there may be conversions of the units of demand to sales revenue, profit margin, costs, pallets, cases, shift hours, and so on.

Role of Planning

Most planning that takes place in an organization is done **hierarchically,** with strategic planning done once a year or less and tactical planning done quarterly, monthly, weekly, or daily. Each of these levels requires forecasts at different levels of aggregation. Through the functional organization, we can view dimensions of demand in terms of marketing (brand-level forecasts by channel and sales currency and margins), sales (account- or regional-level forecasts by product category in sales currency), operations (distribution territory–level forecasts by SKU in cases or plant-level forecasts by SKU in units; incidentally, a SKU is the lowest level in which

EXHIBIT 9.8
Demand
Management and the
Cross-Functional
Business Process

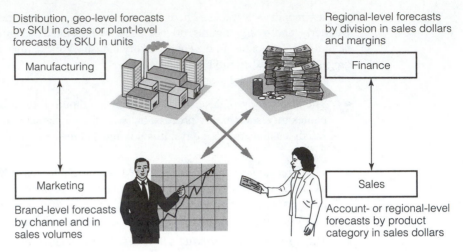

Distribution, geo-level forecasts
by SKU in cases or plant-level
forecasts by SKU in units

Regional-level forecasts
by division in sales dollars
and margins

Manufacturing

Finance

Marketing

Sales

Brand-level forecasts
by channel and in
sales volumes

Account- or regional-level
forecasts by product
category in sales dollars

we might categorize a product, such as a bar code or product code that we might see on a box or the unit itself), and finance (regional-level forecasts by division in sales currency and margins).

Each functional user group has its own requirements (Exhibit 9.8). For example, marketing personnel may want to review the forecast at a brand level in sales and margin rather than at the item level. Similarly, sales personnel prefer reviewing the forecast in currency by region or customer account. To support these related requirements, forecasting approaches may need to be developed for multiple levels; they will also need to reconcile the multiple forecasts.

The level of planning being supported by the forecasting process impacts the data and forecasting models needed.

Moreover, some functions need to view the forecast at the lowest level in a hierarchy, whereas others need to see it at higher levels. However, not all of these functions can necessarily be placed in hierarchies. These different types of planning indicate the key concepts that underlie a forecasting system—aggregation and allocation.

Because these requirements by the functional groups can occur at different levels in the three dimensions of demand, demand analysts have a need for good data and efficient, effective, data-driven forecasting methodologies to support such developments. All this is reflected in an operational demand plan, which assures that the right amount of the right product gets shipped to the right customer or location in the right time (and, of course, at the right price)—a bit of jargon, but nevertheless important to remember.

Reconciling Cross-Functional Forecasting Processes

Most companies go through a process to project their sales and operations plans for the next 1–3 years, which they use to create the budget. The assumptions and analyses are done at the macro level. Because their annual sales levels are used to drive financial planning, the forecasts are usually expressed in dollars (or other currency) and the time granularity is typically expressed in months and quarters. The entities being forecast are often product categories or other product organizational groupings.

As the year progresses, the business also needs sales projections to plan financials as well as operations. During the monthly planning cycle, marketing is responsible for a sales forecast—financial or unit volumes—by brand or product family. At the same time, the operations departments project demand at the SKU level for the next several weeks as inventories are produced, distributed, and delivered to customers.

Each functional group in the company has its own forecast requirements. For example, marketing personnel may want to review the forecast at the brand level in sales and margin currency rather than in SKUs and unit volume level. Sales personnel may find it more useful to see a forecast in currency by region or customer account. In support of these related requirements, forecasting approaches may need to be developed for multiple levels to support the reconciliation of these different forecasts. To work effectively, the forecasting process must generate views that are familiar to each of the functions. These views need to be at different aggregation levels or dimensions as well as in different versions of the measure of demand. Some functions need to view the forecast in currency and some in units.

A "best practice" forecasting process, striving to obtain the "single best number forecast" to drive the business, involves obtaining consensus among different functional organizations.

The differing functional views of a forecast are important for reaching consensus. Each function must first review the forecasts and then approve or modify them based on their view of the business: integrating sales, marketing, operations, and financial plans into a single plan; constantly balancing supply and demand; using company resources effectively throughout the enterprise; and making the results visible in inventory investment, product availability, and customer service.

Market Intelligence and Judgmental Overrides

The multilevel forecasting hierarchy requires demand to be collected and stored at the lowest level of product and location so that appropriate summaries can be obtained for product lines and families by customer locations and segments. Typically, demand is collected by the warehouse or DC so that baseline forecasts can be reviewed and adjusted by sales account, channel, or region. For example, a coupon program for a consumer product may be planned regionally; therefore, the impact on a product's forecast needs to be assessed from a regional perspective. If, for instance, a competitor plans a promotional event, a defensive change to a similar brand's forecast needs to be created. These are generally made as judgmental overrides to a baseline forecast by field forecast managers and consolidated into a consensus forecast.

After a baseline forecast has been produced, other organizations in the company, as well as its trading partners, contribute to refining the forecast through a collaborative process known as Collaborative Planning, Forecasting, and Replenishment (**CPFR**). The American Production and Inventory Society (APICS) dictionary defines CPFR as (1) a collaboration process whereby supply chain trading partners can jointly plan key materials to production and delivery of final products to end customers, where collaboration encompasses business planning, sales forecasting, and all operations required to replenish raw materials and finished goods; (2) a process philosophy for facilitating endorsed by Voluntary Interindustry Commerce Standards (**VICS**). In the CPFR model, retailers and manufacturers extend collaboration from operational planning through execution, enabled by Internet technology.

The result of a collaborative forecasting process is a one-number forecast that becomes the basis for replenishment and production plans to meet customer needs in a timely and cost-effective way.

The Internet is the perfect means of changing conventional industry models because it constitutes an infrastructure that transcends traditional boundaries. Instead of sequential relationships, in which orders are placed with a supplier, inventory is consumed, and another order is placed, web-enabled planning systems now provide a near-instantaneous communications link among trading partners.

New ways to take account of customer and consumer preferences affect the view of point-of-sale replenishment solutions because the use of the Internet means we can communicate changing needs and habits to more companies more effectively than the traditional place-order, consume-inventory cycles.

Customer orders may be placed directly with the supplier without ever leaving home. This is another kind of demand that must be taken into account when developing forecasting solutions. Suppliers can now reside in the center of a web of customers—all communicating needs and information via the Internet.

Analyzing Demand Variability

The starting point in analyzing demand variability is the waterfall chart that we have previously used for evaluating forecasts in earlier chapters. Sometimes, the analysis can be as simple as determining the customers causing the most variability by ranking the annual forecast fluctuations for each item by customer. For example, consider the 48 periods in a monthly time series, called N1410, from the MICRO section (see the CD) of IJF-M3 competition database (Exhibit 9.9). It is a fairly noisy, seasonal time series with minimal trend. The preliminary ANOVA decomposition (not shown) suggests 9% trend, 48% seasonality, and 43% noise. The last six periods are used as a holdout sample and are arbitrarily labeled as July through December 2000. Thus, the first data value in the series is January 1997.

The objective of a demand variability or root cause analysis is to highlight the items that have the greatest impact on the uses of the forecast.

EXHIBIT 9.9 Time Plot of N1410, a 48-Month Time Series from the MICRO Section of the IJF-M3 Competition Database

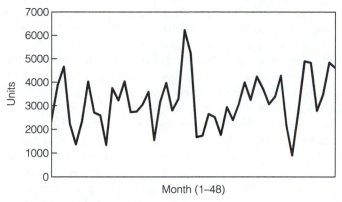

Month (1–48)

(*Source*: PEER Planner FSS on CD)

In Exhibit 9.10, an augmented waterfall chart shows the six values in the hold-out sample in the row labeled Actual = >. "A(YTD) + F(BOY)" stands for Actuals (Year-to-Date) plus Forecasts (Balance-of-Year). In that column, the value 42,500 is the sum of the actuals for the year 2000. For the rows below this, each value represents a partial sum of actuals added to the remaining forecasts in 2000. For example, the first forecast made with actuals through June 2000 results in six forecasts over the 6-month forecast horizon. Thus, A(YTD) + F(BOY) = 38,067 represents the sum of 6 months of actuals (= 16, 520) and 6 months of forecasts. The rows for forecasts 2 through 6 show forecasts with decreasing forecast horizons. For forecast performance to improve, the numbers in this column should gradually get closer to the annual total (= 42,500).

The row labeled 1 step (rolling) average represents a three-period moving average of the one-step-ahead forecasts. These are the forecasts along the lowest diagonal (2661, 3305, 3937, 3181, 5375, 4573), which are useful for determining (along with the corresponding actuals) the behavior of a random error distribution for some models and the associated prediction intervals for forecasts based on the model.

EXHIBIT 9.10 Waterfall Chart for N1410, a 48-Month Time Series from the MICRO Section of the IJF-M3 Competition Database

Start/Horizon	Jul-00	Aug-00	Sep-00	Oct-00	Nov-00	Dec-00	A(YTD) + F(BOY)	Rolling Average	Next Year Cumulative Forecast
Actual = >	4900	4820	2780	3540	4840	5100	42,500	4166	
1. 6/00	2661	3364	3177	3006	5004	4335	38,067	3067	14,663
2. 7/00		3305	3117	2944	4942	4273	40,001	3122	17,389
3. 8/00			3937	3765	5764	5096	44,802	4488	27,482
4. 9/00				3181	5180	4513	41,894	4291	25,993
5. 10/00					5375	4707	42,642	5041	29,942
6. 11/00						4573	41,973	4573	33,273
1 step (rolling) average	2661	2983	3301	3474	4164	4376	0	0	0

(*Source*: PEER Planner FSS on CD)

The column labeled Rolling Average is a three-period moving average of the one-, two-, and three-step-ahead forecasts taken horizontally. This is useful in assessing how forecasts over a 3-month lead time are behaving. The column labeled Next Year Cumulative Forecast shows the cumulative forecast for the following year. It is a way to have some visibility concerning how the model is projecting forecasts past the current year of interest.

What is of interest now is to display the results for the various accuracy measures that are calculated from the forecast errors (Error = Actual − Forecast). The measures of particular interest (see Chapter 5) are derived from errors, absolute errors, percentage errors, absolute percentage errors, percentage variance, and absolute percentage variance. In a percentage error calculation, the actual is the denominator, whereas in a percentage variance calculation, the forecast is the denominator. Exhibit 9.11 displays a waterfall chart using absolute percentage errors and absolute percentage variances. The difference is a matter of interpretation. When the forecast is a plan, most likely expressed in currency, then forecast accuracy tends to be measured with a percentage variance metric.

It may also be necessary to look at forecast performance over the lead time of the product. For example, Exhibit 9.12 displays the waterfall chart for series N1410 by first accumulating the data in 2-month periods. This can be done as a rolling 2-month period or in contiguous 2-month periods. Users should be careful in trying

EXHIBIT 9.11 Waterfall Charts with (a) Absolute Percentage Errors and (b) Absolute Percent Variances for N1410, a 48-Month Time Series from the MICRO Section of the IJF-M3 Competition Database

(a)

Start/Horizon	Jul-00	Aug-00	Sep-00	Oct-00	Nov-00	Dec-00
Actual =>	4900	4820	2780	3540	4840	5100
1. 6/00	45.69%	30.21%	14.28%	15.08%	3.39%	15.00%
2. 7/00		31.43%	12.12%	16.84%	2.11%	16.22%
3. 8/00			41.62%	6.36%	19.09%	0.08%
4. 9/00				10.14%	7.02%	11.51%
5. 10/00					11.05%	7.71%
6. 11/00						10.33%
1 step (rolling) average	45.69%	38.56%	39.58%	27.73%	20.94%	10.51%

(b)

Start/Horizon	Jul-00	Aug-00	Sep-00	Oct-00	Nov-00	Dec-00
Actual =>	4900	4820	2780	3540	4840	5100
1. 6/00	84.14%	43.28%	12.50%	17.76%	3.28%	17.65%
2. 7/00		45.84%	10.81%	20.24%	2.06%	19.35%
3. 8/00			29.39%	5.98%	16.03%	0.08%
4. 9/00				11.29%	6.56%	13.01%
5. 10/00					9.95%	8.35%
6. 11/00						11.52%
1 step (rolling) average	84.14%	64.99%	53.12%	28.84%	16.88%	10.92%

(Source: PEER Planner FSS on CD)

Start/Horizon	Jul-00	Aug-00	Sep-00	Oct-00	Nov-00
Actual = >	9720	7600	6320	8380	9940
1. 6/00	38.01%	13.93%	2.17%	4.42%	6.05%
2. 7/00		15.50%	4.10%	5.89%	7.29%
3. 8/00			21.87%	13.71%	9.26%
4. 9/00				0.23%	2.48%
5. 10/00					1.43%
6. 11/00					
1 step (rolling) average	38.01%	26.76%	25.13%	12.53%	7.84%

EXHIBIT 9.12 Waterfall Chart with Absolute Percentage Errors over 2-Month Lead Time for N1410, a 48-Month Time Series from the MICRO Section of the IJF-M3 Competition Database

(*Source*: PEER Planner FSS on CD)

to interpret these two alternative ways of expressing the data. For example, a quarterly summary is based on calculating sums of contiguous 3-month periods, not rolling 3-month periods.

It may also be important to consider the forecast performance of a product over its lead time.

9.3 AUTOMATED STATISTICAL FORECASTING

Nowadays, forecasting systems frequently incorporate options for automatic forecast method selection. The particular implementation of an automated procedure may vary greatly among the software programs; even within an individual program, it can take on a number of different forms. In this section, we describe an approach for the automatic selection of an exponential smoothing forecasting method found in Chapter 8.

In most forecasting software systems, model choices for exponential smoothing can be made manually or through several levels of automation. In a manual mode, the user makes all choices, including model selections, estimates of smoothing parameters, and initializations. More typically, the forecaster's responsibility is to select a smoothing method, and the software will use an optimization procedure to find the best values to assign to the smoothing weights. The type of exponential smoothing method (Holt, Winters, or damped trend) depicted in a Pegels diagram (Chapter 8) depends directly on the types of trend and seasonal patterns found in the historical data and the forecast profile expected in the forecast horizon. The values given by the smoothing weights determine the relative emphasis given to the immediate and distant past in the historical data. Initial values for level, trend, and seasonal indexes are usually required to start the updating process inherent in most smoothing algorithms.

For the high-volume forecasting needed for demand management, inventory and production planning, a software package in automatic mode is essential to perform the "best" model selection.

Selecting Models Visually

The first step in model selection is to visually display the salient features of the historical data. For well-behaved time series, the appropriate type of trend and/or seasonality pattern may be readily seen in a time plot of the data. In contrast, with volatile time series, it may be more difficult to discern the appropriate type of trend, especially when the data are seasonal. In some cases, it will be difficult to judge even whether the series is seasonal or nonseasonal. This can happen when the series has been affected by special events, such as promotions, which can mask themselves as a seasonal pattern in monthly data. Turning points also complicate model identification, because they may make exponential smoothing result in poor forecasts, because no particular smoothing procedure is designed for it.

Seasonal Model Selection

If at least two to three seasonal cycles are available, seasonality can usually be identified visually. We can check to see if the peaks (troughs) occur during the same quarter or month of every year. Further, we can trace yearly peaks and yearly troughs; if they are widening over time, this signals that seasonality is multiplicative and, if they are relatively constant, this suggests additive seasonality. Other visual displays include box plots by month (each box plot representing the values for a given month), which can point to a pattern of seasonality (Exhibit 9.13a). The medians in the box plots are connected to indicate an average level of the seasonality. The height of the box plot indicates the variability of the seasonal period. On the other hand, box plots by year (each box representing 1 year's values) can be helpful in distinguishing additive from multiplicative seasonality (Exhibit 9.13b). The medians are connected in the box plots, which suggest little or no trend in the data. There is an unusual value in one of the years (1998). If the height of the box plot increases with increasing years, this suggests a multiplicative seasonality. In this case, we should choose additive seasonality.

EXHIBIT 9.13a Box Plots by Year for Series N410 from IJF-M3 Competition, 50 Monthly Values from the MICRO set

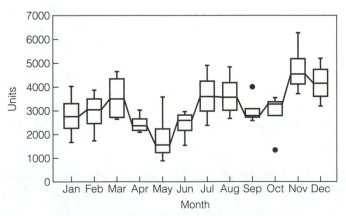

(*Source*: www.maths.monash.edu.au/~hyndman/forecasting/)

EXHIBIT 9.13b Box Plots by Month for Series N410 from IJF-M3 Competition, 50 Monthly Values from the MICRO Set

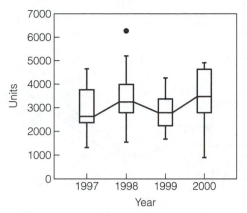

(*Source*: www.maths.monash.edu.au/~hyndman/forecasting/)

The following rules may be found useful when deciding whether to use an additive or multiplicative seasonal model.

1. We recommend using additive seasonal models if the data are inventory demands, intermittent demands, or without trends.
2. Otherwise, multiplicative seasonal models will work quite well.

Sometimes, the appearance of intrayear fluctuations in a time plot may not indicate seasonality, in that the peaks (or troughs) do not recur at the same time each year. Selecting a seasonal model for such a series is not advisable because the underlying source of the fluctuations, miscast as seasonal, go unmodeled. One visual aid to help with this is the tier chart, in which the horizontal axis displays the seasons of the year (in quarters or months) and the values for the seasons of any one year are connected. In Exhibit 9.14a, the seasonality is vivid, whereas in Exhibit 9.14b the intrayear variation is too irregular to be modeled as a seasonal index.

Trend Model Selection

Volatility in a time series makes it difficult not only to distinguish seasonality in a time series but to distinguish trend as well. Smoothing the seasonal fluctuations of a volatile series may be necessary to permit the identification of the underlying trend. The following visual rules can be useful for choosing an appropriate trend for the time series (Tashman and Kruk, 1996):

1. If the recent trend in the time series is unstable (defined as a change in direction from growth to decline or vice versa), use a constant level method.
2. If the recent trend is stable and appears flatter (slower) than the global trend, use a damped trend method.
3. If the recent trend is stable and appears as steep or steeper than the global trend, use a linear or exponential trend method.

If the data are seasonal, these rules can be applied to the deseasonalized data.

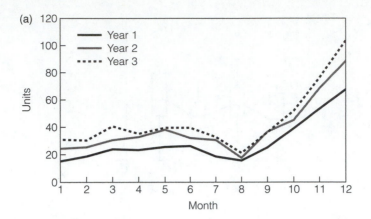

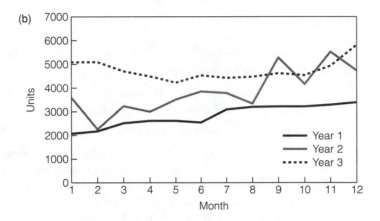

When the main source of variation in a time series is due to seasonality, the time series can be smoothed by seasonal adjustment (see Chapter 6). This suggests that the forecaster should examine a plot of the deseasonalized data or the annualized data. Alternatively, the use of an appropriate moving average of the time series (a four-period moving average for quarterly data or a 12-period moving average for monthly data) may be useful.

Outliers in Model Selection

Finally, there are techniques for dealing with outliers and special events. An outlier may be due to a disruption of business (as a result of a catastrophic act of nature or a work stoppage), to a windfall resulting perhaps from a legal ruling or business restructuring, or to a simple data-entry error. If the outlying values are not identified and, in some way, reduced in influence, the underlying estimates of the level, trend, and seasonal components of a time series can be severely distorted. For example, an extremely high outlier toward the current end of the time series will raise the current trend estimate and result in upwardly biased forecasts for the near future. It can also exaggerate the forecast errors and the width of prediction intervals.

Automatic Method Selection

Most commonly, automatic method selections involve some kind of a contest or tournament among a set of forecasting methods. Each of the included methods is used to forecast a particular time series and the one that does so most accurately is declared the preferred ("best") method. Typically, each of several methods is fit to the entire time series and the procedure that results in the best value of a performance statistic (e.g., MSE or MAD) is declared the best. See Chapter 5 for forecast accuracy measures.

Sliding Simulations

The sliding simulation is an evaluation procedure that involves a three-way split of the time series (Makridakis and Hibon, 1990). First, a subset of the historical data is withheld from a time series to serve as a test period for evaluating forecasting accuracy of a method. Next, the remaining period of fit is divided between the first T_1 observations and the remaining T_2 observations. We call the first T_1 observations the within-sample fit period and the T_2 observations, from $T_1 + 1$ to $T_1 + T_2$, the post-sample fit period.

A sliding simulation then performs, for each method under consideration, a pair of rolling simulations. The first rolling simulation is implemented using the post-sample fit period data to compute forecast error measures, measures that are used as error minimization criteria to optimize the smoothing weights and to select the best-performing method at each lead time. As a result, one method may be chosen to supply one-step-ahead forecasts while another is selected for two-step-ahead forecasts. The second rolling simulation is performed on the test period data and has the traditional purpose of evaluating the accuracy of the forecasts made by the method selected.

The results of the M2-Competition (see Section 5.3) were not so positive for the sliding simulation process. For example, the sliding simulation selection of the best smoothing method (at each lead time) among constant, damped, and linear trend procedures did not systematically outperform any of the individual smoothing methods when these were calibrated either post-sample or within-sample (Makridakis et al., 1993, Exhibit 3; Makridakis and Hibon, 2000). In addition, two of the three smoothing methods performed more poorly when calibrated post-sample (the linear trend being the exception). Nevertheless, the sliding simulation is a practical approach that integrates method selection, coefficient optimization, and forecasting accuracy evaluation.

A Unified Framework

Gardner (1988) proposed and implemented a procedure for selecting exponential smoothing models in an automatic mode. First, a test is performed to determine whether the time series is better modeled as seasonal or nonseasonal. If nonseasonal is chosen, the damped trend, nonseasonal method is selected. If seasonal is chosen, the damped trend, multiplicative seasonal method or additive seasonal is chosen. The justification for this framework lies in the ability of the damped trend model to represent time series with no trends and time series with strong trends as special cases.

If a time series is trending, then detrending the data through differencing will lead to a time series that is less variable—that is, it becomes a time series with a smaller variance. This observation can be used to create a procedure for choosing an appropriate type of trend for the data. In essence, the procedure compares the variability of (1) the original series, (2) the first differences, and (3) the differences of order two of the series (see Section 4.5).

1. If the least volatile series is the original series, then the original series must be without trend and, hence, the most appropriate smoothing procedure is one without a trend. This could be simple smoothing or a constant level model with seasonality.

2. If the first differences reduce the variability, the damped trend is most appropriate.

3. If the least volatile series is the second difference, the strong trend, linear or exponential, is recommended.

The damped trend method includes a trend modification parameter, ϕ (see Chapter 8). With $0 < \phi < 1$, an upward trend gradually decays. However, for the extreme case when $\phi = 1$, the trend is linear and for the other extreme, when $\phi = 0$, the trend is flat. Hence, the damped trend model can emulate a linear trend or constant level in the time series.

One limitation of this framework is that it does not include additive seasonality, an option that can be especially important when the seasonal time series takes on values close to zero. Also, the damped trend optimization algorithm is complex and sensitive to outliers—there can be no guarantee that optimal value for ϕ will properly reflect the nature of the trend in the data.

When tested on 111 time series from the M-Competition, this procedure for selecting exponential smoothing models resulted in a significantly better forecast accuracy than would have obtained by always using a Holt-Winters model (linear trends). However, this procedure achieved about the same forecast accuracy as that resulting from always using a damped trend (Gardner and McKenzie, 1988). Another analysis on the same data suggested substantial agreement between this procedure and the visual selection rules; the same trends were selected 68% of the time (Tashman and Kruk, 1996).

Some Caveats on Automatic Method Selection

The drawbacks of fully automated systems for exponential smoothing can become problematic if the data are very irregular or unusual data characteristics, such as outliers, go undetected. Even a few outliers, especially near the start of the forecast horizon, can have a significant impact on the eventual pattern of the forecasts; for example, an extreme value at the end of a historical period can change the direction of a trend line or distort the presence of a seasonal pattern. Hence, we recommend a visual inspection of data as the first step in model selection.

Automatic methods do not incorporate the power of the visual inspection of data; automatic options in forecasting software merely run a variety of procedures and attempt to identify one, which, by some measure of accuracy, works best for the time series at hand. The number of options itself is limited and may exclude other important modeling options available to the forecaster. Nevertheless, experience shows that a reliance on an automatic procedure for selection of a smoothing method is more beneficial than simply applying the same smoothing procedure to all the data. Moreover, for forecasting large quantities of time series for products and services by customer/location-specific ship-to points, an automatic procedure is a practical necessity.

 Ideally, the criterion for selecting a "best" method should be based on a forecasting evaluation rather than a measure of goodness of fit.

When dealing with exceptional products or aggregates that are of particular importance to the business operation, we recommend some degree of manual control, especially with regard to adjusting outliers and the selection of a particular model. There is very little gained by manually selecting parameter values for the components because their setting has relatively little impact on the forecast profile generated by the smoothing algorithm. Experience suggests the determination of the best smoothing weights should be left to the optimization (search) algorithms embedded in the forecasting software.

A **heuristic** methodology, called Focus Forecasting®, has been a preferred approach in the production and inventory management trade literature (Smith, 1996) in the past. Focus Forecasting is a nonstatistical methodology based on rules about past historical patterns that do not have any theoretical underpinnings to establish optimality properties. Although little has been documented on its comparative accuracy with statistical techniques, a study of cookware demand by Gardner and Anderson (1997) demonstrated that it produces less accurate results than the seasonal, damped trend exponential smoothing models (described in detail in Chapter 8). Nevertheless, it has an intuitive appeal that is often appreciated by practitioners.

Searching for Optimal Smoothing Procedures

Once an exponential smoothing method has been selected, an algorithm is applied to find the optimal values of the smoothing weights. Once the weights are determined, the level, trend, and seasonal components of the forecasting equation are constructed to produce model extrapolations into the future.

The optimization algorithm itself requires some technical choices to be made, although many programs specify default values, often behind the scenes. The technical choices concern: (1) which error-minimization criterion to use, (2) which search procedure to use, and (3) which starting values (also called initial values) to use for initiating the search procedure.

In the simplest case of simple exponential smoothing, the only component of the forecasting equation is the current level; it is an exponentially weighted average of the historical values, with the relative emphasis given to the more recent values versus the distant past. The simple exponential smoothing technique (Chapter 8) has a smoothing equation:

$$L_t = \alpha Y_t + (1 - \alpha) L_{t-1}$$

in which α is the smoothing coefficient ($0 \leqslant \alpha < 1$). The value of the level, L_t, becomes the forecast of the next period's data value ($Y_t(1) = L_t$), so that the error in estimating or forecasting one period ahead is:

$$e_t(1) = Y_{t+1} - Y_t(1)$$
$$= Y_{t+1} - L_t$$

The typical smoothing algorithm seeks to calculate L_t to minimize these one-period-ahead forecast errors. This requires that we first specify an error-minimization criterion—typically, the MAD or MSE. Second, we must calculate the value of α that satisfies this criterion; if it is close to zero, this implies more even weighting of the recent and distant past and, if it is close to unity, this implies that virtually all emphasis is placed on the recent past.

The implementation of exponential smoothing algorithms is complicated by the need for starting values or assumptions about the value of the level/trend/seasonal indexes as of the initial time period in the series.

Error-Minimization Criteria

The error-minimization criterion defines what is best or optimal in an optimization procedure. In principle, any error measure can serve as a basis for optimization. In practice, software programs rely most commonly on a squared error measure: MSE or RMSE. This means that calculations are made to keep the squared one-period-ahead forecast errors to a minimum.

The most common alternative error-minimization measures are the MAD and the MAPE. In both cases, the optimization algorithm seeks to keep the absolute errors to a minimum. In an evaluation of the comparative accuracy of exponential smoothing procedures, Schnaars (1986) designed an optimization algorithm that sought to minimize the MAPE.

The choice of the error criterion, also interpretable as a loss function, could make a difference in practice. In one study, Dielman (1996) showed that if the time series contains outliers or extreme values, then an optimization criterion based on absolute errors is safer than one based on squared errors. However, a specific analysis of exponential smoothing by Makridakis and Hibon (1990) concluded that the choice of error-minimization measure makes very little difference in terms of eventual forecasting accuracy (in a forecasting evaluation).

Searching for Optimal Smoothing Weights

Modern forecasting software can be expected to include at least one search procedure for optimizing the smoothing weights. The most common is the grid search. For example, by selecting a smoothing weight α in simple exponential smoothing, we can implement a crude grid search by setting α equal to designated values between 0 and 1, for example, the 100 values (0.01, 0.02, 0.03, . . . , 0.98, 0.99, 1.00). Then, for each α value, an error measure of choice is calculated. Finally, the α value, which keeps the defined measure to a minimum, can be found. Although this design might be adequate for a simple model, it proves to be unacceptably slow for procedures such as the Winters method, which involves simultaneously optimizing three smoothing weights—one each for the level, trend, and seasonal components. Using the crude grid search for this would require not just 100 comparisons but 1 million. Therefore, certain shortcuts are typically employed to cut down on the number of comparisons that must be considered.

A more refined grid search begins by comparing just a few values—for example, the two values 0.33 and 0.66 in simple exponential smoothing by determining which of these generates the smaller error. This is followed by finer and finer searches about this point. If 0.33 is the initial choice over 0.66, the next step is to compare the values 0.16, 0.33, and 0.49 in the hope of progressively homing in to the overall optimum value.

Much faster than the grid search is the simplex (or hill-climbing) procedure, which presets certain values for the weights and then seeks local (error) minimum points. However, the simplex method carries the risk that there could be more than one local minimum, in which case the smoothing weights that result depend on the preset values. It is recommended that the simplex search be used only for relatively well-behaved (nonerratic) time series.

Starting Values

In order to initiate the grid search or simplex algorithm, starting values must be assigned to the level, trend, and seasonal components. A starting value is an estimate of that component's value during the initial time period in the historical series. The influence of the starting value gradually diminishes as the actual historical data are entered and, if the series is long enough, the impact becomes negligible.

In many time series, especially seasonal series, starting values can make a difference in the smoothing weights and forecasts generated (Chatfield and Yar, 1991). The usual choices for starting values (Gardner, 1985, 1988; Makridakis and Hibon, 1990) distinguish the seasonal component from the level and trend components. For the seasonal component, we can use the classical decomposition seasonal indexes as starting values (see Chapter 6). Because these indexes give the same weight to periods in the recent and distant past, only the first 3 complete years of data should be used to allow the seasonal patterns to evolve over time. An alternative to the classical decomposition is a linear regression model with seasonal dummy variables (Chapter 12).

For the level and trend components, there are several options:

Early values or the very first data point often serves as the starting value for the level component. The change from the first to the second data points (or average of a group of early changes) serves as the starting value for the trend component.

The (global) mean of the time series can be used as the starting value for the level component. If the data are seasonal, the mean of the deseasonalized series can be used.

The slope and intercept of a linear regression model can be used as the starting values for the trend and level components, respectively. This option cannot be applied to damped trends.

Backcasting reverses the time order of the data, using the most recent data point as the start and updating in reverse until an estimate for the earliest time is reached. Unlike regression, backcasting can be applied to damped as well as linear trends.

Many programs do not provide a choice or indicate which specific options are used. According to Makridakis and Hibon (1990), "on the average there are few benefits in attempting to find optimal ways to initialize the values of (non-seasonal) exponential smoothing methods"; however, they admit that for individual series, especially seasonal series, the forecaster might be able to improve forecasting accuracy by trying alternative starting values.

What is the bottom line? Most forecasting practitioners do not wish to take time to adjust the technical settings in their software's computational algorithms to attempt to accommodate the particulars of individual time series. The comforting news is that empirical research suggests that reliance on the program's default settings is unlikely to be very harmful and is certainly cost-beneficial. But, just as the user of an automatic camera might find it rewarding to go for manual overrides on specific occasions, so the forecaster will find occasional rewards for the extra effort in attending to the details.

9.4 DISAGGREGATE PRODUCT-DEMAND FORECASTING CHECKLIST

The following checklist can be used as a scorecard to help identify gaps in the forecasting process that will need attention. It can be scored or color-coded on three levels: green = YES, yellow = SOMEWHAT, and red = NO

_____ Is timely data available for modeling, analysis, and exception handling in:

- POS data for early warning of rapid shifts in demand?
- Electronic promotion calendars for dealing with scheduled marketing/sales events?
- Periodic demand history to establish baseline trends?
- Seasonal and calendar factors to adjust for predictive patterns?

_____ Are demand forecasts adequately segmented and aggregated to serve marketing, financial, and operational needs of the business?

_____ Do customers and business partners collaborate on their business plans and demand projections on a regular basis?

_____ Are there electronic links (Internet, intranet, etc.) established to facilitate data communication with vendors, customers, and suppliers?

_____ Have service and performance measurements been implemented to monitor forecast performance in terms of:

- Equipment change-over times?

- Manufacturing cycle times?
- Work-in-process (WIP) inventory balances?
- Order backlogs?
- Product obsolescence?
- Transshipment costs?
- Revenue objectives?

_____ Has a forecasting system implementation been evaluated in terms of:

- Employee understanding, acceptance, and use?
- System features meeting expectations?
- System performance being acceptable?

9.5 HOW TO CREATE A TIME-PHASED REPLENISHMENT PLAN

The basic function of DRP is to create a recommended order that is sent to manufacturing plants in order to plan production. Exhibit 9.15 shows the spreadsheet calculations involved in forecasting the Planned Orders and Months Supply. There are a number of inventory factors that must be taken into account, such as on-hand, on-order, and backordered quantities.

Basic Distribution Resource Planning

The basic DRP calculation starts with a final forecast. The final forecast (labeled Total Forecast in Exhibit 9.15) is the total forecast of independent demand. There may be other forecasts that need to be included, such as a forecast from a division or region that is not part of the main forecasting system. These are handled in the lines below the final forecast. In this spreadsheet, the gross requirements are the sum of the forecast lines. The gross requirements are determined for any practical number

EXHIBIT 9.15 Basic DRP Calculation with a Minimum Order Quantity of 2000 and Multiple of 50 Units

Min/Mult Example	May-01	Jun-01	July-01	Aug-01	Sep-01	Oct-01	Nov-01	Dec-01	Jan-02	Feb-02	Mar-02	Apr-02	May-02	Jun-02
Total Forecast	1520	1689	1958	2210	3214	3225	3345	3578	3845	3895	3956	4012	4123	4231
Prior Forecast	1500													
Other Forecast														
Gross Requirements	3020	1689	1958	2210	3214	3225	3345	3578	3845	3895	3956	4012	4123	4231
Firm Planned Orders														
Scheduled Receipts														
Planned Receipts		5500	3200	3250	3350	3550	3850	3900	3950	4000	4150			
Planned Orders	5500	3200	3250	3350	3550	3850	3900	3950	4000	4150	4250			
Projected Inventory	380	4191	5433	6473	6609	6934	7439	7761	7866	7971	8165			
OnHand	2000													
MonthsCoverage	2													
LeadTime	1	month												
OnOrder	1400													
Minimum	2000													
Multiple	50													

of periods into the future, from 12 to 18 months being typical. Next, we determine planned receipts. At time $T = 2$ (June 01, in this case), the gross requirements over a lead time of one period are 1689 and the ending (projected) inventory at $T = 1$ (May 01) is 380. The gross requirements over safety time (months coverage) starting at lead-time-period ahead is 4168 ($= 1958 + 2210$). The planned receipts are calculated as:

$$\text{Planned receipts} = \text{Gross requirements summed over the lead times} \\ - \text{Projected inventory} + \text{Gross requirements summed} \\ \text{over the safety times starting at lead-time-period ahead}$$

For example, the planned receipts for period $T = 2$ are $(1689 - 380) + (1958 + 2210) = 5477$. Next, the planned orders are offset one-month back, determined by the lead time, so that the orders can be received as planned. Scheduled receipts already committed. The firm planned orders are the overrideable receipts. The projected (ending) inventory for this period is determined by:

$$\text{Projected inventory} = \text{Previous period ending inventory} \\ + \text{Planned receipts} - \text{Gross requirements}$$

The projected inventory at the end of period $T = 2$ (Jun 01) is 4168 ($= 380 + 5477 - 1689$). Once the projected inventory has been determined, a calculation of months supply can be made as a measure of safety stock.

At the next period, the process repeats itself. Now, for $T = 3$ (Jun 01), the planned receipts are $(1958 - 4168) + (2210 + 3314) = 3214$. These are the planned orders offset to $T = 2$. The ending inventory is 5424 ($= 4168 + 3214 - 1958$).

For the very first period, things are a little different. The initial projected inventory is 380 ($= 2000 + 1400 - 3020$), which is given by

$$\text{Projected inventory} = \text{On-hand} + \text{On-order} - \text{Gross requirements} \\ \text{(initial period)}$$

Minimum and Multiples

When we take minimum quantities and multiples into account, the DRP calculation needs to be augmented. In this example (Exhibit 9.15), a minimum order is 2000 and additional orders are placed in multiples of 50. At $T = 2$ (Jun 01), the planned receipts previously calculated ($= 5477$) become 5500 because of the minimum and multiple conditions. The projected inventory now has 23 additional units and becomes 4191 ($= 4168 + 23$). The remaining calculations remain the same, taking the minimum order quantity and multiples in consideration.

SUMMARY

Demand forecasters are in the business of making statements about future demand for products and services in the face of uncertainty. It is part of an ongoing process affecting sales, marketing, inventory, production, and all other aspects of the supply chain. A bottom-up forecast incorporates a logical and coherent series of steps that, if performed in an organized management-supported fashion, can improve forecasting effectiveness, reliability, and accuracy throughout the supply chain.

In this chapter, we have developed a process for the automated forecasting of large volumes of end items in the supply chain. Some of the functions supported by a forecasting system include statistical modeling and forecasting engine, database management, decision support and analysis, and exception handling and reporting.

An effective forecasting process will result in lower costs and improved customer satisfaction. By making forecasting a vital element in the supply chain, the firm is on the right track to achieving the goal of having the right quantity of the right product on the right truck at the right time!

REFERENCES

Brown, R. G. (1982). *Advanced Service Parts Inventory Control.* 2nd ed. Norwich, VT: Materials Management Systems.

Chatfield, C., and M. Yar (1991). Prediction intervals for the multiplicative Holt-Winters. *Int. J. Forecasting* 7(1), 31–38.

DeLurgio, S. A., and C. D. Bahme (1991). *Forecasting Systems for Operations Management.* Homewood, IL: Business One Irwin.

Dielman, T. E. (1996). *Applied Regression Analysis for Business and Economics.* 2nd ed. Belmont, CA: Wadsworth.

Fildes, R., and C. Beard (1992). Forecasting systems for production and inventory control. *Int. J. Ops. Prod. Manage.* 6(3), 4–27.

Fisher, M., J. H. Hammond, W. R. Obermeyer, and A. Raman (1994). Making supply meet demand in an uncertain world. *Harvard Business Rev.* May–June, 83–93.

Gardner, E. S. (1985). Exponential smoothing: The state of the art. *J. Forecasting* 4, 1–38.

Gardner, E. S. (1988). *Autocast II User Manual.* Bridgewater, NJ: Core Analytic.

Gardner, E. S., and E. A. Anderson (1997). Focus forecasting reconsidered. *Int. J. Forecasting* 13, 501–8.

Gardner, E. S., and E. McKenzie (1988). Model identification in exponential smoothing. *J. Op. Res. Soc.* 39, 863–67.

Makridakis, S., C. Chatfield, M. Hibon, M. Lawrence, T. Mills, K. Ord, and L. Simmons (1993). The M2-competition, a real-life judgmentally based forecasting study. *Int. J. Forecasting* 9, 5–29.

Makridakis, S., and M. Hibon (1990). Sliding simulation: A new approach to time series forecasting. *Manage. Sci.* 36, 505–12.

Makridakis, S., and M. Hibon (2000). The M3-competition: Results, conclusions and implications. *Int. J. Forecasting* 16, 451–76.

Martin, A. J. (1995). *DRP—Distribution Resource Planning.* Rev. ed. Essex Junction, VT: Oliver Wight Limited Publications.

Schnaars, S. P. (1986). A comparison of extrapolation procedures on yearly sales forecasts. *Int. J. Forecasting* 2, 71–85.

Smith, B. T. (1996). *Focus Forecasting®: Computer Techniques for Inventory Control.* Boston: CBI Publishing.

Tashman, L. J., and J. M. Kruk (1996). The use of protocols to select exponential smoothing procedures: A reconsideration of forecasting competitions. *Int. J. Forecasting* 12, 235–53.

PROBLEMS

9.1 Exhibit 1.5 depicts a forecasting problem for the weekly shipment of a canned beverage product.

  a. Create a diagram of an integrated supply chain for a manufacturer of canned beverage products.

  b. Discuss why forecasts are required on a weekly basis.

  c. Who are the principal users of the forecast?

  d. Discuss the benefits that these forecast users would have from an accurate forecast.

  e. Discuss what problems inaccurate forecasts can create in this supply chain.

9.2   Repeat Problem 9.1 for a distributor of automotive replacement parts.

9.3   Repeat Problem 9.1 for a retailer of cosmetic products.

9.4   Forecasts must include estimates for new markets, new products, new labels, and new business conditions. Make some specific suggestions about what a forecaster should take into account when working for:

a. A global manufacturer of construction machinery

b. A private-label manufacturer supplying department stores

c. A distributor of cataloged products

d. A retailer of consumer goods

9.5   a. In making a replenishment of a particular product (see table), determine the planned order (A) for Feb 02, when the procurement lead time is 1 month.

b. If months coverage is set to 3 months instead of 2 months, what will the planned order (A) be?

c. If the user places a firm planned order of 100 units in Feb 02, what will the planned order be (months coverage is 2 months)?

	Jan 02	Feb 02	Mar 02	Apr 02	May 02	Jun 02
Final forecast		129	350	171	150	264	164
Prior forecast		323	0	0	0	0	0
Other forecast		0	0	0	0	0	0
Gross requirements	452	350	171	150	264	164	
Firm planned orders							
Scheduled receipts							
Planned receipts		0	16	(A)	164	343	143
Planned orders		16	(A)	164	343	143	171
Projected inventory	655	321	414	428	507	486	
Months supply		2.89	2	2	2	2	2

9.6   In the accompanying replenishment plan for a particular product, determine the planned order (AA) for Month 2 when the minimum order quantity is 200 units.

	Month 1	Month 2	Month 3	Month 4	Month 5
Gross requirements	100	100	100	100	100
Planned receipts		(AA)		200	
Planned order	(AA)		200		200
Projected inventory	0	100	100	100	100
Lead time	1 month				
Safety stock	1 month				
On-hand	100				
Minimum	200				

9.7   In the accompanying replenishment plan for a particular product, determine the planned order (AAA), when the procurement lead time is 1 month, the minimum order quantity is 20 units, and you must order in multiples of 5 units.

	Month 1	Month 2	Month 3	Month 4	Month 5
Gross requirements	10	20	32	47	51
Planned receipts		20	(AAA)	45	50
Planned order	20	(AAA)	45	50	70
Projected inventory	0	0	3	2	1
Lead time	1 month				
On-hand	15				
Minimum	20				
Multiple	5				

CASE 9A DEMAND FOR ICE CREAM (CASES 1A, 3A–8A CONT.)

The entrepreneur has recently acquired a demand forecasting and replenishment planning system to plan her orders to the new factory. The monthly historical data through August of YR06 is given in Case 3A. The gross requirements (gallons) for September of YR07 through August of YR07 are {66,270, 59,760, 52,404, 51,133, 51,424, 56,473, 67,658, 66,616, 73,819, 80,214, 81,130, 77,610}. The on-hand inventory in the current period is 22,000 gallons and 40,000 gallons are due in (on-order).

(1) She wishes to keep a constant 2 weeks of inventory. Develop a schedule of planned orders for the next 12 months. Assume that the minimum order quantity is 5000 gallons and that orders must be made in multiples of 1000 gallons.

(2) Because it is expensive to store ice cream, the entrepreneur decides to shorten the months coverage to only 1 week. She also plans to order ice cream on a weekly basis. Prepare a schedule of planned orders for the next 8 weeks. Assume that the minimum order quantity is 1000 gallons and that orders must be made in multiples of 500 gallons.

CASE 9B DEMAND FOR SPARE RIBS

You are working as the inventory manager for the Ribby Joe Restaurant Chain. Ribby Joe restaurants are strategically located in busy shopping malls in middle- to upper-middle-class neighborhoods. An unsolicited feature article on Chef Joe by a national newspaper helped boost the popularity of the restaurant. According to top management, spareribs, a finger-licking appetizer, are always in demand by their clientele. Because business is stable at Ribby Joe's, the demand for spareribs is also at a steady level. Management reports the following monthly sales of ribs (in pounds) for last year (January through December) in one of their restaurants: {1198, 979, 1061, 1114, 1139, 1092, 1160, 1102, 1120, 1216, 1123, 1122}. Because there is only 1 year's worth of data, it is not possible to determine if there is seasonality, so your first effort is to use simple exponential smoothing. However, to account for the different number of days and weekends in a month, weights were assigned to the days of the week. Monday through Thursday carry weight 1.0; Friday, Saturday, and Sunday have weights 2.0, 2.5, and 1.5, respectively.

(1) Normalize each month by assuming the data occurred last year.

(2) Plot the historical and normalized data.

(3) Run a simple exponential smoothing forecast for a 3-month horizon.

(4) Plot the historical smoothed data with the forecasts.

(5) What recommendations can you make to management regarding this technique for projecting demand for spare ribs?

You plan to replenish inventory every week. For the first forecast period, you have 40 pounds on hand and 160 pounds on order, lead time is 1.0 week, and you

The Supply Chain

1. Analyzing Customer Demand: What should we make and when?

Based on customer demand, product design, cost, and pricing considerations, the ice cream manufacturer (as in Cases 1A, 3A–8A, 10A, 13A–15A) sets the supply chain in motion. For instance, cocoa beans for making chocolate will be sourced from Africa or South America.

3. Bill of Materials: Are we producing the right amount of the right product?

Pulling together components and knowing exactly how many components are needed for a given product, manufacturers utilize demand signals to assure the most efficient and cost effective manufacturing process.

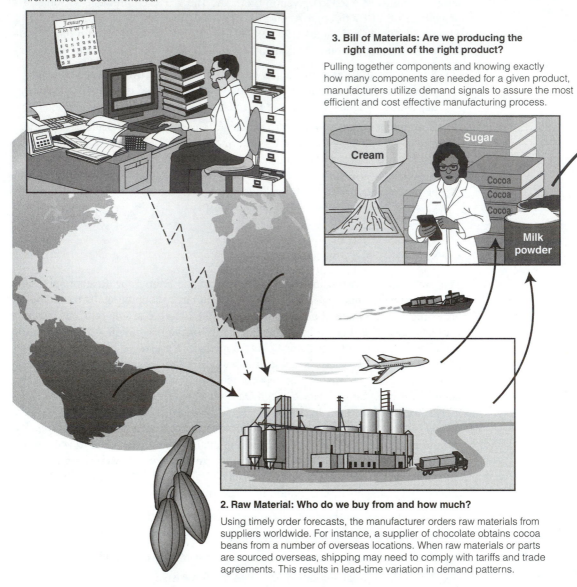

2. Raw Material: Who do we buy from and how much?

Using timely order forecasts, the manufacturer orders raw materials from suppliers worldwide. For instance, a supplier of chocolate obtains cocoa beans from a number of overseas locations. When raw materials or parts are sourced overseas, shipping may need to comply with tariffs and trade agreements. This results in lead-time variation in demand patterns.

4. Assembly: How do we make the final product?

As parts move along the assembly line, subassemblies are transformed into finished goods. Inventory and shipping information is communicated throughout the channel to distribution and retail/wholesale sites.

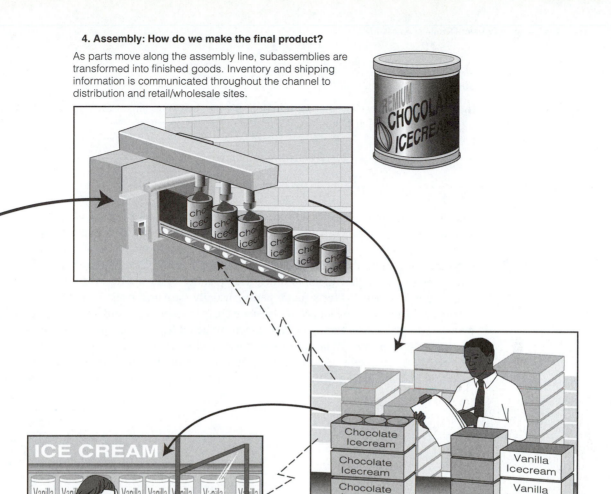

5. Distribution: Where do we distribute product?

Picking, packing, and shipping are the essential warehouse management functions to keep product moving from manufacturer to consumers. Replenishment plans driven by forecasts keep customer orders in sync with optimized inventory positions.

6. Retail/Wholesale: What is the proper assortment and allocation of merchandise in stores?

Accurate forecasts assure that the right amount of the right product is available to consumers when they need it. Poor forecasts can lead to overstocks, out-of-stocks and loss of profit.

plan to maintain a constant 1.5 weeks of inventory. You can order a minimum of 20 pounds and in multiples of 10 pounds.

(6) Develop a time-phased order schedule for the next 2 months. Calculate the planned orders, scheduled receipts, and projected inventory.

CASE 9C FORECASTING FOR INVENTORY CONTROL

Exponential smoothing has found wide applications in inventory control systems. In a production-distribution system, the inventory function can be divided into components, such as pipeline or work-in-process inventory, cycle inventory, seasonal inventory, and buffer inventory. The essence of inventory control is to determine when to order new stock and how much to order. The quantity of most interest is the maximum reasonable demand for a future period, usually measured over a lead time. As a demand planner for a manufacturer, you are charged to place the orders for an item whose historical values for 1 year are shown in the table. (The manufacturer operates on a 4-4-5 pattern.) The lead time is assumed to be 1 week, and you have a requirement to maintain 2.5 weeks of inventory. For the first forecast period, you have four units on hand and two units on order (due in).

	Q:01	Q:02	Q:03	Q:04	Forecast
WK01	7	10	10	5	**2**
WK02	8	9	6	6	**2**
WK03	10	10	3	11	**12**
WK04	12	5	11	7	**6**
WK01	5	6	3	8	**7**
WK02	3	4	8	10	**6**
WK03	8	7	13	10	**10**
WK04	9	12	18	9	**10**
WK01	7	13	12	9	
WK02	4	8	14	12	
WK03	4	10	14	6	
WK04	5	11	10	3	
WK05	5	8	6	3	

(1) Provide a schedule of planned orders for the next 2 months.

(2) Use the same fixed-time order logic as in (1) to determine the schedule of time-phased orders if the minimum order quantity is three.

(3) Use the same fixed-time order logic to determine the schedule of time-phased orders if the minimum order quantity is three and the lead time is 2 weeks.

(4) Use the same fixed-time order logic to determine the schedule of time-phased orders if the minimum order quantity is three, the lead time is 1.5 weeks, and you need to maintain 2 weeks of inventory.

(5) Use the same fixed-time order logic to determine the schedule of time-phased orders if the minimum order quantity is three, the multiples is 2, the lead time is 1.5 weeks, and you need to maintain 2 weeks of inventory.

Your management is concerned about the effectiveness of your forecasting model. You are asked to run some evaluation results to analyze the impact of some systematic changes in the demand pattern, characterized by a ramp, a step, and an impulse.

(6) Use a simple exponential smoothing model on the first 49 values of the data.

(7) For **ramp,** create a one-period-ahead forecast. Assume the historical values for period 50–60 are going to be {11, 10, 9, 8, 7, 6, 5, 4, 3, 2, 1}. Create another one-period-ahead forecast based on the first 50 values. Continue in this pattern for the rest of the historical data. Plot the historical data and the one-period-ahead forecasts for the 11-period forecast horizon.

(8) For **step,** create a one-period-ahead forecast. Assume the historical values for period 50–60 are going to be {14, 14, 14, 14, 14, 14, 14, 14, 14, 14, 14}. Create another one-period-ahead forecast based on the first 50 values. Continue in this pattern for the rest of the historical data. Plot the historical data and the one-period-ahead forecasts for the 11-period forecast horizon.

(9) For **impulse**, create a one-period-ahead forecast. Assume the historical values for period 50–60 are going to be {18, 0, 0, 0, 0, 0, 0, 0, 0, 0, 0}. Create another one-period-ahead forecast based on the first 50 values. Continue in this pattern for the rest of the historical data. Plot the historical data and the one-period-ahead forecasts for the 11-period forecast horizon.

(10) Give an interpretation of the patterns in the one-step-ahead forecasts. Repeat (7)–(9) with the damped trend exponential smoothing model.

Forecasting Models

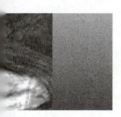

10

Creating and Analyzing Causal Forecasting Models

"Knowledge cannot spring from experience alone, but only from the comparison of the inventions of the intellect with the observed facts."

ALBERT EINSTEIN

IN THIS CHAPTER, WE INTRODUCE the concept of a causal model and describe the steps for creating and analyzing causal models for forecasting purposes. You will learn about a three-stage iterative model building procedure, consisting of

- Identification—using the data to tentatively identify a model
- Estimation—fitting algorithms and inferences about the parameters
- Diagnostic checking—evaluating the model for adequacy

Transformations are important in the preliminary identification and diagnostic checking (residual analysis) stages of the model-building process. Transformations of data should be performed to improve your understanding of what the data are trying to tell you; can make the assumptions of least-squares theory with normality assumptions valid; can have significant value outside the realm of model building, too, that is, in the evaluation and interpretation of data; and can also be applied effectively in presenting management with the results of a forecasting analysis.

The diagnostic checking process, designed to reveal departures from underlying assumptions in a statistical forecasting model, is an important phase for forecasters to learn about. It can be a powerful visual tool for assessing the potential effectiveness of a forecasting model, isolating unusual values, identifying hidden patterns, and understanding the nature of randomness in historical data.

10.1 A MODEL-BUILDING STRATEGY

In previous chapters we have been focused on the extrapolative, self-driven techniques of forecasting, those that make use only of the historical time series of the variable to be forecast. By relying exclusively on self-driven techniques for extrapolation, we may overlook information because we do not consider the behavior of related variables (factors or indicators) and underlying error patterns. Because of their added complexity, causal models require us to pay special attention to a modeling strategy, one that will help to reduce the risk of taking modeling missteps and making erroneous interpretations of results.

Causal models seek to explain past history and forecast future values based on mathematical relationships among variables in conjunction with a specific chance variable.

A general model-building strategy, developed for ARIMA time series models by G. E. P. Box and G. M. Jenkins (1976; Box et al. 1994), can be used for developing causal forecast models. Box and Jenkins introduced an approach, known as the Box-Jenkins methodology, that evolved as a result of their direct involvement with forecasting problems in business, economic, and engineering environments. Their systematic approach resulted in the development of ARIMA models for forecasting, which are covered in Chapters 14 and 15. The modeling procedure consists of the following three stages:

1. *Identification* consists of using the data and any other knowledge that will tentatively indicate whether the forecast variable can be adequately described with a simple or rudimentary model.

2. *Estimation* consists of using the data to make inferences about the parameters that will be needed for the tentatively identified model and to estimate their values.

3. *Diagnostic checking* involves the examination of residuals from fitted models, which can result in either no indication of model inadequacy or a determination of model inadequacy together with information on how the series may be better described.

The procedure is iterative. After each iteration, we examine the residuals for any lack of randomness and, if residuals were found to be serially correlated (a common problem of nonrandomness in time series), we use this information to modify a tentative model. The modified model is then fitted and subjected to diagnostic checking again until an adequate model is obtained.

10.2 WHAT ARE REGRESSION MODELS?

Regression analysis is the principal method of causal forecasting. Forecasters begin a regression analysis by identifying the factors—called independent or explanatory variables—that they believe have influenced and will continue to influence the variable to be forecast (the dependent variable).

It is useful to categorize explanatory variables as internal or external. Internal variables, also called policy variables, can be controlled to a substantial degree by managerial decisions, and their future values can be set as a matter of company policy. Examples include product prices, promotion outlays, and methods of distribution. External or environmental variables are those whose level and influence are outside organizational control. Included here may be variables that measure weather and climate, demographics such as the age and gender of the population in the market area, decisions made by competing enterprises, and the state of the macroeconomy as measured by rates of economic growth, inflation, and unemployment. Normally a regression analysis will attempt to account for both internal and external explanatory variables. Regression modeling for forecasting is described in greater detail in Chapters 11 and 12.

The forecaster's beliefs about the way a dependent variable responds to changes in each of the explanatory variables is expressed as an equation, or series of equations, called a regression model. A regression model also includes an explicit expression of an error variable to describe the role of chance or underlying uncertainty.

A model with a single explanatory variable is called a simple regression model. Multiple regression refers to a model with one dependent and two or more explanatory variables.

A successful regression analysis provides useful estimates of how previous changes in each of the explanatory variables have affected the dependent variable. In addition, assuming that the underlying structure is stable, forecasts of the dependent variable can then be conditioned on assumptions or projections of the future behavior of the explanatory variables. For example, suppose that a regression analysis of the demand for a product indicates that price increases in the past, holding other things constant, have been associated with less than proportional reductions in sales volumes (i.e., demand has been price inelastic). This knowledge may be useful both for forecasting future demand and for adjusting product-pricing policy. The inelastic demand suggests that price increases might be improving profitability. Demand forecasts would then be made in light of the price changes that the company plans to institute.

Similar feedback can be obtained through regression analysis of the influence of external economic variables such as the GDP. Although the firm cannot control the rates of economic growth, projections of GDP growth can be translated via regression analysis into forecasts of product sales growth.

As a forecasting approach, regression analysis has the potential to provide not only forecasts of the dependent variable but useful managerial information for adapting to the forces and events that cause the dependent variable to change.

Indeed, a regression analysis may be motivated as much or more by the need for policy information as by the interest in forecasting. It is important to note that no extrapolative forecasting method can supply policy information, such as how product sales respond to price and macroeconomic variables. When such information is desired, explanations are required, not merely extrapolations. The chapter-opening quotation suggests that Einstein would have preferred the explanatory approach to forecasting, which analyzes the data ("the observed facts") in terms of a model ("an invention of the intellect") that relates the dependent and explanatory variables.

The term regression has a rather curious origin in studies of inheritance in biology by Sir Francis Galton (1822–1911). His studies showed that whereas tall (or short) fathers had tall (or short) sons, the sons were on the average not as tall (or as short) as their fathers. Thus, Galton observed that the average height of the sons tended to move toward the average height of the overall population of fathers rather than toward reproducing the height of the parents. This phenomenon was labeled regression toward the mean, and it has since been observed in a wide spectrum of settings from economic behavior to athletic performance.

The Regression Curve

A regression curve can be used to describe a relationship between a variable of interest and one or more related variables that are assumed to have a bearing on the forecasting problem. If data are plentiful, a curve passing through the bulk of the data represents the regression curve. The data are such that there is no functional relationship describing exactly one variable Y as a function of X; for a given value of the independent variable X, there is a distribution of values of Y. This relationship may be approximated by determining the average (or median) value of Y for small intervals of values of X.

A regression curve (in a two-variable case) is defined as that "typical" curve that goes through the mean value of the dependent variable Y for each fixed value of the independent variable X.

In most practical situations, there are not enough observations to "even pretend that the resulting curve has the shape of the regression curve that would arise if we had unlimited data" (Mosteller and Tukey, 1977, p. 266). Instead, the observations result in an approximation. With only limited data, a shape for the regression curve (e.g., linear, quadratic, or exponential.) is assumed and the curve is fitted to the data by using a statistical method such as the method of least squares. This method is explained shortly.

A Simple Linear Model

Because regression analysis seeks an algebraic relationship between a dependent variable Y and one or more independent variables, the deterministic (or systematic) component of the model describes the mean (expected) value for Y given a specific value of X:

$$\text{Deterministic component} = \text{Mean } Y$$

when Y is some function of X. In practice, there is considerable variability in Y for a given X around a mean value. This mean value is an unknown quantity that is commonly denoted by the Greek letter μ with a subscript $Y(X)$ to denote its dependence on X:

$$\mu_{Y(X)} = \beta_0 + \beta_1 X$$

One key assumption in the linear regression model is that for any value of X, the value of Y is scattered around a mean value.

The straight line may be approximately true; the difference between Y and the straight line is ascribed to random errors. Thus, the observed values of Y will not necessarily lie on a straight line in the XY plane but will differ from it by some random errors.

$$Y = \beta_0 + \beta_1 X + \text{Random errors}$$

Thus, the simple linear regression model for Y can be expressed by the sum of a deterministic component $\mu_Y(X)$ and a random component ε:

$$Y = \mu_{Y(X)} + \varepsilon$$
$$= \beta_0 + \beta_1 X + \varepsilon$$

where the mean (expected) value of random errors ε is assumed to be zero. The intercept β_0 and slope β_1 are known as the regression parameters. The model is linear in the parameters. Both β_0 and β_1 are unknown parameters to be estimated from the data. As a standard statistical convention, it is useful to designate unknown parameters in models by Greek letters, to distinguish them from the corresponding statistics b_0 and b_1 estimated from the data. Regression analysis for simple linear models is discussed in Chapter 11.

As forecasters, we can view the deterministic component as describing a systematic *change*, whereas the random errors depict measured *chance* (in the sense of our notion of forecasting described in the beginning of Chapter 1). In a particular application of the model, the forecaster has data that are assumed to have arisen as a realization of the hypothetical model. The next step is to come up with a rationale for estimating the parameters, β_0 and β_1, in the model from a given set of data.

The Least-Squares Assumption

There are many techniques around for estimating parameters from data, but the method of ordinary least squares (OLS) is the most common and it has a sound basis in statistical theory. This is not to say that other techniques have little merit. In fact, weighted least-squares techniques of several kinds have been found to have increased importance in practical applications such as **robust regression** (Mosteller and Tukey, 1977). We consider some robust regression models for forecasting in Chapter 12.

The method of least squares is the most widely accepted criterion for estimating parameters in a model.

Consider now one reasonable criterion for estimating β_0 and β_1 from data in a simple linear regression model. OLS determines values b_0 and b_1 (because these will be estimated from data, we use small letters b_0 and b_1 to replace the Greek β_0 and β_1), so that the sum of the squared vertical deviations (squared residuals) between the data and the fitted line,

$$\text{Residual} = \text{Data} - (b_0 + b_1 X)$$

is less than the sum of the squared vertical deviations from any other straight-line fit that could be drawn through the data:

$$\text{Minimum of } \sum (\text{Data} - \text{fit})^2 = \sum (\text{Residuals})^2$$

A vertical deviation is the vertical distance from an observed point to the line. Each deviation in the sample is squared and the least-squares line is defined to be the straight line that makes the sum of these squared deviations a minimum. The notation for this is as follows. Consider the data pairs (Y_i, X_i) for $(i = 1, \ldots, n)$. Let $y_i = Y_i - \bar{Y}$ and $x_i = X_i - \bar{X}$, where $\bar{Y} = (\sum Y_i)/n$, and $\bar{X} = (\sum X_i)/n$. The symbol Σ denotes the summation over n values. Then $\Sigma D^2 = \Sigma(Y_i - b_0 - b_1 X_i)^2$ is minimized. Exhibit 10.1 shows the calculations of b_0 and b_1 for a small set of data.

EXHIBIT 10.1
Example Illustrating
the Calculation of
Regression Coefficients
in a Simple Linear
Regression

Y_i	X	$y_i = (Y_i - \bar{Y})$	$x_i = (X_i - \bar{X})$	$y_i x_i$	x_i^2
3	1	−5	−2	10	4
5	2	−3	−1	3	1
7	3	−1	0	0	0
14	4	6	1	6	1
11	5	3	2	6	4
Sum 40	15			25	10

Average: $\bar{Y} = 8 \quad \bar{X} = 3$
$b_1 = 25/10$
$b_0 = \bar{Y} - b_1 \bar{X} = 8 - 2.5(3) = 0.5$
Fitted equation: $\hat{y} = 0.5 + 2.5X$

Linear Regression: One Explanatory Variable

In Exhibit 10.2, the dependent variable Y is the annual sales volume; the explanatory variable X is the advertising cost 1 year earlier. Because X is a lagged value of the data, it allows us to forecast the first period based on an actual value of advertising (see Exhibit 10.3). Note that the slope coefficient in the equation $Y = 18.28 + 0.22\,X$ is 0.22 (Exhibit 10.2). This means that, when the X value for period 36 is 1, the forecast for the next period (37) is 18.5. Thus, a one-step-ahead forecast can be made past the end of the data without having to make a forecast for advertising. It is obtained by calculating $Y_{36}(1) = 18.28 + 0.22 \times (1) = 18.5$. The other parts of the table will be covered more fully in chapter 11.

Exhibit 10.4 shows a scatter diagram of the sales (dependent variable) and lag advertising (independent variable). A good fit is no surprise, because the correlation coefficient between Y and X is relatively high ($r = 0.67$) and the scatter appears linear.

10.3 CREATING MULTIPLE LINEAR REGRESSION MODELS

When we have reasons to believe that more than one explanatory variable is required to solve a forecasting problem, the simple regression model will no longer provide the appropriate representation and we need to develop a linear regression model for forecasting a dependent variable from a projection of more than one explanatory variable. In this case, we are extending the simple linear regression model with one explanatory variable to a model that can encompass multiple explanatory variables. The most direct extension is the multiple linear regression model.

A multiple linear regression model relates a dependent variable to more than one explanatory variable.

EXHIBIT 10.2   Sales versus Advertising (Case 4D)

Regression Statistics	
Multiple R	0.67
R^2	0.45
Adjusted R^2	0.44
Standard error	4.42
Observations	35

ANOVA

	df	SS	MS	F
Regression	1	537.55	537.55	21.49
Residual	33	645.38	19.56	
Total	34	1182.93		

	Coefficients	Standard Error	t Statistic	Lower 95%	Upper 95%
Intercept	18.28	1.42	12.89	15.4	21.17
X variable 1	0.22	0.04	5.26	0.13	0.3

Observation	Y	Predicted Residuals	Standardized Residuals
1	21.52	−1.02	−0.23
2	21.73	−0.73	−0.17
3	22.16	−6.66	−1.53
4	24.10	−8.80	−2.02
5	22.81	0.69	0.16
6	28.85	−4.35	−1.00
7	22.81	−1.51	−0.35
8	23.03	0.47	0.11
9	24.32	3.68	0.84
10	26.04	−2.04	−0.47
11	26.91	−11.41	−2.62
12	18.93	−1.63	−0.37
13	22.81	2.49	0.57
14	24.54	0.469	0.11
15	31.65	4.85	1.11
16	32.30	4.20	0.96
17	28.20	1.40	0.32
18	27.77	2.73	0.63
19	25.40	2.60	0.60
20	31.65	−5.65	−1.30
21	23.03	−1.53	−0.35
22	20.87	−1.17	−0.27
23	23.46	−4.46	−1.02
24	18.93	−2.93	−0.67
25	19.36	1.34	0.31
26	21.30	5.19	1.19
27	26.04	4.56	1.05
28	26.91	5.39	1.24
29	28.85	0.65	0.15
30	19.79	8.51	1.951
31	29.49	1.81	0.41
32	32.30	−0.10	−0.02
33	21.95	4.45	1.02
34	19.36	4.04	0.93
35	21.95	−5.55	−1.27

EXHIBIT 10.3
Output for a Model of Sales versus (Lagged) Advertising

EXHIBIT 10.4
Scatter Diagram of Sales and Advertising Variables

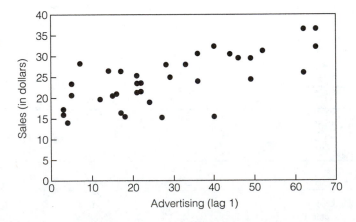

In multiple linear regression, the regression function $\mu_{Y(X)}$ is the deterministic component of the model and takes the form

$$\mu_{Y(X)} = \beta_0 + \beta_1 X_1 + \cdots + \beta_k X_k$$

where $X_1, \ldots, X_k$ are explanatory variables (or regressors) and $\beta_0, \beta_1, \ldots, \beta_k$ are called regression parameters. Explanatory variables are also commonly termed independent variables, and forecasters use these terms interchangeably. This model arises when the variation in the dependent variable Y is assumed to be impacted by changes in more than one explanatory variable. Thus, the average value of Y is said to depend on $X_1, X_2, \ldots, X_k$. The dependence on X is henceforth suppressed in the notation; let $\mu_{Y(X)} = \mu_Y$. In this case, we speak of a multiple regression of Y on $X_1, \ldots, X_k$.

A multiple linear regression model (the analysis is discussed in Chapter 12) may be used to fit a polynomial function, such as

$$\mu_Y = \beta_0 + \beta_1 X + \beta_2 X^2 + \cdots + \beta_k X^k$$

By letting $X_1 = X$, $X_2 = X^2, \ldots, X_k = X^k$, we obtain the form of the linear regression model. Often X is used to represent a time scale (weeks, months, or years). It is worth emphasizing that the regressors may be any functional form; there need only be linearity in the β parameters. What is meant by linearity? Examples of linearity in the parameters include $\mu_Y = \beta_0 + \beta_1 t$, $\mu_Y = \beta_0 + \beta_1 e^t$ and $\mu_Y = \beta_0 + \beta_1 \ln X$. On the other hand, functional forms for a regression model, such as $\beta_1 \sin(\gamma_1 t) + \beta_2 \cos(\gamma_1 t)$ and $\beta_1 \exp(\gamma_1 t) + \beta_2 \cos(\gamma_2 t)$, are *not* linear in the β and γ parameters.

Some Examples

A Consumption Model

Consider an economics illustration of a multiple linear regression model with two explanatory variables. The dependent variable, Y, measures annual consumption in the United States, X_1 measures national disposable income, and X_2 is the value of the nation's wealth in the form of corporate stocks. All variables are expressed per capita (i.e., divided by the U.S. population at end of year). The essence of the model is to see if the dependent variable consumption (what people spend) can be predicted on the basis of income (what people earn) and wealth (what people own).

The model is:

$$\text{Consumption} = \beta_0 + \beta_1 \text{ Income} + \beta_2 \text{ Wealth} + \varepsilon$$

For a sample of recent years, coefficients were determined that resulted in a fitted equation given by:

$$\text{Fitted consumption} = -176 + 0.935 \text{ Income} + 0.047 \text{ Wealth}$$

The estimate of β_1 is $b_1 = 0.935$, and this suggests that, if wealth does not change, then every \$1 increase in income will raise consumption on the average by 93.5 cents. (Keynes called this coefficient the marginal propensity to consume.) So the evidence is that we spend most of our additional earnings; in contrast, from the estimate of β_2 ($b_2 = 0.047$), we seem to save most of our added wealth.

The constant term has no economic significance here, again, because it represents an estimate of consumption during a hypothetical year in which both income and wealth are zero.

For future years, forecasts of consumption levels can be made from projections of income and wealth, as follows:

$$\text{Consumption}_{\text{forecast}} = -176 + 0.935 \times \text{Projected income} + 0.047 \text{ Projected wealth}$$

The accuracy of these forecasts depends on the quality of the income and wealth projections as well as on the magnitude of random variation in consumption.

Log-Linear Models

Multiple linear regression models can be readily adapted to incorporate cases in which the response rate of the dependent variable to one or more explanatory variables is believed to be damping, growing exponentially, or changing in other nonlinear ways. The linear model can also be adapted to suit a common economic assumption of constant elasticities of response. All these can be accomplished by using transformations to the explanatory and/or dependent variables.

One of the most common transformations leads to the log-linear model

$$\ln Y = \ln \beta_0 + \beta_1 \ln X_1 + \beta_1 \ln X_2 + \cdots + \ln \varepsilon$$

where ln refers throughout to a natural logarithm ($\log_e$). Because a one-unit change in the log of a variable can be viewed as a 1% change in the level of the variable, each coefficient represents the $\beta\%$ response in the dependent variable to a 1% increase in an explanatory variable. Economists refer to such ratios of percentage changes as elasticities.

Through this model, the forecaster expresses the belief that each elasticity is a constant (does not vary with the level of X or level of Y). This is in contrast with the assumption in the consumption model that each response rate $-\Delta Y/\Delta X$ is viewed as a constant. For example, if Y denotes sales volume and X is price, the linear model in the same form as the consumption model assumes that there is a single number representing the average response to a \$1 change in price. If, however, we believe that the effect of a \$1 change in price is one thing if the original price is \$10 (and hence the price change is 10%) and quite another if the original price is \$100, then the log-linear model may be more appropriate. In the log-linear model, demand is scaled to the percentage change, not the absolute change, in price.

An example of a log-linear regression model in which the demand for a product Y is related to three explanatory variables, product price X_1, advertising budget X_2, and sales force support X_3, is:

$$\ln \text{Demand} = \beta_0 + \beta_1 \ln \text{Price} + \beta_1 \ln \text{Advertising} + \beta_3 \ln \text{Salesforce} + \varepsilon^*$$

where $\ln \varepsilon$ is rewritten as ε^*. The fitted equation is

$$\text{Fitted ln Demand} = -2.0 \times \ln \text{Price} + 0.125 \ln \text{Advertising} + 0.250 \ln \text{Salesforce}$$

The estimate of the price elasticity β_1 is $b_1 = -2.0$, which indicates that each 1% price increase, holding the advertising and sales force budgets constant, is estimated to reduce demand by 2%. Similarly, the advertising elasticity β_1 is estimated by $b_2 = 0.125$, which denotes an estimated 1/8 of 1% increase in demand per 1% increase in advertising, holding price and sales force budget constant. Alternatively stated, it is estimated to take an 8% increase in advertising to raise demand by 1%.

Forecasts of future demand can use the estimated regression as a forecasting formula and then convert the final result from log to level. For example, if a forecast for ln Demand is calculated to be 8.5, the reversal of the log transformation is $e^{8.5} = 2.7188.5 = 4915$ units. Most forecasting software should make such a reverse transformation unnecessary by presenting the forecasts in the original units of the dependent variable.

During a year, a pizza restaurant used two types of advertising, newspaper ads and radio spots (Exhibit 10.5). The number of ads per month is listed in column D, the number of minutes of radio advertising is shown in column E, the monthly sales in hundreds are shown in column F, and the estimated values of Y based on the regression relationship appear in column G. Column H shows errors (Actual − Estimate).

The multiple linear regression model has two X coefficients. The fitted equation for pizza sales is given by:

$$\hat{Y} = 65.84 + 6.25 \text{ (News ads)} + 21.39 \text{ (Radio time)}$$

The first coefficient (for newspaper ads) is poor because the probability is 0.30 that it is due to chance. The second coefficient (for radio time) is better, but the probability is rather large at 0.09. However, this regression model is not as bad as it looks (Exhibit 10.6). You can also test the effect of both X coefficients considered together. This is really a test on the entire regression equation. The F probability is 0.01 that the regression equation is due to chance. This result is not unusual; often the regression equation looks good, but individual coefficients do not. We discuss the derivation of these test statistics in greater detail in the next chapter. For now, we only need to get a sense as to how they are used in practice.

EXHIBIT 10.5 Pizza Sales versus Newspaper Advertisements and Radio Spots

Y-axis: Sales (000)

A	B	C	D	E	F	G	H
Year	Month	Period	News ads X_1	Radio time X_2	Sales Y	$\hat{Y}$	Error
20XX	1	1	12	13.9	436	438.1	−2.1
	2	2	11	12	380	391.21	−11.21
	3	3	9	9.3	301	320.97	−19.97
	4	4	7	9.7	353	317.03	35.97
	5	5	12	12.3	464	403.88	60.12
	6	6	8	11.4	342	359.64	−17.64
	7	7	6	9.3	302	302.23	−0.23
	8	8	13	14.3	407	452.9	−45.9
	9	9	8	10.2	385	333.97	51.03
	10	10	6	8.4	226	282.98	−56.98
	11	11	8	11.2	376	355.36	20.64
	12	12	10	11.1	352	365.72	−13.72

(*Source*: REGRESS.XLS, The Spreadsheet Forecaster on CD)

EXHIBIT 10.6
Regression Summary
of Pizza Sales Model

Regression Statistics	
Multiple R	0.83
R^2	0.68
Adjusted R^2	0.61
Standard error	39.89
Observations	12

ANOVA

	df	SS	MS	F	Significance F
Regression	2	30998.6	15499.3	9.74	0.01
Residual	9	14320.1	1591.12		
Total	11	45318.7			

	Coefficients	Standard error	t Statistic	p Value
Intercept	65.84	85.42	0.77	—
X_1	6.25	11.2	0.558	0.30
X_2	21.39	14.7	1.45	0.09

(*Source*: Exhibit 10.5)

Why are the p values large when the F probability is small? This usually happens when the X variables are highly correlated with one another (Exhibit 10.7). The correlation coefficient is 0.89. In this case, the variables tend to increase and decrease at the same time. Correlated X variables make it difficult for the regression calculations to isolate the individual effects of each variable. Nevertheless, together the X coefficients do a reasonable job in predicting Y. To improve this model, we would need to look for a variable X that was not highly correlated with radio time but that we have reason to believe might help explain pizza sales.

Exhibit 10.8 shows the pizza sales and fitted values as a time plot. The historical pattern is tracked quite well, but how well will this model forecast? That will depend on the quality and accuracy of the explanatory variables. To forecast, we simply plug in values for X_1 and X_2 in the previous equation and determine the forecast $\hat{Y}$.

EXHIBIT 10.7
Scatter Diagram of
Radio Time versus
News Ads

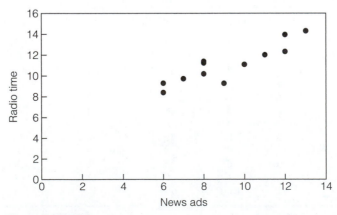

(*Source*: Exhibit 10.5)

EXHIBIT 10.8 Time Plot of Pizza Sales and Fitted Values

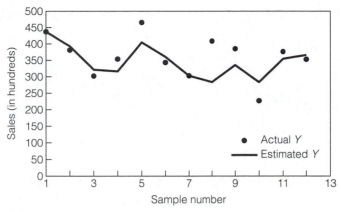

(*Source*: Exhibit 10.5)

10.4 LEARNING FROM RESIDUAL PATTERNS

In Exhibit 4.3 (Chapter 4), we interpret the residuals from the straight-line model for the FRB industrial production series INDPROD as showing a cyclical pattern. This pattern can be related to economic contractions and expansions. The plot suggests that an economic contraction or recession may be underway, as shown by the plunge of the FRB series below the trend level. In addition, the steady recovery to more normal growth is evident in the latest 3 years. Implicit in this analysis of the data (Data = Model + Error) is a model, characterized by the straight-line trend, in which the error describes all the variation not contained in the trend. A summary of the data assigns a fitted value to each value (Data = Fit + Residual), where the fit is the straight line created through the data values. To look at the data in more detail, we examine the residuals. However, Exhibit 4.3 is only a preliminary summary, in that it is based on a simple model that may expose additional patterns for more detailed analysis at a later stage. This iterative process is typical of an exploratory data analysis procedure. Fortunately, much useful information in a residual analysis can be obtained effectively using graphical techniques.

A residual analysis is concerned with revealing departures from the assumptions of a model and suggesting corrective steps.

In other situations, an economic index may not have a persistent trend but, rather, it may meander as it grows for certain fairly long periods but then drop over other long periods. Then, it is the first differences (or period-to-period changes) that may reveal the interesting secondary patterns. For example, the monthly index for a regional commodity (Exhibit 10.9a) can meander, but its month-to-month changes (Exhibit 10.10a) will behave more like a random residual series. The correlogram in Exhibit 10.9b declines gradually from $r_1 = 0.93$ to approximately 0 at lag 14, thereafter

EXHIBIT 10.9
(a) Time Plot and
(b) Correlogram of
100 Monthly Values
of Flour Price Index
at Buffalo, NY
(BUFLOUR.DAT;
Cryer and Miller,
1994)

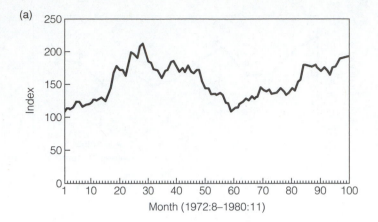

(a)

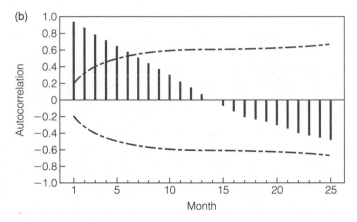

(b)

becoming negative. Note that an r_1 coefficient so close to 1 suggests that the series is highly correlated with its lag 1 values, which is typical of a meandering series.

The sample autocorrelations of the differenced series in Exhibit 10.10b, on the other hand, show the behavior of a more random series; $r_1 = 0.14$, $r_2 = 0.00$, $r_3 = -0.03$, $r_4 = 0.06$, and $r_5 = 0.02$ are close to 0. A series whose period-to-period changes are random is known as a random walk. Random walk models are widely applied in finance for predicting price movements and in the development of an efficient markets theory (Malkiel, 1998).

 A residual analysis may suggest relationships, transformations, or the need for a better understanding of events that may not be apparent in the bulk of the data.

A time plot that has no visible pattern (Exhibit 10.11a) provides no evidence against the assumption that the errors in the model are random, have zero mean, and show constant variance.

EXHIBIT 10.10
(a) Time Plot and
(b) Correlogram of
the Month-to-Month
Changes of a Flour
Price Index at
Buffalo, NY

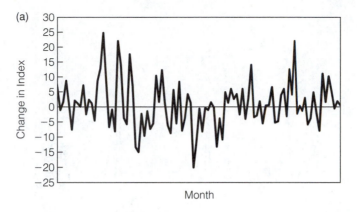

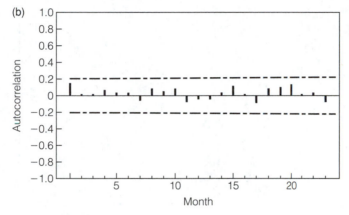

(*Source*: Exhibit 10.9)

It is worth pointing out at this time the random data need not be normally distributed. Note that the correlograms of the differenced flour price index (Exhibit 10.10b)
and the random lognormal data (Exhibit 10.11b) are indistinguishable. They both display the characteristics of a random walk. In fact, the normal probability plot of the
differenced flour price index (Exhibit 10.12a) appears more normal (points lie along a
straight line) than the random lognormal data (Exhibit 10.12b) (lognormal data are
such that the logarithms of the data are normally distributed, not the data itself).

A correlogram in which no pattern is visible is consistent with a basic assumption
about randomness in a regression model.

Next, we discuss some tests that indicate whether a randomness assumption can
be supported by the data.

EXHIBIT 10.11
(a) Time Plot and (b)
Correlogram of Data
without any Pattern—
100 Random Values
Generated from a
Random (Lognormal)
Series

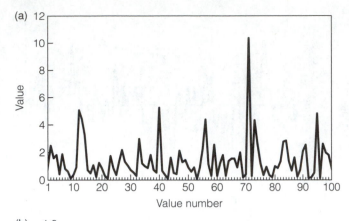

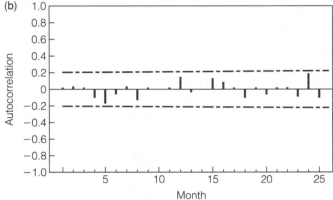

A Run Test for Randomness

Some tests indicative of randomness (or lack thereof) in residuals include tests based on first differences, rank correlation, and runs of signs. The first-differences test assumes that, if there is nonrandomness in the form of trend, the number of positive first differences will be large for a residual series with an upward trend and small for a residual series with a downward trend.

In the rank correlation test, residuals are ranked in order depending on size and then correlated with a straight-line trend. The distribution of the test statistic is difficult to determine, so approximate tests based on the Student's t distribution are used. Details of these tests and similar ones may be found in books on nonparametric statistics (Walsh, 1962).

The runs-of-signs test is a simple way to test for randomness by counting runs. A run is a string of pluses ($+$) or minuses ($-$) that accumulates when a plus ($+$) value is assigned to an observation larger than the average value K of the data and a minus ($-$) value is assigned to a value less than the average. In counting these runs, we let the number of consecutive runs of pluses be denoted by r; n denotes the number of values in the sample. Then, the average number of runs over all possible

EXHIBIT 10.12
Normal Probability
Plots for (a) the
Differenced Flour
Price Index Data and
(b) the Sample of
Random Lognormal
Data

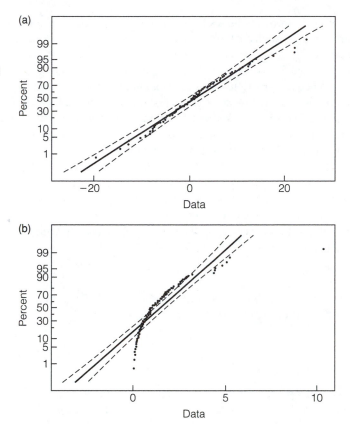

sequences of runs is $1 + [2r(n - r)/n]$. A measure of the expected dispersion of runs among different sequences that could have occurred is the standard error given by

$$\{2r(n - r)[2r(n - r) - n]/n^2(n - 1)\}^{1/2}$$

Thus, if the observed number of runs is within two standard errors of the expected (average) number of runs, then there is little evidence that the series is not random. The runs measure is a measure of conformity of the data to a model specification of randomness.

It is often observed that changes in many stock prices behave essentially like a random series.

The average (expected) number of runs and associated standard errors is 37 (± 3.6) for the differenced flour price index (Exhibit 10.9a) and is 36 (± 3.5) for the random lognormal data (Exhibit 10.11a). The observed values for number of consecutive runs of plusses is 24 for the differenced flour price index and 23 for the random lognormal data. Consequently, in both examples there is evidence that these series are not random according to the runs-of-signs test because the observed *r* is only 24 and 23 for the two series, respectively.

When runs-of-signs counting is applied to residuals, we expect that the number of consecutive positive or negative residuals will be neither small nor large for a random residual series. A straight-line trend with positive slope will have a run of negative signs followed by a run of positive signs. A wildly fluctuating series, on the other hand, will exhibit too many sign changes.

Because of the severe problems created by autocorrelated errors, it is important to be able to detect their presence and have means for correction. (See Chapter 4 for a treatment of autocorrelation analysis.) Other procedures, based on specific assumptions about the nature of the autocorrelation, lead to test statistics (e.g., Durbin-Watson statistic) that are discussed in the context of forecasting with normal linear regression models (Chapter 12).

> The ordinary correlogram is the most widely used graphical tool for detecting the presence of first- or higher-order autocorrelation.

Nonrandom Patterns

Much of a residual analysis for time series models can be carried out effectively through a visual inspection of data patterns and correlograms. If a pattern can be discerned in a residual plot, then these patterns typify a violation of one or more assumptions about randomness in a regression model. Let us discuss the identification of such patterns and some remedies first. Five basic types of patterns are frequently observed in residual plots: no visible pattern, cyclical pattern (positively autocorrelated residuals), nonlinear relationships, increasing dispersion, and linear trend.

A cyclical pattern (Exhibit 10.13) is often evident when we fit linear models to time series data. For example, economic expansions and recessions in the business

EXHIBIT 10.13 Time Plot of Residuals with a Cyclical Pattern

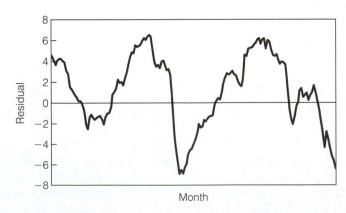

EXHIBIT 10.14 Time Plot of Residuals with a Cup-Shaped Pattern

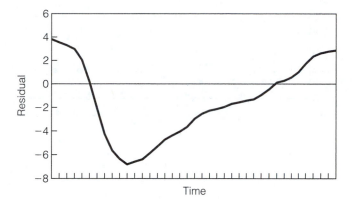

cycle can be visualized in a residual plot. In the special case of time series, it makes sense to connect residuals sequentially in time to expose underlying cyclical patterns. This procedure highlights autocorrelation in residuals.

Another reason for the appearance of nonrandom patterns is that a linear model is being fit to an inherently nonlinear phenomenon. For instance, a plot of sales of a new product may show a rate of growth that is faster than linear growth. Likewise, the income tax rate on individual earnings has a nonlinear relationship with earnings. When we attempt to fit such nonlinear relationships with a linear model, the residuals often appear to have a cup-shape or inverted-cup shape (Exhibit 10.14).

It often happens that the residuals for a nonlinear relationship may not look cup shaped over the entire regression period. However, if we make forecasts from the straight-line model, the forecast errors might show increasing dispersion (Exhibit 10.15), known as heteroscedasticity. If it does not become apparent that the data are nonlinear from analyzing the residuals over the entire regression period, the pattern of over- or underforecasting will certainly exhibit nonlinearity over a long enough period.

 Heteroscedasticity refers to variability in data that is not constant or not homogeneous (nonconstant variance).

It is important to distinguish between nonlinear growth in trend and nonlinear variations as a result of a short-term cycle. In the first case, the nonlinear relationship between two variables will continue in the same direction over a long time. In the case of a short-term cycle, the nonlinear relationship will change direction at the peaks and troughs of each cycle.

Finally, trends, up or down, may be apparent in the residuals (Exhibit 10.16). This pattern can also be the result of a nonlinear relationship between the variables.

EXHIBIT 10.15 Time
Plot of Residuals
with Increasing
Dispersion

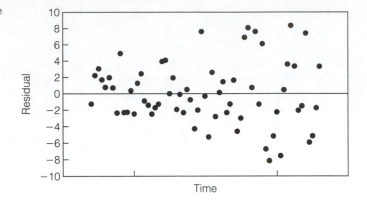

EXHIBIT 10.16 Time
Plot of Residuals
with a Trending
Pattern

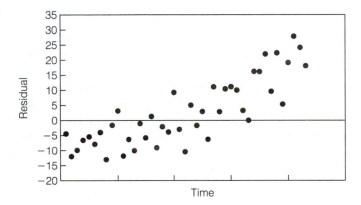

Graphical Aids

In addition to the visual examination of residual plots for nonrandomness, nonlinearities, and outliers, other graphical aids, such as histograms, stem-and-leaf displays, box plots, and probability plots, can prove useful. Exhibit 10.17 shows the box plots of the residuals for two competing ARIMA models. Both sets of residuals appear to be centered on 0, as expected, but the model 1 residuals appear to have a slightly broader distribution in the middle. Model 2 appears to have a longer right tail in the distribution. These two box plots of the residuals turn out to be similar; each shows a nonsymmetrical distribution in the residuals, with a longer tail for positive residuals. Because of the outliers, robust/resistant alternatives should be considered to complement the approach taken for these models.

Box plots offer an effective tool for graphically examining the distribution of residuals.

EXHIBIT 10.17 Box Plots of Residuals for Two Competing Models

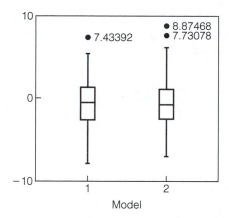

We can also use box plots effectively to look at the distribution of the residuals for a particular period of the year. This may prove useful in determining which months are likely to create more difficulties in forecasting. Exhibit 10.18 displays the residuals for model 1 in a box plot for each month. Note the highly skewed distributions for June and October, which may be interpretable from the underlying date used in the model. January residuals appear to show the greatest variability and September residuals the least. March, July, and September have fairly symmetrically distributed residuals.

The residuals of a model should also be plotted against the explanatory variable to help detect outliers, assess nonhomogeneity (nonconstancy) of variance, and determine if a transformation is required.

Identifying Unusual Patterns

When we look at the residuals, it may become apparent that certain individual observations or small sequences of them are in some sense unusual. When the objective is to extrapolate past results into the future with statistical models, outliers can severely affect the accuracy of forecasts, especially if they occur in the most recent

EXHIBIT 10.18 Box Plots of Residuals for Individual Months, Model 1

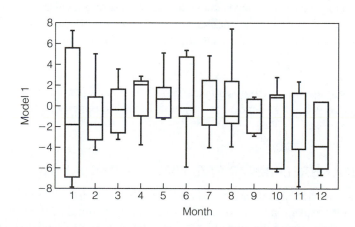

time periods. Outliers are not expected to recur or to influence the data in the same way again, so replacing them with values that are more "typical" should negate their effect on a forecasting model.

Outliers can severely affect the accuracy of forecasts, especially if they appear during the most recent time periods.

For example, severe snowstorms keep many people from getting to work. As a result, telephone calling tends to increase significantly as people call their offices to say they would not be in, call friends and relatives to determine if they were safe, and so forth. For forecasting purposes, it is necessary to adjust the revenues derived from telephone calls for this period downward to more typical revenues, so as not to overstate them under normal circumstances.

When we look at residuals resulting from the transformation and modeling steps, we may be required to prevent distortions caused by outliers. This is especially true if the method used is not resistant to outliers. Here are some examples:

- The outlier may be caused by an administrative decision. If so, the staff of the department responsible for corporate financial results should be consulted. This type of problem is common with revenues, in which retroactive billing distorts revenue data.

- Price changes produce another form of irregularity that requires adjustment. In this case, part of the entire series must be adjusted to put the data on a constant rate base. Alternatively, price indexes can be constructed and used.

- An irregular observation may not be reconcilable with a real event. If so, it should be documented so as to indicate why it was felt to be an outlier.

On the other hand, if we are attempting to build a model to explain past results, we should think long and hard before adjusting past data. The explanatory model may be designed to help us understand how extreme or unusual events affect the process of generating data. If we want to know the impact of severe weather on telephoning habits, the data will provide excellent indications and, as such, should not be adjusted. More often, however, the reasons for unusual values are unknown and must be investigated.

Even with the assistance of statistical methods, human judgment is often required to spot patterns and take innovative action to isolate an underlying problem.

10.5 VALIDATING PRELIMINARY MODELING ASSUMPTIONS

Regression models are based on a number of statistical assumptions that are made more explicit in the next chapter when we deal with the inferences drawn from such models. Transformations of data are often required to validate these assumptions.

Transformations

One of the most important reasons for making transformations of data is that linear relationships among variables are desired. The theory of linear regression is widely used by forecasters to describe relationships among variables. Straight-line fits are the most useful and simplest patterns of this kind to visualize. If more than one variable is assumed to be related to the variable to be forecast, making a transformation that closely approximates a linear relationship among these variables has certain advantages.

Aside from the desire to apply linear regression models, there are other reasons why transformations are useful. By using a suitable transformation, we can find certain desired characteristics of the data, such as additivity of trend and seasonal components. In addition, changes in trend and percentage changes in growth can be better visualized. Many numbers that are reported for public use and to business managers, for example, deal with profits and sales. In reporting results, current sales may be compared with sales in the immediate past and with sales during the same period 1 and 2 years earlier. This involves differencing of data. In addition, the expression of comparisons as percentages is consistent with the use of a logarithmic transformation because growth rates can be viewed as absolute changes in the logarithms of the data. Logarithmic transformations often tend to stabilize variance, in the sense that variability around a model is kept constant, which is an important assumption in many modeling situations.

> By selecting the appropriate transformation, we are assured that the underlying modeling assumptions can be approximated as closely as possible.

Achieving Additivity

Recall that in Chapter 4 we used the ANOVA table as a preliminary tool to gain insight into the nature of trend and seasonality. This approach assumes that trend and seasonal are additive in the sense that Data = Trend + Seasonal + Irregular. An outlier resistant tool for analyzing additivity in two-way tables is the median polish described in Tukey (1977) and Emerson and Hoaglin (1983). The median polish is designed to replace the original table by an associated residual table for which the medians (as opposed to the arithmetic means) for each row and each column are 0. This procedure for taking out medians is an iterative, ad hoc procedure in the sense that the procedure depends on whether we start with rows or columns. However, as a complementary tool to the ANOVA approach, it is effective in prescreening data for unusual values that can severely distort the importance of the underlying additivity of trend and seasonal effects.

In general, the differences between the ANOVA decomposition and the median polish are that an analysis using means tends to produce fewer residuals with large magnitudes, fewer residuals whose magnitudes are close to 0, and more residuals of moderate size. If large errors appear in data that are inconsistent with the additive decomposition of the trend and seasonal components, this will have a large effect on the fitted ANOVA decomposition as well as on the row and column averages. In other words, the ANOVA decomposition has poor resistance to outliers. The advantage of a median polish is that it generally produces substantial residuals for all likely outliers.

Exhibit 10.19 displays a box plot of the differences in the residuals from a two-way median polish of the formal-wear rental sales (Exhibit 4.8) data and the corresponding residuals from the ANOVA decomposition against months. Exhibit 10.19 suggests that, for the months April and August, the residuals from the ANOVA decomposition differ significantly from the corresponding residuals from a median polish. Because the median polish is resistant to outliers, this says that the monthly effect may be biased in the ANOVA decomposition for the trend and seasonal components. If differences were small, we expect to see box plots around a zero line. Furthermore, it appears that Quarter 2 and Quarter 3 are not represented well in the ANOVA decomposition, due to the rather large variation in the differences.

Exhibit 10.20 displays the box plots of these residuals against years, in which the outliers point to Year 7 as the most unusual. In addition, the variations in the box

EXHIBIT 10.19
Box Plot of the Differences in Residuals from a Median Polish and ANOVA Decomposition against Months for Residuals from a Two-Way Table Decomposition for the Formal-Wear Rental Sales Data

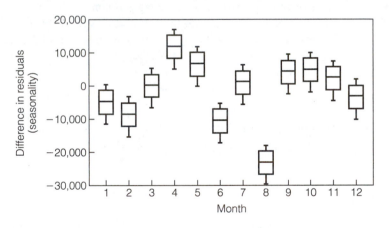

(Exhibit 4.8)

EXHIBIT 10.20
Box Plot of the Differences in Residuals from a Median Polish and ANOVA Decomposition against Years for Residuals from a Two-Way Table Decomposition for the Formal-Wear Rental Sales Data

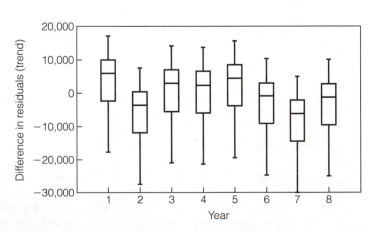

(Exhibit 4.8)

plots are negatively skewed, so that there may be a tendency for trend (annual averages) to be understated. This could be as a result of "trend leakage" into the seasonal pattern (seasonal variation going up with trend).

In addition to having a few unusual observations, the data might also systematically depart from an additive structure (e.g., Data = {Trend + Seasonal} + Error). The diagnostic plot consists of a scatter diagram between the residuals in the table and the corresponding comparison values cv_{ij} defined by

$$cv_{ij} = S_i T_j / m$$

where T_j is the row effect (trend), S_i is the column effect (seasonal), and m is the grand mean (Hoaglin et al., 1983). If the diagnostic plot reveals no consistent trend or pattern, we can conclude that the data do not depart systematically from an additive model. If the slope of the diagnostic plot is k, then simple powers near $(1 - k)$ may be used to provide useful transformations (for details, see Hoaglin et al., 1983, Chap. 6). Exhibit 10.21 displays a scatter diagram of the residuals from a median polish against the corresponding comparison values. This diagnostic plot has a straight-line slope of $k = 0.64$. Hence, we may want to try a cube root transformation ($k = 0.66$) on the data before modeling.

The diagnostic plot is designed to let us judge the extent to which there is any systematic nonadditivity in the data.

We treat residual analysis in the context of normal multiple regression modeling in Chapters 11 and 12.

EXHIBIT 10.21
Scatter Diagram of the Residuals from a Median Polish against Comparison Values for Formal-Wear Rental Sales

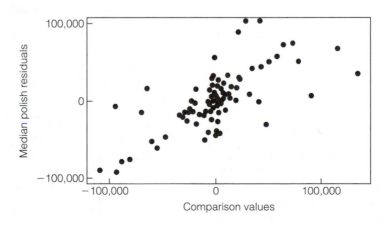

SUMMARY

Many analysts and researchers proceed by seeing how alike things are. Others proceed by trying to understand why things are different. The analysis of residuals is consistent with the latter approach. A residual analysis may suggest relationships, transformations, or the need for a better understanding of events that may not be apparent in the bulk of the data. As discussed earlier, the unplanned findings often yield the most interesting and important results in a forecasting study. Our awareness will also guide us to the next step in the analysis, which deals with the correction and adjustment of a seasonal pattern.

In this chapter, we have introduced several analytical tools that are useful for the preliminary adjustment of data before formal modeling steps are taken. The assumptions that form a basis for selecting an appropriate transformation include the approximate normality of the error distribution in a model, the constancy or uniformity of variability in the errors, and the applicability of linear forms for regression or time series (ARIMA) models.

Residual analysis is a process designed to reveal departures from assumptions about the underlying distribution of random errors and model formulation. Because significance testing and setting confidence limits for model parameters depends on the validity of assumptions about the error distribution, residual analysis is, perhaps, the single most valuable diagnostic tool for evaluating forecasting models.

REFERENCES

Box, G. E. P., and D. R. Cox (1964). An analysis of transformations. *J. Royal Statist. Assoc. B* 26, 211–43.

Box, G. E. P., and G. M. Jenkins (1976). *Time Series Analysis: Forecasting and Control.* San Francisco: Holden Day.

Box, G. E .P., G. M. Jenkins, and G. Reinsel (1994). *Time Series Analysis, Forecasting and Control.* 3rd ed. Upper Saddle River, NJ: Prentice Hall.

Chatterjee, S., A. S. Hadi, and B. Price (2000). *Regression Analysis by Example.* 3rd ed. New York: John Wiley & Sons.

Cryer, J. D., and R. B. Miller (1994). *Statistics for Business—Data Analysis and Modeling.* Belmont, CA: Duxbury Press.

Dielman, T. E. (1996). *Applied Regression Analysis for Business and Economics.* 2nd ed. Belmont, CA: Wadsworth.

Emerson, J. D., and D. C. Hoaglin (1983). Analysis of two-way tables by medians. In *Understanding Robust and Exploratory Data Analysis,* edited by D. C Hoaglin, F. Mosteller, and J. W. Tukey. New York: John Wiley & Sons.

Hanke, J. E., and A. G. Reitsch (1998). *Business Forecasting.* 6th ed. Englewood Cliffs, NJ: Prentice Hall.

Hoaglin, D. C., F. Mosteller, and J. W. Tukey, eds. (1983). *Understanding Robust and Exploratory Data Analysis.* New York: John Wiley & Sons.

Malkiel, B. G. (1998). *A Random Walk down Wall Street: Including a Life-Cycle Guide to Personal Investing.* New York: Norton.

Mosteller, F., and J. W. Tukey (1977). *Data Analysis and Regression.* Reading, MA: Addison-Wesley.

Tryfos, P. (1998). *Methods for Business Analysis and Forecasting: Text and Cases.* New York: John Wiley & Sons.

Tukey, J. W. (1977). *Exploratory Data Analysis.* Reading, MA: Addison-Wesley.

Walsh, J. E. (1962). *Handbook of Nonparametric Statistics.* Princeton, NJ: Van Nostrand.

PROBLEMS

10.1. As a measure of output from the manufacturing sector of the economy, the FRB Index of Industrial Production (FRB) is expected to be strongly negatively correlated with unemployment (UEMP). From the data for FRBP and UEMP in the accompanying table:

Year	FRB	UEMP
1973	130	4.8
1974	129	5.5
1975	118	8.3
1976	131	7.6
1977	138	6.9
1978	146	6
1979	153	5.8
1980	147	7
1981	151	7.5
1982	139	9.5

Source: U.S. Bureau of the Census (1983). *Statistical Abstract of the United States, 1984*. 104th ed. Washington, D.C.: U.S. Government Printing Office.

a. Create and interpret the scatter diagram. Do the variables appear to be negatively related?

b. Label the points on the graph sequentially: A = 1973, B = 1974, and so on, and connect the points in this alphabetical order. Interpret the diagram now in light of the theoretical assumptions about the relationship between FRB and UEMP.

c. Display a scatter diagram of the annual changes (first differences) in the FRB and UEMP. What can you infer from this pattern?

d. If a straight line is fit to the data in (c), resulting in the equation

Fitted (Annual change UEMP)
= 0.66 − 0.138 (Annual change FRB)

how should you interpret the number −0.138 in this relationship?

e. If the predicted change in FRB is 0, what do you expect the change in unemployment to be?

f. If the predicted change in FRB for 1983 is +2, what is your prediction of the level of unemployment in 1982?

10.2. A relationship between sales and advertising (in thousands of dollars) is given by the equation:

Fitted sales = 18.3 + 0.208 (Advertising)

a. Find an interpretation of the number +0.208.

b. If the predicted advertising expenditure is 0, what do you expect the sales to be?

c. If the predicted advertising expenditure for 1983 is +4 (in thousands of dollars), what is your prediction of the sales?

10.3. Since 1960, the annual world crude oil production (in millions of barrels of oil) has grown steadily (OIL, Chatterjee et al., 2000, Table 6.17).

Year	Oil	Year	Oil
1960	7,674	1974	20,389
1962	8,882	1976	20,188
1964	10,310	1978	21,922
1966	12,016	1980	21,722
1968	14,104	1982	19,411
1970	16,690	1984	19,837
1972	18,584	1986	20,246
		1988	21,338

a. Construct the scatter plot of oil versus year, and examine the degree of linearity in the data. Without performing any calculations, do you recommend a transformation of the data?

b. Construct a scatter plot of the logarithms of oil versus year, and examine the degree of linearity in the data. Without performing any calculations, do you recommend a transformation of the data?

c. As another way to look at the data, calculate the percentage changes of the data, and examine the degree of linearity in a scatter plot.

d. In general terms, write down the equations from which you would forecast oil production from linear models obtained from (a), (b), and (c).

10.4. A relationship between housing starts and mortgage rates gave rise to two different models. For each, create the forecasting equation you would use to make projections for the next two periods for housing starts, when $Rates_T(1) = 10.1$ and $Rates_T(2) = 9.7$.

a. Starts = 1808.8 − 29.05 (Rates)

b. Difstart = −21 − 0.000166 (Difrate)

10.5. The relationship between quarterly consumer expenditures (Consexp) and money stock (Moneystock) in the United States was found to be

Consexp = −154.72 + 2.30 (Moneystock)

obtained by fitting a regression model from 1952 to 1956 using data from the accompanying table (Chatterjee et al., 2000, Table 8.1). Both variables are measured in billions of current dollars. This model is of interest to economists performing fiscal and monetary policy analyses.

a. Create a scatter plot of the two variables, and visually examine the degree of linearity in the plot.

b. Examine a time plot of the residuals from this model by first calculating fitted values and subtracting Consexp historical values:

Year	Consexp	Moneystock
1952: I	214.6	159.3
II	217.7	161.2
III	219.6	162.8
IV	227.2	164.6
1953: I	230.9	165.9
II	233.3	167.9
III	234.1	168.3
IV	232.3	169.7
1954: I	233.7	170.5
II	236.5	171.6
III	238.7	173.9
IV	243.2	176.1
1955: I	294.4	178.0
II	254.3	179.1
III	260.9	180.2
IV	263.3	181.2
1956: I	256.6	181.6
II	268.2	182.5
III	270.4	183.3
IV	275.6	184.3

c. To check the model assumptions, perform a run test on the residuals and/or calculate the Durbin-Watson statistic. What are your conclusions about the pattern in the residuals?

d. Plot the sample autocorrelation function (autocorrelogram) of the residuals.

e. What is your projection of consumer expenditures in the 1957:I if the forecast for moneystock for that quarter is 185.2?

f. It was deemed necessary to adjust for autocorrelation in the residuals. Using the Cochrane-Orcutt procedure, the equation was refit and turned out to be:

Consexp = −215.4 + 2.64 (Moneystock)
Repeat (e).

10.6. Analyze the variable Y with X_1 and X_2 from the following table (Chatterjee et al., 2000, Table 4.1).

Y	X_1	X_2
12.37	2.33	9.66
12.66	2.56	8.94
12.00	3.87	4.40
11.93	3.10	6.64
11.06	3.39	4.91
13.03	2.83	8.52
13.13	3.02	8.04
11.44	2.14	9.05
12.86	3.04	7.71
10.84	3.26	5.11
11.20	3.39	5.05
11.56	2.35	8.51
10.83	2.76	6.59
12.63	3.90	4.90
12.46	3.16	6.96

a. Create the three scatter plots for the data, and visually interpret the correlations among these variables.

b. Calculate the product moment correlation coefficients between the pairs of variables.

c. Without running any regressions, how do you interpret the relationship between Y versus X_1 and X_2, individually and collectively? That is, comment on the slope coefficients you would find in the three regression models.

d. Manually draw straight lines through the scatter diagrams. Does X_1 or X_2 help explain the variation in Y? Do X_1 and X_2 help explain the variation in Y?

10.7. Consider the four data sets of pairs of variables $\{(Y_i, X_i), i = 1, 2, 3, 4\}$. (Anscombe, F. J. (1973). Graphs in statistical analysis. *Am. Statist.* 27, 17–21).

Y_1	X_1	Y_2	X_2	Y_3	X_3	Y_4	X_4
8.04	10	9.14	10	7.46	10	6.58	8
6.95	8	8.14	8	6.77	8	5.76	8
7.58	13	8.74	13	12.74	13	7.71	8
8.81	9	8.77	9	7.11	9	8.84	8
8.33	11	9.26	11	7.81	11	8.47	8
9.96	14	8.1	14	8.84	14	7.04	8
7.24	6	6.13	6	6.08	6	5.25	8
4.26	4	3.1	4	5.39	4	12.5	19
10.84	12	9.13	12	8.15	12	5.56	8
4.82	7	7.26	7	6.42	7	7.91	8
5.68	5	4.74	5	5.73	5	6.89	8

a. Create the four scatter plots for the data, and visually interpret the correlations among the pairs of variables.

b. Calculate the product moment correlation coefficients between the pairs of variables.

c. Without running any linear regressions, how do you interpret the relationship between Y_1 versus X_1, Y_2 versus X_2, Y_3 versus X_3, and Y_4 versus X_4.

d. Which dataset(s) most closely adhere to what is assumed for a simple linear regression model?

e. If we assume that a straight-line fit through each data set given by the equation $Y = 3 + 0.5X$, calculate and plot the residuals for each of the four data sets. Comment on the pattern of the residuals.

10.8. An economist wants to predict monthly automobile sales (Y) in millions of dollars based on monthly wages (W), the 3-month Treasury bill rate (R) in percentage terms, and the monthly Consumer Price Index (CPI) based on approximately 10 years of data. The multiple linear regression model to be estimated has the form:

$$Y_t = \beta_0 + \beta_1 W_t + \beta_2 R_{t-1} + \beta_3 \, CPI_{t-1}$$

a. Why are the interest rate and CPI lagged by a month?

b. What are the expected signs of the coefficients in this equation if the data falls in line with theory?

c. If β_3 turns out to be -50, how do you interpret this coefficient?

10.9. In order to predict the monthly sales of durable goods via a linear regression model, a forecaster uses the following variables: open market rate on prime 4–6 month commercial paper (I) lagged 1 month, Consumer Price Index for durable goods (P), monthly sales of durable goods (SD), inventory sales ratios for all durable goods in retail stores (IS) lagged 1 month, average hourly gross earnings of workers (E), and retail inventory of department stores in durable goods (DI), lagged 6 months.

a. Write down the linear regression model for the sales of durable goods.

b. What are the expected signs of the regression coefficients?

c. What is the benefit of lagged dependent variables to forecasting?

10.10. The monthly price of silver per ounce in dollars is assumed to be related to its price the previous month (SILVER, Dielman, 1996, Table 4.8). A simple linear regression model was built to test this theory.

a. Create a time plot of the silver data.

b. Create a scatter plot of $Silver_t$ against $Silver_{t-1}$, and calculate the product moment correlation coefficient.

c. Without running a regression, what is the model you would create for Silver?

d. What is the expected sign of the regression coefficient for the independent variable?

10.11. Repeat Problem 10.10 for one or more of the following monthly time series on the CD:

a. Unemployment rate data (UNEMP, Dielman, 1996, Table 4.7)

b. Prime rate data (PRIME4, Dielman, 1996, Table 4.14)

c. Retail sales for U.S. retail sales stores (RETAIL, Hanke and Reitsch, 1998)

d. Wheat exports (SHIPMENT, Dielman, 1996, Table 4.9)

e. Shipments of spirits (DSHIP, Tryfos, 1998, Table 6.23)

f. Retail sales in a region (RSALES, Tryfos, 1998, Table 6.25)

10.12. In forecasting market share for a bank's products, a forecaster developed a multiple linear regression model for monthly installment loan demand (Y) in the region in terms of installment loan demand lagged 1 month, unemployment rate (U), and an index of economic activity (I).

a. Write down an expression of the model for the forecaster.

b. What are the expected signs of the regression coefficients in the model?

c. If the forecaster's bank consistently had 32% of the total market for installment loans in the region, how should you use the model to come up with a forecast for the bank?

d. Even if the bank's share is relatively constant, what market activity could severely distort the resulting forecast?

CASE 10A Demand for Ice Cream (Cases 1A, 3A–9A cont.)

The entrepreneur has recently acquired a demand forecasting and replenishment planning system to plan her orders to the new factory. Prior to building her own causal models, the entrepreneur would like you to do a little research on what kinds of causal models have already been developed for the industry.

(1) From the literature, select a study in the area of ice cream demand forecasting. Give a description of the problem.

(2) What is the modeling approach taken in the study? What kinds of assumptions were made?

(3) Are data provided with the study? Describe some of the data analysis techniques used.

(4) Give an example of an implementation of a model in the study.

(5) Summarize the recommendations offered in the study.

CASE 10B Demand for Air Travel (Cases 4B–6B cont.)

You are authorized to search for a relationship of the passenger traffic with factors that affect demand. Because the demand for air travel is so sensitive to its price, you decide that investigating own-price elasticities of demand would be desirable. The VP of Operations suggests that you segment the market as follows: (a) long-haul international business, (b) long-haul international leisure, (c) long-haul domestic business, (d) long-haul domestic leisure, (e) short-haul business, and (f) short-haul leisure.

(1) Provide a flowchart describing a systematic approach to estimating elasticities for these six market segments.

(2) Research studies on the international air travel demand that focus on models that provide elasticity estimates.

(3) Summarize and interpret one example of a model that provides elasticity estimates.

(4) Based on your findings, what recommendations should you offer to the VP with regards to price elasticity in the various market segments?

CASE 10C FORECASTING BICYCLE DEMAND (CASES 5C AND 7C CONT.)

The CEO who hired you asks you to research causal modeling approaches that have been used for studying bicycle demand. Forecasting models for bicycle demand can be used for developing models, which can predict both mode and route choice as a function of route characteristics. Discrete choice modeling techniques have been applied to predicting bicycle-route choice as well as mode choice. Discrete choice route models have also been used for estimating elasticities (see http://www.tfhrc.gov/safety/pedbike/vol2/contents.htm).

(1) From the literature, select a study in the area of bicycle demand forecasting. Give a description of the problem.

(2) What is the modeling approach taken in the study? What kinds of assumptions were made?

(3) Are data provided with the study? Describe some of the data analysis techniques used.

(4) Give an example of an implementation of a model in the study.

(5) Summarize the recommendations offered in the study.

CASE 10D FORECASTING FOREIGN IMPORTS

The quantities that influence the import values of goods can be determined employing macroecomic consumption theory. The Chief Financial Officer of WEDOIMPORTS.com is looking for some help in developing a number of models quantifying the import value of goods from Germany to the United States. Your research leads you to something called the Mundell-Fleming model (see http://www.fgn.unisg.ch/eurmacro/tutor/2countrymundellfleming.html).

(1) Give a description of the model in terms of the key assumptions underlying the theory.

(2) What are the key variables in the model?

(3) Illustrate the use of the model with a small-scale application from the literature.

(4) How do you create forecasts with the model?

(5) What are some of the drawbacks and limitations of the model?

CASE 10E THE COMMERCIAL REAL ESTATE MARKET: OFFICE OCCUPANCY RATES

As an intern of a financial services company, you are given the responsibility to analyze the commercial real estate market in a large geographical area. The objective is the evaluation of the performance of a client's investment in class A office buildings. Class A office buildings have the highest quality level in terms of facilities, landscaping, and architecture, and the occupancy rates for class A offices is an important indicator of commercial real estate. The determinants of office occupancy rates have been identified as (a) market size of commercial real estate, (b) corporate income, (c) rent of class A offices relative to classes B and C, and (d) construction activities of office buildings in the area. A proxy for the market size variable is taken to be civilian employment.

(1) Research the definitions of the variables and find sources for the data.

(2) Postulate several linear regression models between office occupancy rates and the related factor(s). Consider the use of leading and lagging indicators and create a rationale for doing so.

(3) Provide an interpretation of the coefficients in your proposed models (without actually producing the models).

(4) Develop a flowchart for the modeling steps to be undertaken.

CASE 10F DEMAND FOR APARTMENT RENTAL UNITS (CASES 6C AND 8B CONT.)

Aside from an interest in projecting rental unit vacancies, the management of the apartment complex produced an econometric model for apartment prices in terms of two macroeconomic variables: personal income (PRICE) and civilian employment (CIVL_EMP). Because these macroeconomic variables are quarterly, the model is based on the first 40 quarterly periods:

$$\log_{10} (\text{PRICE}) = -20.17 + 4.51 \times \log_{10} (\text{CIVL\_EMP})$$

The next step in the development of the model was to relate prices to vacancies. A regression was performed over the same period between the annual change in prices and the average annual number of vacancies with the result:

$$\text{Average monthly vacancies} = 9.20 - 0.0044 \ (\text{Annual change in prices})$$

(1) What is the impact of a percentage change in CIVL_EMP on PRICE?

(2) What is the result of a large change in prices on vacancies? Does this make sense?

(3) If the prices for YR10 and YR11 are $1477 and $1528, respectively, what are the average number vacancies in the final year?

(4) Is this an over- or underforecast and what is the percentage error?

(5) If employment growth for the coming year is assumed to be 153,000, what is the projected price?

APPENDIX 10A ACHIEVING LINEARITY

Rarely do time series have patterns that permit the direct application of standard statistical modeling techniques. Hence, before statistical forecasting models can be introduced, it is important to know how to put data in the proper form for modeling and to transform the data so that they will be consistent with the modeling assumptions of a quantitative forecasting technique.

A very flexible family of transformations on a (positive) variable Y is the power transformations devised by Box and Cox (1964):

$$W = (Y^\lambda - 1)/\lambda \quad \text{for } \lambda \neq 0$$
$$= \ln Y \quad \text{for } \lambda = 0$$

The parameter λ is estimated from the data, usually by the method of maximum likelihood. This procedure involves several steps:

1. Select a value of λ from within a reasonable range-bracketing zero, say $(-2, 2)$. A convenient set of values for λ is $\{\pm 2, \pm 1\ 1/2, \pm 1, \pm 2/3, \pm 1/2, \pm 1/3, \pm 1/4, 0\}$.

2. For the chosen λ, evaluate the likelihood that

$$L_{\max}(\lambda) = -n/2 \ln(\lambda) = (\lambda - 1)\sum \ln Y_i$$
$$= -n/2 \ln(\text{Residual SS}/n) + (\lambda - 1)\sum \ln Y_i$$

where n is the total number of observations, and SS denotes sum of squares.

3. Plot $L_{\max}(\hat{\lambda})$ against $\hat{\lambda}$ over the selected range and draw a smooth curve through the points.

4. Select the value of λ that maximizes $L_{\max}(\lambda)$. This is the maximum likelihood estimate $\hat{\lambda}$ of λ.

5. For applications, round the maximum likelihood estimate $\hat{\lambda}$ to the nearest value that makes practical sense. This should have a minor impact on the results, especially if the likelihood function is relatively flat (as it is in many cases).

6. Determine an approximate $100(1 - \alpha)\%$ confidence interval for λ from the inequality

$$L_{\max}(\lambda) - L_{\max}(\hat{\lambda}) \leq \frac{1}{2}\chi^2(1 - \alpha)$$

where $\chi^2(1 - \alpha)$ is the percentage point of the χ^2 distribution with one degree of freedom, which leaves an area of α in the upper tail of the chi-squared distribution.

7. Draw the confidence interval on the plot of $L_{\max}(\lambda)$ against λ by drawing a horizontal line at the level

$$L_{\max}(\hat{\lambda}) - \frac{1}{2}\chi^2(1 - \alpha)$$

of the vertical scale. The two values of λ at which this line cuts the curve are the end points of the approximate confidence interval. Exhibit 10.23 also shows the 95% confidence interval for λ in the toll revenue data (TOLLREV; see the CD). The plot suggests it may be useful to consider cube root for revenues.

However, for ease of interpretation, the logarithmic transformation ($\lambda = 0$) could also be considered.

Exhibit 10.22a shows the maximum likelihood estimates of the four series in a telecommunications model. The results suggest that for practical purposes it may be useful to consider a cube root for revenues, square roots for messages and business telephones, and no transformation for the employment data.

Box and Cox (1964) gave a somewhat more general version of the Box-Cox transformation:

$$W = [(Y + \lambda_2)^{\lambda_1} - 1)] / \lambda_1 \quad \text{for } \lambda_1 \neq 0$$
$$= \ln(Y + \lambda_2) \quad \text{for } \lambda_1 = 0$$

where $Y + \lambda_2 > 0$. It is assumed that, for some appropriate values of λ_1 and λ_2, the transformed time series can be well described by a model of the type discussed in this book. In the telecommunications example, the data were first detrended with straight-line trends. The resulting parameter estimates, determined by a maximum likelihood method, are shown in Exhibit 10.22b. The logarithmic ($\lambda_1 = 0$) and square root transformations ($\lambda_1 = 0.5$) are generally the most frequently occurring in practice.

EXHIBIT 10.22
(a) Estimate of λ in the Box-Cox Transformation for the Telecommunications Model; (b) Estimate of (λ_1, λ_2) in the Box-Cox Transformation for the Telecommunications Example

(a)

Series Code	$\hat{\lambda}$	Transformation close to
Toll revenues (TOLL_REV)	0.36	Cube root ($\lambda = 0.33$)
Toll messages (MSG)	0.43	Square root ($\lambda = 0.5$)
Business telephones (BMT)	0.57	Square root ($\lambda = 0.5$)
Nonfarm employment (NFRM)	0.93	No transformation ($\lambda = 1.0$)

(b)

Series Code	$\hat{\lambda}_1$	$\hat{\lambda}_2$	Transformation close to
Toll revenues (TOLL_REV)	0.36	−21.4	Cube root ($\lambda = 0.33$)
Toll messages (MSG)	0.43	−7.5	Square root ($\lambda = 0.5$)
Business telephones (BMT)	0.57	−490.8	Square root ($\lambda = 0.5$)
Nonfarm employment (NFRM)	0.93	−6508	No transformation ($\lambda = 1.0$)

11

Linear Regression Analysis

"There is nothing permanent but change."
HERACLITUS, FOURTH CENTURY BC GREEK PHILOSOPHER

IN THIS CHAPTER, WE INTRODUCE the assumptions underlying the method of linear regression analysis with normally distributed errors. We examine the key statistical assumptions required for understanding, interpreting, and evaluating the output of a linear regression model. The material is basic to most of the forecasting techniques used in practice. Linear regression models can be used to

- Predict a dependent variable from one or more (related) independent variables
- Describe a functional relationship with independent variables in which the estimated coefficients provide a business interpretation
- Construct prediction intervals for forecasts
- Develop plans for policy evaluation, where the model serves to express various policy alternatives

11.1 GRAPHING RELATIONSHIPS

In the previous chapter, we introduce regression analysis as the principal tool for creating causal models. But how do we put a model together? We begin a regression analysis by identifying the factors, called independent or explanatory variables, that are believed to have influenced and will continue to influence the variable to be forecast (the dependent variable). The scatter diagram is a useful graphical tool for exploring these relationships. Next, we create a regression model by estimating the coefficients in the model by the method of least squares. This gives us a fitted equation from which we can derive forecasts.

In Chapter 1, we mention relating monthly telephone toll revenues to toll messages in a telecommunications problem. There we note that there is not a one-to-one correspondence between revenues and messages because additional factors, such as the distance between the parties, time of day, and duration of calls, cause variation in the revenue per message. In recent years, due to changes in the telecommunications industry, these factors are disappearing. However, the same kinds of considerations are still important in creating a model for a revenue-quantity relationship in demand planning. For example, variations in room revenues per unit in the hotel industry can be attributed to room discount rates, day of week, length of stay, and so on. Nevertheless, the telecommunications data example will be useful for illustrating the basic ideas behind linear regression models.

In Chapter 3, we introduce the scatter diagram as a visual means of displaying an underlying relationship between two variables. The scatter diagram of toll revenues versus toll messages (Exhibit 11.1) shows a very narrow cluster of points lying along a line with positive slope. This suggests a tight revenue-quantity relationship between toll revenues and toll messages, perhaps described by a simple curve such as a straight line.

In Chapter 1 and again in Chapter 10, we discuss the idea that a model is an abstraction of reality that can be expressed in a simple mathematical relationship. For the revenue-quantity relationship, we can specify such a model as:

$$\text{TOLL\_REV} = \beta_0 + \beta_1 \, \text{MSG} + \varepsilon$$

where ε represents an error term (also called disturbance or noise term), and the coefficients β_0 and β_1 are parameters of the model. Because toll-message volumes

EXHIBIT 11.1
Scatter Diagram of
Monthly Telephone
Toll Revenues
(TOLL_REV) and
Message Volumes
(MSG) over a
10-year Period

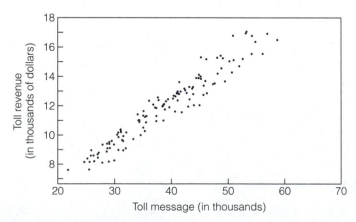

(*Source*: Levenbach and Cleary, 1984)

(MSG) do not fully explain the toll revenues (TOLL_REV), the error term ε must include factors such as time of day, duration of the call, and distance of the call. If these factors average out for any period, β_1 can be interpreted as the average change in toll revenues from an additional message unit. (Here the averaging is over the unaccounted-for times, durations, and distances of the message.)

Forecasting revenues in future periods is based simply on forecasts of volumes (also called units in demand planning applications) in those months. The accuracy of these forecasts depends on the accuracy in forecasting volumes and the relative size of the error term. Exhibit 11.2 shows the results of a forecast evaluation with a holdout sample of the last 12 periods and using actual monthly volumes over the forecast horizon.

> Ex post forecasting uses actual values of explanatory variable(s) in the holdout sample to create a forecast in a regression equation.

Recall that we create a 12-period forecast for the holdout sample, then update historical data with one new period, refit the model, and make forecasts for the remaining 11 months in the holdout sample. This process is repeated until month 11 in the holdout period, at which time there is a single forecast made for period 12.

Exhibit 11.3 displays the key summary statistics from the evaluation. It shows that the estimated slope coefficient varies from 3.37 to 3.43, not a wide range; but what can we infer from the model about the slope parameter? The R^2 statistic, which is a measure of goodness of fit, is high (between 0.91 and 0.93, out of a maximum value of 1). What does this statistic imply, if anything, about the accuracy of the forecast? Another measure is the standard deviation of the residuals, designated as $\sqrt{(MSE)}$, varies between 2124 and 2258 and will be useful in creating probability limits on forecasts. An important measure of forecast performance is the MAPE.

EXHIBIT 11.2 Forecast Performance Evaluations in a Waterfall Chart for Forecasting Revenues (TOLL_REV) from Volumes (MSG) with a 12-Month Holdout Sample

APE	1-Step	2-Step	3-Step	4-Step	5-Step	6-Step	7-Step	8-Step	9-Step	10-Step	11-Step	12-Step	MAPE
Period T	11.9	4.9	9.2	8.9	1.3	5.6	2.5	5.8	1.9	2.1	7.5	5.4	5.6
Period 2	4.4	9.0	8.4	1.7	5.5	2.2	5.7	2.2	2.0	7.6	5.7		4.9
Period 3	9.5	8.6	1.2	5.1	2.3	6.0	2.1	1.7	7.5	5.7			5.0
Period 4	8.2	1.4	5.6	2.6	5.9	2.4	1.8	7.2	5.7				4.6
Period 5	1.0	5.4	2.2	5.5	2.4	1.5	7.3	5.4					3.8
Period 6	5.9	2.4	6.0	1.9	1.6	7.0	5.5						4.3
Period 7	2.0	5.8	2.5	2.0	7.1	5.3							4.1
Period 8	6.4	2.2	1.5	7.5	5.3								4.6
Period 9	2.7	1.7	7.0	5.7									4.3
Period 10	1.2	7.2	5.2										4.5
Period 11	6.7	5.4											6.1
Period 12	5.0												5.0

(*Source*: Exhibit 11.1)

This evaluation suggests that for forecasting purposes, we should be able to predict monthly TOLL_REV values within 5% provided the MSG forecast is without error.

The regression analysis summary output will help answer questions about the reliability of the parameter estimates in a regression equation.

As another example, consider developing a predictive regression model of annual housing starts with mortgage rates. The annual housing starts and annual mortgage rates could be collected for a number of years and the changes in the respective variables plotted in a scatter diagram (Exhibit 11.4). We note in Chapter 3 that a scatter diagram of the level of housing starts and mortgage rates did not seem to show a strong linear pattern. Exhibit 11.4, on the other hand, shows that, on average, a change in annual housing starts tends to decrease with increasing changes in mortgage rates. In this housing starts example, the plot suggests that we should consider a straight-line relationship between the period-to-period changes (first differences) in housing starts and the period-

EXHIBIT 11.3
Summary Statistics from the Forecast Evaluation Results for Forecasting Revenues (TOLL_REV) from Volumes (MSG) with a 12-Month Holdout Sample

N	INT	SLOPE	R^2	$\sqrt{(MSE)}$	MAPE
108	−2185.5	3.45	0.918	2124.4	5.6
109	−1725	3.41	0.914	2174.6	4.9
110	−1929.7	3.43	0.916	2177.9	5
111	−1410.9	3.38	0.916	2211.1	4.6
112	−1805.2	3.42	0.916	2249.4	3.8
113	−1871.8	3.42	0.919	2240.3	4.3
114	−1578.8	3.39	0.921	2246.9	4.1
115	−1653.8	3.4	0.922	2239.9	4.6
116	−1367.5	3.37	0.923	2247.6	4.3
117	−1281.1	3.37	0.926	2239.8	4.5
118	−1378.5	3.38	0.929	2232.7	6.1
119	−1692.4	3.41	0.93	2258.3	5

(*Source*: Exhibit 11.1)

EXHIBIT 11.4
Scatter Diagram and Least Squares Fit of Housing Start Changes (Diff_HOUS) versus Mortgage Rate Changes (Diff_RATE)

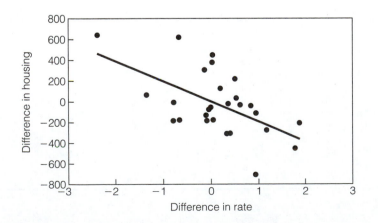

to-period changes in mortgage rates and *not* between the original levels. In Exhibit 11.15, we show (later in the chapter) how to forecast with differenced data.

What Is a Linear Relationship?

A relationship between toll revenues and toll messages can be expressed in a linear model because the equation for the dependent variable (TOLL_REV) is linear in the parameters. For example, $Y = b_0 + b_1 X_1 + b_2 X_2$ is an example of a linear combination, but $Y = b_0 + b_1 X_1^{b_2}$ is not, because the coefficient b_2 does not enter the relationship in a linear fashion. If the message volumes are replaced by the square of the message volumes in the relationship, it is still described by a linear model. What is important is the mathematical form of the relationship, not what it might look like on a graph (in this two-dimensional example).

A linear model can be written as a linear combination (a sum of coefficients b_i multiplied by independent variables).

A linear model (in linear regression) is frequently misinterpreted by forecasters as being just a straight line or linear trend line fit to the data. Of course, a linear trend line fit is a special case of a linear relationship in which the independent variable is expressed by a variable time or a set of ordered equally spaced numbers.

11.2 CREATING AND INTERPRETING OUTPUT

So far, there has not been any information developed about the reliability of the coefficients and how statistically sound the model is. In addition, we need a way to obtain probability limits on forecasts and determine confidence intervals for parameters. This requires the assumptions of normality of the errors in the model.

Normal linear regression models are the most widely known and frequently applied causal methods for forecasting, policy analysis, and econometric model building.

The Precision of the Estimated Regression

Exhibit 11.4 shows a set of data scattered about a straight line. It illustrates the regression relationship between two variables, Y and X. $\bar{Y}_t$ denotes the sample (arithmetic) mean of the observed values of Y, and $\hat{Y}$ are the fitted values. The total sum of squares about the sample mean of $\bar{Y}_t$ is

$$\sum (Y - \bar{Y})^2 = \sum [(Y - \hat{Y}) + (\hat{Y} - \bar{Y})]^2$$
$$= \sum ([(Y - \hat{Y})^2 + 2\sum ([(Y - \hat{Y})(\hat{Y} - \bar{Y}) + \sum (\hat{Y} - \bar{Y})^2$$

With some algebraic manipulations, the middle term equals zero, and hence

$$\sum (Y - \bar{Y})^2 = \sum (Y - \hat{Y})^2 + \sum (\hat{Y} - \bar{Y})^2$$

Total SS	Error SS	+ Regression SS
on $(n-1)$	on $(n-2)$	on 1 degree of
degrees of	degrees of	freedom
freedom	freedom	

Each sum of squares (SS) has degrees of freedom associated with it. The degrees of freedom (df) indicate how many independent pieces of information are associated with the calculation of the sum of squares. This is not generally easy to interpret. We note first that the degrees of freedom for the Total SS is always the sum of the df on the right-hand side: $n - 1 = (n - 2) + 1$; that is, the Total SS needs $n - 1$ independent pieces of information from the n deviations $\{Y_1 - \bar{Y}, Y_2 - \bar{Y}, \dots, Y_n - \bar{Y}\}$ because one piece is known, namely that the sum of these n deviations is always zero (Problem 11.5). The vertical line represents the deviation of a value Y from the mean value $\bar{Y}$. Exhibit 11.6a shows a linear least squares regression line fitted to the observed points.

The total variation can be expressed in terms of (1) the variation explained by the regression and (2) a residual portion called the unexplained variation. Exhibit 11.6a shows the explained variation, which is expressed by the vertical distance between any fitted (predicted) value and the mean, or $\bar{Y} - \hat{Y}_t$. The circumflex (^ or "hat") over the $\hat{Y}_t$ is used to represent fitted values determined by a model. Thus, it is also customary to write $b_0 = \hat{\beta}_0$ and $b_1 = \hat{\beta}$. Exhibit 11.6b shows the unexplained or residual variation—the vertical distances between the observed and the predicted values, $(Y_t - \hat{Y}_t)$.

The total variation in a dependent variable is composed of a component explained by the regression and a residual portion, called the unexplained variation.

EXHIBIT 11.5 Total Variation of Y about the Overall Mean of Y (housing starts changes)

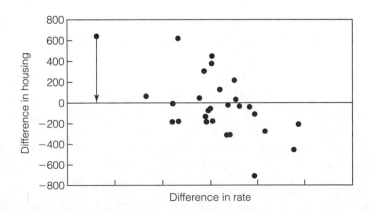

EXHIBIT 11.6
(a) Explained and
(b) Unexplained
Variation in Least-
Squares Regression

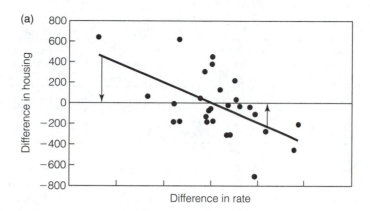

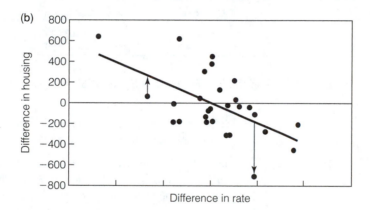

We can construct a table from these sums of squares relationships; this is an ANOVA table, examples of which are shown below in Exhibit 11.7. The first example corresponds to the housing starts example (Diff_HOUS vs. Diff_RATE), and the second example is for the revenue-message relationship (TOLL_REV vs. MSG). Note that the Total SS column adds up as the sum of the Regression SS and the Error (Residual) SS. The Total DF is always 1 less than the number of observations. As we will see shortly, the MS in column 4 is SS/df. This ANOVA table is part of any standard output in a regression-analysis package.

R^2 Statistic

We can see how useful the regression line is as a predictor by seeing how much of the Total SS is given by the Regression SS. The R^2 statistic has the following derivation. As previously shown, the sum of squares about the mean (Total SS) can be expressed as the sum of two other terms, namely, the sum of squares about regression (unexplained variation is the Error SS) and the sum of squares due to regression (explained variation is the Regression SS). Here regression is used in the sense of the

EXHIBIT 11.7
Examples of the ANOVA Portion of a Regression Output (a) Housing Starts versus Mortgage Rates (Diff_HOUS vs. Diff_RATE); (b) Toll Revenue versus Message Volume (TOLL_REV vs. MSG)

(a) ANOVA

	df	SS	MS	F
Regression	1	874316.16	874316.16	13.9
Residual	25	1577595.06	63103.80	
Total	26	2451911.21		

	Coefficients	Standard Error	t Statistic
Intercept	14.47	49.00	0.30
X variable 1	−206.50	55.48	−3.72

(b) ANOVA

	df	SS	MS	F
Regression	1	8218924134	8218924134	1603
Residual	118	605011236.7	5127213.87	
Total	119	8823935371		

	Coefficients	Standard error	t Statistic
Intercept	−1848.93	1035.34	−1.79
X variable 1	3.42	0.09	40.04

fitted equation. The fraction Regression SS/Total SS is known as R^2 and is a measure of goodness of fit.

$$R^2 = \text{(Explained variation)} / \text{(Total variation)}$$
$$= \text{(Total SS} - \text{Error SS)} / \text{Total SS}$$

Practitioners at times rely too heavily on a mistaken interpretation of the appearance of a good result when a model displays a high R^2 for the wrong reasons, as in the case of fitting straight-line regressions to trending data with time as the explanatory variable. On the other hand, we expect a "good" model to have a reasonably high value for R^2. Notice that R^2 can never be negative or exceed unity.

It is not necessarily true that a high R^2 statistic means that we have a good forecasting model.

Continuing with the revenue–quantity demanded relationship, we consider a simple straight-line model between the level of message volumes (MSG) and the level of revenues (TOLL_REV) in the telecommunications example. Suppose we wish to predict the toll revenues from the toll messages. A sample of revenue and message volumes are collected from a billing record and plotted as a scatter diagram (Exhibit 11.1), in which the dependent variable (revenue) is put on the vertical axis and the independent variable (message volume) is put on the horizontal axis. Based on a sample of 120 monthly periods, a least-squares fit for the toll revenue model results in a fitted equation:

$$\text{Predicted TOLL\_REV} = -1849.93 + 3.42 \, \text{MSG}$$

The estimate of the slope coefficient, $b_1 = 3.42$, says that, on average, an additional message unit will generate approximately \$3.42 in added toll revenues. The constant term, $b_0 = -1848.93$, is an estimate of what toll revenues might be during a hypothetical time period when message volume is zero. Because no such time period was observed, the constant term here lacks interpretation.

The equation itself serves as a forecasting formula. An m-period-ahead forecast for toll revenues at time period $t = T$ is calculated from:

$$\text{TOLL\_REV}_T(m) = -1849.93 + 3.42 \times \text{MSG}_T(m)$$

Note that for this equation to work, we will need forecasts for the MSG variable to plug into the formula.

The regression equation serves as the forecasting formula.

Exhibit 11.8 is a computer output showing the summary statistics from a simple linear regression model relating toll revenues as a function of toll messages. Whether it might be appropriate to transform the data first (e.g., with logarithms or the Box-Cox transformation; see Chapter 10) is not considered at this point. Let us examine some details of Exhibit 11.8 that we have not dealt with before.

The sample size (observations = 120) refers to the number of observations used in the regression. In this case, the regression was performed over 128 months, a little more than 10 years.

The standard error (= 2264.34), also called the standard error of the estimate, is a measure of the variability about the fitted regression function. Because this statistic is related to the magnitude of the unexplained variation, a desired objective is to find a model that has the lowest estimated standard deviation of the residuals.

One note of caution is that the standard error can only be used to compare models when the dependent variable is of the same form. For example, the standard deviation of the residuals of a model built on the sales of a product cannot be directly compared to the same statistic for a model built on the logarithms of the sales of the product. The latter statistic will have a different interpretation because of the transformation.

The R^2 statistic (= 0.931), also called the coefficient of determination, is the proportion of the total variation about the mean value of Y that is explainable by performing a linear regression on X. In this case, 93% of the revenue variation about the mean is explained by the message data. This is known as a measure of the goodness of fit of the regression. Although many software programs show many digits after the decimal place, it is not necessary to imply such high precision.

In Exhibit 11.8, the variable intercept represents the constant in the linear equation. The coefficient of this constant is -1848.93. The X variable 1 is the independent variable (MSG); its coefficient is 3.42.

The column labeled standard error represents the estimated standard deviation of the regression coefficients. The ratio of coefficients to estimated standard deviations

EXHIBIT 11.8
Regression Output
for a Simple Linear
Regression Model
Relating Telephone
Toll Revenues
(TOLL_REV) as a
Function of Toll
Messages (MSG)

Regression Statistics

Multiple R	0.965
R^2	0.931
Adjusted R^2	0.931
Standard error	2264.34
Observations	120

ANOVA

	df	SS	MS	F
Regression	1	8218924134	8218924134	1603
Residual	118	605011236.7	5127213.87	
Total	119	8823935371		

	Coefficients	Standard Error	t Statistic
Intercept	−1848.93	1035.34	−1.79
X variable 1	3.42	0.09	40.04

(*Source*: Exhibit 11.1)

produces the t statistics in the next column. These statistics are shown in parentheses beneath the corresponding coefficient estimates in the following equation:

$$\text{TOLL\_REV} = -1848.93 + 3.42\ \text{MSG}$$
$$(-1.79) \qquad (40.03)$$

where TOLL_REV is revenues and MSG is volume of messages.

A t statistic is obtained by dividing an estimated standard error into the parameter estimate.

Interchanging the Role of Y and X

It is not uncommon for a practitioner to think that a simple linear regression of X on Y and the regression of Y on X should give equivalent inferences about the relationships between X and Y and between Y and X. For example, by interchanging the dependent and independent variables and looking at the intercept and slope estimates from the least squares the regression equation (although not realistic, in this case) for toll messages X against toll revenues Y becomes

$$X = a_2 + b_2 Y$$
$$= 1317.7 + 0.27 Y$$

The slope $b_1 = 3.42$ for the revenue equation (Y) is not equal to the reciprocal of the slope $b_2 = 0.27$ for the message equation (X), but why? The reason these two regressions give different results is that the line obtained by minimizing the sum-of-squared vertical deviations is different from the line derived by minimizing the sum-of-squared horizontal deviations. Practitioners (and students) may easily misinterpret the basis of least squares, and this misunderstanding can lead to an invalid use of regression equations.

Linear Correlation

When the scatter diagram between two variables X and Y depicts a band of points lying in a linear pattern, we called this linear association. In the context of a regression relationship, when the two variables are nearly independent, the regression coefficients b_1 and b_2 are very small and the two regression lines are almost at right angles. On the other hand, when they are so closely related that the one can be taken to determine the other absolutely, the two regression lines coincide. In this case $b_1 = 1/b_2$. From this observation arises the notion of linear association or correlation between two variables.

A measure of the strength of the relationship between X and Y seems to depend on the angle between the two regression lines. The most common measure of the strength of the relationship is a statistic called the sample product moment correlation coefficient, which we introduce in Chapter 3. Because $r^2 = b_1 b_2$, it can be seen that $r = 0$ when there is no association and $r^2 = 1$ when one variable determines the other. The statistic r ranges from -1 to $+1$. By interchanging X and Y, we can see from the formula that r is the same no matter which variable is used to predict the other.

It should also be evident that a strong linear association does not imply causality. In the revenue-message example, the correlation coefficient is the same no matter which variable is treated as the dependent variable. However, it would be meaningless to state the revenue volumes are the cause for the demand for messages in a telecommunications business.

A product moment correlation coefficient measures the strength of a linear association between a pair of variables, but a strong linear association (correlation) does not imply causality in the relationship.

When two series have a strong positive association, the scatter diagram has a scatter of points along a line of positive slope. A negative correlation shows up as a scatter of points along a line with negative slope.

In the case of simple linear regression, the relationship between R^2 and the square of the sample product moment correlation coefficient (a measure of association) is quite simple; they are the same ($r^2 = R^2$). However, for multiple linear regressions, it turns out that there is more than one correlation coefficient, so we do not know what to compare R^2 with.

It turns out that for simple linear regression, the least-squares fit is determined by the five summary statistics:

- $\overline{Y}$: sample (arithmetic) mean of Y
- $\overline{X}$: sample (arithmetic) mean of X
- s_Y: sample standard deviation of Y
- s_X: sample standard deviation of X
- r: product moment correlation coefficient

To see this, we note that the point determined by the sample means $(\overline{X}, \overline{Y})$ is always on the least-squares line. To determine a second point on the line, we associate with each increase of one sample standard deviation s_X in X an increase of r sample standard deviations s_Y in Y, where r is the product moment correlation coefficient $(-1 < r < 1)$. Thus, the slope estimate b_1 is the ratio of the Y change divided by the X change:

$$b_1 = rs_Y/s_X$$

and the intercept estimate b_0 is $(\overline{Y} - b_1\overline{X})$.

For the example of changes in housing starts versus changes in mortgage rates, the least-squares regression model is $\hat{Y} = 14.47 - 206.5\,X$, which can be determined directly from the five summary statistics:

- $\overline{Y}$: sample mean of Y $(= -21.02)$
- $s_{\overline{Y}}$: sample standard deviation of Y $(= 302.9)$
- $\overline{X}$: sample mean of X $(= 0.115)$
- $s_{\overline{X}}$: sample standard deviation of X $(= 0.885)$
- r: product moment correlation coefficient $(= -0.597)$

From the scatter plot (Exhibit 11.4) and least-squares fit (Exhibit 11.7), it appears that mortgage rates are potentially helpful in explaining the housing starts data. We can use this relationship to forecast housing starts by plugging in a forecast for the change in mortgage rates for the next period and determining the change that could be expected in housing starts for the next period.

11.3 MAKING INFERENCES ABOUT MODEL PARAMETERS

Certain aspects in a regression output can only be properly interpreted if we make additional assumptions about the random errors. Up to this point, we have not made any assumptions about the distribution of the random error term in the regression model. If the random errors are normally distributed, then an extensive statistical theory is applicable.

The Normality Assumption in Regression

The normal assumption (as in normal distribution) states that, in a random sample of n outcomes $\{Y_1, Y_2, \ldots, Y_n\}$ of a variable Y, the corresponding random error terms $\{\varepsilon_1, \varepsilon_2, \ldots, \varepsilon_n\}$ arise independently from a common normal (also called Gaussian) distribution with mean 0 and (unknown parameter) variance σ^2. A compact notation for the latter statement is:

$$\varepsilon_i \sim N(0, \sigma^2) \quad \text{for } i = 1, 2, \ldots, n$$

In this way, the normal linear regression model can be written as

$$Y_i = \mu_{Y(X)} + \varepsilon_i$$

where $\varepsilon_i \sim N(0, \sigma^2)$, and $i = 1, 2, \ldots, n$. For a multiple linear regression model, the expression of these assumptions summarized by:

- The mean μ_Y of Y is linear in the β values; that is, $\mu_Y = \beta_0 + \beta_1 X_1 + \cdots + \beta_k X_k$ is a linear combination of β values multiplied by X values, where β_0, $\beta_1, \ldots, \beta_k$ are called the regression parameters.

- The ε_i are independent, identically distributed with a standard normal $N(0, \sigma^2)$.

Contrast this with the basic assumptions made for simple linear regression earlier in the chapter. We have simply extended the model to include the important normality assumption for Y (specifically, about the distribution of the error term ε.)

Important Distribution Results

Consider a sample (or time series values Y_t, $t = 1, \ldots, n$) in the array in Exhibit 11.9. The n independent values $(Y_1, \ldots, Y_n)$ of Y, observed together with the values of the corresponding independent variables, will be used to make inferences about the regression parameters $\beta_0, \beta_1, \ldots, \beta_k$ and the error variance σ^2.

References are made throughout this book to statistical significance tests based on the normal, Student's, chi-squared, and F probability distributions. The following results point to why certain probability distributions result for significance tests arising from the normality assumption in regression models.

- If $Y_1, Y_2, \ldots, Y_n$ are normally and independently distributed random variables with mean μ_i and variance σ^2, then the sum $Y = \Sigma k_i Y_i$ (where the k_i are constants), is also distributed normally with mean $\Sigma k_i \mu_i$ and variance $\Sigma k_i^2 \sigma^2$.

- If $Y_1, Y_2, \ldots, Y_n$ are normally and independently distributed standardized variables with mean equal to 0 and variance equal to 1, then the ΣY_i^2 follows a chi-squared distribution with n degrees of freedom.

- If S_1, S_2 are independently distributed random variables, each following a chi-squared distribution with k_1 and k_2 degrees of freedom, respectively, then

$$F = (S_1/k_1)/(S_2/k_2)$$

has an F distribution with (k_1, k_2) degrees of freedom.

Certain confidence intervals for parameters and significance tests for summary statistics and those derived from normal regression theory use these results. They may be found in any book on regression analysis or statistical inference, as well as any of the numerous business statistics textbooks. It is beyond the scope of this book to fully cover the inferential developments here.

EXHIBIT 11.9 Data Array for Multiple Linear Regression Analysis

$$
\begin{array}{ccccc}
Y_1 & X_{11} & X_{12} & \cdots & X_{1k} \\
Y_2 & X_{21} & X_{22} & \cdots & X_{2k} \\
Y_3 & X_{31} & X_{32} & \cdots & X_{3k} \\
\vdots & \vdots & \vdots & \ddots & \vdots \\
Y_n & X_{n1} & X_{n2} & \cdots & X_{nk}
\end{array}
$$

With the normality assumption, it becomes possible to examine the validity of the model in terms of the significance of the parameters.

Significance of Regression Coefficients

Recognizing that there will always be random errors in the observed data, it is important to provide a measure of the precision of the parameter estimates and of the reliability of the fitted coefficients in the regression equation. We need to deal with quantifying the uncertainty in the parameter estimates and the values predicted from the model. We may be able to determine if a specific independent variable X_i is necessary by using the t test on the estimated regression coefficient b_i or, equivalently, establish a confidence interval for the corresponding parameter β_i. We proceed here with the confidence interval approach.

To interpret the results of a confidence interval calculation for a parameter, we observe whether the parameter falls inside the interval or not. If a particular (hypothesized) value of the parameter β_i falls inside the confidence interval, then this implies that, given the effects of all other independent variables, X_i does not explain a (statistically) significant amount of additional variability in Y. On the other hand, statistical significance is recognized if the parameter falls outside the confidence interval. In the telecommunications example, for instance (as shown in Chapter 3), the 95% confidence interval for the slope is given by

$$0 \pm 1.96 \times (\text{Standard error of the slope}) = \pm 1.96(0.09) = 0.17$$

A slope of 0 occurs when X and Y are not correlated. Clearly, the estimated slope coefficient ($=3.42$) is outside this interval, so we conclude that the slope is significantly different from 0.

Cases can occur in which each regression coefficient b_i is not significant and yet the regression as a whole is significant, as indicated by an F test. Hence, an **F test** should always be performed in multiple linear regression analysis.

Inferences from Summary Statistics

Let us discuss summary statistics for simple linear regression models. These statistics include the R^2 statistic, t statistic, F statistic, and Durbin-Watson (DW) statistic. The F and R^2 statistics are generalized to be applicable for models with multiple regressors. Additional statistics, such as the correlation matrix and the incremental F statistics, only have meaning in the context of multiple linear regression models.

Goodness of Fit

In general, when we calculate the multiple correlation coefficient R, or its square, we obtain a measure of the effectiveness of the regression fit

$$R^2 = \text{Regression sum of squares/Total sum of squares}$$

$$= (\text{SST} - \text{SSE})/\text{SST}$$

where $\text{SST} = \Sigma(Y - \overline{Y})^2$ is the total variation, and $\text{SSE} = \Sigma(Y - \hat{Y})^2$ is the unexplained variation. Hence, R^2 is the proportion of the variation of Y that has been explained by including particular independent variables. It is also commonly referred to as the coefficient of determination.

The coefficient of determination R^2 measures the proportion of variation in the dependent variable that can be explained by including explanatory variables in a regression equation.

To compare models with a different number of independent variables, the adjusted R^2, corrected for degrees of freedom, is used. This is given by:

$$\overline{R}^2 = 1 - [(n - 1)(n - R^2)/(n - k - 1)]$$

where k = number of independent variables. Unlike R^2, $\overline{R}^2$ can decrease when a variable is added. In fact, for $k \geq 1$, $R^2 \geq \overline{R}^2$ and, moreover, $\overline{R}^2$ can be negative.

Calculating the **F statistic** carries out an overall test of significance of the regression:

$$F = \frac{\text{Regression SS}/\text{Regression df}}{\text{Residual SS}/\text{Residual df}}$$

$$= \frac{\text{Mean square due to regression}}{\text{Mean square due to error}}$$

An examination of p-values or the F table (found in most business statistics texts) shows that, for most practical problems, an observed value of approximately 4 or more probably points to statistical significance for the appropriate degrees of freedom, provided normality assumptions hold. In general, if the F statistic is significantly greater than unity, it indicates that the data do not support an assumption of zero values for the regression coefficients. Then we are inclined to infer that there is a regression relationship with at least one of the $\beta_0, \beta_1, \ldots, \beta_k$ different from zero.

The t Statistic

The t statistic measures the statistical significance of the regression parameter for a specific independent variable. The t ratio follows a Student's t distribution that looks very similar to the familiar bell curve of the normal distribution. However, a t distribution is shorter and fatter, and its variance $[= v/(v - 2)]$ is larger than that of the standard normal distribution ($= 1$). For each positive integer v (Greek nu), called the degrees of freedom, there corresponds a different t distribution. Under the normality assumption, we can assign $100(1 - \alpha)\%$ confidence limits for β_i by calculating

$$b_i \pm t(v, 1 - \tfrac{1}{2}\alpha) \, \text{SE}(b_i)$$

where $t(v, 1 - \tfrac{1}{2}\alpha)$ is the $100(1 - \tfrac{1}{2}\alpha)$ percentage point of a t distribution with v degrees of freedom. The standard deviation of b_i, denoted by $\text{SE}(b_i)$, is better known as the standard error of b_i. To apply a confidence interval, we plug in the estimated standard error for $SE(b_i)$.

The t value is approximately 2.0 in absolute value for the 95% confidence level and $v > 30$. When this is the case, we may reject a hypothesis that the regression coefficient is 0 when 0 is not in the confidence interval. As we have seen before, a statistically significant value not equal to 0 is said to exist for the estimated parameter. Thus, in Exhibit 11.8, the observed t values in the t statistic column for the slope coefficient (X variable) can be interpreted as statistically significant. This cannot be interpreted as proof of a cause-and-effect relationship between the dependent and independent variables, however. For example, each variable may be related to a third (possibly causally linked) factor and only coincidentally related to one another.

The t statistic measures the statistical significance of the regression parameter for a specific independent variable.

The F Statistic

The analysis of the ANOVA table in Exhibit 11.8 displays a comparison of the average sum-of-squared deviations explained by the regression with the unexplained sum-of-squared deviations. This comparison forms the basis for the F test:

$$F = \frac{\text{Mean square due to regression}}{\text{Mean square due to error}}$$

$$= \frac{8,218,924,134}{5,127,214} = 1603$$

If there is no relationship between Y and X, then $\hat{Y} = \bar{Y}$ and F equals zero. The calculated F statistic is compared to the tabular value for the appropriate degrees of freedom. For a simple linear regression, $F = t^2$, and a value of F greater than approximately 4.0 indicates significance at the 5% significance level. For a multiple linear regression, which we discuss in the next chapter, we must use the F table to determine if the overall regression is significant.

The rationale for the F test is that if there is a relationship between the dependent and independent variables, the variation of the estimated values from the observed values will be less than the variation between the estimated values and the mean value of Y; that is, the F ratio will be significantly different from 1.0.

An Incremental F Test

It is often important to test to see if the inclusion of an additional variable significantly improves the fit of a linear regression model. For example, we may want to see if the inclusion of the FRB index of industrial production or the SRC index of consumer sentiment in a model relating housing starts to mortgage rates results in a significant reduction in the sum of squares due to errors.

In multiple linear regression problems, the significance of the regression coefficient cannot, in general, be assessed with confidence intervals on a one-by-one basis. This is because individual regression coefficients are correlated among themselves. We must perform an incremental F test. The numerator of this F statistic, F^*, is the change in the sum of squares due to error in the old model minus the sum of squares due to error in the new model divided by the difference in error degrees of freedom. The foregoing quantity is divided by the MSE of the new model:

$$F = \frac{(\text{Residual SS}_{\text{old}} - \text{Residual SS}_{\text{new}})/(\text{df}_{\text{old}} - \text{df}_{\text{new}})}{\text{Residual SS}_{\text{new}} / \text{df}_{\text{new}}}$$

When only one variable is added to the model at a time, the incremental F test is equivalent to a t test. If the t statistic for the new variable is significant, so is the incremental F statistic. However, when several variables are added at a time, the group of variables may be significant even though one or more t statistics appear insignificant. For example, **indicator variables** are frequently used in econometrics to account for seasonal variation. In this case, even though one or more of the variables appear insignificant, the seasonal variation is described by the presence of all the indicator variables.

The incremental F test indicates whether the added variables as a group are statistically significant.

11.4 Autocorrelation Correction

In many forecasting applications, autocorrelation arises because of an incorrect specification in the form of the relationship for the variables. Autocorrelation manifests itself in forecasting in a pattern of prolonged overforecasting or underforecasting, period after period, into the future.

The underlying assumptions to be considered are that ordinary least-squares estimation is based on model errors that are uncorrelated and that normality implies that model errors are pairwise independent—this may be unrealistic in practice. Some additional ways of dealing with this problem for time series data are explored in later chapters on ARIMA modeling.

First-Order Autocorrelation

To illustrate the occurrence of autocorrelation in the error term of a forecasting model, consider a simple linear regression model with one independent variable X:

$$Y_t = \beta_0 + \beta_1 X_t + \epsilon_t$$

where ε_t depends on the value of itself one period ago; that is,

$$\varepsilon_t = \rho\varepsilon_{t-1} + v_t$$

when v_t is normally distributed $N(0, \sigma^2)$.

It can be shown that if the true parameter $\rho = 0.8$, the sampling variance of β will be more than four times the estimated variance given by the OLS solution. Because the estimated variance is understated, a t test could falsely lead to the conclusion that the parameter is significantly different from zero.

Two main consequences of using OLS analysis in models in which the errors are autocorrelated are that sampling variances of the regression coefficients are underestimated and invalid and that forecasts have variances that are too large. When OLS is used for estimation, the calculated acceptance regions or confidence intervals will be narrower than they should be for a specified level of significance. This leads to the false conclusion that estimates of parameters are more precise than they actually are. There will be a tendency to accept a variable as significant when it is not, and this may result in a misspecified model.

Testing for Serial Correlation

In time series forecasting, it is not unusual to be in violation of regression assumptions because of autocorrelation in random errors; hence, it is important to be able to test for their presence. The Durbin-Watson (DW) statistic is the conventional statistic used to test for autocorrelation (first-order only!). However, we may find it equally informative to construct a plot of a complete set of sample autocorrelations (correlogram) for assessing the nature of autocorrelation in time series data.

The Durbin-Watson Statistic

The DW statistic is commonly used to test for first-order autocorrelation in residuals (Durbin and Watson, 1950, 1951). The DW statistic is calculated from the residuals e_t in a model, and it has the formula

$$d = \left[\sum (e_t - e_{t-1})^2 \right] \Big/ \sum e_t^2$$

If the residuals $\{e_t,\ t = 1, \ldots, n\}$ are positively correlated, the absolute value of $e_t - e_{t-1}$ will tend to be small relative to the absolute value of e_t. If the residuals are negatively correlated, the absolute value of $e_t - e_{t-1}$ will be large relative to the absolute value of e_t. Therefore, d will tend to be small (near 0.0) for positively correlated residuals, large (near 4.0) for negatively correlated residuals, and approximately equal to 2.0 for random residuals. For large samples, d is approximately equal to $2(1 - \rho)$.

The sampling distribution of d depends on the values of the independent variable X_t in the sample. Therefore, the test is able to provide only upper (d_u) and lower (d_l) limits for significance testing. We either accept the null hypothesis of zero autocorrelation or reject it in favor of first-order positive autocorrelation. If $d < d_l$, the zero autocorrelation hypothesis is rejected in favor of first-order positive autocorrelation. If $d_l < d < d_u$, the test is inconclusive. If $d > 4 - d_l$, the zero autocorrelation hypothesis is rejected in favor of first-order negative autocorrelation. The inconclusive region also includes $4 - d_u < d < 4 - d_l$. Because the DW statistic is meant to test for first-order autocorrelation, an approximate rule is to observe that $\rho = (1 - 0.5d)$ and significant autocorrelation is found when r_1 is outside $\pm 2/\sqrt{n}$, where n is the number of observations in the time series.

In most practical situations, an effective solution can be found by simply calculating the lag 1 corrrelation in the correlogram.

With the advances that have been made in computer processing and graphical analysis, it is often more informative to plot the entire correlogram of the residuals of a model and visually determine the nature of autocorrelation patterns (needed for the identification of ARIMA models). Both the DW statistic and the correlogram are inappropriate if lagged dependent variables are incorporated into the model, however.

The Durbin h Statistic

The h statistic can be used to test for serially correlated residuals when a lagged dependent variable appears as an independent variable in a model. The h statistic is defined by

$$h = r\{n/(1 - nV(b_1))\}^{1/2} \qquad \text{for } nV(b_1) < 1$$

where $r \cong 1 - 0.5d$ and $V(b_1) =$ estimated variance of b_1 (the coefficient of Y_{t-1}).

The h statistic is a large ($n > 30$) sample statistic used to test for serially correlated residuals in a model. This statistic is tested as a standard normal deviate, and, if $h > 1.645$, we reject the hypothesis that the residuals have zero serial correlation (at the 5% percent level). Because only the estimated variance of b_1 is required to compute h, it does not matter how many independent variables or higher-order lagged dependent variables are included in the model.

Adjusting for Serial Correlation

There are several approaches that can be tried to eliminate a problem of serial correlation in regression models. These include modeling the first differences or the percentage changes year over year in the time series, performing a transformation on the data (the transformation should be based on the assumed nature of the autocorrelated error structure), including an autoregressive term (last period's actual) in the model, and building an ARIMA model of the residuals of the regression model. We discuss the first two next.

Taking First Differences to Reduce Serial Correlation in Residuals

Modeling first differences of the data may solve a serial correlation problem. Let the assumed model be

$$Y_t = \beta_0 + \beta_1 X_t + \varepsilon_t$$

where the errors are autocorrelated as follows (i.e., $\rho = 1$):

$$\varepsilon_t = \varepsilon_{t-1} + v_t$$

and where v_t is a random error term that is not correlated with ε_t. Replacing $t-1$ for t in the model gives

$$Y_{t-1} = \beta_0 + \beta_1 X_{t-1} + \varepsilon_{t-1}$$

By subtracting Y_{t-1} from Y_t, we get

$$Y_t - Y_{t-1} = \beta_1(X_t - X_{t-1}) + (\varepsilon_t - \varepsilon_{t-1})$$

Because $\varepsilon_t - \varepsilon_{t-1} = v_t$, this gives the result

$$Y_t - Y_{t-1} = \beta_1(X_t - X_{t-1}) + v_t$$

If the errors in the original model were autocorrelated, modeling first differences of the variables would eliminate this restricted form of autocorrelation. By definition, the $\{v_t\}$ are assumed to be independent, identically distributed random variables.

It is often the case that the presence of serial correlation is so great that it becomes obvious from looking at the correlogram of the residuals or the residual plot. An example of this problem is illustrated in a model relating growth in business telephones to nonfarm (excluding manufacturing) employment. The time plots are shown in Exhibit 11.10. A scatter plot (Exhibit 11.11a) shows a positive relationship because both nonfarm employment and sales of business telephones are increasing. The regression relationship (Exhibit 11.11b) indicates a significant slope coefficient ($=53.1$; $t=7$),

EXHIBIT 11.10
Time Plots of
Monthly (a) Business
Telephone Series
(BMT) and (b) Non-
farm (less manufac-
turing) Employment
(NFMA) over a
4-Year Period

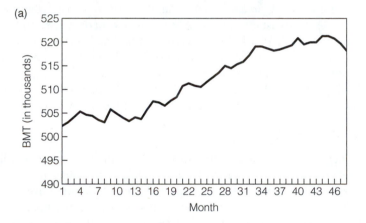

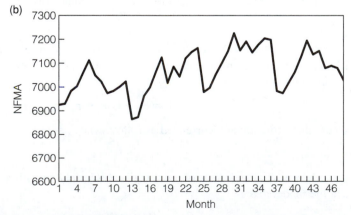

(*Source*: Levenbach and Cleary, 1984)

EXHIBIT 11.11
(a) Scatter Plot of
Business Telephones
(BMT) versus Non-
farm Employment
(NFMA); (b) Regres-
sion Output from
Business Telephone
Model

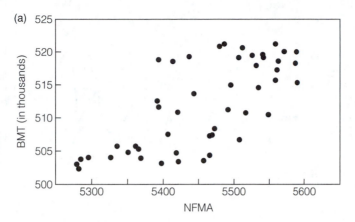

(b)

Regression Model Using	R^2	β (Empl)	t statistic	ρ	DW	Estimated SD
BMT	0.84	0.033	10.6	—	0.52	1.77
First-differenced BMT	0.41	0.024	3.1	1.0	1.01	1.19
Cochrane-Orcutt procedure	0.79	0.032	8.7	0.56	1.39	1.12

and RMSE = 4663. In this case, both variables increase with time. The regression model explains approximately 51% of the variation in the business telephone data. However, the residual plot in Exhibit 11.12a shows that the residuals are serially correlated, which is corroborated by the decaying correlogram (Exhibit 11.12b).

A regression model for the first differences of the variables has an R^2 value of only 0.03. The regression coefficient $(= -3.3; t = 1.3)$ is not significant. However, this is because the changes in the variables, which are very small relative to the level of the original series, are being modeled. The residuals of this model, shown in Exhibit 11.13a, appear almost stationary. The correlogram of the residuals (Exhibit 11.13b) shows no evidence first-order autocorrelation. However, because the regression coefficient is not significant, this suggests that the differenced BMT data behave like a random walk without any relationship to the differenced NFMA data. Note that the lag 12 spike in the correlogram is significant. This suggests an ARIMA structure in the model, which we address in Chapters 12 and 13.

The Cochrane-Orcutt Procedure

A procedure to estimate ρ in the presence of first-order autocorrelation was proposed by Cochrane and Orcutt (1949). This iterative procedure produces successive estimates of ρ until the difference between successive estimates becomes insignificant.

The initial estimate of ρ is derived from the residuals by

$$\rho = \left[\sum e_t e_{t-1}\right] / \sum e_{t-1}^2$$

The value ρ is substituted into the model

$$Y_t - \rho Y_{t-1} = (1-\rho)\beta_0 + \beta_1(X_t - \rho X_{t-1}) + v_t$$

EXHIBIT 11.12
(a) Time Plot of
Residuals Showing
Trending Pattern;
(b) Correlogram of
the Autocorrelated
Residuals

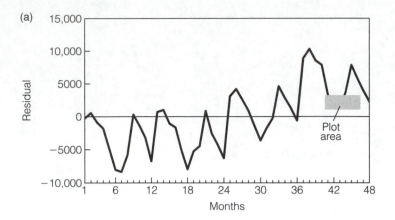

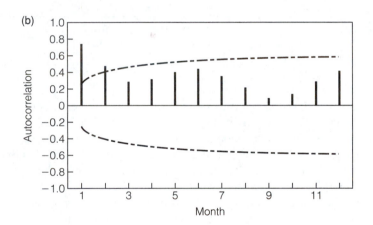

EXHIBIT 11.13
(a) Time Plot of
Residuals That Shows
No Trending Pattern;
(b) Correlogram of
the Residuals from
the Model with
Differenced Data

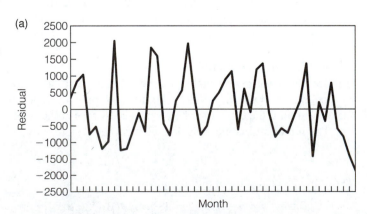

(b)

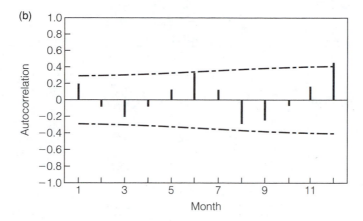

and the transformed equation is solved by using OLS. A second estimate of ρ is made in the same manner:

$$\rho = \frac{\sum v_t v_{t-1}}{\sum v_{t-1}^2}$$

where the v_t values are the residuals from the OLS fit.

The estimate of ρ is compared with the first estimate ρ. If these two values are reasonably close, the second estimate is used. If not, another iteration is made. The Cochrane-Orcutt procedure can also be used to obtain an initial estimate of ρ for use with the **Hildreth-Lu procedure** (Hildreth and Lu, 1960). This can reduce the range of ρ that needs to be searched. The Cochrane-Orcutt procedure cannot be used if there are lagged dependent variables in the model. The details of the Hildreth-Lu procedure are not discussed in this book.

A Comparison of Results Using the Cochrane-Orcutt and Hildreth-Lu Procedures

In a study relating access line gain to the economic variables housing starts (HOUS) and the differenced FRB index of industrial production (DFRB), dummy variables (D_1 and D_2) were used to account for labor strikes. The residuals from this model turned out to be autocorrelated. The model was estimated by using the Cochrane-Orcutt (C-O) and Hildreth-Lu (H-L) procedures, and the results are summarized in Exhibit 11.14.

The H-L procedure results in a slightly lower standard deviation for the residuals than the C-O procedure does, indicating that ρ is closer to 0.40 than to 0.32. It is interesting to note that the lost demand attributable to the strike (D_1) is lowered from −225,270 (which the OLS regression estimates it to be) to −148.140 (the H-L estimate). The coefficient of the housing starts variable shows a decline from 568.5 (OLS) to 493.5 (H-L). The standard error of the access line gain variable increases from 80.6 (OLS) to 106.9 (H-L) when the autocorrelation in the residuals is corrected. This is consistent with the presence of serial correlation resulting in estimates of standard error that are too low.

EXHIBIT 11.14	Regression Model		
Comparison of Serial Adjustment Corrections in a Model	Parameter	OLS	Cochrane-Orcutt	Hildreth-Lu
for Access Line Gain	R^2	0.65	0.59	0.57
(t statistics are	b_1 (HOUS)	568.5	511.9	493.5
shown in brackets		−7.1	−5.2	−4.6
below the coefficient	b_2 (DFRB)	13,185	13,040	12,940
estimates)		−6.3	−5.9	−5.8
	b_3 (D_1)	−246,720	−237,490	−235,740
		(−4.2)	(−4.5)	(−4.6)
	b_4 (D_2)	−225,270	−160,190	−148,140
		(−3.8)	(−3.1)	(−2.9)
	r		0.32	0.4
	Durbin-Watson	1.33	1.92	2.08
	Est. SD	57,980	54,160	54,100

Causal Regression with Prediction Intervals

As we have seen in the modeling process, a forecaster needs to identify an appropriate model for the data and then estimate the parameters of the model. In addition to summary statistics and significance tests, it is important also to consider:

- The forecasts given by the model for future periods. Are they reasonable?

- The comparison of forecasts with actuals. Is the accuracy level acceptable?

- The residual pattern over the fitted and forecast periods. Does it appear to be random?

- Probability limits for the forecast errors and their cumulative sum over the forecast period. These are useful for monitoring the forecasts, as actual results become available. A pattern of overforecasts, underforecasting, or too many values falling outside the limits suggests that the model needs to be reevaluated.

A prediction interval is a range of possible values around the forecast; the width of the range depends on the probability that actual data in the future will fall within the range.

Exhibit 11.15 shows prediction intervals for a simple linear regression model. Prediction intervals for individual (new) values of Y are shown in columns H and I. Confidence intervals for the true mean of Y are shown in columns K and L. These intervals are different because it is easier to forecast the mean than individual values. Hence, the confidence interval for the true mean is tighter than the prediction intervals for the individual Y values corresponding to a given X value.

EXHIBIT 11.15   Prediction Intervals for Individual Y Values and Confidence Intervals for the Mean Value of Y

A	B	C	D	E	F	G	H	I	J	K	L
									Prediction Intervals		
						For Individual Y			For the mean of Y		
Year	Mon	Per	X_1	Y	Y est	Std Error	Y min	Y max	Std Error	Y min	Y max
N/A	N/A	N/A	500	2.4	2.80	0.27	2.14	3.47	0.13	2.49	3.12
			510	2.9	2.86	0.27	2.21	3.51	0.12	2.58	3.14
			520	3.3	2.91	0.26	2.27	3.55	0.10	2.66	3.17
			530	3.0	2.97	0.26	2.34	3.59	0.10	2.73	3.20
			550	3.1	3.07	0.25	2.45	3.69	0.08	2.87	3.28
			570	3.2	3.18	0.25	2.56	3.80	0.09	2.96	3.40
			620	3.3	3.45	0.28	2.77	4.13	0.14	3.10	3.80
			640	3.6	3.56	0.29	2.84	4.28	0.17	3.14	3.98

SUMMARY

This chapter has presented most of the basic ideas we need to apply normal regression theory to forecasting problems. It is worth noting that linear regression theory is basic to forecasting; regression models can be used to describe a relationship between the variable to be forecast and one (or more) related explanatory variable(s); and the method of least squares for parameter estimation, together with normality assumptions, provides the classical statistical formulas from which many forecasting techniques are derived.

The output from a simple linear regression model and interpretation of the most significant quantities (statistics) include the R^2 statistic, which indicates the percentage of the total variation about the mean value of Y that is explained by performing a linear regression on X; the t statistic, which is used to decide if the slope and intercept coefficients are significantly different from zero; the F statistic, which indicates if the overall regression is statistically significant; and the Durbin-Watson statistic d, which is used to test for the presence of first-order autocorrelation in the time series residuals. Because a consequence of autocorrelated errors is that the sampling variances of the regression coefficients are understated, there is a tendency to accept a variable as significant when it may not be.

Correlation analysis is a useful tool for looking for linear associations between pairs of variables, but it cannot replace thoughtful and thorough data analysis on the part of the forecaster; similarly, variables that are otherwise highly correlated may appear to have a low correlation coefficient because of the presence of outliers. In these circumstances, a robust correlation coefficient, r^*, serves as an alternative to the sample product moment correlation coefficient (see Appendix 11a). The differences between the two kinds of correlation coefficients could imply the presence of outliers. If both conventional and robust methods yield similar results, we can quote conventional results with an added degree of confidence because classical assumptions are likely to be reasonable for these data; if results differ significantly, we need to dig deeper into the data source to find reasons for the difference.

REFERENCES

Bowerman, B. L., and R. O'Connell (1993). *Forecasting and Time Series—An Applied Approach*. 3rd ed. Belmont, CA: Duxbury Press.

Box, G. E. P. , G. M. Jenkins, and G. Reinsel (1994). *Time Series Analysis, Forecasting and Control*. 3rd ed. Englewood Cliffs, NJ: Prentice Hall.

Chatterjee, S., A. S. Hadi, and B. Price (2000). *Regression Analysis by Example*. 3rd ed. New York: John Wiley & Sons.

Cochrane, D., and G. N. Orcutt (1949). Applications of least squares to relationships containing autocorrelated error terms. *J. Am. Statist. Assoc.* 44, 32–61.

Cryer, J. D., and R. B. Miller (1994). *Statistics for Business*. Belmont, CA: Wadsworth.

DeLurgio, S. (1998). *Forecasting Principles and Applications*. New York: Irwin/McGraw-Hill.

Devlin, S. J., R. Gnanadesikan, and J. R.Kettering (1975). Robust estimation and outlier detection with correlation coefficients. *Biometrika* 62, 531–45.

Dielman, T. (1996). *Applied Regression Analysis for Business and Economics*. Belmont, CA: Duxbury Press.

Durbin, J., and G. S. Watson (1950). Testing for serial correlation in least squares regression: I. *Biometrika* 27, 409–28.

Durbin, J., and G. S. Watson (1951). Testing for serial correlation in least squares regression: II. *Biometrika* 38, 159–78.

Hanke, J. E., and A. G. Reitsch (1998). *Business Forecasting*. 6th ed. Englewood Cliffs, NJ: Prentice Hall.

Hildreth, G., and J. Y. Lu (1960). *Demand Relations with Autocorrelated Disturbances*. Technical bulletin 276. Lansing, MI: Michigan State Agricultural Experiment Station.

Levenbach, H., and J. P. Cleary (1984). *The Modern Forecaster*. Belmont, CA: Wadsworth.

Makridakis, S., S. C. Wheelwright, and R. J. Hyndman (1998). *Forecasting Methods and Applications*. 3rd ed. New York: John Wiley & Sons.

Tryfos, P. (1998). *Methods for Business Analysis and Forecasting: Text and Cases*. New York: John Wiley & Sons.

Wilson, J. H., and B. Keating (1998). *Business Forecasting*. New York: Irwin McGraw-Hill.

PROBLEMS

11.1. A regression analysis of salaries versus years of employment yielded the following output. Unfortunately, some key pieces of information are missing.

Regression Statistics

R^2	(C)
Standard error of estimate	5.823
Observations	46

ANOVA

SOURCE	df	SS	MS	P Value
Regression	1	(B)	1730.6	0.000
Residual	44		(A)	
Total	45	5444.3		

	Coefficient	Std Error	t Statistic	P Value
Intercept	24.635	2.482	9.93	0.000
YRS EM	0.6565	(D)	4.53	0.000

a. Obtain the following information:

   (A) Residual mean square =

   (B) SSR =

   (C) R^2 =

   (D) Standard error of YRS EM coefficient =

   Also obtain the degree of linear association between SALARY and YRS EM.

b. Based on the results, construct a 99% confidence interval for the slope of the regression line.

c. What information is there in the regression output about whether or not the estimated slope coefficient is significantly different from zero?

d. Supplement your prediction with an approximate 95% confidence interval.

11.2. The regression output for the changes in housing starts versus changes in mortgage rates is given in the accompanying table.

Predictor	Coefficient	SD	t	P
Constant	1.24	48.46	0.03	0.980
X	−194.21	55.28	−3.51	0.002
$S = 254.2$	$R^2 = 32.2\%$		$R^2 = 29.6\%$	

ANOVA

Source	df	SS	MS	F	P
Regression	1	797463	797463	12.34	0.002
Error	26	1680067	64618		
Total	27	2477530			

a. Based on the results, construct a 99% confidence interval for the true slope of the regression line.

b. What information is there in the regression output about whether or not the estimated

slope coefficient is significantly different from zero?

c. Supplement your prediction with an approximate 95% confidence interval.

11.3. The summary statistics for the original housing starts versus mortgage rates relationship is given in the accompanying table. Determine the least-squares fit between housing starts and mortgage rates.

Variable	N	Mean	Median	Tr Mean	SD	SE Mean
Starts	29	1546.3	1507.6	1535.9	337.1	62.6
Rates	29	9.039	8.800	8.959	2.339	0.434

Variable	Min	Max	Q1	Q3
Starts	1014.5	2356.6	1291.9	1747.3
Rates	5.740	14.490	7.520	10.130

Correlation of Starts and Rates = −0.202

11.4. Complete the accompanying ANOVA table for the regression results for Problem 11.3.

Predictor	Coefficient	SD	t	P
Constant	XXXX	XXXX	7.14	0.000
Rates	−29.05	(D)	−1.07	0.294
$S = 336.2$	$R^2 = (C)\%$		$\overline{R}^2 = 0.5\%$	

ANOVA

Source	df	SS	MS	F	P
Regression	1	(B)	129326	1.14	0.294
Error	27	3052370	(A)		
Total	28	3181697			

Unusual Observations

Obs	Rates	Starts	Fit	SD Fit	Residual	St Resid
10	7.5	2356.6	1592.4	75.9	764.2	2.33R
20	14.5	1062.2	1387.9	160.7	−325.7	−1.10X

a. Obtain the following information:

(A) Error mean square =

(B) SSR =

(C) R^2 =

(D) Standard error of Rates coefficient =

Also obtain the degree of linear association between Starts and Rates.

b. Based on the results, construct a 99% confidence interval for the true slope of the regression line.

c. What information is there in the regression output about whether or not the estimated slope coefficient is significantly different from zero?

d. Supplement your prediction with an approximate 95% confidence interval.

11.5. The summary statistics for the changes in housing starts versus changes in mortgage rates is given in the accompanying table. Determine the least-squares fit between DifStart and DifRate.

Variable	N	N*	Mean	Median	Tr Mean	SD	SE Mean
DifStart	28	1	−21.0	−48.4	−20.1	302.9	57.2
DifRate	28	1	0.115	0.035	0.143	0.885	0.167

Variable	Min	Max	Q1	Q3
DifStart	−707.6	640.8	−184.5	110.9
DifRate	−2.380	1.870	−0.207	0.590

Correlation of DifStart and DifRate = −0.567

11.6. The relationship between the mailing list size (in thousands of names) and sales (in thousands of dollars) for a group of catalogs is:

Average list size = 150   Sample SD = 70

Average sales = $5000   Sample SD = $2000

$r = 0.75$

a. Assuming a simple linear regression model, what are the least-squares estimates of the intercept and slope in the equation?

b. Write down the equation of the regression line relating list size to sales.

c. How much would you expect the sales level to be if you did not mail a catalog?

d. What level of sales would you expect for a catalog mailed to 50,000 people?

e. What percentage in the variation in list size can be explained by the fact that some generated more sales than others did?

f. Find the predicted and residual for a situation that leads to sales of $4,000,000 from a catalog mailout to 200,000 names.

g. If the standard error of estimate S_e is $1,330,000, find the 90% confidence interval for the slope coefficient.

11.7. Using the formal-wear rental sales data from Exhibit 4.19, create a scatter diagram in which the tick marks on the X axis correspond to each year in the data (e.g., 1985, 1986, ... , 1992). Observe the pattern of the yearly averages (or medians) in terms of trend as well as the nature of the yearly standard deviations (or similar measure of spread) in the data.

a. Create a set of 11 seasonal dummies and perform a multiple linear regression model with

a constant, a straight line Time variable, and the dummies as independent variables. Ignore the summary statistics, but review a plot of the residuals. Create the residual plot in the same manner as the plot of the original data. Comment on the typical yearly residual as well as a measure of the yearly spread in the residuals.

b. Repeat (a) with the logarithm of the original data.

c. What is the nature of the residual variability in the two data sets, and how do the trend patterns differ?

d. To which data set do you prefer to apply a multiple linear regression model with the seasonal dummies, taking into consideration the basic assumptions underlying a linear regression model? Comment on your assumptions and explain why variability in the data is a more important issue than the pattern of fitted values.

11.8. Repeat Problem 11.7, for one or more of the following monthly time series:

a. Unemployment rate data (UNEMP, Dielman, 1996, Table 4.7)

b. Silver prices data (SILVER, Dielman, 1996, Table 4.8)

c. Prime rate data (PRIME4, Dielman, 1996, Table 4.14)

d. Retail sales for U.S. retail sales stores (RETAIL, Hanke and Reitsch, 1998)

e. Wheat exports (SHIPMENT, Dielman, (1996, Table 4.9)

f. Shipments of spirits (DSHIP, Tryfos, 1998, Table 6.23)

g. Retail sales in a region (RSALES, Tryfos, 1998, Table 6.25)

h. Australian beer production (AUSBEER, Makridakis et al., 1998, Table 2.2)

i. International airline passenger miles (AIRLINE, Box et al., 1994, Series G)

j. Australian electricity production (ELEC-TRIC, Makridakis et al., 1998, Fig. 7-9)

k. French industry sales for paper products (PAPER, Makridakis et al., 1998, Fig. 7-20)

l. Shipments of French pollution equipment (POLLUTE, Makridakis et al., 1998, Table 7-5)

m. Sales of recreational vehicles (WINNEBAG, Cryer and Miller, 1994, p. 742)

n. Sales of lumber (LUMBER, DeLurgio, 1998, p. 34)

o. Sales of consumer electronics (ELECT, DeLurgio, 1998, p. 34)

11.9. Repeat Problem 11.7 for one or more of the following quarterly time series on the CD, using three seasonal dummies to define a seasonal cycle:

a Sales for Outboard Marine (OMC, Hanke and Reitsch, 1998, Table 4.5)

b Shoe store sales (SHOES, Tryfos, 1998, Table 6.5)

c Domestic car sales (DCS, Wilson and Keating, 1998, Table 1-5)

d Sales of The GAP (GAP, Wilson and Keating, 1998, p. 30)

e Air passengers on domestic flights (PASSAIR, DeLurgio, 1998, p. 33)

11.10. Using the accompanying data set for the weekly sales of a baked good, determine if a weekly sales cycle exists apart from the strong growth in demand (Hanke and Reitsch, 1995, Table P.7A)

Week	Q1	Q2	Q3	Q4
1	22.46	30.21	39.29	47.31
2	20.27	30.09	39.61	50.08
3	20.97	33.04	41.02	50.25
4	23.68	31.21	42.52	49
5	23.25	32.44	40.83	49.97
6	23.48	34.73	42.15	52.52
7	24.81	34.92	43.91	53.39
8	25.44	33.37	45.67	52.37
9	24.88	36.91	44.53	54.06
10	27.38	37.75	45.23	54.88
11	27.74	35.46	46.35	54.82
12	28.96	38.48	46.28	56.23
13	28.48	37.72	46.7	57.54

a. Plot the data as a time plot and fit a straight line through the 52 weekly values.

b. Construct a probability plot of the residuals. Can you identify any unusual values?

c. Organize the residuals in a two-way ANOVA table (as described in Chapter 4) and decompose the residuals into a weekly seasonal and quarterly trend components. How do you interpret the 13 seasonal-week averages? Are they additive or multiplicative?

d. Use the 13 weekly seasonal means and modify the straight-line forecasts for periods 53–65 to produce a week-cycle adjusted forecast. Do these forecasts have a commonsense interpretation? Use a 95% confidence interval for the means, if it is available in your software package.

e. Fit a straight line through the first 39 values of the series and project the straight line through the remaining 13 values of the holdout sample. Use the 13 weekly seasonal means determined in (b) and use them to adjust the straight-line projection with the presumed sales cycle. Compare the results with the actual values in the holdout sample, using MAPE, MAE, and so on. Are your conclusions about the value of the sales cycle any different from your conclusion in (c)?

11.11. Repeat Problem 11.10 for one or more of the following sets of weekly data:

a. Shipments of canned beverage (Exhibit 4.11)

b. Sales of clips for sun glasses (Exhibit 6.13)

c. Store sales of a drugstore (DRUGSTOR, Tryfos, 1998, Table 6.24)

d. Sales of absorbent paper towels (TOWEL, Bowerman and O'Connell, 1993, Table 9.1)

11.12. Economists involved in international trade study relationships between agricultural exports and exchange rates (Bessler and Babubla, (1987). Forecasting wheat exports: Do exchange rates really matter? *J. Business Econ. Statist.* 5, 397–406; the data are reported in Dielman, 1996, Table 3.15). The dependent variable Y is monthly U.S. wheat export shipments and the dependent variable X is the real index of weighted-average exchange rates for the U.S. dollar from 1974:1 to 1985:3.

a. Determine a linear relationship between Y and X. Does the sign of the slope coefficient make economic sense? What conclusions do you reach by reviewing a scatter diagram between Y and X?

b. Consider the latest portion of the data from 1982:1 to 1985:3.

   (i) Develop a scatter diagram between Y and X and contrast this with the scatter diagram in (a).

   (ii) Are your conclusions about a possible linear relationship between the variables the same?

   (iii) Verify your conjecture by estimating a linear regression relationship. What is the nature and strength of the relationship?

   (iv) Can you provide an economic rationale for the conclusions drawn from (ii) and (iii)?

   (v) What percentage of the total variation in Y has been explained in the two situations?

c. Construct 95% confidence intervals for the slope parameter in the regression equations in (a) and (b). What inferences can you draw from this?

d. Review the residuals from the model in (b). September 1984 (6605, 151.073) looks suspect in relation to the rest of the data. Visual inspection suggests that 6605 may be an input error. Try to rerun the regression analysis with a "more representative" replacement value of 3305. Repeat (b) and (c).

e. Alternatively, delete September 1984 from the data set. Repeat (b) and (c). Contrast the results with your answer in (d). Do you feel justified in deleting versus replacing an errant observation in this case?

11.13. Since 1960, the annual world crude oil production (in millions of barrels of oil) has grown steadily (OIL, Chatterjee et al., 2000, Table 6.17).

Year	Oil	Year	Oil
1960	7674	1974	20,389
1962	8882	1976	20,188
1964	10,310	1978	21,922
1966	12,016	1980	21,722
1968	14,104	1982	19,411
1970	16,690	1984	19,837
1972	18,584	1986	20,246
		1988	21,338

a. Construct the scatter plot oil versus year, and fit a simple linear (straight-line) regression model. What is the average annual increase in oil production per year?

b. Create 95% confidence limits for the slope coefficient.

c. Plot the residuals against year, and construct the normal probability plot of the residuals.

d. Construct the scatter plot of the logarithms of oil versus year, and fit a simple linear (straight-line) regression model. What is the average annual percentage increase in oil production per year?

e. Create 95% confidence limits for the slope coefficient.

f. Plot the residuals against year, and construct the normal probability plot of the residuals.

g. Contrast the results of the two models constructed for oil. Which will probably produce better forecasts, in your opinion?

11.14. For Problem 11.13, create a holdout sample of the latest three periods in each of the two models for oil.

a. Calculate three one-period-ahead forecasts for each model, and summarize the forecast errors with two accuracy measures (MAPE and MdAPE).

b. Contrast the results and interpret the forecasting ability of each model based on the forecasting performance.

c. Did the fit statistics give you any indication of how the forecasting results will compare in the two models?

CASE 11A DEMAND FOR METHYLENE DIPHEYLENE DIISOCYANATE

Methylene dipheylene diisocyanate (MDI) is a polymer used in rigid foam insulation for energy conservation building materials. As the chemical industry expert for a consulting firm, you need to get a better understanding of the MDI market because the client is interested in commercializing a new proprietary polymer as a better alternative to MDI. It is suspected that energy prices might be a factor influencing the demand for MDI. The energy price to be considered is the composite (oil, gas, coal), weighted price of energy as provided by the U.S. Census Bureau (energy data). The accompanying table contains 21 years of historical data for these variables.

Year	MDI (millions of pounds)	Fuel Price (cents/million BTU)
1	1.8	42.4
2	3.6	41.1
3	3.9	39.7
4	4.8	38.5
5	6.1	37.9
6	5.6	37.7
7	70.4	36.1
8	101.6	35.5
9	132.0	33.3
10	120.0	37.2
11	280.0	36.0
12	290.0	38.0
13	300.0	60.0
14	245.0	67.0
15	310.0	69.0
16	330.0	76.0
17	355.0	78.0
18	410.0	88.0
19	470.0	106.0
20	520.0	140.0
21	500.0	140.0

(1) Create time plots of the data and a scatter diagram of the two variables.

(2) Examining the price history, can you surmise the approximate period of the data? Is it relevant to know this?

(3) Although regression does not imply causation, only association, create a simple linear regression model of MDI demand (dependent variable) versus fuel price (independent variable)

(4) Display and summarize any summary statistics that you feel are relevant and omit any that are not.

(5) Economic theory might suggest taking logarithms of the variables first, so that the (slope) regression coefficient can be interpreted as a constant elasticity. Perform such an analysis, as in (1)–(4).

CASE 11B DEMAND FOR AIR TRAVEL (CASES 4B–6B, AND 10B CONT.)

The VP of Operations has been advised that some simple linear regressions modeling is useful to gain insights into the relationships among the variables, and has requested that you use your newly learned forecasting skills. The quarterly data for 4 years is shown in the table in Case 4B.

(1) Create simple linear regression models of passenger miles versus each of the independent variables.

(2) Perform a preliminary data analysis of the data, including plots, correlograms, and scatter diagrams.

(3) Make an independent forecast of each of the independent variables using an exponential smoothing model, leaving out the last 2 quarters as a holdout sample.

(4) Using the regression model, project the remaining two quarters using the history (ex post) and independent forecasts (ex ante) from (2).

(5) Summarize your results with at least two accuracy measures.

(6) Make a plot of the history, forecasts, and 95% confidence limit on the forecasts.

(7) Compare your results with the 2 quarters (actuals). Are the actuals within your confidence limits?

(8) What are your recommendations on the value and limitations of such forecasts to a practitioner?

CASE 11C ENERGY FORECASTING (CASES 1C AND 8C CONT.)

Before embarking on a full-scale econometric modeling project, you want to gain further insight into the data by running some simple linear regression models and compare them with the exponential smoothing models from Case 8C.

To keep matters simple, you postulate that a factor that could affect consumption is a population variable as well as the GNP (now called GDP). The accompanying table shows energy consumption (in quadrillion BTU), U.S. population (in millions), and GNP (in billions constant 1982 dollars). (www.eia.doe.gov/emeu/aer/ and www.census.gov/statab/).

(1) Create simple linear regression models of consumption versus each of the independent variables, using the first 35 years of the historical data (1949–1983).

(2) Make an independent forecast of the independent variables using an exponential smoothing model (Case 8C).

(3) Using the regression model, project the remaining five years (1984–1988), using the history (ex post) and independent forecasts (ex ante) from (2).

(4) Summarize your results with at least two accuracy measures.

(5) Make a plot of the history, forecasts, and 95% confidence limit on the forecasts.

Year	Energy Consumption	Population (in millions)	GNP (in billions of constant 1982 dollars)	Year	Energy Consumption	Population (in millions)	GNP (in billions constant 1982 dollars)
1949	31.98	149.2	1109.0	1970	67.84	205.1	2416.2
1950	34.62	152.3	1203.7	1971	68.29	207.7	2484.8
1951	36.97	154.9	1328.2	1972	73.99	209.9	2608.5
1952	36.75	157.6	1380.0	1973	75.71	212.0	2744.1
1953	37.66	160.2	1435.3	1974	72.00	213.9	2729.3
1954	36.64	153.0	1416.2	1975	76.01	216.0	2695.0
1055	40.21	165.9	1494.9	1976	78.00	218.4	2826.7
1956	41.74	168.9	1525.6	1977	79.99	220.2	2958.6
1957	41.79	172.0	1551.1	1978	80.90	222.6	3115.2
1958	41.65	174.9	1539.2	1979	78.29	225.6	3192.4
1959	43.47	177.8	1629.1	1980	76.34	227.7	3187.1
1960	45.09	180.7	1665.3	1981	73.25	229.3	3248.8
1961	45.74	183.8	1708.7	1982	73.1	231.5	3166.0
1962	47.83	186.5	1799.4	1983	76.74	234.0	3279.1
1963	49.65	189.2	1873.3	1984	76.47	236.2	3501.4
1964	51.82	191.8	1973.3	1985	76.78	238.7	3618.7
1965	54.02	194.3	2087.6	1986	79.22	241.1	3721.7
1966	57.02	196.6	2208.3	1987	82,84	246.3	3847.0
1967	58.91	198.7	2271.4	1988	84.96	247.4	3996.0
1968	62.42	200.7	2365.6				
1969	65.62	202.7	2423.3				

(6) Compare your results with an updated data source. Are the actuals within your confidence limits?

(7) Make a forecast for 1989–1988, and compare this with an updated source for the independent variable. Are the actuals within your confidence limits?

(8) What are your recommendations on the value and limitations of such forecasts to a practitioner?

CASE 11D NEW YORK METS HOME ATTENDANCE

Professional sports are big business everywhere, and no sport is bigger than major league baseball in the United States. The salaries of professional baseball players have attained levels undreamed of by past generations of players. Their contracts are a major investment for a baseball franchise. Consequently, a ball club must be able to evaluate the return on such investments.

What are the major sources of revenue for a baseball team? The largest is the television stations that pay to broadcast the games. However, television income is generally fixed in advance according to long-term contracts. Another large component of revenues is the receipts from ticket sales. While representing an agency for prominent sports figures, you are asked to investigate a number of potential factors influencing demand and how they might be incorporated in a regression model for ballpark attendance.

Year	Ballpark Attendance	Year	Ballpark Attendance	Year	Ballpark Attendance
1962	922,530	1971	2,266,680	1981	701,910
1963	1,080,108	1972	2,134,185	1982	1,320,055
1964	1,732,597	1973	1,912,390	1983	1,103,808
1965	1,768,389	1974	1,722,209	1984	1,829,482
1966	1,932,693	1975	1,730,566	1985	2,751,437
1967	1,565,492	1976	1,468,754	1986	2,762,417
1968	1,781,657	1977	1,066,825	1987	3,027,121
1969	2,175,373	1978	1,007,328	1988	3,047,724
1970	2,697,479	1979	788,905	1989	2,918,710
		1980	1,178,859	1990	2,732,745

To get started on the problem, you are given the data for the first 29 years of paid attendance at New York Mets home games since the franchise was started in 1962.

(1) Summarize the essential characteristics of the data with time plots, summary statistics, box plots, and correlograms. Are there any outliers, and can you reconcile them?

(2) Update the data to a current year and repeat (1). Do you see any benefit to using all the data rather than a reasonable current period?

(3) Develop a modeling flowchart to forecast ballpark attendance.

(4) Discuss a number of potential factors influencing demand. Research the availability of the data for these factors you have selected.

A previous study resulted in a regression model for Mets home attendance (in thousands) of the form:

$$METS\_ATT = -2742.5 + 2977.3 \, LNMWIN$$
$$- 1893.9 \, LNYWIN + 616.5 \, DUM\_POST$$

where

LNMWIN = ln (Number of games won by Mets in current season)

LNYWIN = ln (Number of games won by Yankees in current season)

DUM_POST = Dummy variable for making playoffs in prior season

(5) In a season in which the Mets win 90 games and the Yankees win 70 games, what would the expected Mets home attendance be with and without making the prior-season's playoffs?

CASE 11E Japanese Automobile Export (Case 7E cont.)

You continue to investigate the econometric modeling approach for this study.

(1) Research the literature or the web for a specific model that has been developed for modeling automobile exports [e.g., Dixit, A. (1988). Optimal trade and industrial policies for the U.S. automobile industry. In *Empirical Methods for International Trade*, edited by Robert Feenstra. Cambridge, MA: MIT Press]. What is the dependent variable? Describe the author's use of the independent variables.

(2) Develop a flowchart of the modeling steps.

(3) Was there a use of the model for forecasting?

(4) Were any data provided in the study? If so, develop plots, data summaries, and correlograms to get some insight into the data patterns. If not, attempt to gather the data for the key variables and document your findings in terms of the quality of the data sources (accuracy, conformity, timeliness, and consistency; see Chapter 1).

CASE 11F Twin Rivers (Case 3B cont.)

The file **EI_GaUse.xls** contains the 152 (EIUse, GaUse) pairs (winter energy consumption in therms; IEUse = therms of electricity; GaUse = therms of gas; 1 therm approximately 30 kilowatt hours and exactly 100,000 BTU) (Tukey, 1977, Chap. 8).

(1) Perform a preliminary data analysis on the data, including box plots, correlation coefficients, and a scatter diagram.

(2) Research the literature or the web for a specific model that has been developed for this kind of problem. Display the model you found with a brief synopsis of the assumptions and findings.

(3) Develop a flowchart of the steps you need to take to update the study.

APPENDIX 11A A ROBUST CORRELATION COEFFICIENT

One robust estimator of correlation that is less sensitive to outliers than the product moment correlation coefficient r, designated $r^*(SSD)$, is derived from the standardized sums and differences of two variables—say Y and X (Devlin et al., 1975). The first step in obtaining $r^*(SSD)$ is to standardize both Y and X robustly by constructing two new variables Y and X:

$$\tilde{Y} = (Y - Y^*)/S_Y^* \quad \text{and} \quad \tilde{X} = (X - X^*)/S_X^*$$

where Y^* and X^* are robust estimates of location and S_Y^* and S_X^* are robust estimates of scale.

Now, let $Z_1 = \tilde{Y} + \tilde{X}$ and $Z_2 = \tilde{Y} - \tilde{X}$, the sum and differences vectors, respectively. Then, the robust variance of the sum vector Z_1 and difference vector Z_2 are calculated; they are denoted by V_+^* and V_-^*, respectively. These variances are used in the calculation of the robust correlation estimate $r^*(SSD)$ given by

$$r^*(SSD) = (V_+^* - V_-^*) / (V_+^* + V_-^*)$$

The justification for this formula can be seen by inspecting the formula for the variance of the sum of two variables:

$$\text{Var}(Z_1) = \text{Var}(\tilde{Y}) + \text{Var}(\tilde{X}) + 2\text{Cov}(\tilde{Y}, \tilde{X})$$

where Cov denotes the covariance between $\tilde{Y}$ and $\tilde{X}$.

Since $\tilde{Y}$ and $\tilde{X}$ are standardized, centered about zero, with unit scale, the expected variance of Z_1 is approximately

$$\text{Var}(Z_1) \approx 1 + 1 + 2\rho(Y, X) = 2(1 + \rho)$$

where ρ is the theoretical correlation between $\tilde{Y}$ and $\tilde{X}$.

Similarly, for Z_2,

$$\text{Var}(Z_2) \approx \text{Var}(\tilde{Y}) + \text{Var}(\tilde{X}) - 2\text{Cov}(\tilde{Y}, \tilde{X})$$
$$= 1 + 1 - 2\rho(\tilde{Y}, \tilde{X}) = 2(1 - \rho)$$

Notice that the expression

$$\frac{\text{Var}(Z_1) - \text{Var}(Z_2)}{\text{Var}(Z_1) + \text{Var}(Z_2)} \approx \frac{2(1+\rho) - 2(1-\rho)}{2(1+\rho) + 2(1-\rho)} = \rho$$

Some robust estimates of the (square root of the) variance, required in the formula for r^*, are discussed in Chapter 5; these include the MdAD and IQR.

12

Forecasting with Regression Models

"Plurality which is not reduced to unity is confusion; unity which does not depend on plurality is tyranny."
BLAISE PASCAL

IN THE PREVIOUS TWO CHAPTERS, we noted how regression models are created for time series forecasting and econometric analysis. These models can:

- Provide superior forecasts (once the accuracy of the forecasts of the independent variables is assured)

- Provide a forecast user with a model that is encompassing and explanatory, because more than one variable is taken into account

- Be used to develop plans for policy evaluation where the model serves to express various policy alternatives

This chapter presents the statistical issues surrounding the use of linear regression models in various applications. In addition, our treatment points out the value of robust/resistant methods of correlation and regression analysis in demand forecasting by considering the effects of outliers on regression parameter estimates and the methods for dealing with nonnormal situations in which classical least-squares techniques can readily lead to inappropriate forecasts.

12.1 MULTIPLE LINEAR REGRESSION ANALYSIS

When a forecaster has reasons to believe that more than one explanatory variable are required to solve a forecasting problem, the simple linear regression model is no longer appropriate. It is necessary to develop a multiple linear regression model for forecasting a dependent variable based on forecasts of *more than one* independent variable. When multiple variables are involved, the modeling effort can easily become more involved; so we need to start with additional preliminary data analyses. Reviewing scatter diagrams between the dependent variable and each of the independent variables, as well as between pairs of independent variables, can save some time and enhance our understanding of underlying relationships. This arrangement is called the *scatter plot matrix*. More simply, we start by creating a matrix of correlations between pairs of variables. In any case, we will be extending the simple linear regression model with one independent variable to a model that can encompass multiple explanatory variables. The most direct extension is the multiple linear regression model.

 A scatter plot matrix is an array of correlations between pairs of variables.

Recall from Chapter 10 that in multiple linear regression analysis the regression function, $\mu_{Y(X)}$, is the deterministic component of the model and takes the form:

$$\mu_{Y(X)} = \beta_0 + \beta_1 X_1 + \cdots + \beta_k X_k$$

where $X_1, \ldots, X_k$ are k independent variables (or regressors), and $\beta_0, \beta_1, \ldots, \beta_k$ are called regression parameters. This model arises when the variation in the dependent variable Y is assumed to be affected by changes in more than one independent variable. Thus, the average value of Y is said to depend on $X_1, X_2, \ldots, X_k$. The dependence on X is henceforth suppressed in the notation; let $\mu_{Y(X)} = \mu_Y$. In this case, we speak of a multiple linear regression of Y on $X_1, \ldots, X_k$.

The formal theory of normal multiple linear regression analysis is very extensive and is dealt with in a number of regression analysis textbooks (e.g., Draper and Smith, 1998; Chatterjee et al., 2000) and numerous business statistics textbooks. Only its use and interpretation in forecasting problems are our focus here. Some examples of multiple linear regression models have already been presented in Chapter 10.

12.2 ASSESSING MODEL ADEQUACY

The adequacy of model assumptions can be examined through a variety of methods, many graphical and involving the analysis of residuals. There is a range of regression pitfalls that we must be aware of and avoid if possible. These pitfalls include trend collinearity, overfitting, extrapolation, outliers, nonnormal distributions, multicollinearity, and invalid assumptions regarding the model errors (e.g., independence, constant variance, and, usually, normality).

Collinearity Due to Trends

When the dependent and independent variables are time series, there are many pitfalls specific to time series to be avoided. Suppose that we are interested in forecasting the cyclical variation in one series on the basis of predictions of a related cyclical time series, such as business telephones and nonfarm employment. In the regression model, a high value of the R^2 statistic may result from what is known as trend collinearity. This often occurs when both series have very strong trends. It is quite possible that the trends are highly correlated but that the cyclical patterns are not. The dissimilarities in cycle may be masked by the strong trends.

Similarly, when a regression model is built on raw time series, it is not clear just what information will result. If both series have rising trend and corresponding strong seasonality, the regression will very probably show a very high R^2 statistic. Alternatively, if there is a strong underlying relationship between variables but their seasonal patterns differ, the regression may appear insignificant.

In the case of a telephone revenue–message relationship, seasonality is not of primary interest, so we can use seasonally adjusted data. A high correlation now means that there is a strong association in trend-cyclical patterns. In order to determine whether there is strong cyclical relationship, the appropriate procedure is to fit preliminary trend lines and correlate residuals from the fitted values.

In time series analysis, eliminating trend from each variable often greatly reduces trend collinearity. In the example in Exhibit 12.1, the product-moment correlations of the differenced data for quarterly access-line gain, housing starts, and FRB index of industrial production are shown below the diagonal in the matrix. The entries above the diagonal in Exhibit 12.1 represent a robust version of the ordinary correlations. The robust correlations are all somewhat lower, possibly due to outliers or a few influential data values.

These can be compared to Exhibit 12.2, the matrix of correlations for the original (undifferenced) data. Although there are some significant correlations among the differenced independent variables, they are probably not large enough to introduce serious collinearity problems. The forecaster should become cautious, however, when simple correlations exceed values from 0.8 to 0.9. As a rule, we should adjust the data for possible sources of variation in which we are not interested in order to study the relationships with respect to those forces whose effects are of primary interest.

EXHIBIT 12.1
Ordinary and Robust
Correlations for the
Access-Line Gain
Model Involving
Housing Starts and
FRB Index (based on
differenced data)

	1 Access-Line Gain	2 Housing Starts	3 FRB Index
1 Access-Line Gain	1. 00	0.32	0.47
2 Housing Starts	0.72	1	0.36
3 FRB Index	0.57	0.42	1

EXHIBIT 12.2		1 Access- Line Gain	2 Housing Starts	3 FRB Index
Ordinary and Robust	1 Access-Line Gain	1	0.07	0.74
Correlations for the	2 Housing Starts	0.01	1	0.14
Access-Line Gain	3 FRB Index	0.47	0.32	1
Model Involving				
Housing Starts and				
FRB Index (based on				
original data)				

Overfitting

Another source of peril in regression analysis is overfitting. This occurs when too many independent variables are used to attempt to explain the variation in the dependent variable. Overfitting may also arise when there are not enough data points. If the number of independent variables is close to the number of observations, a "good fit" to the observations may be obtained but the coefficient and variance estimates will be poor. This often results in very bad estimates for new observations.

Extrapolation

In forecasting applications, regression models are frequently used for purposes of extrapolation—that is, for extending the relationship to a future period of time for which data are not available. A relationship that is established over a historical time span and a given range of independent and dependent variables may not necessarily be valid in the future. Thus, extreme caution must be exercised in using correlation analysis to predict future behavior among variables. There may be no choice in some cases, but the forecaster should recognize the risks involved.

Outliers

Outliers are another well-known source of complexity in correlation analysis. A single outlier can have a significant impact on a correlation coefficient. Exhibit 12.3 shows a scatter diagram of 60 values from a simulated sample, a bivariate normal

EXHIBIT 12.3
Scatter Plot with
Influence Function
Contours for a
Sample of Bivariate
Normal Data with
the Added Outlier
($n = 60$, $r = 0.9$, with
outlier $r = 0.84$)

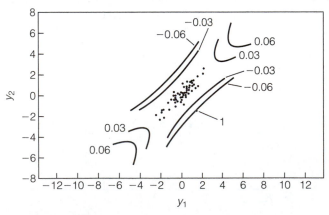

(data from Fisher, 1960).

distribution with population correlation coefficient $\rho = 0.9$. One point was moved to become an outlier. The empirical correlation coefficient is now calculated to be 0.84. Exhibit 12.3 shows that, except for this single point, the scatter is quite linear and, in fact, with this outlier removed, the estimated correlation coefficient is 0.9 (Devlin et al., 1975). Robust alternatives offer some protection in these instances, as well as in others in which there is nonnormality in the error distribution.

Multicollinearity

Be aware of multicollinearity effects when forecasts are based on multiple linear regression models. Models with more than approximately five independent variables often contain regressors that are highly mutually correlated. Because it is likely that there are interrelationships among the variables, we find high pairwise correlations. Simply, one variable may serve as a proxy for another. In such cases, it may be beneficial to seek linear combinations of these variables as regressors, thereby reducing the dimension of the problem. See, for example, Belsley et al. (1980), Draper and Smith (1998), and Chatterjee et al. (2000).

In its simplest form, multicollinearity arises whenever explanatory variables in a regression model are highly mutually correlated. We will find low t values for one or more of the estimated regression coefficients, which suggests that some variable could be dropped from the equation. Multicollinearity is difficult to deal with because it is almost invariably a problem of degree rather than kind. Multicollinearity results in bias, inconsistency, and inefficiency of estimators, which are undesirable, at least from a theoretical viewpoint. In practice, remedies must be found so that the model assumptions can be made approximately true without completely destroying the statistical validity of the model.

Frequently, we would like to retain certain explanatory variables in the equation, even if associated t values suggest that such variables add little to the explanatory power of the equation. The problem is that the variables that are expected to be important have small t values (large standard errors) and lead to unstable estimation of the coefficients.

Multicollinearity arises when the independent variables in a regression model are highly correlated.

Some of the practical effects of multicollinearity include imprecise estimation. Although the concepts of bias, consistency, and inefficiency have very precise theoretical definitions, their effect on the inferences that are drawn from a model may be difficult to determine in every practical application.

Problems of multicollinearity also encourage the misspecification of the model if they are not carefully noted. The stability of coefficients can be severely affected, so that different segments or subsets of the data give rise to vastly different results. This lack of sensitivity precludes the possibility of making sound interpretations of model coefficients.

To examine the relationship among the explanatory variables, we can calculate a variance inflation factor (VIF) associated with each explanatory variable. When a computer package displays these VIFs, we can establish whether a variable is highly correlated with the remaining explanatory variables in the equation and whether dropping any of the variables will eliminate the collinearity effects. (For an extensive discusson and computer example, see Chatterjee et al. (2000), Sec. 9.2, 9.3).

Invalid Assumptions

The standard assumptions discussed in the beginning of this chapter regarding model errors (independence, constant variance, and normality) and the additivity of the model needs to be evaluated. If, for a particular set of data, one or more of these assumptions fail, it may help to transform the data.

Application: Relating Business Telephone Demand to Nonfarm Employment

Growth in the demand for business telephones is known to follow a cyclical pattern. A review of cyclical economic variables and logic suggest that the growth in nonfarm employment can be considered to be an explanatory variable in a simple linear regression model. It is further known that in an economic downturn, for example, the reduction in manufacturing employment is greater than the reduction in business telephone demand. Because most manufacturing employees do not have their own business telephones, this relationship is not surprising. This suggests that nonfarm employment *minus* manufacturing employment should also be considered as an explanatory variable. For aggregate modeling purposes, we have converted the monthly series to quarterly data.

Because both time series are seasonal, we expect that the year-over-year changes in the data will yield a time series that is not seasonal. In fact, to put both time series on a comparable unit-free basis, we choose the percentage changes in the variables as the dependent and independent variables in the linear regression model. The year-over-year percentage changes of the three quarterly series (Exhibit 12.4) suggest that it is preferable to exclude manufacturing employment from the total nonfarm employment.

EXHIBIT 12.4   Time Plot of the Year-over-Year Percentage Changes for Business Telephones (solid line), Nonfarm Employment (dashed line), and Nonfarm Employment Less Manufacturing (dotted line); Nine Years (36 quarters) of Data

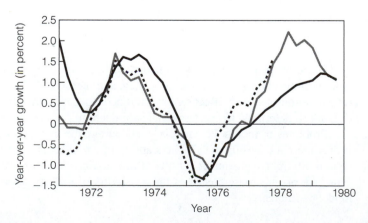

Exhibit 12.5 provides summary statistics for the two models: Model A uses percentage change of nonfarm employment and model B uses percentage change of nonfarm employment less manufacturing employment as its independent variable. Model B shows a 30% improvement in the R^2 statistic and, more important, provides a more accurate 1-year-ahead forecast performance, on average.

The forecast test comparisons were made in the following manner. Regressions were performed over a segment of the data, and forecasts were generated by using actual values for the independent variable (known as ex post forecasting) over the forecast period. Forecast performance was evaluated by calculating the percentage errors for four 1-year-ahead forecasts and three 2-year-ahead forecasts, using the forecast horizon of Year 6 through Year 9. For this particular period, the forecast errors (actual minus forecast) were almost all negative (that is, the model tends to overforecast this period). This approach tests the relative forecasting accuracy of the models.

In a true forecasting environment, forecast errors may also result from inaccurate forecasts of the independent variable as well as misspecification of the model.

If the residuals are independent of one another, have constant variance, and can be assumed to be approximately normally distributed, the coefficient of the employment variable can be interpreted as follows. Assuming $\beta = 0.4$, then, for every 10% increase in nonfarm minus manufacturing employment, there should be an increase of 4% in the demand of additional business telephones.

The low value of the Durbin-Watson statistic (DW $= 0.24$) suggests positive first-order autocorrelation ($r_1 \approx 1 - 0.5$ DW ≈ 0.9) in the residuals. The next step is to reduce or eliminate this problem, which generally involves adding additional variables and building a multiple linear regression model. Alternatively, a number of autocorrelation correction techniques could be tried (see Chapter 11, where we first introduce this problem).

EXHIBIT 12.5
Forecasting Performance Tests and Summary Statistics for Business Telephones–Employment Models; the regression Relationship Is Expressed in Terms of Percentage Changes of the Variables

Summary Statistics

	Model A (business telephones versus nonfarm employment)	Model B (business telephones versus nonfarm employment minus manufacturing)
Number of observations	36	36
R^2	0.37	0.48
F	Significant	Significant
Standard error	0.0069	0.0062
$a \times 10^{-3}$	5.29	3.24
[Std. error $\times 10^{-3}$]	[1.15]	[1.10]
b	0.26	0.38
[Std. error]	[0.06]	[0.07]
Durbin-Watson d	0.16	0.24
MAPE		
one-year-ahead	0.80%	0.60%
two-year-ahead	1.70%	1.70%

The end result is that even a simple model of the type discussed in this example can become a valuable forecasting tool. Because many business and government organizations forecast employment levels, a forecast of a business service, such as telephones, can be readily developed. The predictions obtained from the preliminary model provide a good starting point from which to start improving the forecast, if appropriate.

These model predictions can be (and probably should be) adjusted judgmentally given the shortcomings in the model and the business knowledge surrounding the particular application. It is important to note that even a model with one or more imperfections can be useful if the forecaster takes the time and effort to review the model on an ongoing basis and tries to understand its strengths and weaknesses.

12.3 SELECTING VARIABLES

Graphical methods can play a big role in assessing how the selection of independent variables affects the fitting process and how individual data values affect the least-square estimates of regression coefficients. In some applications, a forecaster may be faced with a relatively large set of independent variables that, on theoretical or practical grounds, should be considered of value in explaining a dependent variable. In such situations, it is clearly impossible to develop a meaningful regression relationship by incorporating all the variables of interest; it is necessary to reduce the set of independent variables for the regression without jeopardizing the usefulness of the model. At such times, the interpretation of the coefficients may be secondary to the need to explain the overall variability with a model.

The set of independent variables for the models fall into three groups:

- Those that for theoretical reasons should be included

- Those that are likely to be of practical benefit, such as proxy and dummy variables (dummy variables are categorical in nature—yes-no, on-off, war-peace—and generally take on the values of 0 or 1)

- Those that may have desirable intuitive or statistical properties

Given a class of models and model assumptions, the next step is to establish rules for accepting or rejecting variables. Generally, it is desirable to estimate lack of fit by an estimate of the expected sum of squared deviations of the fitted values from the true values.

 A good procedure for selecting variables is one that minimizes the lack-of-fit criterion.

For a linear regression model with a constant error variance of σ^2, the ideal quantity to minimize is $\Sigma \text{Var}(Y_i)$, which is σ^2 times the number of independent variables. The variance σ^2 can be estimated as

$$\hat{\sigma}^2 = \left[\sum (Y_i - \bar{Y})^2\right]/n - k - 1$$

where k is the number of independent variables and n is the number of observations.

There are other criteria available for choosing k. These include Mallows's C_p, Anscombe's $s^2/(n - k)$ (as modified by Tukey), and Allen's PRESS (prediction sum of squares). These criteria, which are discussed in greater detail in Draper and Smith (1998), Chatterjee et al. (2000), and Mosteller and Tukey (1977, Chap. 14), should be applied with care; they can serve as a useful guide in making a sensible choice of k.

There are a variety of algorithms for selecting subsets of independent variables. Most of these approaches, although widely used, suffer from a number of theoretical and practical difficulties. Among these are methods using all possible regressions, backward elimination, forward selection, and stepwise regression.

- The method of *all possible regressions* uses all possible regressions for k independent variables. Although everything is covered, the method can be expensive, time consuming, and usually unwarranted.

- The method of *backward elimination* begins with all variables in the model. A partial F test is used to calculate the contribution of each variable. The variable with the lowest F value is removed, and the process is repeated until all variables have F values greater than a preselected value. With this technique, the results for the full model are available and the procedure is cheaper than the all-possible-regressions procedure.

- The method of *forward selection* starts by selecting a variable that best explains the variation in the dependent variable by using the partial correlation coefficient. Using the residuals, a second variable is found that best explains the remaining variation in the dependent variable. The model is then reestimated with the new variable included. The calculations and selection procedure are repeated until no remaining variable has a partial correlation larger than a preselected value. Although this procedure is economical and leads to a continuous improvement in the model, it ignores the possible reduction in importance of earlier variables.

- The method of *stepwise regression* is similar to forward selection. At each step, however, the previously selected variables are examined with a partial F test to determine if any are now not contributing significantly. This method takes into account possible relationships between independent variables but requires more selection criteria. There are also weighted and resistant fitting variations to the stepwise regression method; however, these approaches should never be used without examining numerical and graphical outputs, such as the residual distributions and outliers, and reviewing regression coefficients at each step for possible changes in magnitudes and signs.

Regression by Stages

One approach for identifying variables that should be included in a model is to build a regression model in stages. Using Tukey's notation (Mosteller and Tukey, 1977) for convenience, let $Y_{.1}$ denote the residuals after fitting a model with X_1 only.

1. Instead of plotting the residuals $Y_{.1}$ against a new variable X_2, first regress X_2 on X_1 and denote with $X_{2.1}$ the residuals of X_2 that result after fitting X_1. Then $X_{2.1}$ represents the *additional* information in X_2 that is not already captured by X_1.

2. Plot $Y_{.1}$, against $X_{2.1}$; this will show whether X_2 has a strong apparent dependence with Y, given that X_1 is already in the model. In other words, this plot shows the relationship between the unexplained variation in Y and the additional information in X_2. In the case in which X_1 and X_2 are highly correlated with Y and with one another, the plot of $Y_{.1}$ versus $X_{2.1}$ will show little correlation, because X_2 adds little, given that X_1 is in the model.

3. Regress X_3 on X_1 and X_2; the residuals that result are denoted by $X_{3.12}$.

4. Plot the residuals of the model containing both X_1 and X_2 (thus, $Y_{.12}$) against $X_{3.12}$. The variables that show high dependence are included in the model, and those with little dependence are reserved for future use.

In general, let $X_{i.\text{REST}}$ denote the residuals of X_i on the rest of the independent variables. By regressing Y on $X_{i.\text{REST}}$, the same regression coefficient will result for X_i that results from the regression of Y on all the X values. The remaining regression coefficients are, of course, different. By displaying $Y_{i.\text{REST}}$ (the fit of Y on *all* X_i) against each $X_{i.\text{REST}}$, it is possible to see the impact of each variable and, in some cases, the impact of individual points on the estimate of the regression coefficient.

In order to examine how to combine or compare coefficients in different regressions, Tukey uses the minvar modification of the fitted coefficient b_i, denoted by $b_{i.\text{REST}}$. For each i, $b_{i.\text{REST}}$ is that linear combination of fitted regression coefficients, including unity times the ith coefficient b_i, which has minimum estimated variance. These coefficients can be compared with the estimated variance of b_i itself. As a set of ratios, they may point to a useful combination of independent variables and to important dependencies among estimated regression coefficients.

Application: Building a Cross-Sectional Model for Additional Telephone-Line Development

In this study, an analysis was performed to help identify the potential for increased sales of additional telephone lines within residences in 470 geographic areas. A requirement was that the model should incorporate local economic and demographic data in a formulation understandable and acceptable to the sales personnel responsible for stimulating demand. Areas with below-average development, as predicted by the model, would be candidates for future sales campaigns. Although the particular data set may not be relevant in today's telecom market, the analysis, nevertheless, brings up a number of important points about the modeling process.

Variable Selection

After experimentation with a variety of possible independent variables, several were ultimately selected on theoretical and statistical considerations. Median family income, adjusted for cost-of-living differences among the geographic areas, was an obvious candidate. Of the total work force in each area, the percentage engaged in white-collar employment was selected because these employees are generally intensive users of telephones on the job and tend to take their telephone habits home with them. Of all households, the percentage in which more than one automobile was owned was selected as an indicator of the propensity to consume rather than to save income. A number of other variables could also have served this purpose.

The Model-Building Process

As a first step in building a model, a plot was made of the percentage of additional-line development (additional telephone line per 100 residential access lines) by area. It showed a lack of symmetry in area distribution, particularly where percentages of extension telephone were high.

Scatter diagrams were plotted to investigate the relationships between the dependent and independent variables. Earlier in this chapter, we note that, in a multiple linear regression model, the inclusion of partial residual plots might be helpful, because they show the relationship between the dependent variable and each independent variable, given that all the variables have been entered into the model.

A regression model containing the three independent variables and the partial residual plots was examined. The partial residual plots associated with income, employment, and automobiles are shown in Exhibits 12.6, 12.7, and 12.8, respectively. In Exhibit 12.6, most of the points are concentrated in the $10,000–$18,000 range, with a relatively small number of high-leverage points above $20,000. Exhibits 12.7 and 12.8 show a more uniform distribution. In all three plots, the relationships appear to be positive and linear.

EXHIBIT 12.6   Plot of Partial Residuals versus the Income Variable for the Additional-Line Development Study

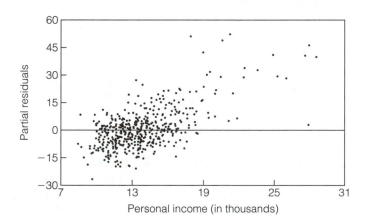

EXHIBIT 12.7   Plot of Partial Residuals versus the Employment Variable for the Additional-Line Development Study

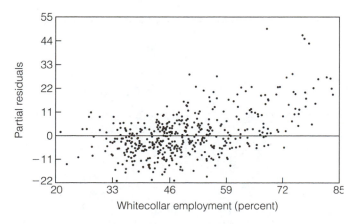

EXHIBIT 12.8   Plot
of Partial Residuals
versus the Automobile-
Ownership Variable
for the Additional-
Line Development
Study

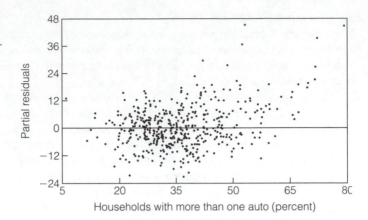

The constancy of variance and normality was checked for in the residuals, which completed the diagnostic residual analysis. Exhibit 12.9 plots the residuals versus the values that were predicted in the model, and Exhibit 12.10 shows the normal probability plot of the residuals of this model. Exhibit 12.9 reveals an apparent problem of increasing residual variance, and most of the residuals associated with development above 85% are positive. The normal probability plot in Exhibit 12.10 shows that the high positive residuals form a longer tail than is present in the normal curve.

In most cases, we can stop at this point, because the departures from the assumptions of the model are not too severe. However, in this case we decided to go further to see whether a more constant residual variance and a more linear probability plot (one closer to normality) can be obtained and whether a better partial residual plot against income can be obtained by transforming the variables. For example, a transformation of the dependent variable might improve the normal probability plot.

Exhibit 12.11 shows a plot of the residuals against the predictions for the model after a logarithmic transformation of the percentage development was taken. In this case, an obvious pattern of decreasing variance results, leading us to conclude that the logarithmic transformation is inappropriate. A similar plot (not

EXHIBIT 12.9
Scatter Plot of
Residuals versus the
Predicted Values for
the Additional-Line
Development Study

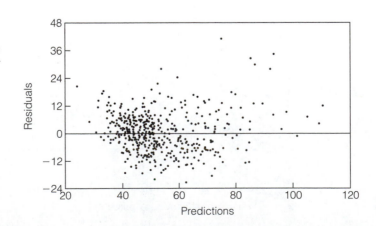

EXHIBIT 12.10
Normal Probability
Plot of the
Residuals for the
Additional-Line
Development Study

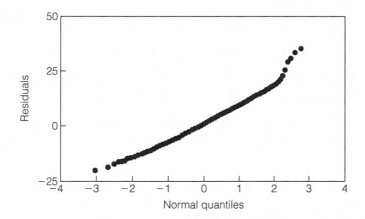

shown) of the residuals versus the predictions for an alternative model was created in which a square-root transformation was applied to the dependent variable and a logarithmic transformation was applied to the income variable. The logarithmic transformation was taken to obtain a linear relationship in the partial residual plot between the two time series. This plot is similar to Exhibit 12.10, but is somewhat improved in that it shows a more uniform residual variance over the range of predictions. In addition, the residuals are not as concentrated at the low end of the percentage-development scale.

Model Summary

In the early stages of the analysis, we failed to notice that, of the 473 geographic areas examined, there were three areas for which no values (observations) were given for the dependent variable, yet there were values for the independent variables for these areas. The software assumed the missing values to be zero, by default, and proceeded with the analysis. Later, a plot of residuals against the predictions offered by the final model (in which zeroes appeared as observed values for the three areas) demonstrated that the three residuals were in fact so large that they dominated all

EXHIBIT 12.11
Time Plot of
Residuals versus the
Predictions after
Taking Logarithms of
the Dependent
Variable for the
Additional-Line
Development Study

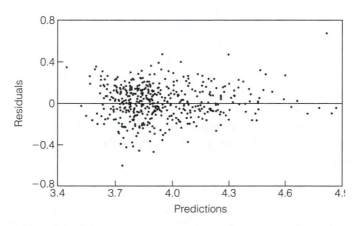

EXHIBIT 12.12
Comparison of OLS
and Robust
Regression Results
for the Final Model
with and without
Outliers

Variable	With Outliers			Without Outliers		
	Coefficient	OLS (SD)	Robust Coefficient	Coefficient	OLS (SD)	Robust Coefficient
Interest	−13.33	−2.26	−13.37	−9.69	−3.47	−13.37
Income	1.93	−0.26	1.94	1.55	−0.39	1.94
White-collar employment	0.032	−0.004	0.031	0.03	−0.006	0.031
Households > 1 auto	0.017	−0.004	0.015	0.017	−0.005	0.016

others and had the effect of lowering the R^2 statistic from 0.69 to 0.43. We then made a probability plot of the residuals against the normal curve, and it also verified the appearance of the three outliers.

A robust regression model could offer some protection against outliers, such as those just described. In Exhibit 12.12, the left-hand part of the table shows that OLS and robust regression analyses yield almost identical results, with no extreme values and approximately normal residuals. A comparison of the left-hand and right-hand sides shows that the outliers distorted the OLS results, but not the robust results. Recall that only three of the 473 observations were extreme, yet these three altered the income coefficient and the constant term significantly. This suggests that a difference between OLS and robust regression coefficients of approximately one standard error is sufficient cause to review the OLS model and the original data in much greater detail.

12.4 INDICATORS FOR QUALITATIVE VARIABLES

There are times a forecaster must apply special treatments to a regression model because the variables represent a specific statistical problem. Solutions to these problems include the use of indicator variables for qualitative factors and for seasonal and outlier adjustment and the use of lagged variables to describe effects that unfold over time.

Use of Indicator Variables

Indicator variables, better known as dummy variables, are useful for extending the application of independent variables to represent various special effects or factors: one-time or fixed-duration qualitative factors or effects, such as wars, strikes, and weather conditions; significant differences in intercepts or slopes for different consumer attributes, such as race and gender; discontinuities related to changes in qualitative factors; seasonal variation; the effects of outliers; and the need to let the intercept or slope coefficients vary over different cross sections or time periods.

Qualitative Factors

In addition to quantifiable variables of the type discussed in earlier chapters, the dependent variable may be influenced by variables that are essentially qualitative in nature. Changes in government or public policy, wars, strikes, and weather patterns are examples of factors that are either nonquantifiable or very difficult to quantify. However, the presence of such factors can influence consumer demand for products and services.

 Dummy, or indicator variables, are used to indicate the existence or absence of an attribute or condition of a variable.

For example, suppose that for any given income level, the sales Y of a product to females exceed the sales of the same product to males. Also, suppose that the rate of change of sales relative to changes in income is the same for males and females. A dummy variable can be included in the sales equation to account for gender. Let $D = 0$ for sales to males and $D = 1$ for sales to females. Then

$$Y_i = \beta_0 + \beta_1 D_i + \beta_2 (\text{Income})_i + \varepsilon_t$$

For this example, the base or control condition will be "males" ($D_i = 0$). The prediction Y of sales to males is therefore

$$Y_i(\text{Males}) = \beta_0 + \beta_2 (\text{Income})_i \quad \text{for } D_i = 0$$

and the prediction of sales to females is

$$Y_i(\text{Females}) = \beta_0 + \beta_1 + \beta_2 (\text{Income})_i \quad \text{for } D_i = 1$$

The coefficient β_1 is called the differential intercept coefficient. It indicates the amount by which sales to females exceed sales to males at a given level of income. A t test can be used to determine whether β_1 is significantly different from zero. Exhibit 12.13 shows a plot of the two regression lines for this example.

Similarly, the mean sales of a product in one geographical area may show the *same* rate of change relative to an economic variable that the sales in another area shows; yet the total sales for each state may be *different*.

EXHIBIT 12.13
Plot of Two
Regression Lines

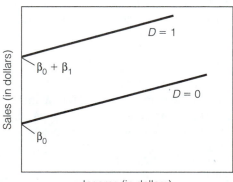

Income (in dollars)

Models that combine both quantitative and qualitative variables, as both of the fore-going examples do, are called analysis-of-covariance models.

We must always be careful to introduce one less dummy variable than the number of categories represented by the qualitative variable. In the last case, the two categories (males, females) can be represented by one dummy variable.

Dummy Variables for Different Slopes and Intercepts

In our example, suppose we want to know whether the intercepts and slopes are different for females and males. This can be tested in a regression model of the form

$$Y_i = \beta_0 + \beta_1 D + \beta_2 (\text{Income})_i + \beta_3 \, D(\text{Income})_i + \varepsilon_i$$

Then

$$Y_i(\text{Males}) = \beta_0 + \beta_2 (\text{Income})_i \quad \text{for } D = 0$$

Likewise,

$$Y_i(\text{Females}) = \beta_0 + \beta_1 + (\beta_1 + \beta_3)(\text{Income})_i \quad \text{for } D = 1$$

It should be noted that the test for the differences between intercepts and slope for males and females are significance tests to see whether β_1 and β_3 are different from zero.

The use of a dummy variable in the additive form allows us to identify differences in intercepts. The introduction of the dummy variable in the multiplicative form allows us to identify different slope coefficients.

Measuring Discontinuities

A change in a government policy or a change in a price may alter the trend of a revenue series. In the case of a price change, the preferable course of action is to develop a price index. Suppose that for any of a variety of reasons, such as lack of time or data, this is not possible. A dummy variable may be introduced into the model as follows. Let

$$Y_t = \beta_0 + \beta_1 X_t + \beta_2 D_t + \varepsilon_t$$

where $D_t = 0$ for $t < T^*$ and $D_t = 1$ for $t \geq T^*$; T^* is the time of the policy or price change; D_t is a dummy variable with a value of 0 for all time less than T^* and a value of 1 for all time greater than or equal to T^*; and X_t is an explanatory variable.

In this example, the predicted values of revenues are:

$$Y_t = \beta_0 + \beta_1 X_t \quad \text{if } t < T^*$$

and

$$Y_t = (\beta_0 + \beta_2) + \beta_1 X_t \quad \text{if } t \geq T^*$$

Another situation that often occurs involves a "yes-no" or "on-off" possibility. For example, the demand for business telephones is very strongly affected by presidential and congressional elections, held in even-numbered years. Storefront campaign offices and candidates' headquarters are established; there is a large increase in the demand for business telephone lines in September and October; and then, in November, the telephones that were related to the election campaign are disconnected. In odd-numbered years, local political elections are held; local politicians do not have the financial resources to establish as many campaign offices. Consequently, the impact on telephone demand is not as great in odd-numbered years. Aside from this, we can be sure that elections occur in odd- and even-numbered years alike. Therefore, it is possible to use a dummy variable that assumes that one-half the telephones access-line gain attributable to an election occurs in September, the other half occurs in October, and it all disappears in November. In this case, for a given year, the important consideration in assessing election-influenced telecommunications demand is whether there is or is not an election. This is what is meant by a "yes-no," "on-off," or categorical variable. The categorical variable does not continue to take on different values for an extended time and, in this sense, is not quantitative.

Adjusting for Seasonal Effects

The forecaster may decide to use dummy variables to account for seasonality in quarterly data. In these situations, a dummy variable is omitted for one of the quarters. The interpretation of the coefficients is most revealing when the omitted quarter is the one that stands out as being different. For housing starts, it makes sense to not exclude a dummy variable for the first quarter. The seasonal effect of the first quarter is then captured by the constant term in the regression equation

$$Y_t = \beta_0 + \beta_1 D_{1t} + \beta_2 D_{2t} + \beta_3 D_{3t} + \beta_4 X_{1t} + \beta_5 X_{2t} + \varepsilon_t$$

where $D_{1t} = 1$ for the second quarter, $D_{2t} = 1$ for the third quarter, and $D_{3t} = 1$ for the fourth quarter. For monthly data, eleven dummy variables would be used (February through December, for example), and then the January seasonality would be captured by the constant term in the regression equation. On the other hand, the fourth quarter should be omitted for most retail sales time series to show the difference between the stronger quarter and the others.

It is possible that only 1 month or quarter has a significant seasonal pattern. However, this is not generally the case in forecasting the demand for sales of telephones (for instance): Usually each month or quarter has a unique seasonal pattern.

 Dummy variables for seasonal adjustment may be of limited value if the season changes over time because the dummy variable approach assumes a constant seasonal pattern.

A useful attribute of the X-11/X-12 computer programs from the U.S. Bureau of Census and X-11-ARIMA/88 seasonal adjustment program from Statistics Canada is that they may be used with data that do not show a constant seasonal pattern. In addition, most series do seem to exhibit changing seasonality over time. The advantage of using dummy variables in a regression is that seasonality and trend-cycle can be estimated simultaneously, in a one-step process; otherwise, a two-step process is necessary: (1) seasonal adjustment of data and (2) performance of the regression. The current ease of using seasonal adjustment computer programs has minimized the one-step advantage of the dummy variable approach. In the one-step process, the dummy variable provides an index of average seasonality that can then be tested for significance by using the t test for an individual dummy variable or the incremental F test for a group of dummy variables.

Eliminating the Effects of Outliers

To illustrate the use of dummy variables to eliminate the effect of outliers, consider the following situation. Let us say that in our data for telephone access-line gain there is a year in which there was a telephone-company strike in April and May and the disconnections of telephone service because of nonpayment of telephone bills were not reported for those months, distorting the annual totals. When we choose a model for predicting telephone access-line gain, we can decide to incorporate a dummy variable into the model. The dummy variable, in this case, is a variable that can generally be set equal to 0 for all observations except an unusual event or outlier. Thus, the values for the dummy variable are equal to 0 for all years except the strike year, when it has a value of 1. Because the dummy variable equals 0 for all periods except that in which the outlier appears, the dummy variable explains the outlier perfectly. That is, the predicted value equals the actual value for the outlier.

The use of such a dummy variable tends to reduce the estimated standard deviation of the predicted errors artificially because what is, in fact, a period of wide variation has been assigned a predicted error of 0. For this reason, we recommend that dummy variables be used very sparingly for outlier correction. They tend to result in a model with a higher R^2 statistic than can perhaps be justified: The model will appear to explain more of the variation than can be attributed to the independent variables. In some cases, the outliers may be a result of random events, and we need to be aware of this. In the case of outliers caused by nonexistent values, it is usually preferable to estimate a replacement value based on the circumstances that existed at the time the outlier occurred.

Be especially cautious when dummy variables and lagged dependent variables occur in the same equation. For example, a dummy variable might be used to correct for a strike that could have a large negative impact on the quantity of a product sold. However, it is then necessary to adjust the value for the subsequent period, or the value of the lagged dependent variable will drive next period's predicted value too low. A preferable alternative is to adjust the original series for the strike effect (if lagged dependent variables are included in the model).

In some circumstances, robust regression techniques may offer a method for estimating regression coefficients so that the results are not distorted by a few outlying values. The variability in the data will not be understated, and the very large residuals will readily indicate that the model, as presently specified, is incapable of explaining the unusual events.

On the other hand, residuals in a model in which dummy variables have been used for outliers suggest that unusual events be perfectly estimated. A robust regression alternative has considerable appeal from a forecasting viewpoint. Because it will not understate the variability in the data, there is less of a tendency to expect a greater degree of accuracy in the forecast period than can be achieved over the fitted period.

At times it may be necessary to introduce dummy variables because it is almost impossible to estimate a replacement value for a missing or extreme data value: There may be too many unusual events occurring at the same time. For example, a strike may coincide with the introduction of wage-price controls or an oil embargo. It would be extremely difficult to determine the demand for a product or service had there been no strike because too many other variables would also have changed.

Moreover, if we attempt to build a model to predict gasoline consumption, based on data that includes the oil embargo period in the 1970s, we will probably incur problems. Because the period encompasses a fuel-supply shortage, the actual consumption in that period is more a function of supply than of demand. Because this situation did not exist in any of the historical data, it might be better to use a robust regression, leaving out the data for the oil embargo, or to use a dummy variable to account for that time period.

12.5 ANALYZING RESIDUALS

What steps should the forecaster follow to sort out all these different ways of analyzing residuals?

1. Review the summary statistics (F, t, incremental F, and R^2 statistics). If the model passes these tests, the residual analysis can begin. Review a residual plot for constancy of variance among residuals; patterns of increasing dispersion with increasing magnitude, as mentioned earlier, suggest that logarithmic or square-root transformations of the dependent variable should be made as a first attempt. Then reestimate the model to accomplish this, and plot the residuals once again against predicted values.

A recommended sequence of residual analysis begins with a plot of the residuals against the predicted values of the dependent variable.

2. Once a satisfactory transformation of the dependent variable has been obtained, if one is indeed required, plot the residuals versus each independent variable. At this time:

   ■ Delete variables with low correlation (< 0.2) and reestimate the model.

   ■ Transform independent variables that exhibit nonlinear relationships (one at a time). After each transformation, reestimate the model and generate a new set of partial residual plots.

   ■ Perform additional analyses if all the plots show linear and significant (not horizontal) relationships.

3. For time series models, plot the residuals and a correlogram of the residuals. Test for serially correlated residuals. The plots may suggest additional variables for consideration. The techniques discussed in the next section may be required if the serial correlation problem remains.

4. Generate percentiles of the residuals versus the percentiles of a standard normal distribution. This will highlight potential outliers (review each end carefully) and indicate if there are departures from normally distributed residuals. Outliers should be investigated and replaced, if replacement is appropriate. Transformations of the dependent variable may be required; alternatively, robust regression may be appropriate if normally distributed residuals cannot be obtained.

Although linear regression models based on nonlinearly related data may appear to give acceptable summary statistics, the inferences drawn from such models can be erroneous and misleading. Residual analysis is an effective tool for graphically demonstrating departures from model assumptions. When looking at regression residuals, keep in mind that:

■ A residual plot of constant variance with no visible pattern is consistent with the basic assumptions of the linear least-squares regression model. If the residuals are also normally distributed, a variety of significance tests can be performed. Also, the run test can be applied to the test for nonrandomness.

■ A residual pattern of increasing dispersion may suggest the need to transform one or more variables. The logarithmic and square-root transformations are the most commonly used to solve this problem. A plot of the residuals versus the dependent variable also highlights the need for transformations.

■ The normal probability plot is a convenient way to decide whether the residuals are normally distributed. In addition, outliers are readily detected at one or the other end of the plot.

12.6 THE NEED FOR ROBUSTNESS IN REGRESSION

In Chapter 11, we show a close relationship between the correlation coefficient as a measure of linear association and the simple linear regression model as a confirmation (goodness-of-fit) of that linearity. Unfortunately, in multiple linear regression models, that sort of simple relationship does not exist. However, correlation coefficients can still play a useful role in sorting through variables that might be usefully paired with the dependent variable in a regression model.

Why Robust Regression?

In least-squares estimation, regression coefficients are derived by minimizing the sum of the squares of the residuals (Residual = Data − Fit). In the minimization process, all residuals are given equal weight. Experience has shown, however, that outliers can have an unusually large influence on least-square estimators. That is, the outliers pull the least-squares fit toward them too much, and the resulting examination of residuals may be misleading because the residuals corresponding to the outliers look smaller than the residuals from a robust fit—they do not appear like outliers.

Robust methods have been created to modify least-squares regression procedures so that outliers have much less influence on the final estimate.

Outliers and other sources of unusual data values should never be ignored. The robust procedures just mentioned operate to reduce their impact on the regression estimates. Other modern forecasting techniques, however, tackle outlier and usual data problems directly—through intervention analysis—rather than through adjustments of a computational algorithm.

M-Estimators

A family of robust estimators, called M-estimators, is obtained by minimizing a specified function of the residuals. Alternate forms of the function produce the various M-estimators. Generally, the estimates are computed by iterated weighted least squares.

One such function $\xi(.)$ of the residuals takes the form:

$$\xi(e) = 0.5e^2 \qquad \text{if } |e| \leq c = Ks$$
$$= c|e| - 0.5e^2 \quad \text{if } |e| > c = Ks$$

where s is a scale estimate, such as UMdAD, the (unbiased) median absolute deviation of the residuals from the median, divided by 0.6745 (see Chapter 3). If the data are normally distributed and the number of observations is large, the divisor 0.6745 is used because then s approximates the standard deviation of the normal distribution. Usually the sample standard deviation $\hat{\sigma}$ is not used as a value because it itself is influenced too much by outliers and thus is not resistant against them. The constant K is chosen to obtain a desired level of efficiency (as compared to least-square regression if the data are normal), and it is often set to between 1 and 2. If K is sufficiently large, the M-estimate will be equivalent to ordinary least-squares estimates.

Minimizing $\xi(e)$ yields an estimate of the regression coefficients; the minimization requires Huber weights, defined by;

$$W_i \begin{cases} = 1 & \text{if } |e_i| \leq Ks \\ = Ks/e_i & \text{if } |e_i| > Ks \end{cases}$$

The statistic s approximates the standard deviation of a normal distribution, and the constant K is chosen as some number close to 1.5 (based on empirical evidence). The iterative procedure has the following steps:

1. Obtain an initial estimate of the regression coefficients. The initial estimates can be obtained in a number of ways such as by OLS.

2. Compute residuals and calculate $Ks = K$ (UMdAD).

3. Compute the weights and perform the method of weighted least squares.

4. Repeat steps 1 through 3 until convergence or a reasonable number of iterations has been performed.

To distinguish robust regression coefficients from OLS coefficients, the robust coefficients for a simple linear regression, denoted by $\beta_0{}^*$ and $\beta_1{}^*$, are calculated using the weighted least squares.

A second M-estimator, called the bisquare estimator, gives zero weight to data whose residuals are quite far from zero (Mosteller and Tukey, 1977). The bisquare weighting function is discussed in Chapter 7 in connection with seasonal smoothing and is defined by:

$$W_i \begin{cases} = 0 & \text{if } |e_i| > Ks \\ = 1 - (e_i/Ks)^2]^2 & \text{if } |e_i| \le Ks \end{cases}$$

It is worth noting that the bisquare-weighting scheme is more severe than the Huber scheme. In the bisquare scheme, all data for which $|e_i| \le Ks$ *will* have a weight less than 1. Data having weights greater than 0.9 are not considered extreme, data with weights less than 0.5 are regarded as extreme, and data with zero weight are, of course, ignored.

To counteract the impact of outliers, the bisquare estimator gives zero weight to data whose residuals are quite far from zero.

A recommended robust regression procedure begins with OLS estimates of the parameters. This is followed by several iterations of bisquare-weighted least squares. To avoid the potential problem of finding local minima, which can result from "bad" initial estimates, we recommend starting the bisquare procedure with a few iterations of the Huber scheme. In practice, we have found that one iteration of the Huber scheme followed by a couple of iterations of the bisquare scheme works quite well (Levenbach, 1982).

Calculating M-Estimates

It is instructive to take a closer look at the weighted least-squares calculations of M-estimates. These estimates can be substantially different when the data contain one or more outliers. For simplicity, consider the simple time series {2.5, 4.0, 5.5, 7.0, 8.5, 10.0, 11.5, 13.0, 14.5, 3.0}, which is of the form $Y = 1.0 + 1.5\,X$ for the first nine points if X represents 1, 2, . . . , 10. Notice that the last value (3.0) is an outlier according to the model.

One iteration of a robust solution for these data, using Huber weights ($K = 1$), is given in Exhibit 12.14. This is analogous to downweighting residuals that are greater than one standard error from their central measure. Notice that the initial estimates are obtained from the OLS solution $\hat{Y}_{(OLS)} = 3.60 + 0.79\,X$. The residuals are

EXHIBIT 12.14
One Iteration of Sample Data, Using Huber Weights ($K = 1$) for Fitting $Y = 1.0 + 1.5X$

(1) X	(2) Y	(3) $e_i = Y_i - \hat{Y}_{(OLS)}$	(4) $|e_i - e_M|$	(5) Huber weights ($K = 1$)
1	2.5	−1.89	2.49	1
2	4	−1.18	1.78	1
3	5.5	−0.47	1.07	1
4	7	9.24	0.36	1
5	8.5	0.95	0.36	1
6	10	1.66	1.07	1
7	11.5	2.37	1.78	1
8	13	3.08	2.49	0.86
9	14.5	3.79	3.2	0.7
10	3	−8.5	9.1	0.31

$\hat{Y}_{(OLS)} = 3.60 + 0.79X$
Median residual e_M (median of column 3) = 0.595
MdAD (median of column 4) = 1.78
$s = $ UMAD = MdAD/0.6745 = 2.64

computed in column 3. Column 4 shows the absolute deviations of the residuals from the median residual ($= 0.595$). The median value in column 4 is the MdAD statistic. Comparing column 3 with the UMdAD statistic ($=$ MdAD/0.6745) derives column 5 ($= 2.64$), using the definition of the Huber weight function. The weight W is 1.0 if the absolute value of the residual in column 3 is less than 2.64; otherwise, the weight is 2.64 divided by the absolute value of the residual. It is evident that the last three points receive a weight less than 1 for this iteration.

The first 13 Huber iterations using $K = 2$ are summarized in Exhibit 12.15. The coefficients are calculated by using the weighted least-squares formula set forth in the preceding section. It can be seen from Exhibit 12.15 that only the last point (the weights column) received a weighting less than 1. With each iteration, the slope of the line gradually increases, as can be seen from the b_1^* column. Notice that in this artificial example, the weight associated with the outlier becomes 0. With real data, however, the Huber weights rarely get smaller than 0.2.

EXHIBIT 12.15
Thirteen Iterations of the Huber ($K = 2$) Solution to the Fit in which the Last Point (W_{10}) is Downweighted (weights are shown only for point W_{10}; weights for W_1 through W_9 are 1 for all iterations)

Iteration	Regression coefficients b_0^*	b_1^*	Weights
1	2.85	1	0.62
2	2.26	1.16	0.38
3	1.83	1.27	0.24
4	1.54	1.35	0.15
5	1.34	1.41	0.09
6	1.21	1.44	0.06
7	1.13	1.46	0.03
8	1.08	1.48	0.02
9	1.05	1.49	0.01
10	1.03	1.49	0.01
11	1.02	1.49	0
12	1.01	1.5	0
13	1.01	1.5	0

EXHIBIT 12.16 Calculation Using Bisquare Weights ($K = 4.2$) for the First Iteration of Sample Data; Bisquare Weights $W_i = [1 - (e_i/Ks)^2]^2$; $s = \text{UMAD} = 2.64$ (residuals are given in Exhibit 12.14, column 3)

Y_i	2.5	4	5.5	7	8.5	10	11.5	13	14.5	3
W_i	0.943	0.977	0.996	0.999	0.985	0.956	0.911	0.852	0.78	0.17

In Exhibit 12.16, a robust solution using bisquare weights ($K = 4.2$) is shown for the first iteration. The bisquare-weighting scheme downweights all but the third (W_3) and fourth (W_4) points in the first iteration (Exhibit 12.17). We can see by looking at the coefficient estimates that the calculations converge after only three iterations to a final fit of the form $\hat{Y}_{BS} = 1.0 + 1.5X$.

EXHIBIT 12.17 Bisquare Solutions to the Fit in Which All but the Fifth and Sixth Values are Downweighted

Regression coefficient ($b*$) and weight (W)		Iteration		
	OLS	1	2	3
b_0*	3.6	1.51	1	1
b_1*	0.79	1.36	1.5	1.5
W_1	1	0.94	0.94	0.83
W_2	1	0.97	0.98	0.91
W_3	1	1	1	0.94
W_4	1	1	1	0.99
W_5	1	0.98	0.98	1
W_6	1	0.95	0.95	1
W_7	1	0.9	0.9	0.99
W_8	1	0.84	0.83	0.94
W_9	1	0.76	0.75	0.91
W_{10}	1	0.12	0.00	0.00

12.7 MULTIPLE REGRESSION CHECKLIST

The following checklist can be used as a scorecard to help identify gaps in the forecasting process that will need attention. It can be scored or color-coded on three levels: Green = YES, Yellow = SOMEWHAT, and Red = NO

_____ Is the relationship between the variables linear?

_____ Have linearizing transformations been tried?

_____ What is the correlation structure among the independent variables?

_____ Have seasonal and/or trend influences been identified and removed?

_____ Have outliers been identified and replaced when appropriate?

_____ Do the residuals from the model appear to be random?

_____ Are any changes in the variance apparent (is there heteroscedasticity)?

_____ Are there any other unusual patterns in the residuals, such as cycles or cup-shaped or trending patterns?

_____ Have F tests for overall significance been reviewed?

_____ Do the t statistics indicate any unusual relationships or problem variables?

_____ Can the coefficients be appropriately interpreted?

_____ Have forecast tests been made?

12.8 HOW TO FORECAST WITH QUALITATIVE VARIABLES

Most business forecasters use regression models based strictly on quantitative independent variables such as dollar expenditures or numbers of customers. However, we can often improve forecast accuracy by adding qualitative independent variables to our regression models. Examples include the location of stores, the condition of real estate property, the season of the year, the set of features in a product, and the age or income categories in a sample of customers.

Qualitative variables let us analyze the effects of information that cannot be measured on a continuous scale.

Qualitative variables are especially easy to use in spreadsheet regression models. The basic idea is to devise independent variables using binary formulas that return only one of two values, 0 or 1. Different combinations of the 0-1 values classify qualitative information to help predict the dependent variable, the one we are trying to forecast.

Modeling with a Single Qualitative Variable

Several examples will help illustrate using qualitative variables. Let us start with a problem in which a single qualitative variable is necessary. The data in Exhibit 12.18 are from a small chain of fast-food restaurants that wanted to predict weekly sales volume based on store location (street or mall) and the population of each store's trading area. Both sales and population data are stated in thousands. One way to make predictions is to run separate regressions for malls and street locations. But this is the hard way. We can make better predictions by consolidating all the data in one model.

Step 1. Setting Up the Worksheet
The first step is to enter formulas in column C that return the value 1 for a mall location and 0 otherwise. Next, execute the regression commands. Include both the mall indicators and the population data. The actual weekly sales in thousands are in column E, and the corresponding fitted values are in column F.

EXHIBIT 12.18
Linear Regression
Model with One
Qualitative Variable:
Analysis of
Restaurant Sales

Regression output:

Constant	9.142
Std error of Y estimate	0.782
R^2	0.997
Number of observations	12
Degrees of freedom	9
X coefficient(s)	10.765 (Male) 0.084 (Area Population)
Std error of coefficient	0.455 0.002

A	B	C	D	E	F
		X_1	X_2	Y Actual	Y Estimate
1	Street	0	151	22	21.8
2	Mall	1	220	39	38.3
3	Street	0	53	14	13.6
4	Mall	1	112	30	29.3
5	Mall	1	332	47	47.7
6	Mall	1	398	54	53.2
7	Street	0	241	29	29.3
8	Mall	1	60	24	24.9
9	Street	0	104	17	17.9
10	Street	0	153	23	22.0
11	Mall	1	162	33	33.5
12	Street	0	410	43	43.5

Step 2. Performing the Analysis

The regression equation is

$$\text{Estimated weekly sales} = 9.143 + 10.765 \times X_1 \text{ (or mall indicator)}$$
$$+ 0.084 \times X_2 \text{ (or area population)}$$

The coefficient 9.143 is the regression constant or intercept value, the base sales estimate when the mall indicator is 0 (when the store has a street location). The mall indicator coefficient is 10.765. We interpret this figure as the average increase in weekly sales produced by a mall location compared to a street location. The area population coefficient is 0.084, meaning that sales should increase by an average of $84 for each thousand population, regardless of store location.

When there is a single qualitative variable, the estimated values in column F can be plotted as two parallel lines (as graphed in Exhibit 12.13). When X_1 is 1, the mall location applies, and the result is the upper regression line. When X_1 is 0, the street location applies, and the result is the lower line. Thus, the actual effect of a qualitative variable is to shift the constant or intercept value. In this example, the constant shifts upward by 10.765 units, the value of the X_1 coefficient. Every point on the upper line is 10.765 units greater than the corresponding point on the lower line.

Modeling with Two Qualitative Variables

Now let us look at the problem in Exhibit 12.19, where two qualitative variables are needed. The aim is to develop an equation to predict the selling price of apartment complexes based on the condition of the property and the number of units in each

EXHIBIT 12.19
Linear Regression
Model with Two
Qualitative Variables:
Analysis of
Apartment-Complex
Sales Prices

Regression output:

Constant	11,357
Std error of Y estimate	11,485
R^2	0.995
Number of observations	13
Degrees of freedom	9
X Coefficient(s)	125,855 (excellent) 81,322 (good) 25,614 (units)
Std error of coefficients	8,160 7,712 637

A	B	C	D	E	F	G
		X_1	X_2	X_3	Y Actual	Y Estimate
Property number	Condition code	Excellent indicator	Good indicator	Number of units	Selling price	Selling price
158	A	1	0	4	230,250	239,668
509	A	1	0	6	295,000	290,896
118	A	1	0	8	345,800	342,124
973	A	1	0	18	599,900	598,263
300	B	0	1	4	185,500	195,135
725	B	0	1	10	335,750	348,819
28	B	0	1	12	419,900	400,047
172	B	0	1	14	449,900	451,275
133	B	0	1	14	455,500	451,275
661	C	0	0	4	118,500	113,813
760	C	0	0	5	143,900	139,427
980	C	0	0	14	377,700	369,952
795	C	0	0	18	455,500	472,408

complex. The condition information is qualitative in nature, with A = excellent, B = good, and C = fair. Again, we could run separate regressions for each condition, but there is an easier way. Formulas in columns C and D convert the condition codes to 0-1 variables. If the property is in excellent condition, column C returns a 1; otherwise it returns a 0. If the property is in good condition, column D returns a 1 and otherwise a 0. We could use a third indicator for fair condition, but it would be redundant; fair condition is already represented by $X_1 = 0$ and $X_2 = 0$.

The regression equation is:

$$\text{Estimated selling price} = 11,357 + 125,855 \times X_1 \text{ (or excellent indicator)}$$
$$+ 81,322 \times X_2 \text{ (or good indicator)}$$
$$+ 25,613 \times X_3 \text{ (or number of units)}$$

The value of the regression constant or intercept value is 11,357. This value is the base price when both X_1 and X_2 are 0, which is the base price for property in fair condition, regardless of the number of units in the complex. Property in excellent condition sells for a premium of 125,855, the X_1 coefficient, compared to property in fair condition. Property in good condition sells for a premium of 81,322, the X_2 coefficient, compared to property in fair condition. The X_3 coefficient, 25,613, is the amount that the price increases for each unit in the complex.

When there are two qualitative variables, the regression equation can always be plotted as three parallel lines (as graphed in Exhibit 12.13). Again, the effect of the qualitative variables is to shift the constant or intercept values.

Modeling with Three Qualitative Variables

When there are three qualitative variables, the regression equation can always be plotted as four parallel lines. Again, the effect of the qualitative variables is to shift the constant or intercept values.

Seasonal patterns complicate forecasting. These patterns can be removed from the data before forecasting by performing a seasonal adjustment. If the range of seasonal fluctuation each year is relatively constant, that is the seasonal pattern is additive, we can also use qualitative variables to handle seasonality within the regression model itself.

An example is given in Exhibit 12.20, where a model is developed to predict the demand for plaster casts at a ski resort. An obvious independent variable is the number

EXHIBIT 12.20   Linear Regression Model with Three Qualitative Variables:
Analysis of Demand for Plaster Casts at a Ski Resort

Regression output:

Constant	14.22			
Std error of Y estimate	2.57			
R^2	0.99			
Number of observations	24			
Degrees of freedom	19			
X coefficient(s)	0.99	6.44	-6.44	-30.59
Std error of coefficient	0.02	1.50	1.49	1.48

A	B	C	D	E	F	G	
		X_1	X_2	X_3	X_4	Y	Y estimate
		Visitors	Winter	Spring	Summer	Number	Number
Period	Year Quarter	(000s)	indicator	indicator	indicator	of casts	of casts
1	1975 Winter	33.63	1	0	0	53	54.0
2	Spring	36.46	0	1	0	41	44.0
3	Summer	41.18	0	0	1	24	24.5
4	Fall	43.16	0	0	0	57	57.0
5	1976 Winter	46.45	1	0	0	70	66.8
6	Spring	50.63	0	1	0	60	58.0
7	Summer	54.41	0	0	1	41	37.6
8	Fall	58.66	0	0	0	77	72.4
9	1977 Winter	62.52	1	0	0	81	82.7
10	Spring	65.55	0	1	0	70	72.8
11	Summer	69.62	0	0	1	50	52.7
12	Fall	72.92	0	0	0	87	86.6
13	1978 Winter	74.64	1	0	0	94	94.7
14	Spring	80.31	0	1	0	86	87.5
15	Summer	80.97	0	0	1	64	64.0
16	Fall	87.75	0	0	0	99	101.3
17	1979 Winter	88.07	1	0	0	109	108.0
18	Spring	94.00	0	1	0	101	101.0
19	Summer	96.16	0	0	1	77	79.0
20	Fall	96.98	0	0	0	110	110.4
21	1980 Winter	103.90	1	0	0	123	123.8
22	Spring	107.77	0	1	0	120	114.7
23	Summer	110.42	0	0	1	95	93.2
24	Fall	114.91	0	0	0	126	128.2

of visitors to the resort. This number is known well in advance because demand exceeds supply and reservations are required. A regression using only the number of visitors to predict the demand of casts (not shown) gives relatively poor results. The problem is that more visitors ski during the fall and winter, so the resort needs variables to take the season of the year into account. In column E, formulas return 1 during the winter, quarter and 0 otherwise. Similar formulas in columns F and G mark the spring and summer quarters. We do not need an indicator for the fall quarter because this is represented by $X_2 = 0$, $X_3 = 0$, and $X_4 = 0$.

The regression equation is:

$$\text{Estimated demand for casts} = 14.22 + 0.99 \times X_1 \text{ (or number of visitors)}$$
$$+6.44 \times X_2 \text{ (or winter indicator)}$$
$$-6.44 \times X_3 \text{ (or spring indicator)}$$
$$-30.59 \times X_4 \text{ (or summer indicator)}$$

SUMMARY

This chapter has explained the normal multiple linear regression model and its assumptions and the interpretation of additional summary statistics. It has provided an overview of common pitfalls to be avoided in building multiple linear regression models (detailed discussions of the pitfalls are presented in the appropriate chapters) and also an overview of regression models that automatically select subsets of independent variables (exercise caution when using automated variable selection programs). We have also discussed modeling assumptions implicit in formulating a forecasting model, including robustness considerations; the importance of the availability of relevant and appropriate independent variables for both the historical and future time periods; and the importance of the interpretation of the results in the light of the assumptions.

Least-squares estimation is sometimes inappropriate because it gives equal weight to outliers, thereby distorting the fit to the bulk of the data. In these circumstances, a robust regression serves as an alternative to the OLS regression. Differences between the two kinds of results may imply the presence of outliers. Robust regression provides a viable alternative if unusual events occurred during the historical period. Weighted least-squares techniques, together with Huber or bisquare weighting schemes, provide the capability to downweight extreme values so that the regression fit is not affected by them. If neither estimation technique will yield satisfactory results, the model chosen is inappropriate for the expected conditions. If both OLS and robust methods yield similar results, quote the OLS results with an added degree of confidence because OLS and normality assumptions are likely to be reasonable for these data. If results differ significantly, dig deeper into the data source or the model specification to find reasons for the differences.

REFERENCES

Bails, D. G., and L. C. Peppers (1982). *Business Fluctuations*. Englewood Cliffs, NJ: Prentice Hall.

Belsley, D. A., E. Kuh, and R. E. Welsch (1980). *Regression Diagnostics: Identifying Influential Data and Sources of Collinearity*. New York: John Wiley & Sons.

Chatterjee, S., A. S. Hadi, and B. Price (2000). *Regression Analysis by Example*. 3rd ed. New York: John Wiley & Sons.

Cryer, J. D., and R. B. Miller (1994). *Statistics for Business—Data Analysis and Modeling*. Belmont, CA: Duxbury Press.

Devlin, S., R. Gnanadesikan, and J. R. Kettering (1975). Robust estimation and outlier detection with correlation coefficients. *Biometrika* 62, 532–45.

Dielman, T. (1996). *Applied Regression Analysis for Business and Economics*. Belmont, CA: Duxbury Press.

Draper, N. R., and H. Smith (1998). *Applied Regression Analysis*. 3rd ed. New York: John Wiley & Sons.

Fisher, R. A. (1960). *The Design of Experiments*. 7th ed. UK: Oliver and Boyd.

Hoaglin, D. C., F. Mosteller, and J. W. Tukey, eds. (1983). *Understanding Robust and Exploratory Data Analysis*. New York: John Wiley & Sons.

Levenbach, H. (1982). Time series forecasting using robust regression. *J. Forecasting* 1, 241–55.

Makridakis, S., S. C. Wheelwright, and R. J. Hyndman (1998). *Forecasting Methods and Applications*. 3rd ed. New York: John Wiley & Sons.

Menzefricke, U. (1995). *Statistics for Managers*. Belmont, CA: Wadsworth.

Mosteller, F., and J. W. Tukey (1977). *Data Analysis and Regression*. Reading, MA: Addison-Wesley.

PROBLEMS

12.1. Data on sales tax revenue are shown in the table below:

Year	Sales Tax Revenues, TAX_REV	Sales Tax Rate, RATE	Retail Sales, SALES (thousands)
1	13,300	6.0	1393
2	18,880	6.0	1528
3	20,800	6.0	1596
4	24,100	6.6	1923
5	27,000	7.0	2163
6	30,300	7.0	2410
7	33,300	7.0	2594
8	34,400	7.0	2735
9	36,600	8.6	2944
10	39,600	9	3206
11	44,400	9	3576
12	48,800	9	3908

Bails and Peppers (1982, Table 7-2).

a. Determine the responsiveness of tax revenues (TAX_REV) to changes in either the tax rate (RATE) or the tax base (SALES) from a constant elasticity model:

$$\log \text{TAX\_REV} = \beta_0 + \beta_1 \log \text{RATE} + \beta_2 \times \log \text{SALES} + \varepsilon$$

b. Compare the regressions in terms of the interpretations for the elasticities in the models:

$$\log \text{TAX\_REV} = \beta_0 + \beta_1 \log \text{RATE} + \varepsilon$$
$$\log \text{TAX\_REV} = \beta_0 + \beta_2 \log \text{SALES} + \varepsilon$$

12.2. Relate revenues from completed projects with the cost of programming time for a project in a computer software company (The data are in AUDIT.DAT on the CD (Menzefricke, 1995, Example 9.1); data from K. W. Stringer and T. R. Stewart, 1986, *Statistical Techniques for Analytical Review in Audition*. New York: John Wiley, pp. 19–22).

a. Draw a scatter plot and describe the nature of the relationship between the two variables.

b. Create summary statistics for central tendency, dispersion, and correlation. What are typical values for the variables? What range do you expect to find 95% of the time for each of the variables? Do any values seem out of line? How do the values appear to be distributed overall?

c. Find the least-squares relationship from the five summary statistics, and plot it on the scatter plot obtained in (a). Interpret the coefficients in the context of this application.

d. Calculate the residuals and the summary statistics for them. Draw a scatter plot of the residuals versus the independent variable. Do the residuals appear to be systematically related to the independent variable? Do the residuals appear to be roughly evenly distributed across the values of the independent variable? Do the residuals display a fairly constant variability across the values of the independent variable?

e. Draw a normal probability plot for the residuals. Are there any serious deviations from the line expected under normality?

f. Try to identify any unusual values such as outliers or influential observations. Describe how each is unusual, so you can look further for possible sources of error or other reasons why the value is different from the bulk of the data.

12.3. Repeat Problem 12.2 using the data for world pulp prices and shipments on the CD (SHIP— pulp shipment in millions of metric tons; PRICE—world pulp price in dollars per ton) (Makridakis et al., 1998, Table 5-2).

12.4. a. Using the data for Problem 12.3, transform the variables to logarithms and create the model:

$$\log \text{SHIP} = \beta_0 + \beta_1 \log \text{PRICE} + \varepsilon$$

b. Repeat Problem 12.2 using this model.

c. Determine the responsiveness of pulp shipments to changes in world pulp prices from this model.

12.5. For a sample of 70 residences sold in a mid-size North American city (REALNEW.DAT on the CD), three variables are selected: (1) MARKET, the amount (in thousands of dollars) for which the residence was sold; (2) AREA, the living area (in square feet) for the residence; and (3) ASSESS, the assessed value in (in thousands of dollars). Create a multiple linear regression model linking variables (2) and (3) to the market value (1) (Menzefricke, 1995, Example 10.1).

a. Draw scatter plots of pairs of variables, and describe the nature of the relationship among the variables.

b. Create summary statistics for central tendency, dispersion, and correlation. What are typical values for the variables? What range do you expect to find 95% of the time for each

of the variables? Do any values seem out of line? How do the values appear to be distributed overall?

c. Find the least-squares regression fit, and plot the residuals versus the fitted values. Interpret the coefficients in the context of this application.

d. Calculate the residuals and the summary statistics for them. Draw a scatter plot of the residuals versus each independent variable. Do the residuals appear to be systematically related to any independent variable? Do the residuals appear to be roughly evenly distributed across the values of the independent variable? Do the residuals display a fairly constant variability across the values of the independent variable?

e. Create a normal probability plot for the residuals. Are there any serious deviations from the line expected under normality?

f. Try to identify any unusual values, such as outliers or influential observations. Describe how each is unusual, so you can look further for possible sources of error or other reasons why the value is different from the bulk of the data.

12.6. Repeat Problem 12.5, adding a categorical variable BASEMENT from the REALNEW .DAT file (on the CD) to the multiple linear regression model for market value (Menzefricke, 1995, Sec. 10.6).

12.7. In an industrial investigation into the performance of a product, the measure of performance Y is assumed to have a curvilinear relationship to an input level X given by the model:

$$Y = \beta_0 + \beta_1 X + \beta_2 X^2 + \varepsilon$$

Using the data stored in INDUSTRY.DAT (on the CD), repeat Problem 12.5 (Menzefricke, 1995, Sec. 10.7).

12.8. A grocery chain is investigating the use of scanners to evaluate its promotional activities. A linearized multiplicative model is postulated to relate SALES (in units) to weekly PRICE (in dollars) in a regression model of the form log SALES = $\beta_0 + \beta_1$ log PRICE + ε. To incorporate promotional effects, the model is augmented by adding two variables, FLYER (flyer promotion) and DISPLAY (display promotion):

$$\log \text{SALES} = \beta_0 + \beta_1 \log \text{PRICE} + \beta_2$$
$$\times \text{FLYER} + \beta_3 \text{DISPLAY} + \varepsilon$$

Complete a regression analysis, using the data in BEVERAGE.DAT (on the CD) and repeating Problem 12.5 (Menzefricke, 1995, Sec. 10.8).

12.9. From an economic database or government website, extract data for consumer price index, overall employment (X_1). Investigate the economic relationship, known as the Phillips curve, with a simple linear regression model of the form Inflation $= \beta_0 + \beta_1 X_1 + \varepsilon$. Inflation is measured by the percentage change in the consumer price index.

a. Place 95% confidence limits on β_1, and give an interpretation of the result.

b. Is there any evidence that a more complex model should be tried?

c. Create a three-period holdout sample for the data, and calculate three one-period-ahead forecasts. How do you assess the forecast-ing ability of this model for projecting inflation one period out?

12.10. Extend the empirical analysis in Problem 12.9 by analyzing a multiple linear regression model involving unemployment for men over 20 years old (X_2) and unemployment for women over 20 years old (X_3).

a. Contrast and interpret the regression models from the fit statistics for

Inflation $= \beta_0 + \beta_2 X_2 + \beta_3 X_3 + \varepsilon$
Inflation $= \beta_0 + \beta_1 X_1 + \beta_2 X_2 + \beta_3 X_3 + \varepsilon$

Are the coefficients reasonable and the equations usable for forecasting?

b. Create residual plots and normal probability plots, and comment on the validity of the normal assumptions.

c. How could you use these models to create inflation forecasts, and what are the caveats for using the equation?

CASE 12A FORECASTING MARKET PRICE FOR RESIDENTIAL HOUSING

A residential real estate market area is being studied by a government agency in terms of stability of housing prices and related demographics. In an initial survey, the agency gathered information on immigration, age, marital status, and family size. As a brand-new employee, your assignment is to make sense out of the data and create some predictive models. In particular, you are asked to create a model for PPRICE, the price per square foot of residential housing. The area is found to be relatively homogeneous in terms of similar type of housing. Selected data for a 15-year period is available for PPRICE, MAR (number of marriages), and HHINC (number of households with annual income greater than $60,000). The data are shown in the accompanying table.

Year	Pprice (dollars)	MAR	HHINC
01	26	274	40,500
02	33	303	41,200
03	38	378	42,100
04	36	363	43,200
05	40	393	44,000
06	40	390	44,400
07	45	452	45,100
08	51	510	46,000
09	50	505	48,800
10	51	515	50,400
11	49	490	51,500
12	53	536	52,300
13	45	452	53,100
14	43	434	53,900
15	42	422	54,500

(1) Create scatter diagrams to assess the degree of association in these variables.

(2) Run a regression of PPRICE on the independent variables.

(3) Consider a first difference on the variables prior to modeling the relationship. Why?

(4) With your best model, take a holdout sample of 2 years and create a 2-year projection of PPRICE:

   a. Using independent trend forecasts for the independent variables

   b. Using actual values of the independent variables in the forecast period

(5) Create a multiple linear regression model with the additional variables MAR and HHINC, using diagnostic tools to assess the appropriateness of the supplementary variables.

(6) Evaluate the performance of your models and make a recommendation for future work.

CASE 12B The Demand for Chicken (Case 8E cont.)

From microeconomic theory it is known that the demand for a commodity generally depends on the income of the consumer, the real price of the commodity, and the real price of complementary or competing products. The first table shows the per capita consumption of chicken (in pounds) in the United States for 1960–1982 along with disposable income per capita (in dollars) and the retail prices for chicken, pork, and beef (in cents); the second table shows chicken consumption for 1983–2001 (Economic Research Service of the U.S. Department of Agriculture).

Year	Chicken (pounds)	Income (dollars)	Chicken Price (cents)	Pork Price (cents)	Beef Price (cents)
1960	28.0	397.5	42.2	50.7	78.3
1961	29.9	413.3	38.1	52.0	79.2
1962	30.1	439.2	40.3	54.0	79.2
1963	30.9	459.7	39.5	55.3	79.2
1964	31.4	492.9	37.3	54.7	77.4
1965	33.7	528.6	38.1	63.7	80.2
1966	35.6	560.3	39.3	69.8	80.4
1967	35.6	624.6	37.8	65.9	83.9
1968	37.1	666.4	38.4	64.5	85.5
1969	38.5	717.8	40.1	70.0	93.7
1970	40.3	768.2	38.6	73.2	106.1
1971	40.2	843.3	39.8	67.8	104.8
1972	41.7	911.6	39.7	79.1	114.0
1973	39.9	931.1	52.1	95.4	124.1
1974	39.7	1021.5	48.9	94.2	127.6
1975	39.0	1165.9	58.3	123.5	142.9
1976	39.1	1349.6	57.9	129.9	143.6
1977	42.8	1449.4	56.5	117.6	139.2
1978	44.9	1575.5	63.7	130.9	165.5
1979	48.3	1759.1	61.6	129.8	203.3
1980	49.0	1994.2	58.9	128.0	219.6
1981	49.4	2258.1	66.4	141.0	221.6
1982	49.6	2478.7	70.4	168.2	232.6

Year	Chicken (pounds)	Year	Chicken (pounds)	Year	Chicken (pounds)
1983	49.8	1990	61.5	1997	72.4
1984	51.6	1991	64.0	1998	72.9
1985	53.1	1992	67.8	1999	77.5
1986	54.3	1993	70.3	2000	77.9
1987	57.4	1994	71.1	2001	77.6
1988	57.5	1995	70.4		
1989	59.3	1996	71.3		

A research firm specializing in agricultural economics has recruited you. Your first assignment is to assist one of the principals in the firm with a study for a chicken franchise on the price elasticities for various meat products. Before delving into a full-fledged study, you perform a number of preliminary analyses to get an understanding of the data.

(1) Create scatter diagrams of chicken consumed against the remaining variables in terms of the original units and log-transformed units. Comment on linearity in the patterns and on whether you notice a high, medium, or low degree of association.

(2) Calculate both the ordinary correlation coefficient and a robust correlation coefficient. Contrast the results. If any appear to be much different, investigate possible causes.

(3) Divide the data in two groups, and repeat (1) and (2).

(4) Run a regression model for the log-transformed variables. Give an interpretation of the coefficients in terms of elasticities.

CASE 12C ENERGY FORECASTING (CASES 1C, 8C, AND 11C CONT.)

An economic research firm is preparing a White Paper to be placed on its website, and you have been providing assistance in the analysis of data sources and the quality of the data, as well as running several univariate exponential smoothing models. Before embarking on a full-scale econometric modeling project, you want to gain further insight into the data by running some simple linear regression models and comparing them with the exponential smoothing models from Case 8C.

To keep matters simple, you postulate that a factor that could affect consumption is a population variable as well as the GNP (or GDP). Case 11C shows a table with the data, consisting of energy consumption (in quadrillion BTU), U.S. population (in millions), and GNP (in billions of 1982 constant dollars) (Selected U.S. Economic, Demographic, and Energy Indicators).

(1) Create multiple linear regression models between energy consumption and two or more of the independent variables, using the first 37 years of the historical data (1947–1983).

(2) Make an independent projection of the independent variables using an exponential smoothing model (Case 8C).

(3) Using the regression model, project the remaining 5 years (1984–1988) using the history (ex post) and independent projections (ex ante) from (2).

(4) Summarize your results with at least two accuracy measures.

(5) Make a plot of the history, forecasts, and 95% confidence limit on the forecasts.

(6) Compare your results with an updated data source. Are the actuals within your confidence limits?

(7) Make a projection for 1989–1998, and compare with an updated source for the independent variable. Are the actuals within your confidence limits?

(8)   What are your recommendations on the value and limitations of such forecasts to a practitioner?

Once we have forecasts for energy consumption, we can create forecasts for energy production based on the relationship between consumption and production (see Case 8C).

(9)   Create a scatter diagram with energy production (dependent variable) and energy consumption (independent) variable.

(10) Create the model for production in terms of consumption by repeating (1)–(6).

(11) Make a forecast for the 1989–1998 period using both ex ante and ex post forecasting.

(12) How many years into the future do you feel comfortable making your projections? Why?

CASE 12D Predicting Academic Performance

The dean of Admissions of your local university is interested in finding out how well an applicant will perform in the graduate program of the Business School. You are asked to develop a relationship between graduate academic performance and some explanatory factors that will give the dean a forecasting equation to estimate an applicant's performance in the graduate school. As a start, you are given the data from a previous study that assumed a multiple linear regression model of the form:

$$\text{Perform} = a + b \times \text{GPA} + c \times \text{Hours} + d \times \text{GMAT} + e \times \text{Work}$$

where Perform = graduate academic performance (3 = Distinguished,
2 = High Pass, 1 = Pass, 0 = No Credit)

GPA = undergraduate GPA score

Hours = average number of study hours per day in the graduate school

GMAT = GMAT score

Work = number of years of work experience

Unfortunately, the original report has been lost, but the data are still available and are provided in the accompanying table.

Performance	GPA	Hours	GMAT	Work (years)
2	3	2.5	622	2
2.8	3.7	6.5	780	6
2.8	3.5	6	650	3
2	3.5	3	630	3
2.8	3.9	4	582	3
1.6	3.7	2.5	750	4
2	3.74	4	687	1.5
2	3.7	3	640	3
1.8	3.5	2	670	3
2.2	3.65	3.5	750	2.5
2	3.1	3.5	640	5
2.4	3.7	3	660	3
1.6	3.1	2	710	2
1.8	3.3	3	720	6
2.4	3.5	3.5	640	2
1.8	3.3	2	680	4
2	3.56	4	750	3
2	2.8	4.5	670	3
2.2	3.5	5	590	1.5
1.8	2.8	6	690	3.5
2.2	3.6	4	580	5
2.4	4.4	4	630	2
1.8	3.1	3	590	2
2.8	3.6	5.5	760	4
2	3.2	3.5	600	2.5

(1) Create time plots of the variables along with appropriate summary statistics to gain a preliminary insight into the data patterns.

(2) Calculate the correlation matrix using the ordinary correlation coefficient and the robust correlation formulae. Contrast any differences you find and interpret them in terms of the underlying data patterns. Are the signs of the coefficients meaningful?

(3) Reconstruct the results of the previous study by running a multiple linear regression model of Performance versus the independent variables.

(4) Interpret the coefficients in the regression model. Are there any variables that do not adequately support the relationship?

(5) Do you have any suggestions for simplifying the relationship? Do you feel that there are omitted variables that may strengthen the relationship?

(6) Create a modeling flowchart to summarize your decisions and suggestions for advancing the study to get improved results.

(7) In examining the residuals, are there any transformations that should be tried?

(8) Summarize your recommendations to the dean in terms of the applicability of this model to helping predict academic performance in a graduate program.

CASE 12E THE COMMERCIAL REAL ESTATE MARKET: OFFICE OCCUPANCY RATES (CASE 10E CONT.)

Review the objectives, indicators, and predictors of the real estate market (Case 10E).

(1) Acquire approximately 3–5 years of quarterly data for the office occupancy rates for class A offices in a state, province, or region.

(2) Characterize the seasonality of the data, and provide illustrative summaries of the historical patterns in the data. Consider transformations of the data, if required.

(3) Collect relevant data for the determinants of office occupancy, including any that you may find appropriate. Create univariate data summaries for these factors.

(4) Create a flowchart of the modeling steps you would use to develop several linear regression models to explain the relationship between office occupancy rates and its determinants.

(5) Calculate the ordinary and robust correlation coefficients for the variables in your analyses. Interpret differences that appear important.

(6) Develop several linear regression models between office occupancy rates and the related factor(s). Consider the use of leading/lagging indicators and provide a rationale for doing so.

(7) Provide an interpretation of the coefficients in your stated models.

(8) Create additional backup summaries, plots, and displays for future reference and in-depth presentations.

(9) Write a brief synopsis summarizing your findings and recommendations for use of the model for forecasting. Add an overview of recommendations to your firm's client.

13

Building ARIMA Models: The Box-Jenkins Approach

"When things look too good to be true, they probably are."
PARAPHRASING DAVID WESSEL (OUTLOOK COLUMN, *WALL STREET JOURNAL*, JUNE 5, 1989)

IN THIS CHAPTER, WE INTRODUCE a class of techniques, called ARIMA, which can be used to describe stationary time series and nonstationary time series with changing levels. For seasonal time series, the nonstationary ARIMA process is extended to account for a multiplicative seasonal component. These models form the framework for

- Expressing various forms of stationary (level) and nonstationary (mostly trending and seasonal) behavior in time series
- Producing optimal forecasts for a time series from its own current and past values
- Developing a flexible modeling methodology

In this chapter, we learn how to select ARIMA models through the three-stage iterative procedure introduced in Chapter 10. This is commonly called the Box-Jenkins procedure and consists of identification (using the data to tentatively identify a model), estimation (fitting algorithms and inferences about the parameters), and diagnostic checking for model adequacy.

13.1 WHY USE ARIMA MODELS FOR FORECASTING?

In Chapter 4, we construct a year-by-month ANOVA table as a data-driven technique for examining trend and seasonal patterns in a time series. This led us to estimate many coefficients for the monthly and yearly means in the rows and columns of a table. Although the results were not intended for generating projections, the ANOVA table provides a useful explorative view of the data. In addition, regression analysis (Chapters 11 and 12) provides a means of fitting trends in which functions of time are treated as independent variables; they often fit well over a limited range of the data, yet frequently forecast poorly when extrapolated. Furthermore, regression models let us incorporate independent variables to add explanatory power (including seasonality) to a forecasting problem. However, independent forecasts of these factors must be added to achieve a comprehensive, final forecast. The ARIMA modeling approach offers an alternative *model-driven* technique to circumvent some of these difficulties within a unified theoretical framework. The meanings of the AR, I, and MA in ARIMA will become clear as the theory unfolds.

In a number of studies, Granger and Newbold (1986) show that forecasts from simple ARIMA models have frequently outperformed larger, more complex econometric systems for a number of economic series. Although it is possible to construct ARIMA models with only 2 years of monthly historical data, the best results are usually obtained when at least 5–10 years of data are available—particularly if the series exhibits strong seasonality.

The ARIMA models have proved to be excellent short-term forecasting models for a wide variety of time series.

A major drawback of the pure ARIMA models is that, because they are univariate, they have very limited explanatory capability. The models are essentially sophisticated extrapolative devices that are of greatest use when it is expected that the underlying factors causing demand for products, services, revenues, and so on will behave in the future much in the same way as in the past. In the short term, this is often a reasonable expectation; however, because these factors tend to change slowly, data tend to show inertia in the short term. However, there are extensions of the ARIMA approach that incorporate explanatory factors for including information such as price, promotions, strikes, and holiday effects. These models are called transfer function (or dynamic regression) models, but are beyond the scope of this text (see Box et al., 1994; Jenkins, 1979; Pankratz, 1991).

A significant advantage of univariate ARIMA models is that plausible forecasts can be developed in a relatively short time. Much more time is usually required to obtain and validate historical data than to build the models. Therefore, a practitioner can often deliver significant results with ARIMA modeling early in a project for which adequate historical data exist. The forecaster should always consider ARIMA models as an important forecasting alternative whenever these models are relevant to the problem at hand.

13.2 THE LINEAR FILTER MODEL AS A BLACK BOX

The application of ARIMA models is based on the idea that a time series in which successive values are highly dependent can also be thought of as having come from a process involving a series of independent errors or shocks, ε_t. The general form of a (discrete) linear process is

$$Z_t = \mu + \varepsilon_t + \psi_1 \varepsilon_{t-1} + \psi_2 \varepsilon_{t-2} + \cdots + \psi_n \varepsilon_{t-n} + \cdots$$

where μ and all ψ_j are fixed parameters and the $\{\varepsilon_t\}$ is a sequence of identically, independently distributed random errors with zero mean and constant variance.

Why is it called a linear filter? The process is linear because Z_t is represented as a linear combination of current and past shocks. It is often referred to as a black box, or filter, because the model relates a random input to an output that is time dependent. The input is filtered, or damped, by the equation so that what comes out of the equation has the characteristics that are wanted.

A linear process is capable of describing a wide variety of practical forecasting models for time series. It can be visualized as a black-box equation transforming random inputs into the observed data.

A linear process can be visualized as a black box as follows (Exhibit 13.1). **White noise** or purely random error $\{\varepsilon_t\}$ is transformed to an observed series $\{Y_t\}$ by the operation of a linear filter; the filtering operation simply takes a weighted sum of previous shocks. For convenience, we henceforth write models in terms of Y_t, which has been mean adjusted, that is, $Y_t = Z_t - \mu$. The weights are known as ψ (psi) coefficients. For Y_t to represent a valid **stationary process,** it is necessary that the coefficients ψ_j satisfy the condition $\sum \psi_j^2 < \infty$.

It can be shown that any linear process can be written formally as a weighted sum of the current error term plus all past shocks. In many problems, such as those in which it is required that future values of a series be predicted, it is necessary to construct a parametric model for the time series. To be useful, the model should be physically meaningful and involve as few parameters as possible. A powerful parametric model that has been widely used in practice for describing empirical time series is called the mixed autoregressive–moving average (ARMA) process:

$$Y_t = \phi_1 Y_{t-1} + \phi_2 Y_{t-2} + \cdots + \phi_p Y_{t-p} + \varepsilon_t - \theta_1 \varepsilon_{t-1} - \theta_2 \varepsilon_{t-2} - \cdots - \theta_q \varepsilon_{t-q}$$

where p is the highest lag associated with the data, and q is the highest lag associated with the error term. The ARMA processes are important because they are

EXHIBIT 13.1 Black-Box Representation of the Linear Random Process

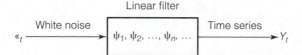

mathematically tractable; moreover, they are flexible enough to describe many time series.

There are some special versions of the ARMA process that are particularly useful in practice. If that weighted sum has only a finite number of nonzero error terms, then the process is known as a moving average (MA) process. It can be shown that the linear process can also be expressed as a weighted sum of the current shock plus all past observed values. If the number of nonzero terms in this expression is finite, then the process is known as an autoregressive (AR) process. The origin of the AR and MA terminology are described a little later with specific examples.

 The ARMA process is important because it is mathematically tractable and can be shown to represent a wide variety of forecasting models for time series.

It turns out that an MA process of finite order can be expressed as an AR process of infinite order and that an AR process of finite order can be expressed as an MA process of infinite order. This duality has led to the principle of parsimony in the Box-Jenkins methodology, which recommends that the practitioner employ the smallest possible number of parameters for adequate representation of a model. In practice, it turns out that relatively few parameters are needed to make usable forecasting models with business data.

It may often be possible to describe a stationary time series with a model involving fewer parameters than either the MA or the AR process has by itself. Such a model will possess qualities of both autoregressive and moving average models: it is called an ARMA process. An ARMA(1,1) process, for example, has one prior observed-value term of lag 1 and one prior error term:

$$Y_t = \phi_1 Y_{t-1} + \varepsilon_t - \theta_1 \varepsilon_{t-1}$$

The general form of an ARMA(p, q) process of autoregressive order p and moving average order q looks like

$$Y_t = \phi_1 Y_{t-1} + \phi_2 Y_{t-2} + \cdots + \phi_p Y_{t-p} + \varepsilon_t - \theta_1 \varepsilon_{t-1} - \theta_2 \varepsilon_{t-2} - \cdots - \theta_q \varepsilon_{t-q}$$

In short, the ARMA process is a linear random process. It is linear if Y_t is a linear combination of lagged values of Y_t and ε_t. It is random if the errors (also called shocks or disturbances) are introduced into the system in the form of white noise. The random errors are assumed to be independent of one another and to be identically distributed with a mean of zero and a constant variance σ_ε^2.

13.3 A MODEL-BUILDING STRATEGY

The Box-Jenkins approach for ARIMA modeling provides the forecaster with a very powerful and flexible tool. Because of its complexity, it is necessary to establish procedures for coming up with practical models. Its difficulty requires a fair amount of sophistication and judgment in its use. Nevertheless, its proven results in terms of forecasting accuracy and understanding the processes generating data can be invaluable in the hands of a skilled user.

A general model-building strategy for ARIMA models is due to Professors G. E. P. Box and G. M. Jenkins (1976), who during the 1960s developed their strategy as a result of their direct involvement with forecasting problems in business, economic, and engineering environments. More extensive treatments of the Box-Jenkins procedure can be found in textbooks such as Bowerman and O'Connell (2004), Box et al. (1994), Delurgio (1998), Makridakis et al. (1998), Newbold and Bos (1994), Pankratz (1983), and Vandaele (1983).

Although a Box-Jenkins methodology is an excellent way for forecasting a time series from its own current and past values, it should not be applied blindly and automatically to all forecasting problems.

The Box-Jenkins procedure consists of the following three stages.

1. *Identification* consists of using the data and any other knowledge that will tentatively indicate whether the time series can be described with a moving average model, an **autoregressive model,** or a mixed autoregressive–moving average model.

2. *Estimation* consists of using the data to make inferences about the parameters that will be needed for the tentatively identified model and to estimate values of them.

3. *Diagnostic checking* involves the examination of residuals from fitted models, which can result in either no indication of model inadequacy or model inadequacy, together with information on how the series may be better described.

The procedure is iterative. Thus, residuals should be examined for any lack of randomness and, if we find that residuals are serially correlated, we use this information to modify a tentative model. The modified model is then fitted and subjected to diagnostic checking again until an adequate model is obtained.

13.4 IDENTIFICATION: INTERPRETING AUTOCORRELATION AND PARTIAL AUTOCORRELATION FUNCTIONS

The primary tool for identifying an ARIMA process is with autocorrelation functions (ACFs) and partial autocorrelation functions (PACFs). ACFs are introduced in Chapter 3 as a way of describing the mutual dependence among values in a time series. Extreme care must be taken in interpreting ACFs, however; the interpretation is complex and requires considerable experience. Attention should be directed to individual values as well as to the overall pattern of the autocorrelation coefficients.

In practice, the autocorrelations of low-order ARMA processes are used to help identify models with the Box-Jenkins methodology.

Autocorrelation analysis can be a powerful tool for deciding whether a process shows pure autoregressive behavior or moving average behavior in ARIMA models.

Autocorrelation and Partial Autocorrelation Functions

The ACF and PACF are widely used in identifying ARIMA models. The corresponding ordinary and partial correlograms are the sample estimates of the ACF and PACF. They play an important role in the identification phase of the Box-Jenkins methodology for forecasting and control applications. Some examples follow, but to simplify writing the model equations, we use a notational device known as the backshift operation (see Appendix 14A for an explanation).

For example, the ACF of an AR(1) process satisfies the simple equation $\rho_k = \phi_1 \rho_{k-1}$, $k > 0$, which, with $\rho_1 = 0$, has the solution $\rho_k = \phi_1^k$, $k \geq 0$. The ACF of an AR (1) process is depicted in Exhibit 13.2. There is a decaying pattern in the ACF; the decay is exponential if $0 < \phi_1 < 1$ (Exhibit 13.2a). For $-1 < \phi_1 < 0$ (Exhibit 13.2b), the ACF is similar but alternates in sign. The PACF shows a single positive value at lag 1 if $0 < \phi_1 < 1$ and a negative spike at lag 1 if $-1 < \phi_1 < 0$.

In general, the autocorrelation function of a stationary autoregressive process will consist of a mixture of damped exponentials and damped sine waves.

The PACF is more complex to describe. It measures the correlation between Y_t and Y_{t-k} adjusted for the intermediate values $Y_{t-1}, Y_{t-2}, \ldots, Y_{t-k+1}$ (or the correlation between Y_t and Y_{t-k} not accounted for by $Y_{t-1}, Y_{t-2}, \ldots, Y_{t-k+1}$). If we denote by ϕ_{kj} the jth coefficient in an AR(k) model, so that ϕ_{kk} is the last coefficient, then it can be shown (Box et al., 1994, Sec. 3.2.5) that the ϕ_{kj} satisfy the set of equations

$$\rho_j = \phi_{k1}\rho_{j-1} + \cdots + \phi_{k(k-1)}\rho_{j-k-1} + \phi_{kk}\rho_{j-k} \qquad j = 1, 2, \ldots, k$$

The ϕ_{kk} expressed in terms of the ρ_j, as a function of the lag k, is the PACF. For example, $\phi_{11} = \rho_1$, and $\phi_{22} = (\rho_2 - \rho_1^2)/(1 - \rho_1^2)$. The PACF ϕ_{kk} will be nonzero for $k \leq p$ and zero for $k > p$, where p is the order of the autoregressive process. Another way of saying this is that ϕ_{kk} has a cutoff after lag p. For example, the PACF of an AR(1) process has one spike at lag 1. It has the value $\rho_1 = \phi_1$.

Another basic process that occurs fairly often in practice is the AR(2) process. In this case there are two autoregressive coefficients ϕ_1 and ϕ_2. Exhibit 13.3 shows the ACF and PACF of an AR(2) model with $\phi_1 = 0.3$ and $\phi_2 = 0.5$. The values in the ACF diminish according to the formula

$$\rho_k = \phi_1 \rho_{k-1} + \phi_2 \rho_{k-2}$$

EXHIBIT 13.2 (a) ACF an AR(1) Process ($\phi_1 = 0.70$); (b) ACF of an AR(1) Process ($\phi_1 = -0.80$)

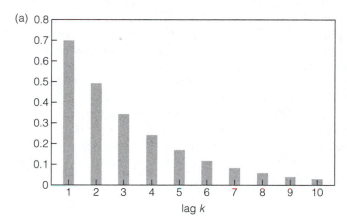

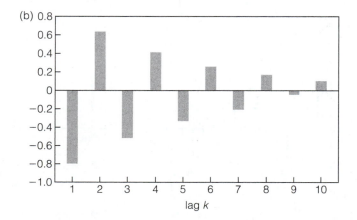

EXHIBIT 13.3 (a) ACF and (b) PACF of an Autoregressive AR(2) Model with Parameters $\phi_1 = 0.3$ and $\phi_2 = 0.5$

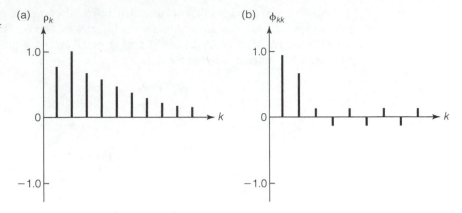

The PACF shows positive values at lags 1 and 2 only. The PACF is very helpful because it suggests that the process is autoregressive and, more important, that it is second-order autoregressive.

If $\phi_1 = 1.2$ and $\phi_2 = -0.64$, the ACF and PACF have the patterns shown in Exhibit 13.4. The values in the ACF decay in a sinusoidal pattern; the PACF has a positive value at lag 1 and a negative value at lag 2. There are a number of possible patterns for AR(2) models. A triangular region describes the allowable values for ϕ_1 and ϕ_2 in the stationary case: $\phi_1 + \phi_2 < 1$, $\phi_2 - \phi_1 < 1$, and $-1 < \phi_2 < 1$. If $\phi_1^2 + 4\phi_2 > 0$, the ACF decreases exponentially with increasing lag. If $\phi_1^2 + 4\phi_2 < 0$, the ACF is a damped cosine wave.

The ACF of a MA(q) process is 0, beyond the order q of the process (i.e., it has a cutoff after lag q). For example, the ACF of a MA(1) process has one spike at lag 1; the others are 0. It has the value $\rho_1 = -\theta_1/(1 + \theta_1^2)$ with $|\rho_1| \le \frac{1}{2}$.

The PACF of the MA process is complicated, so we here simply display the formula for the MA(1) process:

$$\phi_{kk} = [-\theta_1^k/(1 - \theta_1^2)]/(1 - \theta_1^{2(k+1)})$$

EXHIBIT 13.4 (a) ACF and (b) PACF of an Autoregressive AR(2) Model with Parameters $\phi_1 = 1.2$ and $\phi_2 = -0.64$

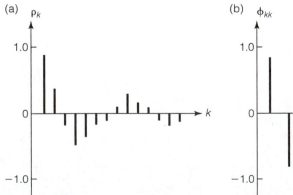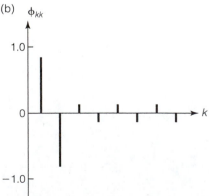

The ACF and PACF of an MA(1) model with positive θ are depicted in Exhibit 13.5. There is a single negative spike at the lag 1 in the ACF. There is a decaying pattern in the PACF. The ACF of an MA(1) process with negative θ (Exhibit 13.6) shows a single positive spike, but the PACF shows a decaying pattern with spikes alternating above and below the zero line.

One important consequence of the theory is that the ACF of an AR process behaves like the PACF of an MA process and vice versa. This aspect is known as a duality property of the AR and MA processes. If both the ACF and the PACF attenuate, then a mixed model is called for.

It turns out that the ACF of the pure MA(q) process truncates, becoming 0 after lag q, whereas that for the pure AR(p) process is of infinite extent. MA processes are thus characterized by truncation (spikes ending) of the ACF, whereas AR processes are characterized by attenuation (gradual decay) of the ACF. Derivations of this kind are beyond the scope of this book, but these important results are derived in Box et al. (1994) and Pankrantz (1983).

EXHIBIT 13.5 (a) ACF and (b) PACF of an MA(1) Model with Positive θ

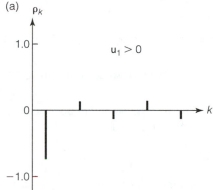
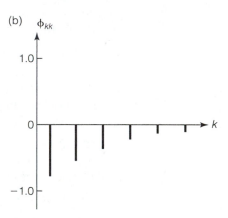

EXHIBIT 13.6 (a) ACF and (b) PACF of an MA(1) Model with Negative θ

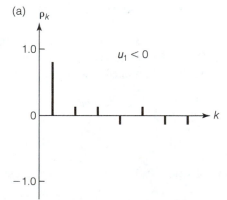
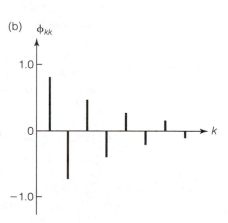

In some situations, it may not be clear whether the ACF truncates or attenuates. Hence, it is useful to supplement the analysis by considering the PACF. Each ARIMA process has an ACF and PACF associated with it.

For an AR process, the ACF attenuates and the PACF truncates; conversely, for an MA process, the PACF attenuates and the ACF truncates.

Once we decide on the degree of differencing required to produce stationarity, we can review a catalog of ACF and PACF patterns to establish the orders p and q of the autoregressive and moving average components of the stationary ARMA model:

$$\phi(B)Y_t = \theta(B)\varepsilon_t$$

The estimated autocorrelation functions, or correlograms, are matched with the cataloged patterns in order to establish a visual identification of the most appropriate model for a given situation. Usually, more than one model suggests itself, so that we may tentatively identify several similar models for a particular time series. Some useful rules for identification are:

- If the correlogram cuts off at some point, say $k = q$, then the appropriate model is MA(q).

- If the partial correlogram cuts off at some point, say $k = p$, then the appropriate model is AR(p).

- If neither diagram cuts off at some point, but does decay gradually to zero, the appropriate model is ARMA(p', q') for some p', q'.

The Mixed ARMA Process

The mixed ARMA(p, q) contains p AR terms and q MA terms. It is given by

$$(1 - \phi_1 B - \phi_2 B^2 - \cdots - \phi_p B^p)Y_t = (1 - \theta_1 B - \theta_2 B^2 - \cdots - \theta_q B^q)\varepsilon_t$$

This model is useful in that stationary series may often be expressed more parsimoniously (with fewer parameters) in an ARMA model than in the pure AR or MA models.

The ACF and PACF of an ARMA(p, q) model are more complex than either the AR(p) or MA(q) models. The ACF of an ARMA(p, q) process has an irregular pattern at lags 1 through q; then the tail diminishes according to the formula

$$\rho_k = \phi_1 \rho_{k-1} + \cdots + \phi_p \rho_{k-p} \qquad k > q$$

The PACF tail also diminishes.

The best way to identify an ARMA process initially is to look for decay or a tail in both the ACFs and PACFs.

In practice, it is sufficient to recognize the basic structure of the ACF and PACF for values of $p = 1, 2$ and $q = 1, 2$. Because of the complexity of the formulae, we illustrate here the typical patterns of the ACF and PACF only for the ARMA(1, 1) process over a range of values of $\phi_1(-1 < \phi_1 < 1)$ and $\theta_1(-1 < \theta_1 < 1)$. The first two values of the ACF are

$$\rho_1 = [(1 - \phi_1\theta_1)(\phi_1 - \theta_1)] / (1 + \phi_1^2 - 2\phi_1\theta_1)$$

$$\rho_2 = \phi_1\rho_1$$

The ACF and PACF of an ARMA(1,1) process with $\phi_1 = 0.7$ and $\theta_1 = -0.6$ are shown in Exhibit 13.7. The value at lag 1 in the ACF is high. The remaining values show an exponential decay. The PACF also shows a decay or tail.

The ACF and PACF of the same ARMA(1, 1) process, but with the signs of ϕ_1 and θ_1 reversed, shows alternating decaying values in the ACF (Exhibit 13.8). In the PACF, there is a large negative value at lag 1 followed by an exponential decay in the remaining values.

EXHIBIT 13.7 (a) ACF and (b) PACF of an ARMA(1, 1) Process with Parameters $\phi_1 = 0.7$ and $\theta_1 = -0.6$

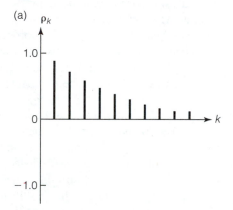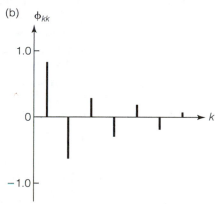

EXHIBIT 13.8 (a) ACF and (b) PACF of an ARMA(1, 1) Process with Parameters $\phi_1 = -0.7$ and $\theta_1 = 0.6$

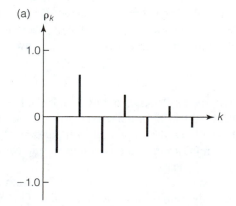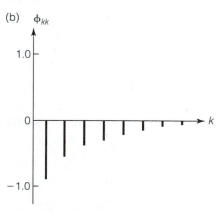

In practice, it can be challenging to identify the process just by looking at empirical ACF and PACF based on real data. There are many possible parameter values to choose from that result in similar-looking patterns in the ACF and PACF. Even when performed on a computer, the range of permissible values for the coefficients are limited. Next, we discuss the restrictions on these parameters.

Invertibility and Stationarity

The ACF and PACF are the two principal tools used to describe the structure of a theoretical ARMA model. The ACF and PACF have patterns that are useful for identifying the order and lag structure of an ARMA model.

The moving average model MA(q), of order q, with lag structure given by the presence and absence of the θ coefficients, is (in backshift notation):

$$Y_t = (1 - \theta_1 B - \theta_2 B^2 - \cdots - \theta_q B^q)\varepsilon_t$$

where the current value Y_t of the time series is assumed to be a linear combination of the current and previous error terms ε_t. In practice, these models are useful for describing events that are affected by random events, such as strikes and policy decisions. Many economic and planning series exhibit behavior that can be reasonably described by a model containing MA components.

No restrictions on the size of θ are required for an MA process to be stationary. However, to avoid certain model estimation problems an invertibility condition needs to be imposed. This arises because, for an MA(1) model, for example, if the coefficient is either θ or $1/\theta$, the ACF will be the same. A MA(1) model is said to be invertible if $|\theta_1| < 1$.

For the general ARIMA process, it is required that the roots of the two polynomial equations in B, namely, $\phi(B) = 0$ and $\theta(B) = 0$, all lie outside the unit circle. The first condition ensures the stationarity of W_t, that is, the statistical equilibrium about a fixed mean; this means that the sum of the coefficients (ϕ) must always be less than 1. The second condition, known as the invertibility requirement, guarantees uniqueness of representation (the weights applied to the past history of W_t to generate forecasts die out). This means that the sum of the coefficients (θ) of a moving average model must always be less than 1.

The multiplicity of a model is not the only thing that determines its invertibility or noninvertibility. Models are called noninvertible because it is impossible to estimate the errors ε_t. Because the starting values $\varepsilon_0, \varepsilon_{-1}$, are unknown, there will be errors in estimating the early ε_t. With noninvertible models, these errors grow instead of decay.

Seasonal ARMA Process

For the pure seasonal AR process $(1, 0, 0)_s$, the ACF and PACF have similar patterns to the corresponding ACF and PACF for the regular AR process $(1, 0, 0)$, except the spikes are s lags apart. Similarly, the autocorrelation patterns of the pure seasonal MA process $(0, 0, 1)_s$ is similar to the patterns found for the corresponding regular MA process $(0, 0, 1)$ with s lags apart.

A more useful model that we look at later is the general multiplicative model, in which nonstationarity is removed through regular $(1 - B)^d$ and seasonal

EXHIBIT 13.9
ACF of a Combined
Regular and
Seasonal AR Model:
$(1 - \phi_1 B)(1 - \phi_{12} B^{12})Y_t = \varepsilon_t$; $Y_t =$
$\phi_1 Y_{t-1} + \phi_{12} Y_{t-12} -$
$\phi_1 \phi_{12} Y_{t-13} + \varepsilon_t$

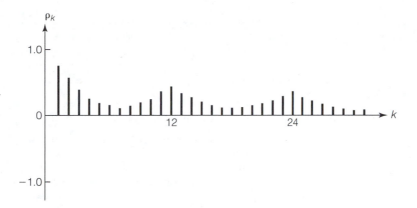

differencing $(1 - B^s)^D$, where $d = 0$, 1, or 2 and $D = 0$, 1, or 2. The ACF and PACF for these models have a mixture of the regular and seasonal patterns in them and, hence, tend to be somewhat complex to construct and identify. For monthly time series, $s = 12$ and the various seasonal spikes appear at lags 12, 24, and so on, requiring a fairly long time series for reliable identification. Hence, in practice, we encounter seasonal ARIMA models with very low AR and MA structure (usually $p = 0$, 1 and/or $q = 0$, 1).

Another process that occurs in practice is the combination of regular and seasonal autoregressive components. Exhibit 13.9 shows the ACF of a particular regular AR(1) and a seasonal AR(12) process. A pattern in which the values reach a peak at multiples of 12 is noticeable, as are buildups to and decays from that peak at the other lags.

There are a large variety of patterns that can emerge in the modeling process. The Box-Jenkins approach is so general that it is impossible to catalog all possibilities. Therefore, it is essential to follow an iterative procedure in developing successful forecasting models.

In later sections, we treat the identification of an appropriate ARIMA model based on real data. The first step is to transform and/or difference the data to produce a stationary series, thereby reducing the model to one in the ARMA class. Then, the ordinary and partial correlograms for the various patterns of differencing that are found in the adjusted data are displayed and compared to the basic catalog of theoretical autocorrelation functions already described. We first look at identifying the regular ARIMA models that are appropriate for nonseasonal data.

13.5 IDENTIFYING NONSEASONAL ARIMA MODELS

This section describes the identification of nonseasonal ARIMA models, also known as regular ARIMA models. These models are appropriate for many business and economic time series that are nonseasonal, are seasonally adjusted, have strong trends, or exhibit a random walk pattern (e.g., changes in stock prices).

In most cases, seasonal time series are modeled best with a combination of regular and seasonal parameters.

Identification Steps

In the previous section we describe the main features of the ARIMA(p, d, q) process:

$$(1 - \phi_1 B - \phi_2 B^2 - \cdots - \phi_p B^p)(1 - B)^d\, Y_t = (1 - \theta_1 B - \theta_2 B^2 - \cdots - \theta_q B^q)\varepsilon_t$$

where the ε_t values are white noise (a sequence of identically distributed uncorrelated errors). Using the backshift operator B again (see Appendix 14A), this expression can be simplified to:

$$\phi(B)(1 - B)^d\, Y_t = \theta(B)\varepsilon_t$$

where the AR(p) terms are given by the polynomial $\phi(B) = (1 - \phi_1 B - \phi_2 B^2 - \cdots - \phi_p B^p)$ and the MA(q) terms are $\theta(B) = (1 - \theta_1 B - \theta_2 B^2 - \cdots - \theta_q B^q)$. For example, the ARIMA(1, 1, 1) process takes the form

$$(1 - \phi_1 B)(1 - B)^d\, Y_t = (1 - \theta_1 B)\varepsilon_t$$

To identify one of these ARIMA models we need to:

- Determine the order of differencing ($d = 0$, 1, or at most 2) required to produce stationarity. This occurs when the ACF of the differenced series decays relatively quickly.

- Determine preliminary estimates of the ϕ and θ coefficients in the resulting ARMA model. Viewing the cutoff and decay patterns of the ACF and PACF does this.

13.6 ESTIMATION: FITTING MODELS TO DATA

The second stage of the model-building strategy is the estimation or fitting stage. Least-squares or maximum likelihood methods can fit ARMA models. An iterative nonlinear least-squares procedure is used to obtain the $p + q + 1$ parameter estimates of an ARMA(p, q) model. The estimates minimize the sum of squares of the residuals e; that is,

$$S(\phi, \theta, \sigma) = \sum e_t^2$$

given the form of the model and the data. Because the procedure is, in general, nonlinear (because of the moving average terms), the initial values for the parameters must be determined so that minimization can proceed.

There are a variety of parameter-estimation methods available in the literature (see Box et al., 1994, Chap. 7) as well as a number of software programs that implement these models with automated algorithms. Hence, the details of an estimation procedure are nowadays primarily of academic interest. Experience with many kinds of business and economic time series has convinced us that parsimonious models are

 When estimating models, few general guidelines exist, except those directing us toward a preference for the use of parsimonious models ("the fewer parameters the better").

generally better forecasting models. Parsimony is a practical rather than a theoretical consideration, however.

Once parameters have been estimated, it is also useful to determine a matrix of estimated variances and covariances of the coefficients. Based on large-sample theory, the standard deviations of the parameter estimates (usually referred to as standard errors) are obtained for this and used to determine whether the parameters are (statistically) significantly different from 0. Some care must be taken here because of the need to establish the appropriateness of the quadratic approximation in the log-likelihood function.

There are two goodness-of-fit statistics that are most commonly used for model selection, analogous to the R^2 statistic for regression models. They are the Akaike information criterion (**AIC**) and the Schwarz Bayesian information criterion (**BIC**). The AIC and BIC are similar in that they are based on a likelihood function, but need to be calculated by a computer program (for details, see Box et al., 1994; Brockwell and Davis, 1996).

Consider Exhibit 13.10. In this example, the Y variable is the quarterly housing permits issued in the United States from 1947:Q1 to 1967:Q2 (Pankratz, 1991, Series #9), the X_1 variable is the housing permits 1 quarter earlier, and the X_2 variable is the housing permits 2 quarters earlier. Formulas in column D repeat the Y values (column F) one row up and column E repeat the values two rows up.

The model is fit from 1947:Q1 to 1964:Q4 (Exhibit 13.11). Note that the X_1 coefficient in this example is 1.20, the X_2 coefficient is -0.54, and the constant is 36.85. With these values for the model parameters, it turns out that the housing-permits data generate a stochastic cycle. A nonseasonal AR(2) model generates a stochastic cycle when the following condition is met:

$$\phi_1^2 + 4\phi_2 < 0$$

Using the estimated model coefficients, we find that the condition is met ($= -0.72$).

The average period p^* of the stochastic cycle is defined by

$$p^* = 360°/\cos^{-1}[\phi_1 \, / \, 2(-\phi_2)^{1/2}]$$

where the cosine inverse is stated in degrees. Because the data are quarterly and $p^* = 4.4$, the cycle repeats every 4.4 quarters on average. In reality, a quarterly series has a cycle every 4 quarters. Estimated Y values and fitting errors are computed in columns G and H. One forecast is made past the end of the data for 1964:Q4 ($= 114.5$).

EXHIBIT 13.10
AR(2) Model for
U.S. Housing Permits

Autoregression, 2 Lagged variables
U.S. Housing permits
Quarterly, seasonally adjusted 1947Q I − 1967Q II

Year	Quarter	Period	Lag1 X_1	Lag2 X_2	Housing permits Y	Y estimate	Error
A	B	C	D	E	F	G	H
1947	# N/A	1	#N/A	#N/A	83.3	#N/A	#N/A
1947		2	83.3	#N/A	83.2	#N/A	#N/A
1947		3	83.2	83.3	105.3	91.7	13.6
1947		4	105.3	83.2	117.7	118.3	−0.6
1948		5	117.7	105.3	104.6	121.2	−16.6
1948		6	104.6	117.7	108.8	98.8	10.0
1948		7	108.8	104.6	93.9	110.9	−17.0
1948		8	93.9	108.8	86.1	90.8	−4.7
1949		9	86.1	93.9	83.0	89.5	−6.5
1949		10	83.0	86.1	102.4	90.0	12.4
1949		11	102.4	83.0	119.6	114.9	4.7
1949		12	119.6	102.4	141.4	125.1	16.3
1950		13	141.4	119.6	158.6	141.9	16.7
1950		14	158.6	141.4	161.3	150.8	10.5
1950		15	161.3	158.6	158.2	144.8	13.4
1950		16	158.2	161.3	136.1	139.6	−3.5
1951		17	136.1	158.2	121.9	114.7	7.2
1951		18	121.9	136.1	97.7	109.6	−11.9
1951		19	97.7	121.9	103.3	88.3	15.0
1951		20	103.3	97.7	92.7	108.1	−15.4
1952		21	92.7	103.3	106.8	92.3	14.5
1952		22	106.8	92.7	102.1	115.0	−12.9
1952		23	102.1	106.8	110.3	101.7	8.6
1952		24	110.3	102.1	114.1	114.1	0.0
1953		25	114.1	110.3	109.1	114.2	−5.1
1953		26	109.1	114.1	105.4	106.2	−0.8
1953		27	105.4	109.1	97.6	104.4	−6.8
1953		28	97.6	105.4	100.7	97.1	3.6
1954		29	100.7	97.6	102.7	105.0	−2.3
1954		30	102.7	100.7	110.9	105.7	5.2
1954		31	110.9	102.7	120.2	114.5	5.7
1954		32	120.2	110.9	131.3	121.2	10.1
1955		33	131.3	120.2	138.9	129.5	9.4
1955		34	138.9	131.3	130.9	132.6	−1.7
1955		35	130.9	138.9	123.1	118.9	4.2
1955		36	123.1	130.9	110.8	113.9	−3.1
1956		37	110.8	123.1	108.8	103.3	5.5
1956		38	108.8	110.8	103.8	107.6	−3.8
1956		39	103.8	108.8	97.0	102.7	−5.7
1956		40	97.0	103.8	93.2	97.2	−4.0
1957		41	93.2	97.0	89.7	96.3	−6.6
1957		42	89.7	93.2	89.9	94.2	−4.3
1957		43	89.9	89.7	90.2	96.3	−6.1
1957		44	90.2	89.9	89.6	96.5	−6.9

			Lag1	Lag2		Housing permits	
Year	Month	Period	X_1	X_2	Y	Y estimate	Error
A	B	C	D	E	F	G	H
1958		45	89.6	90.2	85.8	95.7	−9.9
1958		46	85.8	89.6	96.9	91.4	5.5
1958		47	96.9	85.8	112.7	106.8	5.9
1958		48	112.7	96.9	122.7	119.8	2.9
1959		49	122.7	112.7	119.8	123.2	−3.4
1959		50	119.8	122.7	117.4	114.4	3.0
1959		51	117.4	119.8	111.9	113.0	−1.1
1959		52	111.9	117.4	104.7	107.7	−3.0
1960		53	104.7	111.9	98.3	102.1	−3.8
1960		54	98.3	104.7	94.9	98.3	−3.4
1960		55	94.9	98.3	93.3	97.6	−4.3
1960		56	93.3	94.9	90.9	97.6	−6.7
1961		57	90.9	93.3	91.9	95.5	−3.6
1961		58	91.9	90.9	97.2	98.0	−0.8
1961		59	97.2	91.9	104.7	103.9	0.8
1961		60	104.7	97.2	107.7	110.0	−2.3
1962		61	107.7	104.7	108.2	109.6	−1.4
1962		62	108.2	107.7	110.7	108.5	2.2
1962		63	110.7	108.2	113.2	111.3	1.9
1962		64	113.2	110.7	114.6	112.9	1.7
1963		65	114.6	113.2	112.2	113.2	−1.0
1963		66	112.2	114.6	120.2	109.6	10.6
1963		67	120.2	112.2	122.1	120.5	1.6
1963		68	122.1	120.2	126.6	118.5	8.1
1964		69	126.6	122.1	122.3	122.8	−0.5
1964		70	122.3	126.6	115.9	115.2	0.7
1964		71	115.9	122.3	116.9	109.9	7
1964		72	116.9	115.9	110.1	114.5	
1965					110.4		
1965					109.8		
1965					112.1		
1965					117.6		
1966					112.1		
1966					96.0		
1966					78.0		
1966					66.9		
1967					83.5		
1967					95.8		
1967					107.7		
1967					113.7		

(#N/A=not available)
(*Source*: Pankratz, 1991, Series #9)

EXHIBIT 13.11
Model Fit of AR(2)
Model for Housing
Permits (Fit:
1947:I–1964:IV; fore-
cast: 1965:I–1967:IV)

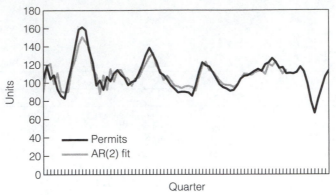

(*Source*: Exhibit 13.10)

It does not necessarily follow that an adequate (good-fitting) model will lead to an accurate forecasting model. However, to have confidence in a forecasting model, we want to be sure that we did the best possible job in identifying an adequate representation of the patterns in the data. In this example, we will examine the adequacy of our model for U.S. housing permits (Exhibit 13.10; from Pankratz, 1983, Case 4; 1991, Series 9), an example of a stationary nonseasonal time series producing stochastic cycles in the forecasting pattern. The model coefficients for the ARIMA (2, 0, 2) model are shown in Exhibit 13.12. Based on the estimated coefficients and their estimated standard errors, a confidence interval for the MA1 parameter can be approximated by $0.11 \pm 2(0.16)$, or $(-0.21, 0.43)$, which suggests that the evidence is lacking for a nonzero MA1 parameter. The ARIMA(2, 0, 2) model estimated by Pankratz (1991) with $\theta_1 = 0$ for these data is:

$$1.20B + 0.54B^2 Y_t = 36.85 + (1 + 0.47B^2)e_t$$

We first examine the residuals (Exhibit 13.13). The residual plot suggests that observation #5 (1984:Q1 = 117.5) is unusual, although this may not be evident from a time plot of the data. The outlier is further confirmed with the probability plot (Exhibit 13.14). Here, observation #5 is the most negative standardized residual and departs from the overall straight-line appearance of the plot. How does one outlier affect the adequacy of the model? We hope it is inconsequential.

The outlier can be corrected by replacing it with its fitted value from the original model (replacing 104.6 with 127.7). The results are summarized in Exhibit 13.15 along with those for the original model for comparison. The outlier-adjusted model has coefficients that differ substantially from the original model. There is also evidence now

EXHIBIT 13.12
Model Summaries for
U.S. Housing Permits,
ARIMA(2, 0, 2)

Model	Coefficient	Standard error	t value
AR1	1.330	0.150	9.000
AR2	−0.630	0.130	−4.800
MA1	0.110	0.160	0.700
MA2	−0.400	0.130	−3.100
Constant	32.580	1.000	32.600
Mean	107.900	3.297	

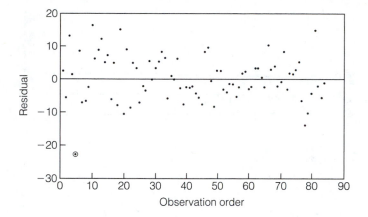

EXHIBIT 13.13
Time Plot of
Residuals from an
ARIMA(2, 0, 2)
Model for Housing
Permits

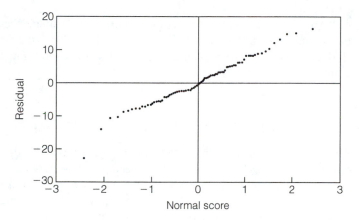

EXHIBIT 13.14
Probability Plot of
Residuals from an
ARIMA(2, 0, 2)
Model for Housing
Permits

EXHIBIT 13.15
Model Summaries for
U.S. Housing Permits;
Outlier Adjusted
ARIMA(2, 0, 2)

Type	Coef Outlier Adjustment	SD	Coef Original
AR1	0.92	0.14	1.33
AR2	−0.3	0.13	−0.63
MA1	−0.5	0.1	0.11
MA2	−0.75	0.08	−0.4
Constant	41.80	1.55	32.58
Mean	108.6	4.03	107.9
MSE	39.8 on 79 df		49.8 on 79 df
p^*	11.1		10.8

that the MA1 parameter should not have been set to 0 in the original model. Whether these differences in parameter estimates have a direct impact on the forecasts generated from these two models is unknown at this stage but that will be examined in the next chapter. It is worth noting, however, that the MSE, a common measure of comparison, is 20% smaller for the adjusted model than for the original model. On the other hand, there is not a noteworthy difference between the stochastic cycle estimates (p^*), both being approximately 11 weeks.

EXHIBIT 13.16
Correlogram of
Residuals from Original ARIMA(2, 0, 2)
Model for Housing
Permits (95% confidence limits)

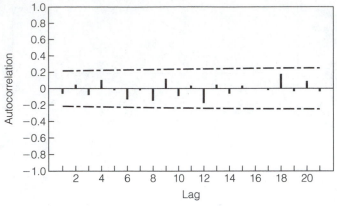

(*Source*: Exhibit 13.13)

There is no evidence in the correlogram of the residuals that an outlier is present or may have an impact on the adequacy of the fit. Both the ordinary correlogram (Exhibit 13.16) and partial correlogram (not shown) show satisfactory support for assuming random residuals.

13.7 DIAGNOSTIC CHECKING: VALIDATING MODEL ADEQUACY

The third and last stage of the model-building strategy is called diagnostic checking, in which analytical techniques are used to detect inadequacies in the model and to suggest model revisions. It may be necessary to introduce new information into the model after this stage and to initiate another three-stage procedure for assessing the new model.

Diagnostic checking involves examining the statistical properties of the residuals. Tests should be designed to detect departures from the assumptions on which the model is based and to indicate how further fitting and checking should be performed. When the model is adequate, the residual series should be independently and randomly distributed about 0. Any dependence or nonrandomness indicates the presence of information that could be exploited to improve the accuracy of the model.

In the diagnostic checking phase, analytical techniques are used to detect inadequacies in the model and to suggest model revisions.

To ensure that the best forecasting model has been obtained, it is important to examine the residuals from the fitted model. If the model is ultimately selected for use as a forecasting tool, the performance of the model should also be monitored periodically during its use so that it can be updated when appropriate.

We consider here three ways in which a fitted model can be checked for adequacy of fit. First, the model can be made intentionally more complex by overfitting; those parameters whose significance is in question can be tested statistically. Second, in a more technical test, correlogram estimates can be tested individually or in an overall chi-squared test. A third check involves a cumulative periodogram test.

Overfitting

When a tentatively identified model is to be enhanced, an additional parameter can be added to the model (overfitting) and the hypothesis that the additional parameter is 0 can be tested by a t test. Thus, overfitting with an AR(2) model or an ARMA(1,1) model, for instance, could test an AR(1) model.

The estimate s_ε^2 of the square of the residual standard error can also be used for diagnostic checking. A plot of s_ε^2 (adjusted for degrees of freedom) against the number of additional parameters should decrease with improved fitting. When overfitting, the improvement, if any, in decreased s_ε^2 may be inconsequential.

Chi-Squared Test

Correlograms and partial correlograms are probably the most useful diagnostic tools available. If the residuals are not random, these diagrams may suggest residual auto-correlations that can be further modeled.

A chi-squared test can be used to evaluate whether the overall correlogram of the residuals exhibits any systematic error. When the chi-squared statistic exceeds the threshold level, the residual series contains more structure than would be expected for a random series. The test statistic, due to Box and Pierce (1970) and modified by Ljung and Box (1978), is given by the formula.

$$Q = n(n+2)\sum (n-k)^{-1}r_k^2$$

where $r_k(k=1, \ldots, m)$ are residual autocorrelations selected (typically $m=20$) n is the number of observations used to fit the model, and m is usually taken to be 15 or 20. Then, Q has an approximate chi-squared distribution with $(m-p-q)$ degrees of freedom. For an ARIMA(p, d, q) model, n is the number of terms in the differenced data. This is a general test of the hypothesis of model adequacy, in which a large observed value of Q points to inadequacy. Even if the statistic Q is not significant, a review of the residual time series for unusual values is still appropriate.

A chi-squared test statistic is used to evaluate whether the overall correlogram of the residuals exhibits any systematic error.

It may also appear practical to examine individual values in the correlogram of the residuals relative to a set of confidence limits. Those values that fall outside the limits are examined further. Upon investigation, appropriate MA or AR terms can be included in the model.

Just as there are for the ordinary correlation coefficient, there are approximate confidence limits for the theoretical autocorrelation coefficients that establish which autocorrelation coefficient can reasonably be assumed to be 0. As a rough guide for determining whether or not theoretical autocorrelations are 0 beyond lag q, Bartlett (1946) has shown that, for a sample of size n, the standard deviation of r_k (an estimated standard error to be plugged into a confidence interval statement) is approximately

$$n^{\frac{1}{2}}\left[1 + 2(r_1^2 + r_2^2 + \cdots + r_q^2)\right]^{\frac{1}{2}} \qquad \text{for } k > q$$

Quenouille (1949) has shown that, for a pth-order autoregressive model, the standard errors of the partial autocorrelogram estimates ϕ_{kk} are approximately $n^{\frac{1}{2}}$ for $k > p$. Assuming normality for moderately large samples, as shown by Anderson (1942), the limits of plus or minus two standard errors about 0 should provide a reasonable guide in assessing whether or not the autocorrelation coefficients are effectively different from 0.

Periodogram Analysis

In case of periodic nonrandom effects, the cumulative periodogram can be an effective diagnostic tool. The test has its basis in frequency-domain analysis of time series (Jenkins and Watts, 1968).

For a residual series $\{e_t\}$ a periodogram $I(f_I)$ is defined by

$$I\left(f_I\right) = \frac{1}{2}n\left[\left(\sum e_t \cos 2\pi f_i t\right)^2 + \left(\sum e_t \sin 2\pi f_i t\right)^2\right]$$

where $f_i = i/n$, $I = 0, 1, \ldots, n/2$, is the frequency. Large values of $I(f_I)$ are produced when a pattern with given frequency f_i in the residuals correlates highly with a sine or cosine wave at the same frequency. Then the normalized cumulative periodogram is defined by

$$C\left(f_i\right) = \left[\sum I\left(f_i\right)\right]/ns^2$$

where s^2 is an estimate of σ_ε^2.

For a white-noise residual series, the plot of $C(f_i)$ against f_i, $j = 0, 1, \ldots, n/2$, is scattered about a straight line joining the coordinates (0, 0) and (0.5, 1). Inadequacies in the fit to a model will show up as systematic deviations from this line. A standard statistical test, called the Kolmogorov-Smirnov test, uses the maximum vertical deviation of the plot from the straight line as the statistic for determining unsuspected periodicities in the data (Box et al., 1994, Chap. 8).

13.8 IMPLEMENTING NONSEASONAL ARIMA MODELS

Let us now consider some examples of real data and their associated correlograms. It is helpful to make some observations regarding the displays.

- A time plot is useful to establish what data adjustments and transformations may be needed. The appropriate amount of differencing should be performed to achieve stationarity, and the differenced data should be plotted.

- The ordinary correlogram should be inspected for pure AR or MA structure.

- It is helpful to inspect the partial correlogram to confirm or supplement the information derived from the ordinary correlogram.

The inspection of the ordinary and partial correlograms together should suggest a preliminary model to be fitted in the first iteration of the model-building process. The most significant patterns in the correlograms should be documented for future reference.

The theoretical forms of an ACF and PACF are the reference tools for visualizing the structure of ARIMA models. The interpretation of the correlograms (estimated ACF and PACF patterns with real data) is more of an art than a science. To learn how various time series models can be identified by these patterns in empirical versions of the ACF (correlogram) and PACF (partial correlogram), see examples using real data in Bowerman and O'Connell (2004), Box and Jenkins (1976), Gaynor and Kirkpatrick (1994), Granger (1989), Granger and Newbold (1986), Hoff (1984), Makridakis et al. (1998), and Pankratz (1983).

The quickest way to acquire the expertise necessary to identify ARIMA models is by looking at examples.

At the early modeling stages, it is generally simpler to attempt to identify only the most obvious patterns. Once the parameter estimates have been obtained for a tentative model, the correlogram of the residuals derived from this model should be inspected for any significant secondary patterns. With experience, it is possible to identify complex patterns relatively quickly. For the beginner, however, learning by example is recommended. Exhibit 13.17 provides general guidance for selecting starting models. For example, if the ACF and PACF both decay, we recommend using an ARIMA(p, 0, q) model. If the ACF and PACF both truncate, try the simpler AR(p) and MA(q) models, especially if one cuts off more quickly.

Index of Consumer Sentiment

The University of Michigan's Survey Research Center (SRC) publishes a quarterly index of consumer sentiment. The index of consumer sentiment is published in the *Business Conditions Digest*, a monthly publication of the U.S. Department of Commerce. The data come from a survey that attempts to ascertain the anticipations and intentions of consumers. Although they reflect only the respondents' anticipations (what they expect others to do) or expectations (what they plan to do), and not firm commitments, such information is nevertheless useful as a valuable aid to economic forecasting. A plot of monthly values was shown in Exhibit 5.5.

EXHIBIT 13.17 Model Identification for Nonseasonal Stationary Time Series		Autocorrelation Function	
	Partial Autocorrelation	Pattern Decays	Truncates
	Truncates	AR	Mixed (ARMA)
	Decays	Mixed (ARMA)	MA

The quarterly index (58 quarterly values from 1978: 1 to 1992: 2) is shown in Exhibit 13.18, and the ordinary and partial correlograms are shown in Exhibit 13.19. The spike at lag 1 (= 0.9) in the partial correlogram and the slowness of decay in the ordinary correlogram suggest that the series may be neither stationary in level nor seasonal, so that a first difference should be taken.

EXHIBIT 13.18
Time Plot of the University of Michigan Survey Research Center Index of Consumer Sentiment (quarterly)

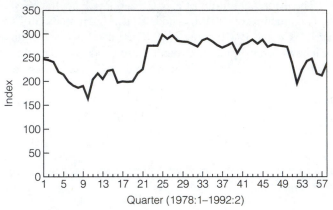

(*Source*: University of Michigan, SRC)

EXHIBIT 13.19
Consumer Sentiment Index Series: (a) Ordinary Correlogram; (b) Partial Correlogram

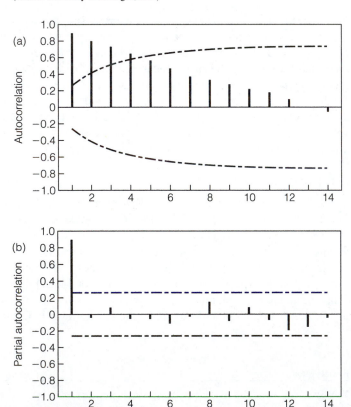

(*Source*: Exhibit 13.18)

Exhibit 13.20 shows the corresponding ordinary and partial correlograms of the first-differenced data. These do not show any significant spikes, so a tentative model turns out to be the simple ARIMA(0,1,0) model (based on 54 values with a holdout sample of 4, it has an estimated $\alpha = -0.2$):

$$Y_t - Y_{t-1} = \alpha + \varepsilon_t$$

An alternative stationary AR(1) model based on 54 values

$$Y_t = 25.1 + 0.9Y_{t-1}$$

A four-period forecast was derived from the two models and compared the actual values by using a holdout sample of the last four values (#58, #57, #56, and #55) of the data. The holdout sample and the accuracy statistics are summarized in Exhibit 13.21. Although the differences are not great, the direction of the AR(1) forecasts appear to take on a different slope than the direction of the data, at least in the very near term.

EXHIBIT 13.20
First Differences
of the Consumer
Sentiment Index
Series: (a) Ordinary
Correlogram; (b)
Partial Correlogram

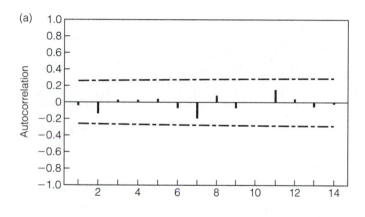

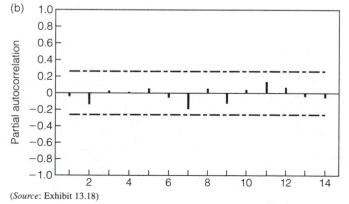

(Source: Exhibit 13.18)

EXHIBIT 13.21	Actual	Percentage Error of AR(1) Model	Percentage Error of First Difference Model
Comparison of the Forecast Performance of AR(1) and ARIMA(0,1,0) Models for SRC Consumer Sentiment Index	#55: 242.2	2.1	2.4
	#56: 247.9	−12.7	−12.2
	#57: 215.6	−14.6	−13.8
	#58: 212.3	−2.8	−1.9
	MPE	−7.0	−6.4
	MAPE	8.1	7.6

(*Source*: Exhibit 13.18)

Seasonally Adjusted U.S. Money Supply

The monthly seasonally adjusted U.S. money supply (M2) series (Exhibit 4.26) is not stationary, as is clear from the slow decay in the ordinary correlogram in Exhibit 13.22. Hence, it requires a first difference. After differencing the series once, the plot does not indicate trend but, rather, shows increasing variability over time (Exhibit 4.27). Because the original money-supply series probably requires a variance-stabilizing transformation, logarithms may indeed give better results. The first differences of the log-transformed data (Exhibit 4.30) appear stationary. The correlograms (Exhibit 13.23) suggest the ARIMA(1,1,0) model on the log-transformed data because of the

EXHIBIT 13.22
(a) Ordinary and (b)
Partial Correlograms
of the Money Supply
Data

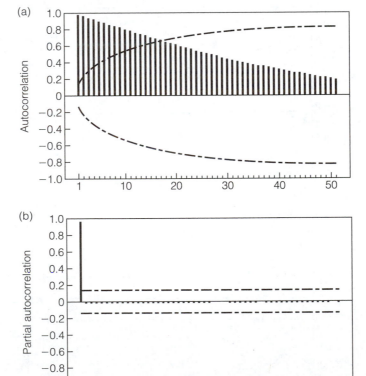

(*Source*: Exhibit 4.26)

EXHIBIT 13.23
(a) Ordinary and (b) Partial Correlograms of the First Differences of Log-Transformed Money-Supply Data

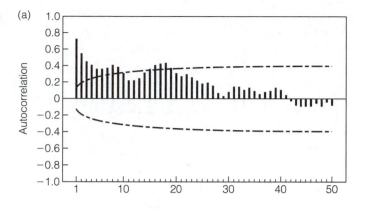

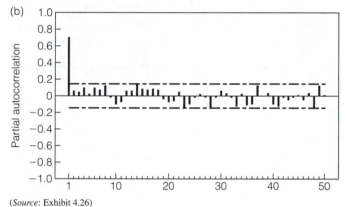

(*Source*: Exhibit 4.26)

EXHIBIT 13.24
Fitted AR(1) Model to the First Differences of Log-Transformed Money Supply

Final Estimates of Parameters

Type	Coef	SD	t
AR1	0.7315	0.0489	15.0
Constant	0.0011	0.0001	8.3

Differencing: 1 regular difference
Number of observations: Original series 204, after differencing 203
Residuals: SS = 0.000775890 (backforecasts excluded)
          MS = 0.000003860 df = 201

Modified Box-Pierce (Ljung-Box) Chi-Squared Statistic

Lag	12	24	36	48
Chi-squared	12.4	31.5	51.0	64.2
df	10	22	34	46

(*Source*: Exhibit 4.26)

decay in the ordinary correlogram and the spike at lag 1 in the partial correlogram. Thus, a tentative model is an ARIMA(1, 1, 0) for the log-transformed money-supply data. The results (Exhibit 13.24) after fitting this model show statistical significance for the parameter (AR1 = 0.7315; $t = 15$) and the constant term (= 0.0011; $t = 8.3$).

EXHIBIT 13.25
(a) Ordinary and (b)
Partial Correlograms
of the Second
Differences of the
Money-Supply Data

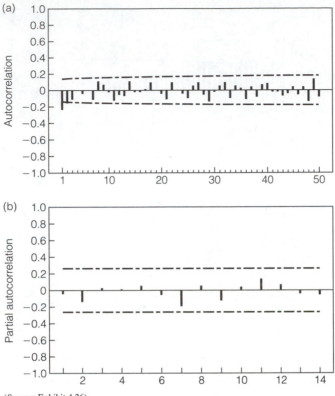

(*Source*: Exhibit 4.26)

Another possibility that could be tried on these data is to take two successive first differences—that is, $(1 - B)^2$—of the original data (Exhibit 4.28). The ordinary and partial correlograms (Exhibit 13.25) suggest a low-order MA model for the second differences, such as an ARIMA(0, 2, 2) model. This is based on the significant negative spikes at lags 1 and 2 in the ordinary correlogram.

The results after fitting this model show statistical significance for the moving average parameters MA1 ($= 0.47$; $t = 6.8$) and MA2 ($= 0.31$; $t = 4.5$). Exhibit 13.26 shows the ordinary and partial correlograms of the residuals of this tentative model. The increasing variability with time may still be a problem, but only a forecast test with holdout samples can resolve which model is most suitable. Although a forecast simulation will provide one of the more valuable inputs to the model-selection process for forecasting, we will pass on taking such a step at this time. In any case, it cannot hurt to have a couple of competing models for ongoing monitoring and analysis in any forecasting application.

The FRB Index of Industrial Production

The FRB index of industrial production is not stationary because level and slope both change over time (see Exhibit 4.2). The very gradual decay in the pattern of correlations in the ordinary correlogram in Exhibit 13.27a suggests taking a first difference.

EXHIBIT 13.26
(a) Ordinary and
(b) Partial Correlo-
grams of the Residu-
als of a Tentative
ARIMA(0, 2, 2)
Model for the Money-
Supply Series (95%
confidence limits)

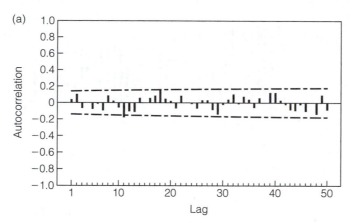

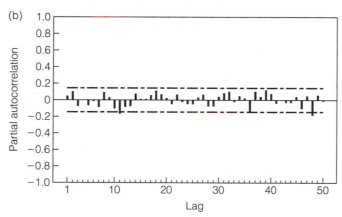

(*Source*: Exhibit 4.26)

The first differences of the FRB data (not shown) do appear stationary. The ordinary and partial correlograms in Exhibit 13.28 illustrate the difficulty in interpreting them precisely. One interpretation (Granger, 1980, p. 74) says there is a decaying pattern in the ordinary correlogram and a truncation at lag 2 in the partial correlogram. This interpretation is based on an AR(1) model on the first differences for the time period January 1948 to October 1974. On the basis of the patterns in Exhibit 13.28, we could also try an AR(2) on the first differences; this is an ARIMA(2, 1 ,0). In fitting this model, the coefficients turn out to be significant for both the AR1 parameter ($\phi_1 = 0.33$; $t = 7.2$) and the AR2 parameter ($\phi_2 = 0.20$; $t = 4.3$).

An alternative interpretation is a decaying pattern in the partial correlogram and a truncation at lag 3 in the ordinary correlogram. This suggests an ARIMA(0, 1, 2) or ARIMA(0, 1, 3). The latter model yielded statistically significant results: MA1 parameter ($\theta_1 = 0.29$; $t = -6.2$), MA2 parameter ($\theta_2 = -0.21$; $t = -4.5$), and MA3 parameter ($\theta_3 = -0.20$; $t = -4.3$).

For forecasting purposes, it is prudent to perform some forecast simulations with holdout samples before finalizing the model. Once again, having a couple of competing models at hand will always be very useful.

EXHIBIT 13.27
(a) Ordinary and (b)
Partial Correlograms
of the FRB Index of
Industrial Production

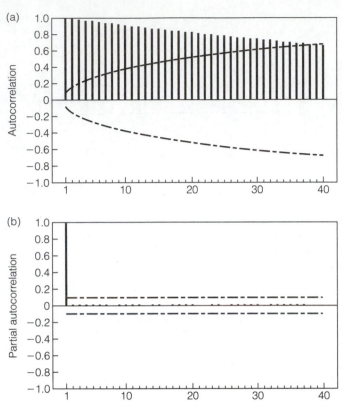

(*Source*: Exhibit 4.2)

13.9 IDENTIFYING SEASONAL ARIMA MODELS

As with regular ARIMA models, seasonal ARIMA models can be classified as autoregressive and moving average. The seasonal MA process is analogous to the regular MA process. It differs from a regular MA process of order 12 where there could be significant values at lags 1 *through* 12 in the ACF. If there is a single dominant value at 12, this indicates a seasonal MA process.

In practice, it is not unusual to find significant spikes at other lags of, say 5 or 9 in the correlogram of a residual series. Generally, these are spurious and may not suggest a secondary seasonal pattern. Their removal through a lag structure generally has little impact on forecasts generated from such a model. It is important, however, to isolate and interpret seasonal lags corresponding to realistic periodicities.

The ACF of a pure seasonal MA process has a single value at the period of the seasonality. The PACF shows a decaying pattern at multiples of 12 if the seasonality has a 12-month period.

The seasonal AR process is also analogous to the regular AR process. However, the pattern of decay that is evident in the ACF is noticed at multiples of the period.

EXHIBIT 13.28
(a) Ordinary and (b)
Partial Correlogram
of First Differences
of the FRB Index of
Industrial Production

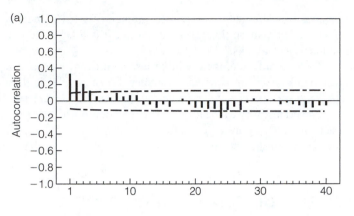

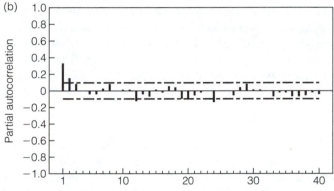

(*Source*: Exhibit 4.2)

For example, the ACF of a first-order seasonal (monthly) AR process has a decaying pattern in the values at multiples of 12. The PACF has a single value at lag 12. It is worth noting that pure monthly seasonal models look like 12 independent series, so that the ACF and PACF are approximately 0, except at multiples of 12.

The ACF of a particular simple combined regular and seasonal moving average process is depicted in Exhibit 13.29. On expanding the factors in the model, it

EXHIBIT 13.29
ACF of a Combined
Regular and Seasonal
MA Model: $Y_t = (1 - \theta_1 B)(1 - \theta_{12} B^{12})\varepsilon_{t-1}$;
$Y_t = \varepsilon_t - \theta_1 \varepsilon_{t-1} - \theta_{12} \varepsilon_{t-12} + \theta_1 \theta_{12} \varepsilon_{t-13}$

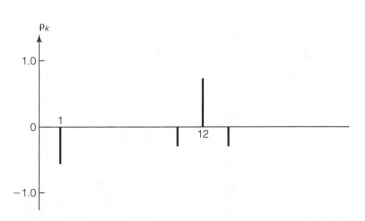

becomes apparent that the current error and errors 1, 12, and 13 periods back in time affect Y_t. The pattern that results in the ACF is a large value at lags 1 and 12 and smaller values at lags 11 and 13.

As a general rule, there will be less significant values at lag 12 plus or minus each regular moving average parameter. For example, if the process is a regular MA(2), there will be smaller values at lags 10, 11, 13, and 14.

The easiest way to identify the order of combined MA processes is to introduce a seasonal moving average parameter in the model. The order of the regular MA parameter will then be apparent from the correlogram of the residuals.

13.10 IMPLEMENTING SEASONAL ARIMA MODELS

Let us examine a more detailed example to illustrate the use of ARIMA models. This analysis will result in a forecast of hotel/motel demand (DCTOUR on the CD); the data have been previously plotted in Exhibit 1.4 and analyzed in Chapter 6 (seasonal decomposition) and Chapter 8 (exponential smoothing). A preliminary ANOVA decomposition (not shown) suggests that the variation due to trend, seasonality, and irregularity accounts for 12.1%, 86.6%, and 1.3%, respectively, of the total variation about the mean.

Preliminary Data Analysis

A time plot of the historical data and various differenced and transformed data are shown in Exhibits 13.30–13.33. These give us a basis for identifying several competing models for the data. Series A is the first difference of the basic data. Series B is the first difference of the log-transformed data. Series A behaves like changes in the data, whereas series B acts much like percentage changes. We will seek seasonal ARMA models for series A and B.

Exhibit 13.30 illustrates an upward trend in the historical data and slightly increasing dispersion in the seasonal variation over time. The seasonal volatility can be validated in the first differences of the data (Exhibit 13.31a). Note that the seasonality appears slightly more uniform in the first differences of the log-transformed data

EXHIBIT 13.30
Time Plot of 8 Years
of Monthly Hotel/
Motel Demand

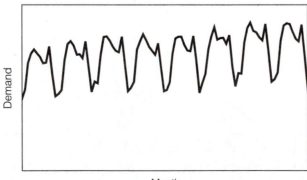

(*Source*: Exhibit 1.4)

EXHIBIT 13.31
Time Plots of (a)
First Differences of
the Hotel/Motel
Demand (series A)
and (b) First
Differences of the
Log of Hotel/Motel
Demand (series B)

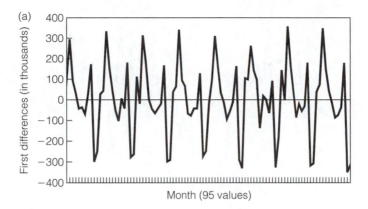

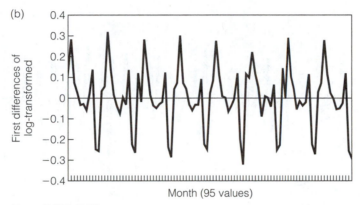

(*Source*: Exhibit 13.30)

(Exhibit 13.31b) than in the untransformed version. The period 12 differences show a cyclical pattern (Exhibit 13.32a). The combination of differences of period 1 and 12 finally appear almost stationary (Exhibit 13.32b); the latter is designated series C.

Taking logarithms can reduce the increasing dispersion in the data over time. Another stationary series (series D) is obtained by taking differences of periods 1 and 12 after log-transforming the data. For purposes of the analysis, it is sufficient to model the differenced data (of periods 1 and 12 together) with and without the logarithmic transformation.

The next step is to examine the ordinary and partial correlograms and match them with a catalog of ACF and PACF patterns; a wide variety of these patterns can be found in the textbooks. Exhibit 13.33 shows seasonality in the pattern and, hence, the underlying data are seasonal (not surprising). Along with the partial correlogram, we may conclude that series B is much like a seasonal AR(1) model. The significant spikes at lag 1 in both correlograms can be interpreted as either a regular AR or MA term. For balance in the model, we will opt for the MA(1) term. The same result applies to series A because the ordinary correlogram for series B (Exhibit 13.33a) and series A (not shown) look essentially the same.

There are some noteworthy spikes, surrounding lag 9 (spikes 8 and 10 appear significant). This suggests that a 'secondary' seasonality may be present, perhaps

EXHIBIT 13.32
Time plots of (a) the
Period 12 Differences
and (b) the First and
Period 12 Differences
of the Hotel/Motel
Demand (series C)

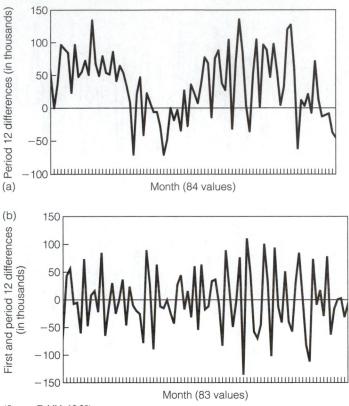

(*Source*: Exhibit 13.30)

attributable to the average period of "long" months in a year. For a tentative model, this should not be of concern.

A tentative (seasonal autoregressive) model with a regular MA(1) term was fit to series A and B, and the correlograms of the residuals were plotted (not shown). The coefficient for the seasonal AR1 parameter ($= 0.99$; $t = 35$) in series B suggests that a seasonal difference is in order—when expressed in backshift notation, a seasonal AR1 parameter is $(1 - 0.99B^{12})$, which is very close to a seasonal difference $(1 - B^{12})$. The regular MA1 was significant ($\theta_1 = 0.75$; $t = 10.4$).

The ordinary and partial correlograms of series D (the first and period 12 differences of the log-transformed hotel/motel demand) was examined next (Exhibit 13.34). These patterns can have a couple of interpretations. Recall that we are not necessarily interested in only a single "best" model. It serves us better to have a couple of different, but competing, models for developing and tracking forecasts. Consequently, the decaying pattern in the estimated PACF suggests an ARMA(0, 2) model, assuming the estimated ACF truncates at lag 2. On the other hand, a decaying pattern in both the estimated ACF and PACF suggests an ARMA(1, 1) model.

The resultant estimated parameters for the first instance turn out to be MA1 = 0.3 and $t = 5.7$, and MA2 = 0.3 and $t = 1.1$. Apparently, the MA2 parameter is not

EXHIBIT 13.33
(a) Ordinary and (b)
Partial Correlograms
of Series B: First
Differences of Log-
Transformed of the
Hotel/Motel
Demand Series

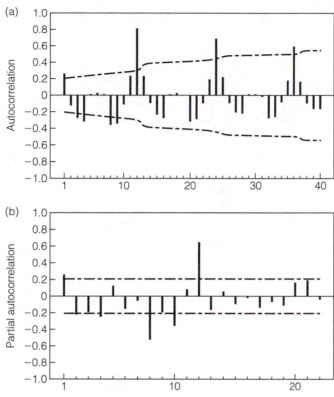

(Source: Exhibit 13.30)

necessary, so the ARMA(0, 1) is sufficient. The ARMA(1, 1) model turns out to have estimated parameters AR1 = 0.08 and $t = 0.5$ (not significant), and MA1 = 0.77 and $t = 7.7$. We end up with an ARMA(0, 1) model for series D with estimated MA1 coefficient = 0.75 and $t = 9.8$. This result is consistent with our analysis of series B.

Taking logarithms turns out to be useful for these data; it eliminates the increasing variability found in the time series. Series C was constructed without first taking logarithms as a means to compare its impact on forecasting.

After fitting the ARMA(0, 1) model to series D, we plotted ordinary and partial correlograms of the residuals. These are shown in Exhibit 13.35. There is still a significant (negative) spike at lag 12 in both correlograms. We will opt for a seasonal MA(1) term to account for it. Exhibit 13.36 displays the ordinary and partial correlograms after taking this last modeling step. There is no more evidence of any useful structure to be extracted from the data.

Model Summary

We have tried and built a number of models. In the backshift notation, these models are estimated from the ARIMA processes shown in Exhibit 13.37. Model E is attributed to Frechtling (1996) and comes from the source of the hotel/motel demand

EXHIBIT 13.34
(a) Ordinary and (b)
Partial Correlograms
of Series D: First and
Period 12 Differ-
ences of the Log-
Transformed
Hotel/Motel
Demand Series

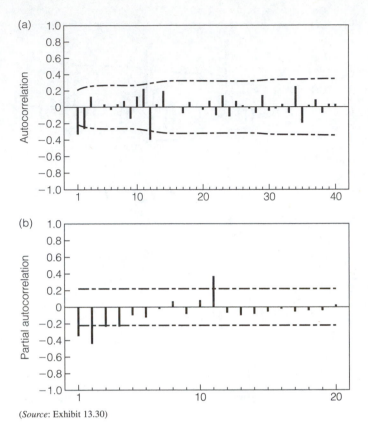

(*Source*: Exhibit 13.30)

(DCTOUR) data. How different are these models in terms of forecasting performance? Although summary statistics can be helpful, the acid test is a forecast evaluation with a holdout sample.

Some Forecast Test Results

Forecast tests were performed for all the models. By holding out Year 8, 12 one-month-ahead forecasts were generated in Year 8, 11 two-month-ahead forecasts, and so on, until we have one 12-month-ahead forecast. The waterfall chart is shown for model D in Exhibit 13.38, displaying the percentage errors ($PE = 100 \times$ (Actual − Forecast)/Actual) and the MAPE (average of absolute PE) for each row.

Exhibit 13.39 summarizes the results of the tests using the MAPE as the criteria. Model D had the best results (when measured in terms of average absolute percent of forecast error), followed by Model C. Models A, B, and E can be safely discarded because they are consistently outperformed and do not appear very competitive. That leaves the two models based on the regular and seasonal differencing. Some other observations can be made about these results:

- All the models exhibit a seasonal profile over the forecast horizon.
- All models tended to overforecast Year 8.

EXHIBIT 13.35
(a) Ordinary and (b)
Partial Correlogram
of the Residuals from
the ARMA(0,1)
Model for Series D
(95% confidence
limits)

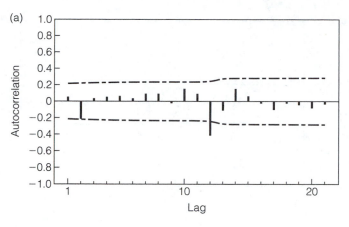

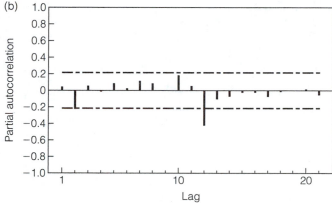

(*Source*: Exhibit 13.30)

- Rival models C and D are versions of the so-called airline model, a model commonly found in seasonal forecasting applications with the Box-Jenkins method.

- Model D (log-transformed) was outperformed by model C only during July through October.

- Log-transformed data reduced seasonal volatility and yielded the best results.

- It should be possible to forecast each of the first six periods within 3% and the remaining six periods individually within 6 or 7%, on average.

If model D is true, 95% confidence limits can be constructed for the forecasts in the holdout period, based on a fitted model up to the holdout period. Exhibit 13.40 displays the forecasts and lower and upper limits calculated as a percentage deviation from the actual hotel/motel demand for the holdout period. As noted before, the forecasts are greater than the actuals. However, the actuals are within the 95% confidence limits for the forecast (except for December 1994). Also note that, because of the log transformation, the confidence limits are asymmetric and are wider for larger values of the forecast.

EXHIBIT 13.36
(a) Ordinary and (b) Partial Correlogram of the Residuals from a Seasonal MA(1) and ARMA(0,1) Model for Series D (95% confidence limits)

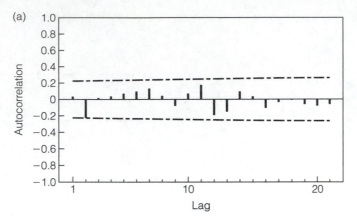

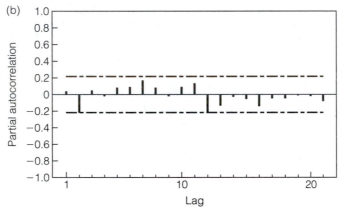

(*Source*: Exhibit 13.30)

EXHIBIT 13.37
ARIMA Model for Hotel/Motel Demand (DCTOUR) in Backshift Notation

Series A	$(1 - \phi_{12} B^{12})(1 - B)Y_t = (1 - \theta_1 B)\varepsilon_t.$
Series B	$(1 - \phi_{12} B^{12})(1 - B) \ln Y_t = (1 - \theta_1 B)\varepsilon_t.$
Series C	$(1 - B^{12})(1 - B)Y_t = (1 - \theta_1 B)(1 - \theta_{12} B^{12})\varepsilon_t$
Series D	$(1 - B^{12})(1 - B) \ln Y_t = (1 - \theta_1 B)(1 - \theta_1 B^{12})\varepsilon_t.$
Frechtling (1996)	$(1 - B^{12})(1 - B)(1 - \phi_1 B)Y_t = (1 - \theta_1 - \theta_2 B)\varepsilon_t$

13.11 ARIMA MODELING CHECKLIST

The following checklist can be used to help identify gaps in the forecasting process that need attention. It can be scored or color coded on three levels: Green = YES, Yellow = SOMEWHAT, and Red = NO.

_____ Does the time plot show recognizable trend, seasonal, or cyclical pattern?

_____ Is the series stationary?

_____ Do seasonal and regular differencing help to create stationarity?

_____ Does the trending series show increasing volatility in the seasonal pattern? Consider taking log transformations of the data first.

EXHIBIT 13.38   Waterfall Chart with Holdout Sample (Year 8 = 1994) for Model D for Hotel/Motel Demand

Hold-Out	1	2	3	4	5	6	7	8	9	10	11	12	MAPE (%)
						PE (%)							
1,001,666	−5.38	−2.22	−1.28	−0.70	−2.33	2.53	−0.36	−2.60	−1.72	−0.06	−2.12	−4.07	2.1
1,073,196	−3.85	−1.95	−1.08	−2.55	1.86	0.41	−2.72	−2.52	−0.59	−2.14	−4.73		2.2
1,421,423	−3.58	−1.75	−2.94	1.65	−0.27	−1.92	−2.62	−1.36	−2.66	−4.75			2.4
1,577,321	−3.37	−3.62	1.28	−0.49	−2.64	−1.83	−1.48	−3.46	−5.29				2.6
1,600,991	−5.27	0.62	−0.88	−2.85	−2.54	−0.70	−3.58	−6.12					2.8
1,594,481	−0.96	−1.55	−3.24	−2.75	−1.39	−2.77	−6.23						2.7
1,510,052	−3.16	−3.93	−3.16	−1.61	−3.49	−5.41							3.5
1,436,164	−5.58	−3.84	−2.00	−3.71	−6.14								4.3
1,404,978	−5.49	−2.68	−4.11	−6.36									4.7
1,585,409	−4.32	−4.80	−6.78										5.3
1,234,848	−6.46	−7.49											7.0
923,115	−9.20												9.2

(*Source*: Exhibit 13.30)

EXHIBIT 13.39   Forecast Test Results with Holdout Sample (Year 8 = 1994)
for the Models A–E for Hotel/Motel Demand (DCTOUR)

MAPE

Model A	Model B	Model C	Model D	Model E
3.2	3.1	2.8	**2.1**	3.0
2.8	2.8	2.6	**2.2**	3.0
3.0	3.1	2.5	**2.4**	2.8
3.1	3.3	2.8	**2.6**	3.3
3.6	3.6	3.3	**2.8**	3.4
3.2	3.4	2.7	**2.7**	3.4
4.2	4.4	**2.9**	3.5	4.1
5.0	5.1	**3.8**	4.3	5.1
5.6	5.7	**3.9**	4.7	5.6
6.6	6.8	**4.6**	5.3	6.8
9.2	8.8	7.7	**7.0**	9.1
12.9	11.4	12.4	**9.2**	12.9

(*Source*: Exhibit 13.30)

EXHIBIT 13.40   Model D 95% Confidence Interval for Forecasts in Holdout Sample (Year 8 = 1994); Forecasts, Lower Limit
(LCL), and Upper Limit (UCL) Expressed as Percentage Deviation from the Actual Hotel/Motel Demand (DCTOUR)

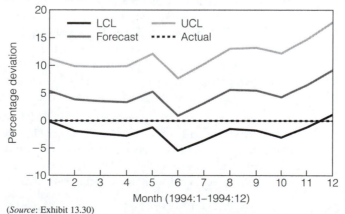

(*Source*: Exhibit 13.30)

_____ Are you watchful of overdifferencing? Be alert to the requirement that moving average parameters corresponding to the differencing that has been taken must achieve stationarity. For example, first differences may induce the need for a moving average parameter of order 1. Differencing with period 4 may induce the need for a seasonal moving average parameter in quarterly series. A check of the correlogram helps here. Overdifferencing may be apparent if regular and seasonal moving average parameter estimates are very close to unity.

_____ Create ordinary and partial correlograms. Are patterns recognizable?

_____ Can correlogram patterns be matched to a theoretical ACF and PACF catalog?

_____ Have all the model parameters (i.e., terms) been included in the model as required?

_____ Have you examined the correlation matrix? There should *not* be a high degree of correlation between parameter estimates (e.g., over 0.7).

_____ Have you examined parameter estimates and their standard errors? The confidence interval for each parameter (including the trend constant if there is one) should not span zero but should be either positive or negative. If the confidence interval does include zero and the interval is basically symmetric about zero, then consideration should be given to eliminating the parameter.

_____ If a first-order regular autoregressive or seasonal autoregressive term is included in the model, the parameter estimates should not be close to 1.0 (e.g., over 0.90 or 0.95). If either is close to 1.0, a regular or seasonal difference should be tried in the model and a forecast test performed to determine whether the differenced model or the autoregressive model is better.

_____ Do the sum of squares of the errors and the standard error of the residuals become smaller as fits of the model improve?

_____ Does the chi-squared statistic fall below the critical value for the associated degrees of freedom? If so, this indicates white noise. (*Note*: A quick, conservative check for white noise is whether the model has a chi-squared statistic below the number of degrees of freedom.)

_____ Are there significant patterns in the correlograms of the residuals? Review the ordinary and partial correlograms for any remaining pattern in the residuals. Give primary emphasis to patterns in the correlogram. (*Note*: If the confidence limits on the correlogram and partial correlogram are approximately 95%, 1 spike in 20 can be expected to be outside the confidence limits due to randomness alone. Decaying spikes in the correlogram, which by visual inspection appear to originate at 1.0, indicate a regular autoregressive process; decaying spikes that originate at a lower level indicate a mixed autoregressive moving-average model.)

_____ Has a deterministic trend constant been estimated? It is suggested that a deterministic trend constant not be added until the very last step. Then, if the mean of the residuals of the final model is more than one standard deviation from zero, put a trend constant in the model and test it for significance. (There may be times when we do not want the model to put a trend in the forecasts.)

_____ There are some combinations of parameters that are unlikely to result within the same model. If these combinations are present, reevaluate the previous analysis of seasonal differences and seasonal autoregressive parameters, seasonal autoregressive and seasonal moving average parameters, and seasonal moving average parameters at other than lags 4 or 12 (or multiples of 4 and 12) for quarterly and monthly series, respectively.

SUMMARY

The Box-Jenkins methodology presents a formally structured class of time series models that are sufficiently flexible to describe many practical situations. Known as ARIMA models, these are capable of describing various stationary and nonstationary (time-dependent) phenomena (unlike ARMA models, which require stationary data) in a statistical rather than in a deterministic manner. This chapter has introduced a three-stage Box-Jenkins modeling process; the three stages are (1) identification (or specification) of forecasting models by using data analysis tools (plotting of raw, differenced, and transformed data), correlograms, and partial correlograms, for the purpose of making tentative guesses at the order of the parameters in the ARMA model; (2) estimating parameters for tentative models; and (3) diagnostic checking, which is a critical step in looking for model inadequacies or for areas where simplification can take place. The rules for identification are (1) if the correlogram cuts off at some point, say $k = q$, then the appropriate model is $MA(q)$; (2) if the partial correlogram cuts off at some point, say $k = p$, then the appropriate model is $AR(p)$; and (3) if neither diagram cuts off at some point, but does decay gradually to zero, the appropriate model is $ARMA(p', q')$ for some p', q'.

REFERENCES

Anderson, R. L. (1942). Distribution of the serial correlation coefficient. *Ann. Math. Stat.* 13, 1–13.

Bartlett, M. S. (1946). On the theoretical specification and sampling properties of autocorrelated time-series. *J. Roy. Statist. Soc.* B8, 27–41.

Bowerman, B. L., and R. O'Connell (2004). *Forecasting and Time Series—An Applied Approach.* 4th ed. Belmont, CA: Duxbury Press.

Box, G. E. P., and G. M. Jenkins (1976). *Time Series Analysis: Forecasting and Control.* Rev. ed. San Francisco. CA: Holden-Day.

Box, G. E. P., G. M. Jenkins, and G. Reinsel (1994). *Time Series Analysis, Forecasting and Control.* 3rd ed. Upper Saddle River, NJ: Prentice Hall.

Box, G. E. P., and D. A. Pierce (1970). Distribution of residual autocorrelations in autoregressive integrated moving-average time-series models. *J. Am. Statist. Assoc.* 65, 1509–26.

Brockwell, P. J., and R. A. Davis (1996). *An Introduction to Time Series and Forecasting.* New York: Springer-Verlag.

DeLurgio, S. A. (1998). *Forecasting Principles and Applications.* New York: Irwin-McGraw Hill.

Frechtling, D. C. (1996). *Practical Tourism Forecasting.* Oxford: Butterworth-Heinemann.

Gaynor, P., and R. C. Kirkpatrick (1994). *Introduction to Time-Series Modeling and Forecasting in Business and Economics.* New York: McGraw-Hill.

Granger, C. W. J. (1980). *Forecasting in Business and Economics.* New York: Academic Press.

Granger, C. W. J. (1989). *Forecasting in Business and Economics.* 2nd ed. New York: Academic Press.

Granger, C. W. J., and P. Newbold (1986). *Forecasting Economic Time Series*. 2nd ed. Orlando, FL: Academic Press.

Hoff, J. (1984). *A Practical Guide to Box-Jenkins Forecasting*. Belmont, CA: Lifetime Learning Publications.

Jenkins, G. M. (1979). *Practical Experiences with Modeling and Forecasting Time Series*. Jersey, UK: GJP (Overseas) Ltd.

Jenkins, G. M., and D. G. Watts (1968). *Spectral Analysis and Its Applications*. San Francisco: Holden-Day.

Makridakis, S., S. c. Wheelwright, and R. J. Hyundman (1 Forecasting Methods and Applications, 3rd ed. New York and Sons.

Newbold, P., and T. Bos (1994). Introductory Business F ed., Cincinnati, Ohio: South-Western Publishing Co.

Pankratz, A. (1983). Forecasting with Univariate Box-Jenki Concepts and Cases. New York: John Wiley & Sons.

Vandaele, W. (1983). Applied Time Series and Box-Jenki Orlando, FL: Academic Press.

Pankratz, A. (1991). Forecasting with Dynamic Regression York: John Wiley & Sons.

Makridakis, S., S. C. Wheelwright, and R. J. Hyndman (1998). *Forecasting Methods and Applications*, 3rd ed. New York: John Wiley and Sons.

Newbold, P., and T. Bos (1994). *Introductory Business Forecasting*. 2nd ed., Cincinnati, Ohio: South-Western Publishing Co.

Pankratz, A. (1983). *Forecasting with Univariate Box-Jenkins Models: Concepts and Cases*. New York: John Wiley & Sons.

Pankratz, A. (1991). *Forecasting with Dynamic Regression Models*. New York: John Wiley & Sons.

Quenouille, M. H. (1949). Approximate tests of correlation in time-series. *J. Roy. Statist. Soc.* B11, 68–84.

Vandaele, W. (1983). *Applied Time Series and Box-Jenkins Models*. Orlando, FL: Academic Press.

Problems

13.1. Using a random normal N(0,1) distribution for the random errors, generate a time series of at least length 50 for the following models:

   a. A sequence of random numbers

   b. An AR(1) model with $\phi_1 = 0.7$

   c. An AR(1) model with $\phi_1 = -0.4$; contrast the patterns with (a) and (b)

   d. An MA(1) model with $\theta_1 = 0.7$; contrast with (b)

   e. An MA(1) model with $\theta_1 = -0.4$; contrast with *I* and (d)

   f. An ARMA(1, 1) model with $\phi_1 = 0.7$ and $\theta_1 = -0.7$

   g. An ARMA(1, 1) model with $\phi_1 = 0.7$ and $\theta_1 = 0.4$; contrast with (e)

13.2. Using a random normal N(0, 1) distribution for the random errors, generate a time series of at least length 50 and construct a plot of the theoretical autocorrelation function for the following models:

   a. An AR(2) model with $\phi_1 = 1.0$, $\phi_2 = -0.75$

   b. An AR(2) model with $\phi_1 = 4.0$, $\phi_2 = -0.5$

   c. An MA(2) model with $\theta_1 = -0.7$, $\theta_2 = 0.2$

   d. An MA(2) model with $\theta_1 = 0.4$, $\theta_2 = 0.2$

13.3. Consider the ARIMA(1, 1, 0) process: $Y_t = Y_{t-1} + \phi_1(Y_{t-1} - Y_{t-2}) + \varepsilon_t$. Show that:

   a. The forecasts are given by the recurrence relations:

$$Y_t(1) = (1 + \phi_1)Y_t - \phi_1 Y_{t-1} + \alpha$$
$$Y_t(2) = (1 + \phi_1)Y_t(1) - \phi_1 Y_t + \alpha$$
$$Y_t(l) = (1 + \phi_1)Y_t(l-1) - \phi_1 \times Y_t(l-2) + \alpha \qquad l > 2$$

   b. Show that the forecast profile of the ARIMA(1, 1, 0) process approaches $\alpha/(1 - \phi_1)$.

13.4. Using the results from Problem 13.1, consider a model fit to a quarterly economic time series to be:

$$Y_t = 2.69 + Y_{t-1} + 0.62(Y_{t-1} - Y_{t-2})$$

   a. What is the mean of the first difference process $W_t = Y_t - Y_{t-1}$ for this example?

   b. If at time $t = T$, $W_T = -5.4$, what are the forecasted *changes* for the subsequent 4 quarters?

c. Calculate the forecasts for the economic time series for the next 4 quarters, assuming $Y_T = 258.5$.

d. Describe the pattern of the forecast for this example.

13.5. Weekly shipments of a consumer product are found to be represented by the model (based on fitting 20 weeks of historical data):

$$Y_t = 0.5Y_{t-1} + 2.0 + \varepsilon_t \qquad \sigma_\varepsilon^2 = 0.9$$

This week's shipments are $Y_{21} = 300$ units.

a. Compute the forecasts for the next 5 weeks.

b. Determine the variance of the process.

c. What is the variance of the five-step-ahead forecast error?

d. Provide approximate 95% prediction interval for the first five forecasts under the assumption that the error process is normally distributed.

13.6. Determine Y_1, Y_2, ..., Y_{14} from the following set of random shocks: $\varepsilon_t = \{-0.6, -1.4, 0.0, 0.9, -0.4, -0.6, -1.3, -0.9, 1.7, -0.6, 0.1, 0.2, 0.9, 0.6, -0.3\}$. Assume $Y_0 = 10$, $Y_{-1} = 9$, and apply to the following models:

a. IMA(1, 1) with $\theta_1 = 0.5$

b. IMA(1, 1) with $\theta_1 = 0.2$

c. ARI(1, 1) with $\phi_1 = 0.5$

d. ARI(1, 1) with $\phi_1 = 0.2$

e. ARIMA(1, 1, 1) with $\phi_1 = 0.2$ and $\theta_1 = 0.5$.

13.7. For the formal-wear rental sales (on the CD), perform the following analyses and provide an interpretation of your findings:

a. Compute the sample ACF and PACF of the original data. Relate the patterns in the correlograms with the patterns of the original data? What is the nature of the nonstationarity?

b. Take a first difference and a seasonal difference of the original data and plot the sample ACF and PACF of your results. Have you adequately removed nonstationarity in your data?

c. Run an ARMA(1, 1) model on the differenced series (first and seasonal). What have you assumed about the differenced series?

d. Run an ARMA(2, 2) model on the differenced series. Is there evidence of overfitting and how did you determine that?

e. Perform various diagnostic checks on the residuals available in your software program. Can you suggest a more parsimonious (simpler) ARMA representation of the differenced data?

13.8. Repeat Problem 13.9 for the hotel/motel room demand series (DCTOUR).

13.9. Do a logarithmic transformation of the formal-wear rental sales data. Repeat Problem 13.7 for the logged data.

a. Can you distinguish the transformed and original series with the correlogram patterns?

b. Which diagnostic is the most helpful in establishing a preference for one of the two forms of the data?

13.10. Repeat Problem 13.9 with the hotel/motel room demand data (DCTOUR).

CASE 13A DEMAND FOR ICE CREAM (CASES 1A, 3A–10A CONT.)

The entrepreneur has offered you the opportunity to apply some advanced modeling techniques to the data. The monthly historical data through August of YR06 is given in Case 3A.

(1) Develop a flowchart for modeling the data using the Box-Jenkins methodology. The steps should include data gathering, model building, and forecast evaluation (although the latter is done in Case 14A).

(2) Perform and document the identification phase on the data. Summarize your findings for a nontechnical audience.

(3) Based on your findings, create at least three different ARIMA models with the data. These should include transformations, if necessary.

(4) Perform and document a diagnostic checking phase on the models.

CASE 13B DEMAND FOR APARTMENT RENTAL UNITS (CASES 6C, 8B, AND 10B CONT.)

The Cozy House Apartments management has supplied you with data covering a period of 11 years, shown in Case 6C.

(1) Develop a flowchart for modeling the data using the Box-Jenkins methodology. The steps should include data gathering, model building, and forecast evaluation (although the latter is done in Case 14B).

(2) Perform and document the identification phase on the data. Summarize your findings for a nontechnical audience.

(3) Based on your findings, create at least three different ARIMA models with the data. These should include transformations, if necessary.

(4) Perform and document a diagnostic checking phase on the models.

CASE 13C ENERGY FORECASTING (CASES 1C, 8C, AND 11C CONT.)

An economic research firm is preparing a White Paper to be placed on its website, and you have been providing assistance in the analysis of data sources and the quality of the data, as well as running several univariate ARIMA models. Now you would like to show what kind of forecasts you can produce with these models.

You are aware that the data, shown in Case 8C, is annual and is simply trending. No seasonal ARIMA models need to be considered.

(1) Develop a flowchart for modeling the data using the Box-Jenkins methodology. The steps should include data gathering, model building, and forecast evaluation (although the latter is done in Case 14C).

(2) Perform and document the identification phase on the data. Summarize your findings for a nontechnical audience.

(3) Based on your findings, create at least three different ARIMA models with the data. These should include transformations, if necessary.

(4) Perform and document a diagnostic checking phase on the models.

CASE 13D U.S. AUTOMOBILE PRODUCTION (CASES 1D, 3D, 4D, 6D, AND 7D CONT.)

Because you have done such good work up to this point, your new assignment as an analyst for an industry association is to develop several ARIMA forecasting models with older data and make comparisons for similar models with more current data. The older data are shown in Case 3D and represent U.S. automobile production by month from 1977 to 1985, which includes a recession in the early 1980s. (*Automotive News*, 1982 and 1986 Market Data Book issues).

(1) Develop a flowchart for modeling the data using the Box-Jenkins methodology. The steps should include data gathering, model building, and forecast evaluation (although the latter is done in Case 14D).

(2) Perform and document the identification phase on the data. Summarize your findings for a nontechnical audience.

(3) Based on your findings, create at least three different ARIMA models with the data. These should include transformations, if necessary.

(4) Perform and document a diagnostic checking phase on the models.

(5) Gather U.S. automobile production data for a more current period, and repeat (2)–(4).

(6) What is your recommendation as to the value of having older data for comparison? Consider, for example, the comparison of the industry in different business cycles.

CASE 13E DEMAND FOR CHICKEN (CASES 8E AND 12B CONT.)

A research firm specializing in Agricultural Economics has recruited you and provided data on per-capita chicken consumption in Case 8E.

(1) Develop a flowchart for modeling the data using the Box-Jenkins methodology. The steps should include data gathering, model building, and forecast evaluation (although the latter is done in Case 14E).

(2) Perform and document the identification phase on the data. Summarize your findings for a nontechnical audience.

(3) Based on your findings, create at least three different ARIMA models with the data. These should include transformations and differencing, if necessary.

(4) Perform and document a diagnostic checking phase on the models.

CASE 13F ARIMA MODELS FOR BALLPARK ATTENDANCE OF A SPORTS TEAM (CASE 11D CONT.)

In Case 11D, you are given the data for the first 29 years of paid attendance at New York Mets home games since the franchise was started in 1962.

(1) For these data or a comparable set of annual, quarterly, or monthly ballpark attendance data, develop the modeling steps required for building ARIMA models using the Box-Jenkins methodology.

(2) Perform and document the identification phase on the data. Summarize your findings for a nontechnical audience.

(3) Based on your findings, create at least three different ARIMA models with the data. These should include transformations and differencing, if necessary.

(4) Perform and document a diagnostic checking phase on the models.

(5) Prepare a short presentation on how you could best sell your approach to the franchise's management.

14

Forecasting with ARIMA Models

"Prediction is very difficult, especially if it's about the future."
NIELS BOHR, NOBEL LAUREATE IN PHYSICS

LINEAR FILTERING TECHNIQUES, WIDELY USED in control engineering, have gradually found their way into mainstream demand forecasting practice over the past 3 decades. This chapter develops forecasts for a class of model-driven forecasting techniques (known as ARIMA models) that can be used effectively to produce forecasts based on a theory of linear filtering. We rely on the concept of stationarity for time series, introduced in Chapter 4, to provide a description of the class of ARMA processes. The construction of these models applies a single, unified theory to the description of a wide range of stationary and nonstationary (e.g., seasonal and trending) time series, allows a linear representation of a stationary time series in terms of its own past values and a weighted sum of a current error term and lagged error terms, and provides a systematic procedure for forecasting a time series from its own current and past values. Once a tentative ARIMA model has been identified and estimated, forecasts can be computed. This requires:

- Generating minimum MSE forecasts from the difference equation of an ARIMA model
- Developing expressions for prediction variances and prediction limits for forecasts
- Utilizing forecast errors as a way of monitoring forecasts

14.1 ARIMA MODELS FOR FORECASTING

We have seen in the previous chapter how autoregressive integrated moving average (ARIMA) models come from a specialized but highly powerful class of **stochastic processes** called linear filters. The general theory of linear filters (Whittle, 1963) is not new; much of the original work was done in the 1930s by the renowned mathematicians Kolmogorov and Wiener for automatic control problems. However, as a tool for forecasting, these techniques became popular in the 1960s and beyond. A special kind of linear filter, called the autoregressive (AR) process, goes back a little further: AR models were first used by Yule (1927). Slutsky (1937) introduced another kind of filter, called the moving average (MA) process. Autoregressive moving average (ARMA) theory, in which these processes are combined, was developed in 1938 by Wold (1954). A unifying modeling strategy was developed during the 1960s by Box and Jenkins (1970). This strategy is the result of their direct experience with forecasting problems in business, economic, and control engineering applications. We treat the specifics of this modeling strategy for forecasting in this chapter for a variety of practical time series.

The development of the Box-Jenkins method for fitting ARIMA models to nonstationary as well as stationary time series provides for a sound framework for obtaining optimal forecasts of future values of time series data.

When we refer to the theoretical aspects of a linear filter we are dealing with a stochastic process. When we describe a particular formula based on the available time series, we refer to it as a model. Although this distinction is not hard and fast, it is at times convenient to differentiate the theoretical from the empirical aspects of a forecasting problem.

14.2 MODELS FOR FORECASTING STATIONARY TIME SERIES

What kinds of historical patterns do we get with ARMA models and what sort of forecast profiles (projections) do they produce? Once a time series is assumed stationary, it can be used with a linear filter in order to devise a forecast. Once an appropriate ARMA model (linear filter) has been fit to the series, an optimal forecasting procedure follows directly from the theory.

Creating a Stationary Time Series

The ARMA process is designed for stationary time series, that is, series whose basic statistical properties (e.g., means, variances, and covariances) remain constant over time. Thus, in order to build ARMA models, nonstationarity must first be identified and removed (see Chapter 4).

The technical definition of stationarity is a complex one. For practical purposes, a stochastic process is said to be weakly stationary if the first and second moments of the process are time-independent. This assumption implies, among other things, that the mean, variance, and covariances of the data are constant and finite.

Another critical assumption in describing stationary time series comes as a result of the chronological order of the data. The difference between conventional regression methods and time series modeling of the sort possible with ARMA models is that independence of the observations cannot be assumed in practice; in fact, in ARMA, modeling derives its strength from the mutual dependence among observations that is found in time series data.

Nonstationarity typically includes periodic variations and systematic changes in mean (trend) and variance.

White Noise and the Autoregressive Moving Average Model

Consider a process $Y_t = \varepsilon_t$, $t = 0, \pm 1, \pm 2$, where ε_t is independent of all other values $\varepsilon_{t-1}, \varepsilon_{t-2}, \ldots, \varepsilon_{t+1}, \varepsilon_{t+2}, \ldots$ This process is called purely random. If all ε_t are normally distributed as well, the process is said to be white noise. The reason is that the next value, ε_t, of white noise is unpredictable even if all previous values ε_{t-1}, $\varepsilon_{t-2}, \ldots, \varepsilon_{t-q}$ and subsequent values $\varepsilon_{t+1}, \varepsilon_{t+2}, \ldots$ are known. The counts of radioactive decay from a long-lived sample and even the changes in the level of stock prices are examples of data that are stationary and can be regarded as white noise.

The concept of white noise is of central importance to forecasting time series, just as independent random error is of central importance to forecasting with regression models.

One simple linear random process that we encounter in practice is the MA process. For example, the formula for a first-order MA process (MA(1)) is

$$Y_t = \varepsilon_t - \theta_1 \varepsilon_{t-1}$$

where θ_1 is a parameter and σ_ε^2 is the variance of all ε_t. A computer-generated example of an MA(1) process with $\theta_1 = 0.25$ is given in Exhibit 14.1. There is no discernable pattern in the numbers.

In general, a qth-order MA process, abbreviated as MA(q) has the form

$$Y_t = \varepsilon_t - \theta_1 \varepsilon_{t-1} - \theta_2 \varepsilon_{t-2} - \cdots - \theta_q \varepsilon_{t-q}$$

where the model is specified by $q + 2$ parameters σ_ε^2, μ, θ_1, $\theta_2, \ldots, \theta_q$. The parameter μ determines the level of the process and is assumed to be accounted for in Y_t. In practice, the parameters are estimated from the data. This model states that the

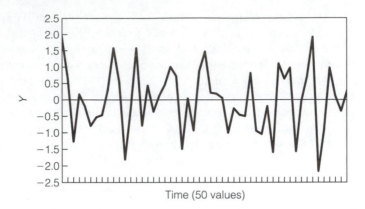

Time (50 values)

values of Y_t, already mean adjusted, consist of a moving average of the errors ε_t reaching back q periods. The coefficients (parameters) of the error terms ε_t are designated by θ values and the minus signs are introduced by convention. It is still assumed that the errors are independent but that the observed values of Y_t are dependent, being a weighted function of prior errors.

The term moving average has been a convention for this model and should not be confused with the moving average MAVG introduced in Chapter 2. In fact, the θ values need not be positive, nor need they add up to unity.

Another simple linear random process is the AR process. The formula for a first-order AR process AR(1), also known as a Markov process, is

$$Y_t = \phi_1 Y_{t-1} + \varepsilon_t$$

where σ_ε^2 and ϕ_1 are parameters. This model states that the value Y_t of the process is given by ϕ_1 times the previous value plus a random error ε_t. Computer-generated realizations of an AR(1) process with $\phi_1 = +0.55$ and $\phi_1 = -0.55$ are shown in Exhibit 14.2. The corresponding correlograms are shown in Exhibit 14.3. Both correlograms have significant spikes (outside the bands) at lag 1 ($= 0.47$ and -0.38). Theoretically, these two spikes are $+0.55$ and -0.55. Other spikes appear to be within the confidence bands. We describe the theoretical patterns of autocorrelation functions (correlograms as they should appear in theory) in the next chapter. Note that the AR process with positive ϕ_1 coefficient has some positive memory of itself in that the plot displays a tendency to retain the pattern of its own immediate past. There are practical examples of data that follow a Markov process.

In general, a pth-order AR process has the form

$$Y_t = \phi_1 Y_{t-1} + \phi_2 Y_{t-2} + \cdots + \phi_p Y_{t-p} + \varepsilon_t$$

where the $p + 2$ parameters, μ, σ_ε^2, and the ϕ values are parameters to be estimated from the data. The Y_t is assumed to be mean adjusted with μ in the equation.

The term autoregression that is used to describe such a process arises because an AR(p) model is much a like a multiple linear regression model.

EXHIBIT 14.2
Realizations of an
AR(1) Process:
(a) $\phi_1 = 0.55$;
(b) $\phi_1 = -0.55$

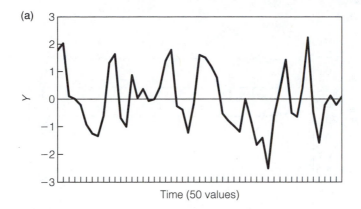

(a)

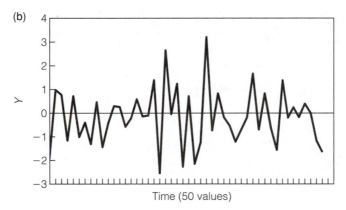

(b)

The essential difference between an autoregression and multiple linear regression is that Y_t is not regressed on independent variables but rather on lagged values of itself—hence the term *auto*regression. A realization of an AR(2) process, with $\phi_1 = +1.0$ and $\phi_2 = -0.5$, is shown in Exhibit 14.4. This one shows an almost regular cyclical pattern. In fact, pure cycles can be described by AR(2) processes.

One-Period-Ahead Forecasts

What are the forecast profiles for these models? As a first example, consider an AR(1) process with $\phi_1 = 0.5$ given by

$$Y_t = 0.5Y_{t-1} + 2.0 + \varepsilon_t$$

We have generated an artificial sample (Exhibit 14.5) and noted the current value at time $t = T$ as well as the previous four observed values. The random errors are calculated from $\varepsilon_t = Y_t - 0.5Y_{t-1} - 2.0$ for $t = T-3$ to $t = T$. We assume now that the true parameters of the model are known and that the current time is $t = T-4$ so that we can examine the pattern of the one-step-ahead forecasts. The forecast error at time $t = T-3$ is $Y_{T-3} - \hat{Y}_{T-4}(1) = 4 - [0.5 \times 10 + 2.0] = -3.0$. At the next period, when the current time is $t = T-3$, the forecast error at time $t = T-2$ is

EXHIBIT 14.3
Correlograms of the
Data in Exhibit 14.2:
(a) $\phi_1 = 0.55$;
(b) $\phi_1 = -0.55$

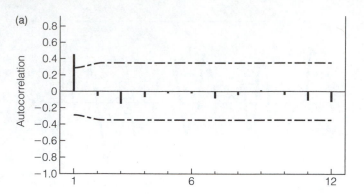

(a)

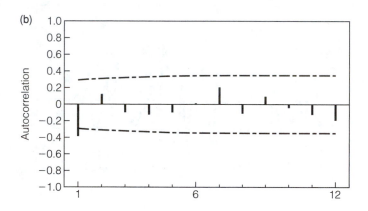

(b)

EXHIBIT 14.4
Realization of an
AR(2) Process ($\phi_1 =$
1.0, $\phi_2 = -0.50$)

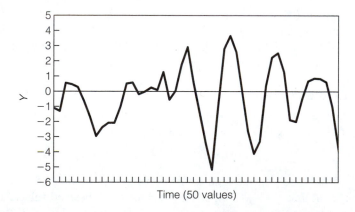

Time (50 values)

$Y_{T-2} - \hat{Y}_{T-3}(1) = 5 - [0.5 \times 4 + 2.0] = 1.0$. Finally, when the current time is $t = T - 1$, the forecast error at time $t = T$ is $Y_T - \hat{Y}_{T-1}(1) = 6 - [0.5 \times 8 + 2.0] = 0$. Note that the random errors are the same as the one-step-ahead forecast errors.

Time Origin	Data Y_t	Residuals
$T-4$	10.0	-3.0
$T-3$	4.0	2.0
$T-2$	5.0	1.0
$T-1$	8.0	0
T	6.0	2.0

EXHIBIT 14.5
Sample Data (hypothetical) for Forecasting with $Y_t = 0.5Y_{t-1} + 2.0 + \varepsilon_t$

The one-step-ahead forecast is given by

$$\hat{Y}_T(1) = 0.5Y_T + 2$$

Because the random errors ε_t are assumed to have zero mean and constant variance σ_ε^2, our best estimate of ε_{T+1} is 0.

l-Step-Ahead Forecasts

The two- and three-step-ahead forecasts are derived through the following procedure.

1. Replace T by $T+2$ in the formula $Y_t = 0.5Y_{t-1} + 2.0 + \varepsilon_t$; then

$$Y_{T+2} = 0.5Y_{T+1} + 2.0 + \varepsilon_{T+2}$$

2. Use the one-step-ahead forecast $\hat{Y}_T(1)$ in place of Y_{T+1} because the latter is not known.

3. Assume $\varepsilon_{T+2} = 0$. Then the two-step-ahead and three-step-ahead forecasts are, respectively,

$$\hat{Y}_T(2) = 0.5\hat{Y}_T(1) + 2.0 + 0 = 4.5$$
$$\hat{Y}_T(3) = 0.5\hat{Y}_T(2) + 2.0 + 0 = 4.25$$

Notice that the forecast of Y_{T+h} with $h \geq 1$ is made from a time origin, designated T, for a lead time or horizon, designated h. This forecast, denoted by $\hat{Y}_T(h)$ is said to be a forecast at origin T, for lead time h. It is the minimum mean square error (MSE) forecast (Box et al., 1994). When $\hat{Y}_T(h)$ is regarded as a function of h for a fixed T, it is referred to as the **forecast function** or **profile** for an origin T.

The forecast error for the lead time h,

$$e_T(h) = Y_{T+h} - \hat{Y}_T(h)$$

has zero mean; thus, the forecast is unbiased.

It is also important to note that any linear function of the forecasts $\hat{Y}_T(h)$ is a minimum MSE forecast of the corresponding linear combination of future values of the series. Hence, for monthly data, the best year-to-date forecast can be obtained by summing the corresponding 1-, 2-, . . . , h-step-ahead forecasts.

A forecast function or profile can be generated by plotting the h-step-ahead forecasts for a fixed time origin, $h = 1, 2, 3, \ldots$

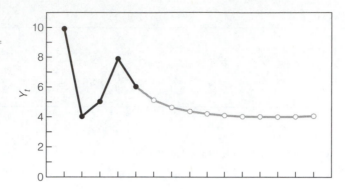

It can be seen from Exhibit 14.6 that for this example there is a geometrical decay to a mean value; the mean equals the constant term divided by $(1 - \phi_1)$. In this case, the mean is 4.0. The equation for the h-step-ahead forecast $\hat{Y}_T(h)$ is then given as

$$\hat{Y}_T(h) = 4.0 + 0.5^h(Y_T - 4.0)$$

In effect, forecasts from AR(1) models with positive parameter will decay (up or down) to a mean level and be level thereafter. AR(1) models with a positive parameter occur much more frequently in practice than ones with a negative parameter.

In the same manner as the AR(1) model, the forecasts for the MA(1) model can be developed. Suppose the MA(1) model is given by

$$Y_t = 3.0 + \varepsilon_t - 0.5\varepsilon_{t-1}$$

Once again $t = T$ is replaced by $T + 1$, and all future errors are set to zero. Then, $Y_{T+1} = 4.0$ and all Y_{T+l} $(l = 2, 3, \ldots)$ are equal to 3.0. The forecast profile for an MA(1) model is rather simple. It consists of one estimate based on the last period's error ($= 4.0$ in the example), followed by the constant mean ($= 3.0$ in the example) for all future periods. This makes sense intuitively because the ACF of the MA(1) model has nonzero correlation only at lag 1 and hence a memory of only one period. In effect, the MA(1) process will yield level forecasts.

The forecast profile for a MA(q) can be generalized rather easily. It consists of values at $h = 1, 2, \ldots, q$ that are determined by the past errors and then equals the mean of the process for periods greater than q.

The Black-Box Representation

In Chapter 13, we state that any linear process can be written formally as a weighted sum of the current error term *plus* all past random error terms. This gives it the representation as a black box, in which random error terms are shown as inputs to a black box, the output of which is the observed Y_t. We have seen in the previous chapter how a linear process, called the ARIMA model, is capable of describing a wide variety of time series for forecasters. This ARIMA process is shown in Exhibit 14.7 as a block diagram to express various forms of stationary and nonstationary behavior in a time series.

EXHIBIT 14.7
Black-Box
Representation of an
ARIMA Process

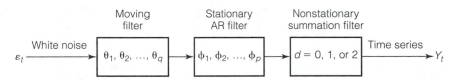

In many problems, such as those in which it is required that future values of a series be predicted, it is necessary to construct a parametric model for the time series. A powerful parametric model that has been widely used in practice for describing empirical time series is the mixed ARMA process:

$$Y_t = \phi_1 Y_{t-1} + \phi_2 Y_{t-2} + \cdots + \phi_p Y_{t-p} + \varepsilon_t - \theta_1 \varepsilon_{t-1} - \theta_2 \varepsilon_{t-2} - \cdots - \theta_q \varepsilon_{t-q}$$

where p is the highest lag associated with the data, and q is the highest lag associated with the error term. For convenience, we henceforth assume that Y_t has been mean adjusted with α, that is, $Y_t = Z_t - \alpha$. If that weighted sum has only a finite number of nonzero error terms, then the process is a moving average process. An ARMA(1, 1) process, for example, has one prior observed-value term of lag 1 and one prior error term:

$$Y_t = \phi_1 Y_{t-1} + \varepsilon_t - \theta_1 \varepsilon_{t-1}$$

In short, the ARMA process is a linear random process. It is linear if Y_t is a linear combination of lagged values of Y_t and ε_t. It is random if the errors (also called shocks or disturbances) are introduced into the system in the form of white noise. The random errors are assumed to be independent of one another and to be identically distributed with a mean of zero and a constant variance σ_ε^2.

 To be useful, an ARMA model should be physically meaningful and involve as few parameters as possible.

The linear process can also be expressed as a weighted sum of the current shock plus all past observed values. If the number of nonzero terms in this expression is finite, then the process is autoregressive. Thus, an MA process of finite order can be expressed as an AR process of infinite order, and an AR process of finite order can be expressed as an MA process of infinite order. This duality has led to the principle of parsimony in the Box-Jenkins methodology (Chapter 13) which advocates that the practitioner employ the smallest possible number of parameters for adequate representation of a model.

14.3 MODELS FOR NONSTATIONARY TIME SERIES

We have noted before that most time series encountered in demand forecasting applications are not stationary. If a nonstationary time series can be made stationary by taking d successive differences (usually of order 0, 1, or 2), this gives an ARMA model for the differenced series. This is called an ARIMA model for the original series—hence the term integrated to suggest "undifferencing." As in calculus, integration is the "opposite" of differentiation.

The three letters p, d, and q give the order of an ARIMA model. By convention, the order of the autoregressive component is p, the order of differencing needed to achieve stationarity is d, and the order of the moving average part is q.

The ARIMA(p, d, q) model is the most general model considered here. It is also the most widely used model. Many times it is necessary to take differences to achieve stationarity, but the resulting series may require only an autoregressive (p) or a moving average (q) component. These models are called autoregressive integrated (ARI) or integrated moving average (IMA) models. The term integrated is used when differencing is performed to achieve stationarity because the stationary series must be summed ("integrated") to recover the original data. The number of differencing is generally 0, 1, or at most 2 in practice (Exhibit 14.7).

Many economic forecasting methods use exponentially weighted moving averages, and they can be shown to be appropriate for a particular type of nonstationary process. Indeed, the stochastic model through which the exponentially weighted moving average produces an optimal forecast is a member of the class of ARIMA models. A range of ARIMA models exists to provide stationary and nonstationary treatments of the time series met in practice.

Forecast Profile for ARMA(1, q) Models

Consider an ARMA(1, 1) model of the form

$$Y_t = 0.5Y_{t-1} + 2.0 + \varepsilon_t - 0.5\varepsilon_{t-1}$$

By following the same procedure of replacing $t = T$ by $T + 1$, assuming ε_{t+1}, $\varepsilon_{t+2}, \ldots = 0$, and using estimates $\hat{Y}_T(1)$, $\hat{Y}_T(2), \ldots$, we can obtain successive forecasts $\{6.5, 4.5, \ldots, 4.0\}$. The forecast for the first period is a function of last period's actual observation and last period's error. All later forecasts are based only on predicted values of Y_t. As in the AR(1) case, it shows a geometrical decay to the mean after the first forecast period.

For the forecast profile for ARMA(1, q) model, the forecasts of the first q periods are a function of observed values, estimated values, and errors. After q periods ahead, there is a geometrical decay to the mean.

Forecast Profile for an ARIMA(0, 1, 1) Model

An example of an ARIMA(0, 1, 1), also known as the IMA(1, 1) model for lack of an AR component, is

$$Y_t = Y_{t-1} + 3.0 + \varepsilon_t - 0.5\varepsilon_{t-1}$$

By following the procedure of replacing $t = T$ by $T + 1$, assuming ε_{t+1}, $\varepsilon_{t+2}, \ldots = 0$, and using estimates $\hat{Y}_T(1)$, $\hat{Y}_T(2), \ldots$, the forecasts $\{10, 13, 16, 19, \ldots\}$ are obtained.

The forecast profile for this model is a straight line after the one-period-ahead forecast. The slope of the line is equal to the constant μ in the model. If $\mu = 0$, the forecast profile is a horizontal line at the level given by $\hat{Y}_T(1)$.

Forecast Profile for an ARIMA(1, 1, 1) Model

An example of an ARIMA(1, 1, 1) model is a special case of a nonstationary ARMA(2, 1) model

$$Y_t = 1.5Y_{t-1} - 0.5Y_{t-2} + 2.0 + \varepsilon_t - 0.5\varepsilon_{t-1}$$

To generate forecasts, the following calculations result:

$$\hat{Y}_T(1) = 1.5(6) - 0.5(8) + 2.0 + 0 - 0.5(-2) = 8.0$$
$$\hat{Y}_T(2) = 1.5(8) - 0.5(6) + 2.0 + 0 - 0 = 11.0$$
$$\hat{Y}_T(3) = 1.5(11) - 0.5(8) + 2.0 + 0 - 0 = 14.5$$

After one period, the moving average term no longer affects the forecasts. The forecast profile for this model trends upward. In this example, it approaches a straight line with slope = Constant/$(1 - \phi_1)$ = $2/(1 - 0.5)$ = 2.

Actually, this is a special case of a stationary ARMA(1, 1) model for the first-differenced data, which is similar to an AR(1) model of the first differences. We have shown before that the forecast profile for an AR(1) model tends geometrically to a mean value [Constant/$(1 - \phi_1)$]. Thus, if forecasts for first differences approach a constant, the forecasts for the original series will grow by this constant amount each period. The forecast profile of the time series becomes a trend line with a constant slope.

Three Kinds of Trend Model

Many time series exhibit trends. It is instructive to compare the characteristics of several simple forecasting models that deal with trends. Exhibit 14.8 shows three alternative approaches to forecasting a trending series. The first model uses time as an independent variable in a simple linear regression model. As new actuals are added, the forecasts do not change, unless the model is updated. The slope and the intercept of the line of forecasts are constant.

The second model is an extension of the first:

$$Y_t - Y_{t-1} = \mu + \varepsilon_t$$

EXHIBIT 14.8
Three Models for
Forecasting a
Trending Time
Series: Straight-Line
Trend versus
ARIMA(0,1,0)

Model Type	Equation	Characteristic of updated forecasts	
		Intercept	Slope
Deterministic trend	$Y_t = \alpha + \beta t + \varepsilon_t$	Constant	Constant
ARIMA *with* deterministic trend constant	$Y_t - Y_{t-1} = \alpha + \varepsilon_t$ or $(1 - B)Y_t = \alpha + \varepsilon_t$	Varies	Constant
ARIMA *without* deterministic trend constant	$(1 - B)^2 Y_t = \alpha + \varepsilon_t$	Varies	Varies

This model has a slope μ, which must be estimated from the data. The forecasts from this model are updated as new actuals become available. This has the effect of changing the intercept of the line of future forecasts but with a constant slope.

The third model can be written

$$Y_t = 2Y_{t-1} - Y_{t-2} + \varepsilon_t$$

Both the slope and the intercept of the lines of updated forecasts will change as new observations become available. The trend can change direction with this model.

Unless there is reason (business or theoretical) to believe that a deterministic relationship exists, autoregressive models are more adaptive or responsive to recent observations than straight-line models are. The choice of taking first differences with trend constant or second differences without a trend constant should depend on the data.

Because it is not helpful to overdifference a time series, second differences should be used only when first differences do not result in a stationary series.

A Comparison of an ARIMA(0, 1, 0) Model and a Straight-Line Model

We now show how the forecasts of an ARIMA(0, 1, 0) model differ from straight-line models, which are commonly used to model trends. Consider a (seasonally adjusted) quarterly economic time series that is generally increasing in trend. Moreover, the series contains several up- and downturns corresponding to economic business cycles.

The correlogram (not shown) of the time series has a gradual decay that confirms nonstationarity in the data due to the trend. Thus, first differences of the quarterly data are recommended; a plot of the first differences (not shown) suggests a constant mean and there is no sign of increasing variability. The correlogram and partial correlogram (not shown) of the first differences also confirm that there are no significant patterns in either diagram. Therefore, an ARIMA(0, 1, 0) model of the seasonally adjusted quarterly data seems appropriate as a starting model:

$$Y_t - Y_{t-1} = \alpha + \varepsilon_t$$

The first differences have a mean value that is 1.35 standard deviations from zero, so a deterministic trend constant is included; the fitted model is given by

$$Y_t = Y_{t-1} + 332.0$$

This is perhaps one of the simplest time series models covered so far. Note that a one-period-ahead forecast is simply the data value for the current quarter plus 332.0. All information before the last observation has no effect on the one-period-ahead forecast. Past data were used to determine that the trend constant should be 332.0.

To compare the forecast profile of the ARIMA(0, 1, 0) model with a simple linear regression model with time as the independent variable, a straight-line fit to the data results in:

$$Y_t = 7165.2 + 60.8t \qquad t = 1, 2, \ldots$$

We can see that the forecast profile of this model is a straight line with a slope = 60.8. Because the slope is positive in this model, the forecasts are always increasing with time.

To compare the response of new data on the forecasts from the two models, four straight-line regressions were performed. The first regression was performed through the fourth quarter of Year 5 (next to the last year); in other words, 4 quarters of the data were kept as a holdout sample for evaluating forecast accuracy. Next, four quarterly forecasts were produced for the holdout sample (i.e., for Year 6). Subsequent regressions and forecasts were also generated by adding another period before fitting the model, and so on. In this manner, ten forecasts were produced, four for quarter 1, three for quarter 2, two for quarter 3, and one for the last quarter of the holdout sample.

The effect of new data on the predictions of the two models can be seen in Exhibit 14.9, which shows the one-step-ahead predictions by the two models for the 4 quarters of Year 6. The predictions show that the straight-line regression model is closer to the actuals for each of the 4 quarters when the fit period ends in the fourth quarter of Year 5. When two more quarters of data are included and the third and fourth quarters of Year 6 are predicted, the straight-line model predictions are again closer to the actual values.

When the first two quarters of Year 6 are added, the changes in the predictions made by the models for the last two quarters of Year 6 are as shown in Exhibit 14.10a. Clearly, the ARIMA model responds to the new data (upturn in Q2 of Year 6) by a much greater amount than does the straight-line model. In fact, for this ARIMA model, the forecasts equal the latest data value plus a constant. Notice that the second quarter of Year 6 is somewhat unusual; this influenced the forecasts for the remainder of Year 6.

Next, two more quarters (Q3 and Q4 of Year 6) are added to the fit period and the 4 quarters of Year 7 are predicted. Now the predictions of the straight-line model are once again superior to the ARIMA model (Exhibit 14.10b). Last, a fit with actuals through the second quarter of Year 7 is done. This time, the plot of the data shows

EXHIBIT 14.9
Comparison of
Forecasts Generated
by the ARIMA(0, 1, 0)
Model and the
Straight-Line
Regression Model
for Year 6

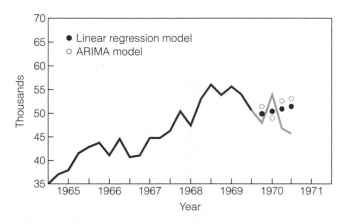

EXHIBIT 14.10
Changes in Model
Predictions for Q3
and Q4 of (a) Year 6;
(b) Year 7

(a)

Year 6	ARIMA	Straight Line
Quarter 3	2756	96
Quarter 4	2807	100

(b)

Year 7	ARIMA	Straight Line
Quarter 3	233	116
Quarter 4	11,499	118

(*Source*: Exhibit 14.9)

a sharp turn upward. The model forecasts are shown in Exhibit 14.11. Once again, the ARIMA model reacts much more quickly to changing conditions. Exhibit 14.11 shows that this time the ARIMA model is correct because the data were indeed continuing upward.

The forecast test just described is not done to claim that either model is superior. Rather, it shows how differently the ARIMA(0, 1, 0) model and straight-line regression models react to new data. There is almost no change in the forecasts from a straight-line model (vs. time) when new data are introduced. However, ARIMA models are affected significantly by recent data. Which model we should select depends more on the particular circumstances of the application and less on the summary statistics from fitting the models.

- If future demand in a given year is expected to be substantially above or below trend (say, based on economic rationale), then the forecast of a straight-line model must be modified. A turning-point analysis or a trend-cycle curve-fitting approach may be the best way of supplementing our judgment about the cycle to a straight-line model (see Chapter 7).

- If we assume that the current short-term trend in the data is likely to continue, an ARIMA model should be seriously considered. Univariate ARIMA models

EXHIBIT 14.11
Comparison of Fore-
casts Generated by
the ARIMA(0, 1, 0)
Model and the
Straight-Line Regres-
sion Model for the
Last 2 Quarters of
Year 7

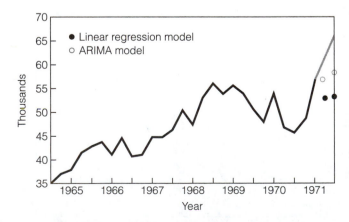

cannot predict turning points, however. If the economy is near a turning point, extreme care should be exercised in using the ARIMA models. Regression models with explanatory variables, econometric models, and transfer function models may be a more appropriate way of sensitizing forecasts to take into account changes in the business cycle.

14.4 SEASONAL ARIMA MODELS

Many of the economic time series reported in business periodicals and online news services are already seasonally adjusted. Corporate and government forecasters use seasonally adjusted data in models for items strongly influenced by business cycles in which a seasonal pattern would otherwise mask the information of primary interest. Because seasonal variation is seldom unchanging, seasonal adjustment based on simplistic smoothing procedures may leave some seasonal pattern of unknown nature in the data. In these situations, it may be preferable to take a model-based approach to seasonal adjustment and forecasting. Seasonal ARIMA models are well suited for these situations.

Other times, forecasts of seasonal data are required. In airline-traffic design, for example, average traffic during the months of the year when total traffic is greatest is of central importance. Clearly, monthly traffic data rather than the seasonally adjusted traffic data are required.

Sometimes seasonality shifts with time. For example, changes in school openings and closings during the year can affect a variety of time series, such as energy demand. If the seasonal change persists, models based on unadjusted data rather than seasonally adjusted data are likely to be more flexible and useful. If changing seasonality is expected, it is therefore better to take account of it through the development of a properly specified seasonal ARIMA model.

A Multiplicative Seasonal ARIMA Model

Models incorporating both regular and seasonal components can be used for describing a typical 12-month seasonal series with positive trend (Y_t = Hotel/motel demand; Exhibit 14.12a). To remove the linear trend from such a series requires at least one first difference (Diff 1 = $Y_t - Y_{t-1}$; Exhibit 14.12b). The resultant series still shows strong seasonality, with perhaps some residual trend pattern. This may call for taking a first-order seasonal difference for the residual series (series $Y_t - Y_{t-12}$; Exhibit 14.12c). A first difference followed by a seasonal difference often give rise to stationary data (free from the trend and seasonal pattern).

It turns out that the order in which the differencing operations are applied to Y_t in Exhibit 14.12a is immaterial and that the operation of successive differencing is multiplicative in nature. That is, series $Y_t^* = Y_t - Y_{t-12}$ in Exhibit 14.12c is the first-order seasonal difference of Y_t. Then, the first difference of Y_t^* is the time plot in Exhibit 14.12d.

EXHIBIT 14.12
(a) Hotel/Motel
Demand; (b) First
Difference; (c)
Seasonal Difference;
(d) First and
Seasonal Differences

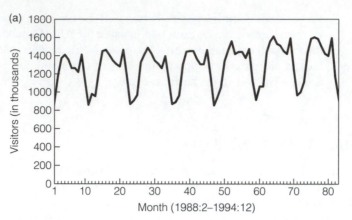

(Exhibit 6.17)

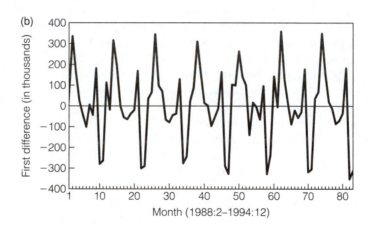

For seasonal ARIMA models, seasonality is introduced into the model multi-plicatively. Although this may seem arbitrary, it can be intuitively explained by our example; if a month—say January—is related to the December preceding it and to the previous January, then it is also related to the December 13 months previous.

The combined operation suggests that if the residual series resembles a random process, or white noise, then a model of the original data could have the form

$$Y_t = Y_{t-1} + Y_{t-12} - Y_{t-13} + \varepsilon_t$$

A forecast from this model is the sum of the values for the same month (Y_{t-12}) in the previous year and the annual change in the values of the previous months $(Y_{t-1} - Y_{t-13})$. This is a special case of an autoregressive model in that past values have been given a constant weight of ± 1. It also shows that those weights given to Y_{t-1} and Y_{t-13} are equal but have opposite signs.

Frequently both seasonality and trend are present in data. Hence, an AR(13) model of the form

$$Y_t = \phi_1 Y_{t-1} + \phi_2 Y_{t-12} + \phi_3 Y_{t-13} + \varepsilon_t$$

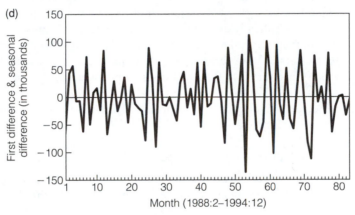

(c)

12 difference (in thousands)

Month (1988:2–1994:12)

(*Source*: Exhibit 14.12a)

(d)

First difference & seasonal difference (in thousands)

Month (1988:2–1994:12)

(*Source*: Exhibit 14.12a)

can be used to interpret this underlying structure. If the coefficients at lags 1 and 13 are approximately equal and have opposite sign and ϕ_2 equals approximately 1, then the resultant model

$$Y_t = \phi_1 Y_{t-1} + Y_{t-12} - \phi_1 Y_{t-13} + \varepsilon_t$$

is a more parsimonious model. These are good representations for strongly seasonal series of period 12:

$$Y_t - Y_{t-12} = \phi_1 (Y_{t-1} - Y_{t-13}) + \varepsilon_t$$

For time series that contain a seasonal periodic component that repeats every s observations, a supplement to the nonseasonal ARIMA model can be applied. In a seasonal process with seasonal period s, values that are s time intervals apart are closely related. This is analogous to the regular ARIMA process in which adjacent ($s = 1$) values are closely related. For a time series with seasonal period s, $s = 4$ for quarterly data and $s = 12$ for monthly data. In a manner very similar to the regular ARIMA process, seasonal ARIMA(P, D, Q) models can be created for seasonal data (to differentiate between the two kinds, it is customary to use capital letters P, D, and Q when designating the order of the parameters for the seasonal models).

14.5 FORECAST PROBABILITY LIMITS

Probability limits are used in monitoring the forecast in the following ways:

- If a high proportion of the forecast errors begin to fall outside the appropriate limits, the values of the model coefficients may have changed or, at least, they should be reestimated. However, the new model might produce significantly different forecasts.

- If systematic patterns appear, the structure of the process may also have changed and additional terms (or transformations) may be required in the model.

- A time plot of the cumulative sum of the forecast errors may indicate changes in the structure of the process as a result of changes in external factors.

One of the goals of forecasting with quantitative models is to be able to make probability statements about the forecasts. This is normally accomplished by calculating forecast error variances under the assumption that the forecast errors are independent, identically distributed random variables. It is also assumed that the number of observations used to fit any model is sufficiently large that errors in estimating the parameters will not seriously affect the forecasts: It is assumed the model is known exactly and that it will remain essentially unchanged for the period for which the forecast is being made.

To examine probability limits for ARIMA models, let us first illustrate their use for some elementary versions.

Probability Limits for an MA(2) Model

Consider the following model $Y_t = \varepsilon_t - 0.6\varepsilon_{t-1} - 0.4\varepsilon_{t-2}$, in which $\theta_1 = 0.6$ and $\theta_2 = 0.4$. For this model, future values are generated from the expressions:

$$Y_{T+1} = \varepsilon_{T+1} - 0.6\varepsilon_T - 0.4\varepsilon_{T-1}$$
$$Y_{T+2} = \varepsilon_{T+2} - 0.6\varepsilon_{T+1} - 0.4\varepsilon_T$$
$$\vdots$$
$$Y_{T+h} = \varepsilon_{T+h} - 0.6\varepsilon_{T+h-1} - 0.4\varepsilon_{T+h-2}$$

As time passes and the actual observations at times $t = T+1$, $T+2$, $T+3$, . . . become available, there is a discrepancy between the one-step-ahead forecast and the observed value. For the one-step-ahead forecast, $\hat{Y}_T(1)$, the forecast error $\mathbf{e}_T$ will be

$$\begin{aligned}
\mathbf{e}_T(1) &= Y_{T+1} - \hat{Y}_T(1) \\
&= [\varepsilon_{T+1} - 0.6\varepsilon_T - 0.4\varepsilon_{T-1}] - [-0.6\varepsilon_T - 0.4\varepsilon_{T-1}] \\
&= \varepsilon_{T+1}
\end{aligned}$$

If ε_{T+1} is not zero, the forecast will be the result of the value of ε_{T+1}. Likewise, for the two-period-ahead forecast $\hat{Y}_T(2)$, the forecast error will be:

$$\begin{aligned}
\mathbf{e}_T(2) &= Y_{T+2} - \hat{Y}_T(2) \\
&= [\varepsilon_{T+2} - 0.6\varepsilon_{T+1} - 0.4\varepsilon_T] - [-0.4\varepsilon_{T-1}] \\
&= \varepsilon_{T+2} - 0.6\varepsilon_{T+1}
\end{aligned}$$

The forecast error now is a result of the two unobserved errors (ε_{T+2}, ε_{T+1}). For the l-step-ahead forecast, the forecast error is equal to $\varepsilon_{T+h} - 0.6\varepsilon_{T+h-1} - 0.4\varepsilon_{T+h-2}$. So, the forecast error l steps ahead will be the result of three unobserved errors for the MA(2) model.

For forecasting purposes, it is assumed that ε_{T+h}, ε_{T+h-1}, and ε_{T+h-2} will be equal to their expected values (namely, 0). In actuality, they will differ from 0 by some amount that cannot be known at the time of the forecast. For all forecasts more than three steps ahead, the forecast error will be the result of three unknown future errors that are assumed to be 0 but will actually be somewhat different from 0.

Now we wish to relate the variance of the l-step-ahead forecast to the variance Var(Y_t) of the process. For the example of the MA(2) model we have been using,

$$\text{Var}(Y_t) = [1 + (0.6)^2 + (0.4)^2]\sigma_\varepsilon^2 = 1.52\sigma_\varepsilon^2$$

Thus, the variance of the process Y_t is 1.52 times the variance of the error term (Box, et al., 1994, Chap. 5). The variance of the one-step-ahead forecast is σ_ε^2, which is the same as the variance of the error terms. This is less than the variance of the series—just determined to be $1.52\sigma_\varepsilon^2$. Consequently, we expect to have less variability in the one-step-ahead forecast than in the series as a whole.

The two-step-ahead forecast error was shown to be equal to $\varepsilon_{T+2} - 0.6\varepsilon_{T+1}$. Using the assumption that the ε_t values are independent of one another and have a constant variance σ_ε^2, the variance of the two-step-ahead forecast error is $[1 + (0.6)^2]$ $\sigma_\varepsilon^2 = 1.36\sigma_\varepsilon^2$. That is, the variance of the two-step-ahead forecast is still less than the variance of the process Y_t. Therefore, the model estimates future observations in a smaller range than just the variance of the series. The variance of the three-step-ahead forecast error is $1.52\sigma_\varepsilon^2$, the variance of the process; after three or more periods are forecast with the MA(2) model, the variance of the l-step-ahead forecast equals the variance of the process Y_t.

All forecasts beyond the three steps ahead into the future for Y_t will give the mean of Y_t, or 0 in this case. To summarize, for the MA(2) model, the best forecast of the distant future is the mean of the process. The variability about that mean (the variance of the forecast) can be estimated from the available data. In Exhibit 14.13,

EXHIBIT 14.13
Forecasts Generated by the MA(2) Model with Associated Probability Limits: $Y_t = \varepsilon_t - \theta_1\varepsilon_{t-1} - \theta_2\varepsilon_{t-2}$

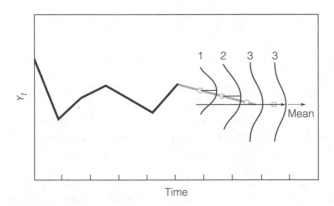

a solid line connects the actual observations. The dotted line represents the forecasts for one, two, three, and four steps ahead. The probability limits expand from the one-step-ahead to the three-step-ahead forecast. The variance of the forecast error equals the variance of the series for all forecasts three or more periods into the future.

Now, if we have an estimate s_ε^2 of σ_ε^2, the variance of the error process, we can obtain the prediction around the forecasts. The first three approximate $(1 - \eta)$ probability limits are:

$$\hat{Y}_T(1) \pm U_{\eta/2} s_\varepsilon$$
$$\hat{Y}_T(2) \pm U_{\eta/2} (1 + \theta_1^2)^{1/2} s_\varepsilon$$
$$\hat{Y}_T(3) \pm U_{\eta/2} (1 + \theta_1^2 + \theta_2^2)^{1/2} s_\varepsilon$$

in which each $U_{\eta/2}$ is the deviation exceeded by a proportion $\eta/2$ of the standard normal distribution. Here the mean was assumed to be 0. From here on out, the probability limits (and the forecast) remain the same for the MA(2) model.

We have looked at some of the properties of the forecasts for one of the simpler ARIMA models, namely ARIMA(0, 0, 2). For a MA(q) model—or ARIMA(0, 0, q)— the model has a memory of only q periods, so the observed data will affect the forecast only if the lead time l is less than q. The following conclusions are reached for the MA(q) model:

- The probability limits expand up to the h-period-ahead forecast if $h < q$.

- The probability limits become constant after the h-step-ahead forecast if $h > q$.

Probability Limits for ARIMA Models

The general procedure for calculating probability limits for ARIMA models follows the same pattern as the derivation of probability limits we have given for the MA(2) model. First, an expression for the l-step-ahead forecast error $e_T(h) = Y_{T+h} - \hat{Y}_T(h)$ is written down. Next the variance of the forecast error is expressed in terms of the error variance σ_ε^2 and the ψ weights, which are computed from the estimated AR coefficients ϕ and MA coefficients θ.

Let s_ε^2 denote an estimate of the error variance σ_ε^2. Then, approximate $(1 - \eta)$ probability limits $Y_T^+(h)$ and $Y_T^-(h)$ for Y_{T+1} are given by

$$\hat{Y}_{T+h}(\pm) = \hat{Y}_T(h) \pm U_{\eta/2} [1 + \sum \psi_j^2)]^{1/2} s_\varepsilon$$

where $U_{\eta/2}$ is the deviation exceeded by a proportion $\eta/2$ of the standard normal distribution.

For a simple linear regression model, the probability limits for the regression line expand in both directions away from the mean value of the independent variable. For a stationary ARIMA model, the concern is not so much with the value of the independent variables. The probability limits for the forecasts of a stationary series will initially expand but will become constant. From this point on, they stay the same for all values into the future.

For a nonstationary ARIMA model, the probability limits will continue to expand into the future. This is similar to the pattern displayed by the probability limits of simple linear regression models as the independent variable takes values further away from its mean.

In Appendix 14A, we can see how to express an ARIMA model in terms of the backshift operator B. This is a more efficient way of writing down the model and deriving general formulae for a forecast function or profile for each model as well as the associated probability limits (Box et al., 1994).

14.6 ARIMA Forecasting Checklist

_____   Have you examined time plots and correlograms for nonstationary behavior in the data?

_____   Considering all the autocorrelations in a correlogram of residuals, does a portmanteau test (e.g., Box-Pierce Q or Ljung-Box) show them to be significantly different from a zero set?

_____   Have you identified seasonality in the data by identifying large seasonal peaks in the correlogram?

_____   If there is evidence of nonstationarity, have you performed differencing on the data?

_____   If there is evidence of seasonality, have you performed seasonal differencing of the data?

_____   If the data is rapidly growing, have you considered taking a logarithmic transformation before differencing?

_____   If there is more than one plausible model identified, have you used the AIC and/or BIC statistic in your model selection?

_____   Is the model parsimonious? Or have you created an overly complex model?

_____   Has a model been forecast-tested over a sufficient period of time (e.g., several business cycles) to determine its forecasting capabilities?

_____   Have you performed a forecast test for any alternative models too? How rapidly does the model respond to new data?

_____   Does the model predict turning points well (or at all)?

Summary

This chapter has demonstrated the feasibility of generating accurate forecasts with ARIMA models. We have shown examples of forecast profiles for several simple model types and compared a specific ARIMA model and a simple straight-line regression model to see how responsive each is to new data. It is evident that the ARIMA model's forecasts are greatly influenced by actual current data observations. This has been shown to be advantageous during certain time periods but disadvantageous in others. We have also emphasized the importance of understanding models and their limitations so that the forecaster can select the appropriate model for the application at hand.

This chapter has dealt with using ARIMA models for forecasting seasonal and nonseasonal data. The choice of ARIMA models or, indeed, of any sufficiently flexible, physically interpretable class of models for describing time series can be achieved through a three-stage iterative procedure, consisting of (1) identification, (2) estimation, and (3) diagnostic checking. This general model-building strategy for constructing ARIMA models with real data is due to Box and Jenkins (1970).

REFERENCES

Box, G. E. P., and G. M. Jenkins (1970). *Time Series Analysis, Forecasting and Control*. San Francisco, CA: Holden-Day.

Box, G. E. P., G. M. Jenkins, and G. Reinsel (1994). *Time Series Analysis, Forecasting and Control*. 3rd ed. Upper Saddle River, NJ: Prentice Hall.

Granger, C. W. J., and P. Newbold (1977). *Forecasting Economic Time Series*. New York: Academic Press.

Slutsky, E. (1937). The summation of random causes as the source of cyclic processes. *Econometrica* 5, 105–46.

Whittle, P. (1963). *Prediction and Regulation by Linear Least-Squares Methods*. London: English University Press.

Wold, H. O. (1954). *A Study in the Analysis of Stationary Time Series*. 2nd ed. Uppsala, Sweden: Almquist and Wicksell.

Yule, G. U. (1927). On a method of investigating periodicities in disturbed series, with special reference to Wolfer's sunspot numbers. *Philos. Trans*. A 226, 267–98.

PROBLEMS

14.1   For the AR(2) model with $\phi_1 = 1$, $\phi_2 = -0.75$, construct a time series of 100 observations, using a random normal $N(0,1)$ distribution for the random errors.

   a. Plot of the sample autocorrelation functions.

   b. How well do the first few spikes in the sample autocorrelation function line up with the corresponding theoretical ones?

   c. How well do the spikes for the longer lags line up? Draw a set of parallel lines at $\pm 2/\sqrt{n}$, where n is the number of terms in a sample time series. Contrast the corresponding spikes that fall outside the band in the theoretical and sample autocorrelation functions. Is there a close correspondence?

14.2   Repeat Problem 14.1 for the AR(2) model with $\phi_1 = 4$, $\phi_2 = -0.5$.

14.3   The model identification and checking methods described in this chapter produced the following AR(2) model for a set of data. Show that this model has a stochastic cycle component and determine its period.

$$(1 - 0.75B + 0.50B^2)Y_t = e_t$$

14.4   The model identification and checking methods described in this chapter produced the following AR(3) model for a set of data. Show that this model has a stochastic cycle component and determine its period.

$$(1 + 0.60B + 0.58B^2 + 0.20B^3)Y_t = e_t$$

14.5   Show that the MA(1) process $Y_t = (1 - \theta_1 B)\varepsilon_t$ can be expressed as an autoregressive process of infinite order. If $\theta_1 = 0.7$, what are the corresponding ϕ values? If the value of θ_1 is -0.7 instead of $+0.7$, how do you expect the appearance of the time series to differ?

14.6   Recall that a second difference is the result of a first difference on the first differences of a series. Contrast this operation with a seasonal difference with $s = 2$. Would you say that a twelfth difference gives the same result as a seasonal difference with $s = 12$?

14.7 Contrast the results of a seasonal difference ($s = 4$) on a second difference of a series to a second difference on a seasonal difference ($s = 4$) of a series.

14.8 a. Sketch the ACF and PACF of an AR(1) process with (i) $\phi_1 = 0.2$, (ii) $\phi_1 = -0.3$, (iii) $\phi_1 = 0.8$, and (iv) $\phi_1 = -0.6$.

   b. For each, contrast the behavior of each pattern in the ACF and PACF.

   c. For (i) and (iii), contrast the ACFs in terms of decay (attenuation).

   d. For (i)–(iii), contrast the PACFs in terms of truncation and sign of spike.

   e. For (ii) and (iv), contrast the ACFs for pattern and decay.

14.9 a. Sketch the ACF and PACF of an MA(1) process with (i) $\theta_1 = 0.2$, (ii) $\theta_1 = -0.3$, (iii) $\theta_1 = 0.8$, and (iv) $\theta_1 = -0.6$.

   b. For each, contrast the behavior in each pattern of the ACF and PACF.

   c. For (ii) and (iv), contrast the PACFs in terms of decay (attenuation).

   d. For (i)–(iii), contrast the ACFs in terms of truncation and sign of spike.

   e. For (ii) and (iv), contrast the PACFs for pattern and decay.

14.10 Given an AR(3) model with parameters $\phi_1 = 0.60$, $\phi_2 = 0.58$, and $\phi_3 = 0.20$, show that this model can be expressed as a product of an AR(1) and an AR(2) component.

14.11 Express the following ARMA(p, q) models in short-hand notation (using the backshift operator B).

   a. ARMA(1, 0)

   b. ARMA(0, 2)

   c. ARMA(1, 2)

   d. ARMA(2, 0)

14.12 Express the following ARIMA(p, d, q) models in short-hand notation (using the backshift operator B).

   a. ARIMA(0, 1, 1)

   b. ARIMA(0, 1, 2)

   c. ARIMA(2, 1, 0)

   d. ARIMA(0, 2, 2)

14.13 Express the following seasonal ARIMA(P, D, Q) models in short-hand notation (using the backshift operator B).

   a. ARIMA(0, 1, 1)$_4$

   b. ARIMA(0, 1, 1)$_{12}$

14.14 The following 12 ARIMA models are commonly found for economic time series. Express these models in the short-hand notation for multiplicative ARIMA models. The seasonal period s is usually $s = 4$ (quarterly) or $s = 12$ (monthly).

   a. (111)(111)s

   b. (212)(011)s

   c. (201)(012)s

   d. (112)(012)s

   e. (200)(011)s

   f. (112)(102)

   g. (211)(012)s log

   h. (012)(112)s log

   i. (011)(011)s log (the "airline" model)

   j. (011)(022)s log

   k. (022)(011)s log

   l. (211)(011)s log

14.15 Using the ARIMA models in Problem 14.14 and software available to you, perform as many models as you can using the seasonal data on the CD. Evaluate your results by summarizing the different models using the diagnostics in the output.

CASE 14A PRODUCTION OF ICE CREAM
(CASES 1A, 3A–10A, AND 13A CONT.)

Because you have performed a thorough preliminary data analysis and developed several exponential smoothing models, the entrepreneur feels encouraged to round out her arsenal of univariate ARIMA series models for forecasting the monthly shipments data.

(1) Reviewing what was learned in Case 13A, select three of the most appropriate ARIMA model formulations for the ice cream data.

(2) Run the models using the 5 five years (60 months), and project the remaining 8 months.

(3) Summarize and interpret the performance measures over the fit period.

(4) Plot a correlogram of the residuals. Is there any suggestion of nonstationarity?

(5) Use at least two forecast accuracy measures and evaluate the performance of the models over the 8-month holdout period.

(6) Summarize your results and make a recommendation as to which model(s) should be retained for forecasting in the future.

(7) With your best model(s), create forecasts for the 16 months following the last data period.

(8) Create a time plot of the history, forecasts, and upper and lower probability limits for presentation to the entrepreneur.

CASE 14B DEMAND FOR RENTAL UNITS
(CASES 6C, 8B, 10B, AND 13B CONT.)

Following your modeling steps in Cases 8B and 13B, you now want to examine forecasts from the ARIMA models. The data are presented in Case 6C.

(1) Holding out YR11 from the analysis, create several ARIMA models for the data.

(2) Summarize and interpret the performance measures over the fit period.

(3) Plot a correlogram of the residuals. Is there any suggestion of nonstationarity?

(4) Use at least two forecast accuracy measures and evaluate the performance of the models over the holdout period.

(5) Accumulate the data into quarterly periods, and repeat (1)–(4).

(6) Using your results from (4) and (5), what are your recommended forecasts for the first quarter of the holdout period? For the whole year?

CASE 14C ENERGY FORECASTING (CASES 1C, 8C, 11C, AND 13C CONT.)

You have been providing assistance in the analysis of data sources and the quality of the data, as well as running several univariate ARIMA models. Now you would like to show what kind of forecasts you can produce with these models. You are aware that the data, shown in Case 8C, is annual and is simply trending. No seasonal ARIMA models need to be considered.

(1) Holding out the latest 5 years in the data (1984–1986) from the analysis, create at least three competitive ARIMA models for the energy consumption and energy production data.

(2) Summarize and interpret the relevant performance measures over the fit period. Avoid extraneous output.

(3) Plot a correlogram of the residuals for each model. Is there any suggestion of nonstationarity? Do you detect any differences in the residual patterns from the models you developed?

(4) Use at least two forecast accuracy measures and evaluate the performance of the models over the holdout period. Do you detect any differences in the forecast error patterns in the models you developed?

(5) If you ran Case 8C (exponential smoothing), run the same scenarios with your ARIMA models and contrast the similarities and differences in terms of (a) effort, (b) cost, (c) expertise required, and (d) significant performance differences. (Suggestion: Use confidence intervals in your comparisons.)

(6) Acquire comparable data for a more recent 10-year time span and try to validate your previous analysis. Are model structures much different? If not, are the estimated values of the coefficients comparable? Do you get similar forecasting performance (including confidence intervals) between the two sets of data?

CASE 14D U.S. AUTOMOBILE PRODUCTION (CASES 1D, 3D, 4D, 6D, 7D, AND 13D CONT.)

Your new assignment as an analyst for an industry association is to develop several ARIMA forecasting models to make some production forecasts. The original data are shown in Case 3D.

(1) Create a strategy for evaluating and summarizing your three best-fitting ARIMA forecasting models for the automobile production data.

(2) Using a holdout sample of 12 months, create a performance evaluation using two forecast accuracy measures and a fixed fit period.

(3) Repeat (2) using all available data as the fit period, excluding the holdout sample.

(4) Create a quarterly series for the data and build the analogous quarterly versions of the monthly ARIMA models.

(5) Repeat (2) and (3) with the quarterly data.

(6) Make an assessment as to the practical value/benefit of (a) fixed vs. rolling fit period, (b) different accuracy measures, and (c) monthly vs. quarterly modeling. Do not simply display numerous tables or look for "hard" answers.

CASE 14E THE DEMAND FOR CHICKEN (CASES 8E AND 12B CONT.)

The data appear in Case 8E.

(1) Create a time plot of the data and the first differences of the data. Comment on the nature of the nonstationarity.

(2) Create a correlogram of the first and second differences of the data. What is the order of differencing required to achieve stationarity?

(3) Divide the data into three groups. Use two or three competing ARIMA models on each group, and create forecasts for the next 12 periods. Evaluate the forecasts with the actuals provided in the following group. How would you use the performance results obtained from the first two groups to tune the forecasts for the third group? (Consider using forecast probability limits).

(4) Repeat (3) with the log-transformed data. What are the qualitative differences in the two analyses? Do you recommend transforming the data in this case?

CASE 14F FORECASTING BALLPARK ATTENDANCE FOR A SPORTS TEAM (CASES 11D AND 13F CONT.)

In Case 11D, you are given the data for the first 29 years of paid attendance at New York Mets home games since the franchise was started in 1962.

(1) Create a strategy for evaluating and summarizing your three best-fitting ARIMA forecasting models for the data.

(2) Using a holdout sample of 5 years, create a performance evaluation using two forecast accuracy measures and a fixed fit period.

(3) Repeat (2) using all available data as the fit period, excluding the holdout sample.

(4) Make an assessment as to the practical value/benefit of (a) fixed vs. rolling fit period, (b) different accuracy measures, and (c) monthly vs. quarterly modeling. Do not simply display numerous tables or look for "hard" answers.

APPENDIX 14A EXPRESSING ARIMA MODELS IN COMPACT FORM

The ARIMA model, in its fullest generality, is cumbersome to write down. It relates a dependent variable to its lagged terms and to lagged error terms. Fortunately, there is a convenient notational device for expressing an operation in which a variable is lagged or shifted; it is known as a **backshift** (backward shift) **operator.**

The Backshift Operator

The backshift notation makes the expression and manipulation of a model much simpler and more like an algebraic operation. The backshift operator B is a convenient notational device for expressing ARIMA models in a compact form. It is defined to be B operating on the index of Y_t, so that BY_t produces Y_{t-1}, which is the value of Y_t shifted back in time by one period (say one month). Hence,

$$B^2Y_t = B(BY_t) = BY_{t-1} = Y_{t-2}$$

The B^2 operation shifts the subscript of Y_t by two time units. Hence, $B^kY_t = Y_{t-k}$.

In the backshift notation, a first difference is simply

$$Y_t - Y_{t-1} = Y_t - BY_t = (1 - B)Y_t$$

At times it may be convenient to use the backshift difference operator ∇ for $(1 - B)$, so that

$$\nabla Y_t = Y_t - Y_{t-1}$$

This backshift notation makes the model look like a polynomial in B "operating" on Y_t. Hence, an AR(1) process becomes

$$(1 - \phi_1 B)Y_t = \varepsilon_t$$

where $Y_t = Z_t - \alpha$, α is the parameter that determines the level of the process, and ε_t is a white-noise process.

Notice that the first difference corresponds to the special AR(1) model in which $\phi_1 = 1$. The real advantage of using the dreaded backshift notation becomes more evident when we want to write down expressions for complex multiplicative ARIMA models.

Note that we have suppressed the constant term α in the model. Redefining the process variable Z_t to include μ does this, so that Y_t henceforth is $(Z_t - \mu)$. In practice, this is accomplished by modeling $Z_t - \bar{Z}$, where $\bar{Z}$ is the arithmetic mean of observed values of the time series.

The AR(1) model can now be written as $(1 - \phi_1 B)Y_t = \varepsilon_t$. Dividing by $(1 - \phi_1 B)$ gives Y_t in terms of ε_t:

$$\frac{Y_t}{(1 - \phi_1 B)}$$

If we use

$$\frac{1}{(1 - \phi_1 B)} = 1 + \phi_1 B + \phi_1^2 B^2 + \cdots \qquad (\text{for } |\phi_1| < 1)$$

it turns out that

$$Y_t = (1 + \phi_1 B + \phi_1^2 B^2 + \cdots)\varepsilon_t = \varepsilon_t + \phi_1 \varepsilon_{t-1} + \phi_1^2 \varepsilon_{t-2} + \cdots$$

Hence, the AR(1) model is equivalent to an MA model of infinite order. Similarly, the MA(1) model, $Y_t = (1 - \theta_1 B)\varepsilon_t$, can be regarded as an autoregressive model of infinite order.

The higher-order AR, MA, and ARMA models can be written as special cases of

$$(1 - \phi_1 B - \phi_2 B^2 - \cdots - \phi_p B^p)Y_t = (1 - \theta_1 B - \theta_2 B^2 - \cdots - \theta_q B^q)\varepsilon_t$$

The series Y_t is assumed to be mean-adjusted, so that the μ term is suppressed in the preceding representation. If we write

$$H(B) = \frac{(1 - \theta_1 B - \theta_2 B^2 - \cdots - \theta_q B^q)}{(1 - \phi_1 B - \phi_2 B^2 - \cdots - \phi_p B^p)}$$

it is evident that Y_t can be viewed as the output from a linear filter whose input is a random series ε_t, with zero mean and constant variance, and whose filter transfer function H(B) is a ratio of two polynomials in the backshift operator B. The purpose of modeling linear models of the ARMA class is to identify and estimate H(B) with as few parameters as possible (parsimonious representation). Once this is done, the representation can be used for forecasting.

Regular ARIMA Models

Let $W_t = (1 - B)Y_t$, so that W_t represents the first difference of Y_t. An ARMA model for W_t is an ARIMA model for Y_t. Assume that a series can be reduced to stationarity by differencing the series some finite number of times (possibly after removing any deterministic trend). The order of differencing is denoted by d. Then,

$$W_t = (1 - B)^d Y_t = \nabla^d Y_t$$

is stationary. For a regular ARIMA(p, d, q) model, the general form is assumed to be

$$(1 - \phi_1 B - \phi_2 B^2 - \cdots - \phi_p B^p)(1 - B)^d Y_t = (1 - \theta_1 B - \theta_2 B^2 - \cdots - \theta_q B^q)\varepsilon_t$$

where the ε_t values are white noise—a sequence of identically distributed uncorrelated errors.

It is required that the roots of the two polynomial equations in B—namely, $\Phi(B) = 0$ and $\Theta(B) = 0$—all lie outside the unit circle. The first condition ensures the stationarity of W_t, that is, the statistical equilibrium about a fixed mean; this means that the sum of the coefficients (ϕ values) must always be less than 1. The second condition, known as the invertibility requirement, guarantees uniqueness of representation (the weights applied to the past history of W_t to generate forecasts die out). This means that the sum of the coefficients (θ values) of a moving average model must always be less than 1. This notation is frequently simplified to:

$$\phi(B)(1 - B)^d Y_t = \theta(B)\varepsilon_t$$

where the AR(p) terms are given by the polynomial $\phi(B) = (1 - \phi_1 B - \phi_2 B^2 - \cdots - \phi_p B^p)$ and the MA(q) terms are $\theta(B) = (1 - \theta_1 B - \theta_2 B^2 - \cdots - \theta_q B^q)$. In short, the ARIMA(1, 1 ,1) model takes the form

$$(1 - \phi_1 B)(1 - B)Y_t = (1 - \theta_1 B)\varepsilon_t$$

APPENDIX 14B FORECAST ERROR AND FORECAST VARIANCE FOR ARIMA MODELS

Consider the general form of the ARIMA(p, d, q) model given by

$$(1 - \phi_1 B - \phi_2 B^2 - \cdots - \phi_p B^p)(1 - B)^d Y_t = (1 - \theta_1 B - \theta_2 B^2 - \cdots - \theta_q B^q)\varepsilon_t$$

This can be expanded as

$$(1 - \phi_1 B - \phi_2 B^2 - \cdots - \phi_{p+d} B^{p+d}) Y_t = (1 - \theta_1 B - \theta_2 B^2 - \cdots - \theta_q B^q)\varepsilon_t$$

An observation at time $t + 1$ generated by this process can be written as:

$$Y_{t+l} = \phi_1 Y_{t+l-1} + \phi_2 Y_{t+l-2} + \cdots + \phi_{p+d} Y_{t+l-p-d} + \varepsilon_{t+1} - \theta_1 \varepsilon_{t+l-1}$$
$$- \theta_2 \varepsilon_{t+l-2} - \cdots - \theta_q \varepsilon_{t+l-q}$$

Alternatively, Y_{t+l} can be written as an infinite weighted sum of current and previous errors ε_t:

$$Y_{t+l} = \varepsilon_{t+1} + \psi_1 \varepsilon_{t+l-1} + \psi_2 \varepsilon_{t+l-2} + \cdots + \psi_l \varepsilon_t + \psi_l + 1\varepsilon_{t-1} + \cdots$$

Consider the l-step-ahead forecast, $\hat{Y}_T(l)$, which is to be a linear combination of current Y_T, previous observations $Y_{T-1}, Y_{T-2}, \ldots$, and errors $\varepsilon_{T-1}, \varepsilon_{T-2}, \ldots$. It can also be written as a linear combination of current and previous errors ε_T, $\varepsilon_{T-1}, \ldots$; thus,

$$Y_T(l) = \psi_l^* \varepsilon_t + \psi_{l+1}^* \varepsilon_{t-1} + \psi_{l+2}^* \varepsilon_{t-2} + \cdots$$

where the ψ_j^* are to be determined. An important result in ARIMA time series theory is that the MSE of the forecast is minimized when $\psi_{l+j}^* = \psi_{l+j}$ (Box et al., 1994).

The variance of the forecast error $e_T(l)$ is given by

$$\mathrm{Var}(Y_{T+l} - \hat{Y}_T(l)) = (1 + \psi_1^2 + \psi_2^2 + \cdots + \psi_{l-1}^2)\sigma_\varepsilon^2$$

These estimates of variance are based on the assumption that the ψ_l are correct. That is, the error in estimating parameters is assumed to be negligible relative to the successive one-step-ahead prediction error.

As an example in using the formula, consider the simple AR(1) model: $Y_t = \alpha + \phi_1 Y_{t-1} + \varepsilon_t$. The forecast errors are

$$e_T(1) = Y_{T+1} - \hat{Y}_T(1)$$
$$= [\alpha + \phi_1 Y_T + \varepsilon_{T+1}] - [\alpha + \phi_1 Y_T]$$
$$= \varepsilon_{T+1}$$
$$e_T(2) = Y_{T+2} - \hat{Y}_T(2)$$
$$= [\phi_1 \varepsilon_{T+1} + \varepsilon_{T+2}]$$
$$\vdots$$
$$e_T(l) = Y_{T+l} - \hat{Y}_T(l)$$
$$= \sum \phi_1 \varepsilon_{T+l-j}$$

The variance of the forecast error is

$$\text{Var}[e_T(l)] = \sigma_\varepsilon^2 \sum \phi_1^{2j}$$
$$= \sigma_\varepsilon^2 (1 - \phi_1^{2j})(1 - \phi_1^2)$$

For a stationary AR(1) model, with $-1 < \phi_1 < 1$, the variance increases to a constant value $\sigma_\varepsilon^2 / (1 - \phi_1^2)$ as the lead time (or horizon) h tends to infinity. For nonstationary models, on the other hand, the forecast variances increase without bound with increasing values of h.

An important property of forecast errors, worth noting here, is that, whereas errors in one-step-ahead forecasts are uncorrelated, the errors for forecasts with longer lead times are in general correlated. It is worth remembering in practice that there are two kinds of correlations to be considered:

- The correlation between forecast errors $e_T(h)$ and $e_{T-j}(h)$ made at the *same* lead time h from *different* time origins T and $T-j$
- The correlation between forecast errors $e_T(h)$ and $e_T(h+j)$ made at *different* lead times from the *same* origin T

General expressions for these correlations can be found in Box et al. (1994).

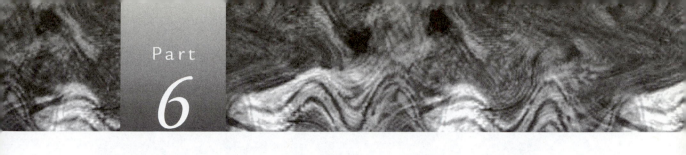

Improving Forecasting Effectiveness

15

Selecting the Final Forecast Number

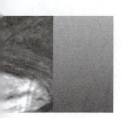

"It is necessary to be humble and accept the probabilistic environment."
HOGARTH AND MAKRIDAKIS (1981)

THIS CHAPTER DEALS WITH DETERMINING the reliability of forecasts generated with forecasting models and using that information to support a recommended forecast for approval. This involves preparing forecast scenarios, establishing credibility with accepted performance standards, evaluating the reliability of the forecasts using forecasting simulations, and reconciling sales force and customer inputs where appropriate. By rigorously adhering to these steps, you will be able to confidently recommend a "final forecast number," which deals with knowing how much of what product will be needed in what place, at what time, and at what price; and packaging and presenting the forecast for approval.

We advocate using a forecaster's checklist to measure a specific forecast relative to generic standards. The purpose of the checklist is to establish standards for the forecasting organization. Both the forecaster and the forecast manager can use the checklist in the preparation and subsequent review of the forecast. By establishing meaningful forecasting standards, forecast evaluation can be greatly simplified.

15.1 PREPARING FORECAST SCENARIOS

One of the most perplexing problems that forecasters face is how to tell a good forecast from a bad one at the time it is presented for approval. Certainly, after the forecast time period has elapsed, anyone can look back and determine how closely the forecast predicted the actual results; but this is after the fact. The forecaster wants to be confident that the forecast is reasonable and credible at the time it is prepared.

After forecasting models have been developed, the business forecaster reaches the stage where actual forecasts are produced, tested, and approved. This effort begins with a generation of scenarios from the models that have survived the selection process.

The philosophy of forecast evaluation is one in which primary emphasis is placed on the forecasting process rather than on the numbers. If the forecaster has meticulously followed a proper forecasting process, the end result will be as good a forecast as can be delivered.

In addition to creating projections from models, the forecaster needs to create scenarios that provide estimates of the reliability of the forecast in terms of limits around the forecast at specified levels of confidence. Alternatively, reliability can be expressed as the likely percentage (amount) of deviation between a forecast and actual performance. For example, suppose that new car purchases for the year are forecast to be 10 million ± 700,000 at a 90% confidence level. Another way of stating this is that, in a particular forecasting model, the average annual deviation (absolute value) between what is forecast and the actual new car sales is approximately 7%. We recommend that forecasters test the validity of their models by simulating a forecast over a holdout period and generating scenarios from the models over time periods for which the actual results are known. In this way, it is possible to establish the likely forecast accuracy.

Exhibit 15.1 illustrates how a forecaster could summarize forecast errors in a model with actual data from Year 1 through Year 15. A projection from the model for Year 11 is generated. The actual data through Year 10 show that the projection is 8.7% greater than the Year 11 actual value. An additional year of actuals is then added to the model, and Year 12 is predicted. This time the projection is only 1.2% greater than the actual. This process can be continued and some average performance can be calculated. In this hypothetical example, the average absolute 1-year-ahead forecast error for five periods is 4.3%.

It might be also be useful to consider the MdAPE (here 1.8%) as well to ensure that a very large miss in one year does not unduly distort the average value. With this approach, the forecaster might expect the 1-year-ahead prediction to be within 4% of the actual value, on average.

Historical Fit	Percentage Error
Year 1–Year 10	−8.7
Year 2–Year 11	−1.2
Year 3–Year 12	5.7
Year 4–Year 13	3.8
Year 5–Year 14	−2.3
MAPE	4.3
MdAPE	3.8

Exhibit 15.1 Summary of 1-Year-Ahead Forecast Errors from a Hypothetical Model with Data for 15 Years (Percentage error = (Actual − Forecast)/Actual)

Combining Forecasts and Methods

The combining of forecasts is treated extensively in the forecasting literature (see Clemen, 1989; Armstrong, 2001). Given that it is frequently difficult to know which smoothing technique to select, it may be useful to bypass a specific model selection procedure entirely. For instance, by automating the modeling process, we can fit a variety of exponential smoothing models and then average their individual forecasts to come up with a combined forecast as a forecast scenario. Different approaches for combining two or more forecasts into a composite forecast have led to some surprising improvements in getting more accurate forecasting results.

Combining forecasts is a technique that can be implemented completely automatically, making it useful for multiseries forecasting.

Averaging Forecasts

In a study of 1001 time series in the M1-competition (Makridakis et al., 1982), a simple average of forecasts of exponential smoothing methods was shown to be more accurate than the best of the individual forecasts. A decade later in the M2-competition study (Makridakis et al., 1993), averaging of forecasts from exponential smoothing methods also turned out to be advantageous, but not quite as good as choosing the best of the individual methods for each time series. These and other competitions have given the technique of combining forecasts widespread acceptance. Combining forecasts spreads the risk that any individual procedure will go far off the mark. As Winkler (1989) noted, "Just as investors create diversified portfolios to reduce risk, a combined forecast can be viewed as having a smaller risk of making an extremely large error than an individual forecast."

One alternative to simple averaging is to base the weights on the accuracy of past forecasting or fitting performance. Another suggestion uses subjective weightings to form a composite forecast. Practitioners might like weighing the forecast based on their personal judgments about which methods most closely reflect reality at the time. Unfortunately, there are no pat answers to combining forecasts, although some promising improvements in accuracy have been reported in the literature.

15.2 ESTABLISHING CREDIBILITY

What is needed is a process that, if followed, will increase the likelihood of good forecasting performance. In other words, it is necessary to establish credibility by setting standards of performance for forecasters that will increase the likelihood of improved forecast accuracy. A checklist that can be used by both forecasters and forecast managers to measure a specific forecast relative to some established standards is included at the end of this chapter, and we recommend its use. The checklist is general in nature and covers the essential elements of an effective forecasting process. It is designed to assist the new forecaster in getting started and to remind the experienced forecaster to cover all steps in the process. Let us review each part of the checklist in turn.

It is necessary to establish credibility by setting standards of performance for forecasters.

Setting Down Basic Facts—Forecast-Data Analysis and Review

To be satisfied that basic facts have been adequately researched, a forecaster should expect to produce tables and plots of historical data. The data should be adjusted to account for changes in geographic boundaries, organizational changes, product groupings and customer segmentations, or other factors that will distort analyses and forecasts. If appropriate, the data should be seasonally adjusted to give a better representation of trend-cycle patterns. Outliers or other unusual data values should be explained and replaced, if this is warranted. It may be useful to indicate the NBER reference dates for the peaks and troughs of business cycles that affect the company. This provides the forecaster with an indication of the extent to which a client's data are impacted by the national business cycles. Knowing this relationship will be helpful when the forecaster reviews the assumptions about the future state of the economy and assesses how these assumptions are reflected in the forecast.

Tables and plots of annual percentage changes provide an indication of the volatility of a series and are useful later in checking the reasonableness of the forecast compared to history. If possible, ratios should be developed between the forecast series and other stable data series that are based on company or regional performance. These ratios should be shown in tables or plots. Once again, these ratios provide reasonableness checks. If some major change is expected in the forecast period, these ratios should help identify the change.

Whenever possible, there should be at least a decade of data available for the forecaster's review. It may not be necessary to show this much history when presenting the forecast to the clients; but it is necessary to have this much data available to analyze the impact of business cycles. If possible, data going back several recessions should be available. However, in many forecasting circumstances, data this old may no longer be relevant. Data back to the most recent recession is desirable because this will reflect the timing, impact, and duration of the economic cycle on the business data.

Documentation of history is an important step that can serve as reference material for all future forecasts and forecasters.

Causes of Change

The next segment of the checklist deals with the causes of changes in past demand trends or levels. The first step is to identify the trend in the data. Regression analysis is an excellent tool for this. A straight-line regression against time, as a starting point, will provide a visual indication as to whether the trend is linear or nonlinear. There should be a plot of the series and its fitted trend on a scale of sufficient breadth to clearly identify deviations from trend. The reasons for the deviations should then be identified and explained in writing. These explanations must be specific. Was there any unusual competitive activity? Was there a change in a promotion program or prices? Did the deviation correspond to a regional or national economic pattern? What was the source of explanation—the forecaster or someone else? Finally, how certain is the forecaster that the reason or explanation stated is correct? Is the forecaster reasonably certain of the cause, or is there insufficient evidence to be confident that the true cause has been or can be identified? Documenting this is particularly helpful to a new forecaster and improves productivity.

There should also be available a record of forecasts and actual performance for at least the previous 3 years. This allows the forecaster and forecast manager to know how well the organization has done in the past and to gauge the possible reaction of the clients to changes in the forecast. It is also possible to determine from these data if any or all of the forecasters on the staff have a tendency to be too optimistic or too pessimistic over time.

A record of forecast performance with actuals should help us gauge the possible reaction of clients to changes in the forecast and reduce the tendency organizations to over- or underestimate demand.

Analyzing Forecast Errors

The next segment of the checklist is concerned with the reasons for the differences between previous forecasts and actual results. This form of results analysis is useful for uncovering problem areas, for identifying the need for new or improved methods, and for determining the quality of the prior forecasts.

At this time, the forecaster or forecast manager is looking for a pattern of overforecasting or underforecasting. The key to identifying the reasons for forecast deviations is to have written records of basic assumptions, which should be reviewed. These assumptions should then be tested for specificity against the standards shown on the checklist. Do the assumptions relate only to the future time periods? Has the forecaster used an assumption that states a positive assertion of facts that may hold true during the forecast period? Was the direction of expected impact stated? Presumably, there are both positive and negative assumptions in terms of their impact on the series being forecasted. Do the assumptions indicate the amount or rate of expected impact, the timing of the initial impact, and the duration of the expected impact?

Accompanying each assumption should be a rationale indicating why the assumption is necessary. The source of an assumption might be the forecaster, company economists, industry forecasters, government publications, or newspaper clippings and journal articles. The forecaster may be absolutely certain that the assumption will prove correct. On the other hand, the forecaster may indicate that it is necessary to make the assumption but that considerable doubt exists as to whether the future will be as assumed.

> The source of the assumption should be identified and the degree of confidence in the assumption should be stated.

Factors Affecting Future Demand

The next segment of the checklist is concerned with the factors likely to affect future demand and, therefore, the forecast. Assumptions have to be made about factors such as income, habit, price of a company's product, price of competing goods, availability of supply, and market potential. In addition, the forecaster should check to see that there is a logical time integration between historical demand and the short- and long-term forecasts. Time plots are very useful here. In addition, there should be logical time integration between related forecast items. The forecast should also be reasonably related to forecasts produced by other organizations in the company (if the forecasting function is not centralized in one organization) such as forecasts of economic conditions, revenues, and expenses.

> Assumptions should be made about factors influencing demand and the time integration between historical demand and the short- and long-term forecasts.

Creating the Final Forecast

At this time, the forecaster and forecast manager can decide whether the forecasting methodologies used present the best methods available at the time. The methods presented in this book have been tested extensively in business applications and have proven to be practical.

Within limits, it is very difficult for a forecaster or forecast manager to fine-tune the numbers presented and have any degree of confidence that the changes are appropriate.

However, the forecaster or manager can carefully review the forecast assumptions for reasonableness. The assumptions are the heart of the forecast, and considerable probing of these assumptions can satisfy the forecaster as to their appropriateness for the forecast period.

The use of multiple methods to arrive at the final forecast is highly recommended.

The forecaster and forecast manager can also review the technical soundness of the analysis and be satisfied that no errors were made. Having performed these forecast evaluations, the forecaster and forecast manager can discriminate between a good forecast and a bad one at the time they are asked to have it approved. This level of managerial involvement is generally required only for extremely sensitive or important forecasts, for training new forecasters, for reviewing exceptional cases, and for spot-checking to ensure that the processes are being followed.

After narrowing down the possible alternatives, the forecaster must determine the parameters of selected forecasting models. These parameters may be ratios such as housing units per acre or market penetration rates. Similarly, the potential car buyer we introduce in Chapter 1 double checks the parameters that make a certain car seem right, including the specific model, color, engine size, and extras such as power steering, power brakes, and a stereo radio.

Model parameters may also be coefficients estimated statistically from data by computer. Computer packages provide immediate access to the flexibility and breadth of potentially useful forecasting techniques.

Verifying Reasonableness

Continuing the analogy in Chapter 1, the car buyer attempts to validate the manufacturer's claims concerning ease of parking, comfort, noise level, braking, and acceleration. Taking a road test does this. For the business forecaster, there are a number of analytical tools available (e.g., residual analysis and forecast simulations) to determine whether it is possible to improve on an initial model. These tools have been presented throughout the book. As a result of the road test, the car buyer may decide to try a different make, model, or engine.

Likewise, the forecaster might decide to add new data, try a different method, or replace a one-time series such as the reciprocal of an unemployment rate with another series such as total employment in a revenue model.

Forecasters validate models through diagnostic checking.

The diagnostic checking stage usually requires a number of iterations. New variables can be considered, transformations of variables can be made to improve the models, and some techniques should be rejected at this stage because of their inability to provide statistically significant results and their inability to achieve the desired objectives of accuracy.

In evaluating alternatives, the forecaster will find patterns or characteristics of the models that will influence the final selection of the models for use. For example, which techniques are more accurate in predicting turning points? In predicting stable periods? Which techniques have the best overall accuracy in the forecast test mode? Do some techniques tend to overpredict or underpredict in given situations? Are the short-term projections of one technique better than another? Do we actually need long-term predictions? Do the coefficients of one model seem more reasonable than those of another either in sign or magnitude?

Selecting a Final Forecast Number

The final steps of the process entail relating projections from various models to the final forecast. The forecaster has several decisions to make. The forecaster recognizes that the various models are abstractions from the real world. The future will never be exactly like the past; the projections from the models must be viewed as job aids in making a subjective judgment about the future. In most cases, that advice is provided in the form of a single "best-bet" number, which represents either the value at some specific time or the cumulative value of a series of data points at the end of a specific period of time.

The "best bet" number can be illustrated best with the median of a hypothetical frequency distribution. There is an even chance that the future outcome will fall above or below that median. Its primary weakness, however, is that the planning and decision-making processes assume that the forecast precisely describes the future when, in fact, it cannot perform such a feat.

Because future events and conditions cannot be predicted consistently with complete accuracy, the end product of the forecasting process can best be described as giving advice.

Decision making involves the assessment and acceptance of risk. Therefore, forecasters can assist decision makers by providing forecast levels and associated probability limits that indicate the chance of each of those levels being exceeded. This does not mean that the forecaster takes a "shotgun" approach to predicting the future by incorporating the extreme alternatives at either end of the range. It simply means that the forecaster should provide the "best-bet" figure and state the associated risk levels or range on each side of the "best bet." If a view of the future is presented in this format, the decision maker has much more information on which to assess the risk associated with decisions.

Creating probabilities associated with alternative views of the future are, of course, highly subjective. In the social sciences, probabilities are developed through some form of scientific sampling process over a long period of time. Such is not the case when it comes to quantifying the probabilities of future events or conditions in a business forecast. A multitude of influencing factors can enter into the picture after the forecast has been made and thereby completely change the course of future events.

The principle of using more than one extrapolative technique can again provide substance to the forecasting process by giving a certain degree of objectivity in the development of risk levels. It is clear, therefore, that the extrapolative techniques play an important role in the decisions on the forecast level that is ultimately produced. This level of managerial involvement is generally required only for extremely sensitive or important forecasts, for training new forecasters, for reviewing exceptional cases, and for spot-checking to ensure that the processes are being followed.

The amount of analysis may seem overwhelming. However, most of the analyses are performed the first time for a given time series and are then updated as new results become available. Often, much of the data and graphic displays can be generated and maintained with spreadsheet software packages.

Role of Judgment

Continuing with the analogy, once the car buyer has purchased the car, subjective judgment comes into play if the car buyer realizes that the purchase was not a good decision. For example, during verification and confidence checks, suppose that the car buyer discovers a flaw so great that the dealer agrees either to repair the car or to exchange it for a slightly different model—the buyer needs to exercise judgment not called for in the original forecast in order to reconcile expectations and reality.

In an actual forecasting situation, it may become apparent that the actuals have exceeded the estimates for several successive periods. Experience may suggest that a model's projections be modified upward by a given amount to account for the current deviation and the forecaster's expectation of whether that pattern will continue.

Subjective judgment in forecasting should be based on all available information, including changes in company policy, changes in economic conditions, contacts with customers, and government policy considerations. This judgment is a real measure of the skill and experience of the forecaster. For this reason, data and processes are only as good as the person interpreting them. This judgment operates on many inputs to reach a final forecast.

Informed judgment plays a critical role in the determination of the final forecast number and, later on, in the determination of when a forecast should be revised.

Judgment is, by far, the most crucial element when we are trying to predict the future. Informed judgment is what ties the forecasting process and the extrapolative techniques into a cohesive effort that is capable of producing realistic predictions of future events or conditions. Informed judgment is an essential ingredient of the selection of the forecasting approach; the selection of data sources; the selection of the data collection methodology; the selection of analysis and extrapolative techniques; the use of analysis and extrapolative techniques during the forecasting process; the identification of influencing market and company factors that are likely to affect the future of the item to be forecast; the determination of how those factors will affect the item in terms of the direction, magnitude (amount or rate), timing, and duration of the expected impact; and the selection of the forecast presentation methodology.

Informed judgment plays a significant role in minimizing the uncertainty associated with forecasting.

Automatic processes, models, and statistical formulas are sometimes used in computing future demand from a set of key factors. However, no such approach is likely to reduce substantially the reliance on sound judgment. Judgment must be based on a comprehensive analysis of market activities and a thorough evaluation of basic assumptions and influencing factors.

Statistical approaches can provide a framework of information around which analytical skills and judgment can be applied in order to arrive at and support a sound forecast. To quote from Butler et al. (1974, p. 7): "In actual application of the scientific approaches, judgment plays, and will undoubtedly always play, an important role. The users of econometric models have come to realize that their models can only be relied upon to provide a first approximation—a set of consistent forecasts which then must be 'massaged' with intuition and good judgment to take into account those influences on economic activity for which history is a poor guide."

The limitations of a purely statistical approach should be kept clearly in mind. Statistics, like all tools, may be valuable for one job but of little use for another. An analysis of patterns is basic to forecasting and a number of different statistical procedures may be employed to make this analysis more meaningful. However, the human element is required to understand the differences between what was expected in the past and what actually occurred and to predict the likely course of future events.

15.3 USING FORECASTING SIMULATIONS

Choosing the Holdout Period

The first step in a forecasting simulation is to divide the historical data between a fit period and a test period. Typically, the fit period starts at the beginning of the historical data and continues to some point in the recent past. For example, the historical data could span periods 1–24 and the fit period could be chosen to be periods 1–20. The last four periods of data are treated as the holdout sample to provide a test of the model's forecasting accuracy over the recent past.

Here are some guidelines to decide how much of the historical data should be held out from the fit period to use in evaluating forecasting accuracy. Suppose the forecasting horizon (h) is three periods, so that we wish to make projections for periods

(25–27). We should aim, at minimum, to withhold the previous three periods (22–24) from the fit period. The accuracy with which a model will forecast the next three periods (25–27) can then be gauged by measuring how accurately the model forecasts the previous three periods (after giving that model only the data from the fit period, through period 21).

Forecasting horizon: From the present, how long into the future do we wish to forecast?

Now consider a situation in which the forecasting horizon is one time period ahead. If we hold out from the model fit just a single time period, then accuracy will be measured based on only one forecast and its corresponding forecast error. If we wish to forecast no farther ahead than period 25, we would certainly not wish to limit our test period to one period, period 24. Doing this means that our gauge of a method's forecasting accuracy for period 25 would be how well that method performed in forecasting only the most recent period, period 24, from the vantage point of period 23. If period 24 is atypical in some sense, our gauge will prove misleading. Prudence in the design of a forecasting simulation requires diversification.

Depth of the simulation: The forecasting horizon sets a logical minimum to the test period; however, we may wish to deepen the simulation—to increase the number of forecasts on which accuracy is being evaluated—by holding out more data.

In statistical studies based on a cross section of data (e.g., 100 test subjects observed at the same time), we frequently find up to a 50-50 split between the sample used for model development and the sample used for validation. In a forecasting context, however, a 50% holdout sample is practically unheard of. As applied to our forecasting example, doing this means that we would use the first 12 periods to fit a model and the remaining 12 periods to test the model's forecasting accuracy. Although such a forecasting simulation would have adequate depth, we might question the timeliness of the results. The model, after all, was developed without knowing anything that happened in the last dozen periods. Events of the more recent past might warrant an alternative manner of accounting for trend or seasonality or might dictate new explanatory variables that should be considered.

Moreover, if we shorten the fit period too much, we may not leave ourselves an adequate basis for developing a plausible forecasting method. In principle, the precision of statistical estimates tend to improve with increases in sample size; hence, too small a fit period may preclude our obtaining a good fit for any forecasting method.

Timeliness: The number of time periods held out from the model fit should be at least as large the length of our forecasting horizon.

If the forecaster is fortunate to have a lengthy time series and a stable forecasting environment, a deep simulation is feasible. When the forecasting environment is more volatile, depth may have to be sacrificed to preserve timeliness. In addition, if the time series is very short to begin with, a forecasting simulation may not be feasible at all. We describe some alternatives to forecasting simulation later in this chapter.

The choice of how much data to use in a holdout sample is a matter of achieving an acceptable balance between depth and timeliness.

Fixed-Origin Simulations

The origin of a simulation is the point from which forecasts are generated. Simulations may be carried out using either a single origin or multiple origins. The former is called a fixed-origin simulation or, alternatively, a static simulation. Starting at the origin (call it time $t = T$), we generate forecasts for time periods $T + 1$, $T + 2$, . . ., where T is the most recent value in the time series.

Recall that the time span from the forecast origin to the end of the forecast period is called the lead time of the forecast. The forecast with a lead time of one-period is called a one-step-ahead forecast, that with a lead time of two periods is a two-step-ahead forecast, and so forth. We can substitute the actual time interval (period, month, etc.) for the word *step* to refer to a one-step-ahead forecast or a 2-month-ahead forecast. Sometimes, the term trace forecasts is used for forecasts derived from a fixed origin.

The origin of a simulation is the final time period in the fit period.

In the previous example, we use period 20 as the simulation origin. In a fixed-origin simulation, we start making projections for periods 21–24 in sequence. The resulting trace forecasts include one forecast each at one period ahead, two periods ahead, three periods ahead, and four periods ahead. Using known actual values, forecast errors are calculated along with measures of bias, precision, and relative error (see Chapter 5).

Fixed-origin simulations have several shortcomings.

Inadequate depth. A fixed-origin simulation yields only one forecast (and hence one forecast error) at each lead time. We should base accuracy judgments on more than a single forecasting situation at each lead time.

Sensitivity to the origin. Forecasts generated from a single origin are liable to be misleading because of occurrences unique to that origin. If period 20 serves as the sole origin for the forecasting simulation, the forecasts issued for periods 21–24 are based on the forecast for period 20. It is possible that a potentially good model will have the misfortune of going astray in its estimate for period 20 and, as a result, perform poorly at all steps ahead. Fildes (1989) has pointed out that the potential for misjudging forecasting models is high when the simulation is based on a single origin, due to high sampling variability across time.

Confounding different lead times. A fundamental problem exists with the interpretation of error measures from a fixed-origin simulation. Each measure is derived from an average of errors at different lead times. The result is an assortment of near-term and long-term forecast errors. For example, a MAPE could be calculated as an average of eight errors: the one for 1 quarter ahead, 2 quarters ahead, and so forth through 8 quarters ahead.

In the averaging process, we may lose sight of individual lead times, so make the nature of the averaging explicit. For example, on determining a MAPE = 24.5%, say, we should state that from a fixed-origin simulation with a horizon of 8 quarters, the errors in forecasting each of 8 quarters ahead from quarter $t = T$ average 24.5%. The near-term errors (e.g., 1 and 2 quarters ahead) are likely to be less than 24.5% whereas the longer-term errors (7 and 8 quarters ahead) are liable to be greater than this figure.

When performing a fixed-origin simulation it is appropriate to *plot forecasts for presentation purposes* and *compare the accuracy* of a new method against a series of judgmental forecasts made at one point in the past. Suppose that, at quarter $t = T$, there are forecasts recorded for each of the succeeding 8 quarters. We may not be able to reproduce these forecasts because there is no record of the methodology used. We would like, however, to determine whether our new method will issue more accurate forecasts if put to the test in quarter $t = T$. Now, the fixed-origin simulation would provide a fair comparison. We choose quarter $t = T$ as the origin and forecast each of the next 8 quarters, thus recreating the exact conditions under which the previous forecasts were made.

Rolling-Origin Simulations

A rolling simulation, or a rolling-origin simulation, involves successively updating the forecasting origin and creating forecasts from each new origin. The procedure goes a long way toward addressing the three deficiencies of the fixed-origin

Exhibit 15.2
Waterfall Chart with
Forecasts Generated
by a Rolling
Simulation

		Periods		
20	21	22	23	24
Origin	F1	F2	F3	F4
	Origin	F1	F2	F3
		Origin	F1	F2
			Origin	F1

simulation—it increases the depth of the forecasting simulation, perhaps substantially; it reduces the sensitivity of the simulation results to the choice of origin; and it provides the means for distinct assessments of forecasting accuracy at each lead time.

Rolling simulations begin the same way as fixed-origin simulations: Data from periods 1 through T are used to fit a model. Then, starting at the end of period $t = T$, the origin, forecasts are generated for periods $T + 1$ through $T + h$. The rolling simulation, depicted in Exhibit 15.2, takes the data value for period $T + 1$ and adds it to the fit period. The model is updated and revised based on the data, which now includes period $T + 1$. (The precise nature of the revision is discussed next.) Using period $t = T + 1$ as the origin, new forecasts are made for periods $T + 2$ through $T + h$. The updating process continues. The data value for period $T + 2$ is added to the fit period, the model revised once again and, from the new origin, $T + 2$, additional forecasts are generated. The process stops after the next to last period $(T + h - 1)$ has done its stint as the forecasting origin.

 Whenever possible, it is preferable to use dynamic (rolling) simulations rather than static (fixed-origin) simulations.

We can use our example to contrast the results of a rolling simulation with the fixed-origin simulation procedure. Period 24 $(T + h = 24)$ is the most recent time period. Period 20 is the simulation origin $(T = 20)$. Hence, the simulation horizon is $h = 4$ periods. The fixed-origin simulation generates four forecasts from origin 20. The rolling simulation also generates four forecasts from origin 20, but then supplies an additional three from origin 21, two from origin 22, and one from origin 23, for a total of 10 forecasts for the assessment of forecasting accuracy. The depth of the simulation has been enhanced substantially, from four to ten forecasts. To illustrate this, Exhibit 15.3 shows a waterfall chart for an ARIMA model for hotel/motel demand (DCTOUR) series.

Analyzing Simulation Errors by Lead Time

In contrast to a fixed-origin simulation, a rolling simulation results in multiple forecasts at virtually every lead time. For the forecasting example shown in Exhibit 15.2, the rolling simulation resulted in 4 one-period-ahead forecasts, 3 two-period-ahead forecasts, 2 three-period-ahead forecasts, and 1 four-period-ahead forecast. To assess

Exhibits 15.3.   Waterfall Chart with Forecast Percentage Errors Generated by an ARIMA Model for Hotel/Motel Demand (DCTOUR)

DCTOUR

Holdout Data	PE (%)												MAPE (%) Model D
	1	2	3	4	5	6	7	8	9	10	11	12	
1001666	−5.38	−2.22	−1.28	−0.70	−2.33	2.53	−0.36	−2.60	−1.72	−0.06	−2.12	−4.07	2.1
1073196	−3.85	−1.95	−1.08	−2.55	1.86	0.41	−2.72	−2.52	−0.59	−2.14	−4.73		2.2
1421423	−3.58	−1.75	−2.94	1.65	−0.27	−1.92	−2.62	−1.36	−2.66	−4.75			2.4
1577321	−3.37	−3.62	1.28	−0.49	−2.64	−1.83	−1.48	−3.46	−5.29				2.6
1600991	−5.27	0.62	−0.88	−2.85	−2.54	−0.70	−3.58	−6.12					2.8
1594481	−0.96	−1.55	−3.24	−2.75	−1.39	−2.77	−6.23						2.7
1510052	−3.16	−3.93	−3.16	−1.61	−3.49	−5.41							3.5
1436164	−5.58	−3.84	−2.00	−3.71	−6.14								4.3
1404978	−5.49	−2.68	−4.11	−6.36									4.7
1585409	−4.32	−4.80	−6.78										5.3
1234848	−6.46	−7.49											7.0
923115	−9.20												9.2

forecasting bias and precision at each lead time, we can average all the one-period-ahead forecast errors, percentage errors, or absolute errors to depict the accuracy of forecasting with a lead time of one. The same evaluation can be carried out for the two-period-ahead errors, the three-period-ahead errors, and so forth. The value for each selected error measure (e.g., MAD, MAPE, and RMSE) should be reported separately for each lead time.

 Multiple error measure should be reported separately for each lead time.

15.4 DESIGNING FORECASTING SIMULATIONS

Is there a minimum depth that the forecaster should strive for? A practical suggestion is to try to achieve a minimum depth of three forecasts at each lead time. This will result in each error measure being based on a minimum of three previous forecasting situations. Although not a rigid requirement, this does discourage the forecaster from placing too much credence in accuracy measures calculated from only one or two forecasting situations.

There is more than one way to achieve adequate depth. In addition to lengthening the simulation horizon, which may not always be feasible, the forecaster can increase the length of the historical time series, the number of time series investigated, the number of forecasting origins investigated, and the number of forecast lead times investigated. Also, by examining other series, we can compensate for lack of depth when using only one time series.

There are other considerations about which test periods to include. Should the test period be constant or decreasing depth? In the standard implementation of a forecasting simulation, the test period ends with period $T + h$, the most recent time period for which data are available. Hence, in a rolling simulation, as we update the fit period, the test period becomes successively shorter. This is why the depth of the simulation decreases as lead time increases. We can call this characteristic of the rolling simulation a decreasing-depth test period.

Some forecasters may find it more appealing to maintain a constant-depth test period, in which the same depth is maintained at all lead times. If the simulation horizon is $h = 4$ time periods, for example, a constant-depth test period results in four forecasts at each lead time $T + 1$ through $T + 4$. The four-period-ahead MAPE will be based on the same number of forecasting situations as the one-period-ahead MAPE. Keeping a constant-depth test period, however, requires the forecaster to hold out twice as many time periods (from h to $2h$) from the fit period.

One of the drawbacks of the constant-depth test period is that it doubles the number of time periods that must be held out of the initial fit period. In addition, the constant-depth test period is wasteful of data. With $2h = 8$ time periods withheld, we could have 8 one-period-ahead, 7 two-period-ahead, 6 three-period-ahead, and 5 four-period-ahead forecast errors, in place of just four at each lead time. To limit the wasting of data, we recommend a compromise between the constant- and decreasing-depth test period. As usual, the forecaster begins the forecasting simulation by choosing the desired simulation horizon (h). This horizon is then incremented by two time periods to ensure that the minimum depth at the lead time h is 3. The forecaster uses all forecasts at lead times up to h so that the test period is decreasing in depth down to a minimum depth of 3.

Pruning the fit period is yet another consideration. As a rolling simulation successively updates the simulation horizon, it is, in turn, extending the fit period. Initially our forecasting simulation example used a 20-period fit period. In the updating, periods 21–23 entered the fit period one at a time, so that the final fit period incorporated 23 periods. An alternative is to maintain a constant fit period by pruning the earliest time period as the origin is updated. So, as period 21 enters the revenue forecasting simulation, period 1 is pruned from the fit period; when period 22 comes in, period 2 departs; and so forth. The pruning procedure is similar to a moving average. Its appeal, like that of the moving average, is that it cleanses the model fit of old data and ensures that we are consistently generating forecasts from fit periods of the same length. But the value of pruning may be suspect. Some univariate forecasting techniques, such as exponential smoothing, already deemphasize older data, making pruning unnecessary.

Single versus Multiple Fit Periods

The multiple origins of a rolling simulation are preferable to the single origin of a fixed-origin simulation. However, even rolling simulations may not go far enough to promote diversity. Jenkins (1979) suggested that, if the forecasts from two different methods are to be compared, then a considerable number of forecasts from different forecasting origins are needed.

The emphasis should be on different, nonoverlapping forecasting periods. Many more forecasting situations can be analyzed, and the variety of fit periods reduces the sensitivity of the results to unusual events. The forecaster should select more than a single fit period. So far, we have chosen forecasting simulations around a single split of the data into fit period and test period. However, for example, our forecasting example data could be sliced as follows:

Fit period 1:   Periods 1–10,   Test Period 11–14
Fit period 2:   Periods 1–15,   Test Period 16–19
Fit period 3:   Periods 1–20,   Test Period 21–24

Forecasting simulations that incorporate multiple fit periods gain both depth and diversity.

Updating versus Recalibrating

In a rolling simulation, each updating of the fit period leads to some revision in the equation used to generate forecasts. The revisions to the forecasting equation take one of two forms. Updating the model involves a revision because new data are brought into the fit period; however, the form of the model (i.e., coefficients) is not revised. In contrast, recalibrating the model incorporates a revision of the coefficients as well. Recalibrating is the more desirable alternative. To update the fit period without recalibrating the coefficients is akin to stepping forward on one foot while locking the other in place: We get a better jump on the future when we jump (update) with both feet. Recalibration, however, is much more computationally intensive.

Dealing with Short Time Series

It is generally difficult to identify clearly patterns of trend and seasonality in short time series. Hence, to identify an appropriate forecasting method, it may be impossible to perform a forecasting simulation. *Short* is a relative term. With forecasting techniques, such as exponential smoothing, a time series with as few as seven consecutive time periods may be long enough for adequate modeling, whereas a time series with 40 observations may be too short for the proper application of ARIMA models.

If time series are too short for a forecasting simulation, the forecaster should reconsider some of the points made earlier. With a rolling simulation and multiple fit periods, even a short series can be manipulated to provide some depth and diversity of forecasting situations. If we have another time series available that is similar in nature, we may do a cross validation, in which we use the second series to test the accuracy of any model developed for the original series. We can reverse these steps as well, using the second series to develop a model and the first to test for forecast accuracy.

15.5 Reconciling Sales Force and Customer Inputs

In many situations, the company's sales force and its trading partners can provide input to the forecasting process that can be of value to the forecaster. This is particularly true when a limited number of customers account for a large share of the total business. It is likely that the sales management has assigned a salesperson to an account and that a sales force composite process is used to identify and quantify future business opportunities. There may be specific situations that support an adjustment to the proposed forecast.

By gaining access to the sales forecasts, the forecaster can determine whether the proposed forecast is consistent with sales plans.

Although the existence of such a sales force input is of great value, there are factors that may diminish its usefulness. At the beginning of the planning process for a given year, the sales quotas, on which sales compensation will be based, have not yet been set. The sales force is usually reluctant to identify all of its most likely opportunities in hopes that the quota will be set somewhat lower. Similarly, even after the quotas are set, the sales force is often reluctant to identify new opportunities later in the year for fear that the quota will be raised.

If they believe the quota will not be revised, the sales force and the distributors have a tendency to forecast high to make sure that product will be available to their customers. Prompt delivery will increase customer satisfaction and make sure that business is not lost to competitors because of a delivery delay.

On the other hand, when the sales force is unlikely to meet its quota and management finds out, it may get more help than it wants from management. In addition, it may get this help every subsequent month. Therefore, there may be a tendency to hold to the business-plan target as the sales forecast until the last moment. The sales force is naturally optimistic and is unwilling to signal an inability to meet the plan until all possible options have been explored.

For these reasons, a statistical model may actually provide better forecasts even though the model does not have the intelligence that resides in the sales force. Some managers have attempted alternative sales force compensation systems that encourage better forecasts by maximizing a bonus based on increased sales and forecast accuracy (Gonik, 1978; Mantrala and Raman, 1990).

15.6 GAINING ACCEPTANCE FROM MANAGEMENT

When a car buyer presents his or her selected car to friends and family, the buyer tries to convince them that the car is a beauty and that its cost does not exceed what had been planned for—besides, won't the neighbors be jealous? The buyer hopes to receive approval, especially when the relatives agreed to help in financing the purchase.

It is necessary to obtain user acceptance and higher management approval of a forecast.

The forecaster must also present a forecast for approval. With pride of authorship, the forecaster thinks the forecast is a beauty and is worth what it cost to produce. The managers will approve, hopefully, but never with much enthusiasm. After all, if things go wrong, it is your forecast!

The Forecast Package

After the forecast has been developed, it must then be documented and communicated to the people who need the information. The forecast package should include:

- The forecast
- A display of the forecast that analytically relates it to the past data (through a graphical and/or tabular display of the historical data and the forecast on the same page)
- Appropriate documentation of the rationale and assumptions regarding external and company factors that are likely to influence the item under study during the forecast period
- Appropriate documentation on the approach that was used to make the forecast and on the extrapolative techniques used during the forecasting process
- A delineation of specific potential decision points related to risk levels and the significance of particular assumptions

The purpose of the forecast package is to communicate the forecast to others and, at the same time, provide credibility to the forecast in the form of supporting documentation.

The value of a forecast is a function of its usefulness to decision makers in the face of future uncertainty. Therefore, merely developing the forecast does not complete the forecaster's job. The product must also be sold to the decision maker. The supporting documentation should emphasize the quality of the process, the inputs used during the process, and the judgment that was applied throughout the process.

The chore of documentation can be minimized by advance planning and continuous record keeping. The forecaster's documentation is as essential as the proof that a car has been serviced in compliance with a warranty is for someone wanting to buy or sell a used car. The forecaster must write down all the specific steps taken and the assumptions made. Only then can the forecaster and the manager have a meaningful analysis of results when the actual data are compared.

If work has been documented, it will be possible to specify a reason or set of reasons for the forecast's differing from actual accomplishments. These reasons will go a long way toward helping the manager evaluate the forecaster's performance. Moreover, without documentation it will not be possible to learn from past experiences and to determine where some problem lies.

From a manager's viewpoint, documentation also simplifies staff turnover problems. If the original forecaster is unavailable, a new forecaster will not have to reconstruct a forecast from scratch—a model or case study will already exist and a body of information will be available for use.

The users of the forecast will also appreciate the additional documentation. Instead of simply having a set of numbers, they will have the kind of information they need to assist them in making decisions about their area of responsibility.

A vital part of gaining credibility with management is quality documentation of the work at various stages of the process.

Forecast Presentations

In practice, there are normally far too many individual forecasts to review and present for approval. We suggest an approach that includes the following.

- Identify major changes from the last forecast.

- Review the forecasts for the strategic items that are most important to success. These are generally those items with the greatest revenue. Pareto's law normally applies where approximately 20% of the items account for 80% of the business. However, there is generally a need to review the items that were only recently introduced and have not generated that much demand to date but that are expected to be the future flagship products for the company. This is particularly true for high-technology products, where the product life cycles are short.

- Review the top 10 improvement opportunity items. These are the items that will contribute to the greatest improvement in overall forecast accuracy. Because there are so many items, it may be desirable to weight the unit forecast miss (Forecast − Actual expressed in percent) by the percentage that the actual revenue for the item is of the total actual revenue. This places the greatest weight on the forecast accuracy of the items having the largest impact on total revenue. These items often have the most impact on the operation of the business.

- Review other major revenue items. These are the items that represent a large percentage of the total revenue and that have been forecast with a high degree of accuracy. Because they represent a large part of the revenue, they should not be overlooked.

- Compare the overall revenue forecast that results from multiplying the unit forecasts by their prices with overall top-down revenue forecasting. This will demonstrate the degree of forecast alignment between the forecasts driving factory and purchasing decisions and the forecasts driving financial planning. This approach can also be used for major product groups.

- Summarize the overall forecast accuracy trends. The use of a revenue-weighted accuracy measure is often used to summarize the performance of the numerous item forecasts. This will provide an indication of how well the forecaster, forecasting system, and other relevant individuals (salesperson, sales manager, sales director, etc.) have done and where improvements are needed.

15.7 THE FORECASTER'S CHECKLIST

The following checklist can be used as a scorecard to help identify gaps in the forecasting process that will need your attention. It can be scored or color-coded on three levels (Green = YES, Yellow = SOMEWHAT, and Red = NO).

Step 1. Setting down basic facts about past trends and forecasts

_____ Are historical tables and plots available?

_____ Are base-adjusted data available? (A constant base is needed. For example, have historical revenues been adjusted to today's base price? Have data been adjusted for mergers and acquisitions?)

_____ Are seasonally adjusted data available?

_____ Have outliers been explained? (As discussed in the treatment of ARIMA modeling, they may significantly affect the forecasts.)

_____ Have NBER cyclical reference dates been overlaid?

_____ Are percentage changes shown in tables and plots?

_____ Have forecast-versus-actual comparisons been made for one or more forecast periods?

Step 2. Determining causes of change in past demand trends

_____ Is a trend identified?

_____ Is it linear or nonlinear?

_____ Are there plots of data and fitted trends?

_____ Is the scale of sufficient breadth to see deviations?

_____ Have the deviations been explained in writing?

_____ Are the explanations of causes specific?

_____ Has the source of the explanations been identified?

_____ Is the degree of certainty about the explanations noted?

Step 3. Determining causes of differences between previous forecasts and actual data

_____ Are differences explained?

_____ Are there any patterns to the explanations?

_____ Are there basic assumptions that can be reviewed?

Step 4. Determining factors likely to affect future demand

_____ Do factors relate to the future?

_____ Do factors indicate the direction of impact?

_____ Do factors indicate the amount or rate of impact, the timing of the impact, and the duration of the impact on demand?

_____   Are there rationale statements for each factor?

_____   Are the sources of any rationale statement identified?

Step 5. Making forecasts for future periods

_____   Time integration: Are the long-term forecast, short-term forecast, and history all shown on one chart?

_____   Item integration: Are the ratios of related items shown, as well as their history through the long-term forecast?

_____   Functional integration: Are related forecasts identified and the relationships quantified?

_____   Have multiple methods been used for key items and have the results been compared?

_____   Has impact on the user of the forecast been considered?

SUMMARY

This chapter addresses the issue of determining the reliability of model forecasts using empirical forecast simulations. The forecast simulations provide an understanding of how well the models would have predicted past demand history. Probability limits can provide a forecast range that can be expected at a given level of confidence. In the case of regression models, this assumes that it is possible to obtain accurate forecasts of the independent variables. Even when this is not feasible, the model can be used to evaluate alternative scenarios. This information may help us in selecting from among alternative models and in advising forecast users about the accuracy they can expect given past performance.

Forecasters and forecast managers can increase the likelihood that the best forecast has been developed by using a forecast checklist similar to the one presented in Section 15.7. We use the checklist to validate that a general forecasting process is practical and comprehensive. We have discussed several issues and considerations in dealing with and using sales force projections effectively, and have offered suggestions in terms of the material that is appropriate for the presentation of the forecast. The interests, requirements, and background of the audience must also be taken into account in deciding on the final presentation format.

REFERENCES

Armstrong, J. S. (2001). *Principles of Forecasting: A Handbook for Researchers and Practitioners.* Boston, MA: Kluwer Academic Publishing.

Butler, W. E. R., A. Kavesh, and R. B. Platt, eds. (1974). *Methods and Techniques of Business Forecasting.* Englewood Cliffs. NJ: Prentice-Hall.

Clemen, R. T. (1989). Combining forecasts: A review and annotated bibliography. *Int. J. Forecasting* 5, 559–638.

Fildes, R. (1989). Evaluation of aggregate and individual forecast method selection rules. *Manage. Sci.* 35, 1056–65.

Gonik, J. (1978). Tie salesmen's bonuses to their forecasts. *Harvard Business Rev.* May–June, 116–23.

Hogarth, R, and S. Makridakis (1981). Forecasting and planning: An evaluation. *Manage. Sci.* 27, 115–38.

Jenkins, G. M. (1979). *Practical Experiences with Modeling and Forecasting Time Series.* Jersey, UK: GJ&P (Overseas) Ltd.

Makridakis, S., A. Andersen, R. Carbone, R. Fildes, M. Hibon, R. Lewandowski, J. Newton, E. Parzen, and R. Winkler (1982). The accuracy of extrapolation (time series) methods: Results of a forecasting competition. *J. Forecasting* 1, 111–53.

Makridakis, S., C. Chatfield, M. Hibon, M. J. Lawrence, T. Mills, K. Ord, and L. F. Simmons (1993). The M2-competition: A real time judgmentally based forecasting study (with comments). *Int. J. Forecasting* 9, 5–30.

Mantrala, M. K., and K. Raman (1990). Analysis of sales force incentive plan for accurate sales forecasting and performance. *Int. J. Res. Marketing* 7, 189–202.

Winkler, R. (1989). Combining forecasts: A philosophical basis and some current issues. *Int. J. Forecasting* 5, 605–9.

PROBLEMS

Background

The Forecasting Environment. GLOBL (a fictitious company) is one of the leading international companies providing consumer technology products to a broad range of worldwide customers. GLOBL's mission is to provide:

- Development, manufacturing, and sales of educational technology products
- Development and sales of the hardware and software systems to support these products
- A broad range of customer-support services ranging from installations, training, consulting, and ongoing maintenance

You have just joined GLOBL as a forecaster. You have received some onsite training and have visited various overseas offices to learn about the scope of the job, which is extensive. Your responsibilities are to provide forecasting services to all GLOBL business areas. You must appropriately serve all aspects of GLOBL business, including planning for demand and supply, marketing, sales and operations, finance, new product development and introduction, and corporate strategy. Your manager has observed that, as with any business function, there are not nearly enough forecasting resources to address all the potential needs at GLOBL. Thus, careful evaluation and prioritization of forecasting work activities must be done. Also, there is a great opportunity to become more efficient by better coordinating some of the forecasting services now separately performed for each GLOBL business area.

Your initial assignment is to the demand/supply planning area of GLOBL. And your first product forecasting responsibility is a set of high-tech consumer products. However, over the first 5 years in their careers, it is usual in GLOBL for forecasters to be rotated through several diverse product and services assignments—as well as different aspects of particular business areas.

Your job description as GLOBL demand forecaster includes:

- Forecasting the demand for a group of GLOBL's products
- Providing regular coordinated communications with the development, sales, and marketing groups
- Developing reliable modeling approaches to predict sales volumes
- Providing periodic, objective, defensible forecasts to the sales and operations (S&OP) process, which will use this forecast for production and capacity planning over the subsequent 6 months
- Providing monthly forecast updates and related information for revenue planning
- Presenting and defending forecasts to senior management, as required
- Reviewing forecasting performance on a regular basis with your user groups and information sources to identify areas needing improvement

Although GLOBL develops, manufactures, and sells a broad range of consumer products, you have three product lines for which you will develop forecasts.

Product Line A: This is a family of consumer products for early childhood development. The customers for these products are preschool children who are physically challenged. Their needs are for educational toys, games, and devices that allow them to better adapt to their environment and enhance their growth potential within the community.

Product Line B: This is a family of consumer products for academia and institutions of higher learning. The customers for these products are students requiring specialized learning devices and educational materials to allow them to cope more effectively and competitively in a general academic environment.

Product Line C: This is a family of consumer products for the occupationally challenged. The customers for these products are adults in the workforce requiring customized aids for enhancing their productivity in the workplace.

The Marketplace for GLOBL Products.   There are five major players in the worldwide educational technology marketplace, plus another dozen niche players. GLOBL has a centralized market intelligence staff that is responsible for overall marketplace trends and outlooks, keeping track of competitie activities and market share, and performing specialized marketplace studies as required by sales, marketing, and product development.

GLOBL Product Development.   GLOBL does all development work on the three products you will be forecasting. This means that GLOBL maintains a development staff whose responsibilities include evaluating and tracking customer requirements for educational products, determining and prioritizing what needs may be best pursued by GLOBL, designing and developing products to meet these needs, determining go-to-market strategies for these products, tracking GLOBL product performance versus objectives, and enhancing products as required to meet GLOBL objectives.

GLOBL Sales Force and Channel Strategy.   GLOBL has a worldwide team of dedicated product sales specialists. There are also a number of business partners who sell GLOBL products, often along with other products and services. There is a strong focus on increasing the use of web-based facilities to exploit e-business sales.

GLOBL Manufacturing.   GLOBL performs manufacturing activity for the products you forecast. Worldwide manufacturing supply/demand planning is performed centrally for all products, although there are several manufacturing sites for each product.

GLOBL Product and Strategy Details

Product Line A: Product Line A sells into the preschool market. In recent years, Product Line A has seen dramatically increased use to support web-based applications.

Product Line B: Product Line B sells into the academic market and institutions of higher learning. Although GLOBL has been in the market for quite a few decades, the original versions of this line were introduced just over 36 months ago. Sales have been normal for the past year. Two years ago, there was an unexpected upswing in demand in Quarter 3, which caused big manufacturing problems. A dedicated sales force does over 90% of sales and has grown significantly in size over the past 3 years. There are currently plans for a further strengthening of the sales budget due to concern that GLOBL is still number 3 in this marketplace. This sales force operates off a quota system with sales contests scheduled approximately once a year, usually in the last quarter. Selling in this marketplace depends on establishing good relationships with the educational institutions. GLOBL's competitors appear to be more successful at this. You have difficulties getting solid information on the product line's sales activities from the sales force.

Product Line C: Product Line C sells into the commercial workplace market. It spans a wide range of occupational functions in industry, supporting complex needs across many business applications. Product Line C has seen modest growth over the past several years; new and esoteric applications in many commercial marketplaces are driving a niche market. GLOBL has divided its sales efforts between its own sales force and its business partners in roughly a 30–70 split. Good contacts with traditional customers have been key to sales success. However, forward-looking strategists are beginning to be concerned regarding the trends to mutual e-procurement initiatives in these customers.

Developing Factors

Step 1: Prepare the Factors.   GLOBL has determined a number of factors that influence the demand for its products. These factors may be different across the product lines, but six factors appeared to have some common value. Unfortunately, these factors often display some similarities in their strength of impact on the demand. It is your responsibility to sort out the behavior of these factors and their influence on demand in order to obtain some quantification of this for presentation to your management. You came up with the idea of an association matrix for the factors. In this exercise, we create such a matrix.

Factor A. **Opportunity momentum index:** Because each sales situation can be characterized by its sell cycle status, win probability, revenue, and so forth, an index of opportunity momentum was created to measure the goodness of a sales opportunity by the end of its sales cycle.

Factor B. **Competitive pricing index:** Price levels for the product were tracked and compared against similar competitive product.

Factor C. **Product attractiveness index:** Product functionality was indexed with those of its major competitors.

Factor D. **Channel investment index:** This is the percentage of advertising/marketing spending on GLOBL product line as a percentage of total spend by business partners.

Factor E. **Industry investment index:** This is the industry expenditures on web-based applications as a percentage of total industry expenditures on applications.

Factor F. **Gross internet product:** This is a measure, like the GDP, measuring e-industry output.

15.1 Suggest and define three factors for each product line to supplement factors A–F.

Step 2: Execute an Impact Change Matrix for the Factors Influencing Product Demand.   An *impact change matrix* is a table in which a measure of the impact on demand (during a time period relative to the current time period) is provided for each factor over a number of periods in the product's life cycle. This measure is on a scale of 1 to 5 (1 = weaker, 2 = moderate, 3 = average, 4 = positive, 5 = stronger). The table is completed as follows.

1. In Current Period, place the measure of impact the factor has on demand during the current period.
2. In Prior Year column, place the measure that reflects the change of the factor's impact on demand a year ago relative to its impact in the current period.
3. In Immediate Past column, place the measure to reflect the change of impact the factor has had on demand in the past 3–6 months relative to its impact in the current period.
4. In the Immediate Future column, place the measure to reflect the change of impact the factor is expected to have on demand in the *next* 3–6 months relative to its impact in the current period.
5. In Next Year column, place the measure that reflects the change of the factor's expected impact on demand next year relative to its impact in the current period.

The accompanying table is an example of an impact change matrix for a product line under a given set of assumptions about the business environment for GLOBL.

Factor	Prior Year	Immediate Past	Current Period	Immediate Future	Next Year	Comments
A	3	3	4	4	5	Index becoming more reliable
B	3	4	4	5	5	Index becoming more important as products are less differentiated
C	5	3	4	3	4	Competition has substantially caught up with GLOBL
D	1	2	3	3	3	Importance of larger customer is diminishing
E	4	4	3	4	3	Gradually decreasing importance as product life cycle is ending
F	4	4	4	4	5	Index is expected to increase in importance as the e-market matures

15.2 Using assumptions about the GLOBL environment for the product lines, create impact change matrixes. There are no unique answers to this.

Step 3: Evaluate the Results in the Association Matrix for the Chosen Factors.   An association matrix is a table of correlations of the sequences in the impact change matrix. Because these sequences are very short, the ordinary correlation coefficient may not be adequate. Hence, we have used a robust correlation coefficient that is analogous to the ordinary correlation coefficient. It is described in Appendix 3A. Note that the association of a factor with itself is always unity ($= 1$) and that the values can range between -1 and $+1$. The completed association matrix for the example matrix in Step 2 is given next.

	A	B	C	D	E	F
A	1					
B	0.80	1				
C	0.0	-0.72	1			
D	0.80	0.95	-0.28	1		
E	-0.80	-0.34	-0.47	-0.70	1	
F	0.60	0.38	0.18	0.18	-0.18	1

Aside from the direction of impact (positive or negative) on demand, there is a changing pattern of impact over time. When two factors display a similar pattern of impact on demand, these factors will be strongly associated. This high association can be positive or negative depending on whether the similarities in patterns move in the same or opposite directions. The importance of these associations may play out when you create models for demand over time using one or more of these factors. Having an early understanding of and insight into this will help you with the modeling issues.

15.3 Using assumptions about the GLOBL environment for your product lines, evaluate the association matrixes for the factors of your choice. This is a quantification of a subjective process to give specificity to your assumptions.

Step 4: Reconcile the Association Matrix Results with Your Plans for Developing Demand Models for Your Product Lines Based on These Factors. Factor F, showing low association with several other factors, might be used together with one or more good factors in a regression model. If this is successful, Factor F will be a good factor to use because its impact changes appear to be dissimilar from the others, thus providing a more independent influence on demand.

15.4 a. Reconcile the results found in step 3 with the assumptions you made about the role of the factors in forecasting the demand for your product lines.

b. Factors A and E are strongly negatively associated. What implication does this have for their future inclusion in modeling demand for the product?

c. Factor A is strongly associated with most of the other factors. Why might this be so? Does it suggest that the other factors should not be included in a demand model for the product that contains Factor A?

Modeling Demand

Step 1: Prepare a Preliminary Analysis of the Product Line and Factors Influencing Demand.

15.5 **A seasonal decomposition.**

a. Create an analysis of a product line series by identifying the trend and seasonal components with the technique(s) of your choice. Summarize your results in a data sheet like the one given here for future use in reconciling multiple projection techniques.

b. Answer the following questions and in doing so apply a projection technique to come up with a four-period forecast (horizon $= 4$ months).

   i. Find the twelve seasonal factors.

   ii. Determine the peak seasonal month.

   iii. Determine the lowest seasonal month.

   iv. Do you notice any subpatterns (e.g., quarter by quarter)?

    v. What is the average monthly change in trend (in level and percent)?

    vi. Determine a projected trend over the forecast horizon.

    vii. Determine the projected seasonal factors for the months in the forecast horizon.

    viii. Calculate a projection of demand over the forecast horizon.

Sample Data Sheet: A Trend × Seasonal Decomposition for Product XYZ

Seasonal Indices (Multiplicative)

Period	Index	Period	Index
1	0.279606	7	0.921522
2	0.708010	8	1.05525
3	1.30809	9	0.928185
4	0.794871	10	0.579209
5	1.09598	11	0.781690
6	1.87311	12	1.67447

Accuracy of Model

MAPE:	75.6
MAD:	88.9
MSD:	17831.7

Row	CombSale	TREN1	SEAS1	DETR1	DESE1	FITS1	RESI1
⋮							
31	230	514.440	0.92152	0.44709	249.59	474.07	−244.068
32	605	533.780	1.05525	1.13343	573.32	563.27	41.726
33	965	553.119	0.92819	1.74465	1039.66	513.40	451.603
34	450	572.459	0.57921	0.78608	776.92	331.57	118.427
35	460	591.798	0.78169	0.77729	588.47	462.60	−2.603
36	765	611.138	1.67447	1.25176	456.86	1023.33	−258.331

Model Projections

Row	Period	Projection
1	37	176.29
2	38	460.08
3	39	875.32
4	40	547.27

15.6 Correlation analysis with the factors.

Answer the following questions for each product line in the order that they appear below. In doing so, you will be creating scatter plots and the correlation measures for these factors.

    a. Create time plots for the factors.

    b. Create scatter plots for the factors.

Sample Table

Scatter Plot	Degree of Scatter	Direction of Scatter	Visually Interpret Linear Association (−1, +1)
Factor A	Very narrow	Positive	0.8
Factor B	Narrow	Negative	−0.6
Factor C	Broad		0.3
Factor D	—		
Factor E		—	

c. Calculate the correlation coefficient as a measure of association using the sample table as a guide.

	Factor A	Factor B	Factor C	Factor D	Factor E
Factor A	XXXXXXX				
Factor B		XXXXXXX			
Factor C			XXXXXXX		
Factor D				XXXXXXX	
Factor E					XXXXXXX

d. Calculate the robust alternative as a measure of association. The robust alternative is like an insurance policy. If the robust alternative is equal or very close to the correlation coefficient, then we have assurance of the validity of the correlation coefficient. However, when there are unusual values in the scatter diagram, the correlation coefficient is a deficient and unreliable measure of association. In such situations, the robust alternative and the correlation coefficient are not close. This can provide a valuable warning to the analyst to probe deeper into the nature of the deficiency.

Step 2: Execute Univariate and Multivariable Models for Product Lines.   An analysis of each product has been run creating time plots and scatter plots for factors A–E. Answer the following questions in the order that they appear. In doing so, you will be creating scatter plots and the association/correlation measures for these factors with each product.

15.7 **Regression analysis.** Regression analysis determines whether one set of data (one or more independent variables) has any relationship, or correlation, to another set of data (dependent variable). You can make predictions once you calculate these relationships.

a. Collect data for a period of time or from multiple sites so you can perform a regression analysis. Predict future sales (dependent variable) based on the values specified for the key factors (independent variables) if the correlation between the key factors and sales is strong enough.

b. What is your dependent variable and which factors will you be using for independent variables?

c. Build simple linear regression models for each product with one of its factors. Determine the equation and use the equation to project product demand for a particular value of the factor. Estimate a range.

Example: Building a simple linear regression model for Product XYZ demand versus Factor A. A regression with Product XYZ demand (dependent variable) against Opportunity Pipeline factor (independent variable) was run with the following results.

Regression output summary	
Constant	−46.9
Std error of Y Estimate	120.5
R^2	0.79
Number of observations	34
X coefficient(s)	1.20

i. Determine the equation.

Answer: The equation is **ProdXYZ demand = − 46.9 + 1.20 (Factor A)**

ii. Use the equation to project ProdXYZ Demand if Factor A = 600.

Answer: **ProdXYZ demand = −46.9 + 1.20 (600) = 673.1 = 673**

iii. What is the estimated range on this projection?

Answer: Formula for calculating is:

Projection ±2 (Std error of Y Estimate)

Range: 673 − 2(120.5) to 673 + 2 (120.5) = **[432,914]**

d. Build multiple linear regression models for each product with two of its factors. Determine the equation and use the equation to project product demand for a particular value of the factor. Estimate a range.

Example: Building a multiple linear regression model for Product XYZ demand versus Factor B and Factor C.

Regression output summary:

Constant	3515.4
Std error of Y estimate	195.0
R^2	0.47
Number of observations	34
X coefficient(s)	99.4 and −2899.5

i. Determine the equation.

Answer: The equation is **ProdXYZ demand = 3515.4 + 99.4 (Factor B) − 2899.5 (Factor C)**

ii. Use the equation to project ProdXYZ demand if Factor B = 1.4 and Factor C = 1.1.

Answer: **ProdB demand = 3515.4 + 99.4 (1.4) − 2899.5 (1.1) = 465. 2 = 465**

iii. What is the estimated range on this projection?

Answer: Formula for calculating is:

Projection ±2 (Std error of Y Estimate)

Range: 465 − 2(195) to 465 + 2(195) = **[75,855]**

e. **Summarize the projections with ranges.** Complete the following table based on your results. (We will use the results in step 4 to reconcile ranges and recommend a final forecast.)

Model Projections and Range: XYZ versus	Lower Limit of Range	Projection	Upper Limit of Range
Factor A	Jan: 432 Feb: Mar:	673	914
Factor B	Jan: 378 Feb: Mar:	762	1,146
Factor C	Jan: 0 Feb: Mar:	280	778
Factor D	Jan: 105 Feb: Mar:	491	877
Factor B and Factor C	Jan: 75 Feb: Mar:	465	855

Step 3: Evaluate Model Performance Summaries. You are now approaching the close of another forecasting cycle and you are requested to evaluate the forecasting models for ProdXYZ created during this period.

15.8. a. **Summary of projections with ranges from exponential smoothing models.** You are asked to recommend a projection and range from the models for Product XYZ in a summary table.

The following results for exponential smoothing (ES) models based only on the historical data were determined for ProdXYZ. Place your recommended projection and range in the table for each period in the forecast horizon.

ES Model	MAPE	MAE	RMSE	Lower Limit	Projection	Upper Limit
Simple	174	147	204	203	607	1,011
Linear Trend	54	129	188	279	645	1,015
Damped Trend	52	125	189	284	641	1,021
Recommended						

b. **Summarize the projections with ranges from regression models.** The following results with the multi-variable regression models were found for ProdXYZ (Problem 15.7). Fill in the missing values.

Model Projections and Range: XYZ versus:	Lower Limit of Range	Projection	Upper Limit of Range
Factor A	Jan: 432 Feb: Mar:	673	914
Factor B	Jan: 378 Feb: Mar:	762	1,146
Factor C	Jan: 0 Feb: Mar:	280	778
Factor D	Jan: 105 Feb: Mar:	491	877
Factor B and Factor C	Jan: 75 Feb: Mar:	465	855

Step 4: Reconcile Model Projections with Structured Judgment to Arrive at a Final Forecast Number.

15.9 Fill in the following table combining the results of the two major modeling approaches performed for this forecast: univariate exponential smoothing techniques based exclusively on the historical information in ProdXYZ and multivariable regression models based on the inclusion of factor information. Apply informed judgment to arrive at a recommended forecast number.

Judgment Factors	Projection	Range

	Final Forecast Number	Range
	Jan:	(,)
	Feb:	(,)
	Mar:	(,)

CASE 15A DEMAND FOR ICE CREAM
(CASES 1A, 3A–8A, 10A, 13A, AND 14A CONT.)

The entrepreneur has come to the point where she wants you to evaluate the various methodologies and come up with an approach to forecasting the demand for ice cream in her business. The monthly historical data through August of YR06 is given in Case 3A.

(1) Use the first 5 years of the historical data to develop an exponential smoothing model (Case 8A) and an ARIMA model (Cases 13A and 14A).

(2) Use two different groups of weights to combine the forecasts into a single forecast for the holdout period: (a) equal weight and (b) weights proportional to the inverse of the MSE.

(3) Evaluate the accuracy of the forecasts over the holdout period with two accuracy measures.

(4) What recommendations do you make for combining forecasts in this context?

CASE 15B DEMAND FOR APARTMENT RENTAL UNITS
(CASES 6C, 8B, 13B, AND 14B CONT.)

The owner of an apartment complex wants to improve the efficiency of his business and has asked you to create predictions of future vacancies. The management has supplied you with data covering a period of 11 years, shown in Case 6C.

(1) Use the first 10 years of the monthly data to develop an exponential smoothing model (Case 8B) and an ARIMA model (Cases 13B and 14B).

(2) Use two different groups of weights to combine the forecasts into a single forecast for the holdout period: (a) equal weight and (b) weights proportional to the inverse of the MSE.

(3) Evaluate the accuracy of the forecasts over the holdout period with two accuracy measures.

(4) What recommendations do you make for combining forecasts in this context?

CASE 15C ENERGY FORECASTING
(CASES 1C, 8C, 11C, 13C, AND 14C CONT.)

An economic research firm is preparing a White Paper to be placed on its website, and you have been providing assistance in the analysis of data sources and the quality of the data, as well as running several univariate exponential smoothing and ARIMA models. Now you would like to show what kind of improvements in forecast accuracy you could produce by combining models.

(1) Use the 10 years of the monthly data to develop an exponential smoothing model (Case 8C) and an ARIMA model (Cases 13C and 14C).

(2) Use two different groups of weights to combine the forecasts into a single forecast for the holdout period: (a) equal weight and (b) weights proportional to the inverse of the MSE.

(3) Evaluate the accuracy of the forecasts over the holdout period with two accuracy measures.

(4) What recommendations do you make for combining forecasts in this context?

CASE 15D U.S. AUTOMOBILE PRODUCTION
(CASES 1D, 3D, 4D, 6D, 7D, 13D, AND 14D CONT.)

Your latest assignment as an analyst for an industry association is to evaluate the forecasting accuracy of a combined forecast consisting of an exponential smoothing model and an ARIMA model. The original data is shown in Case 3D and represents U.S. automobile production by month from 1977 to 1985, which includes a recession in the early 1980s. (*Automotive News*, 1982 and 1986 *Market Data Book* issues).

(1) Use the 20 years of the monthly data to develop an exponential smoothing model (Case 8D) and an ARIMA model (Cases 13D and 14D).

(2) Use two different groups of weights to combine the forecasts into a single forecast for the holdout period: (a) equal weight and (b) weights proportional to the inverse of the MSE.

(3) Evaluate the accuracy of the forecasts over the holdout period with two accuracy measures.

(4) What recommendations do you make for combining forecasts in this context?

CASE 15E DEMAND FOR CHICKEN
(CASES 8E, 12B, 13E, AND 14E CONT.)

Your assignment now is to assist one of the principals in the firm with a study for a chicken franchise on some long-term trend forecasts for various meat products.

(1) Use the first 20 years of the historical data to develop an exponential smoothing model (Case 8E) and an ARIMA model (Cases 13E and 14E) and forecast the remaining 5 years.

(2) Use two different groups of weights to combine the forecasts into a single forecast for the holdout period: (a) equal weight and (b) weights proportional to the inverse of the MSE.

(3) Evaluate the accuracy of the forecasts over the holdout period with two accuracy measures.

(4) What recommendations do you make for combining forecasts in this context?

CASE 15F COMBINING FORECASTS

A number of techniques for combining forecasts have been suggested in the literature. But how do you determine which is the best for a particular situation? There is probably no final answer to this, so the next best thing is to look at a number of different examples to see if any have characteristics similar to your own forecasting situation.

Consider as many of these studies as is feasible and summarize your own conclusions about the promise of these approaches in terms of:

(1) Multideterminant approaches

(2) Interpretations of combined forecasts

(3) Combining exponential smoothing with regression models

(4) Combining exponential smoothing and ARIMA models

(5) Combining regression with ARIMA models

16

Implementing the Forecasting Process

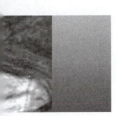

"Those who have knowledge don't predict. Those who predict, don't have knowledge."
LAO TZU, 6TH CENTURY BC CHINESE POET

SUCCESSFUL FORECASTING ORGANIZATIONS are those that have discovered how to apply effective management practices and processes to what is essentially a nontraditional business discipline. In the forecasting discipline, we do not have the power to change future demand, so our role is to change the target or forecast to better align it with actual future demand.

In this chapter, we emphasize the importance of improving forecasting effectiveness by considering forecasting approaches, implementing change, forecasting systems, measuring performance, monitoring forecasts, forecast integration, and virtual forecasting services.

16.1 PEERING INTO THE FUTURE: A FRAMEWORK FOR PROCESS IMPROVEMENT

Recall that we have followed four key steps in our approach to solving forecasting problems in this book. These same steps are key to forming an effective framework for improving the forecasting process.

Prepare: Define the purpose and role of the job or organization, define the major areas of responsibility, set objectives, and establish indicators of performance.

Execute: Define short-term goals and action plans, and carry out a plan for each area of responsibility.

Evaluate: Perform forecast monitoring (define objectives for the forecaster), know what to monitor, develop a measurement plan, develop indicators of forecast accuracy, and develop scores for performance.

Reconcile: Select the best approach, integrate forecasts, support forecasting systems, and get top management involvement.

Prepare

Purpose and Role

What is the purpose of a forecasting organization? This requires considerable thought because it is difficult to be effective unless we know what it is that needs to be accomplished. One role of a forecast manager is, thus, as an advisor to that company's senior management and managers of end-user organizations. To fulfill the other part of the role, a forecast manager must manage colleagues and their work. Here is a mission statement.

The responsibility of a forecasting organization is to provide top-quality advice— primarily advice about future demand for a firm's products and services.

We can spend many hours wrestling with the purpose of the job. Developing meaningful indicators of performance can consume a great deal of time and result in many debates. Experience will cause us to reject some indicators and replace them with others that are more relevant. Naturally, both the indicators and attendant levels of performance will change over time as the business evolves.

Major Areas of Responsibility

Next, we try to define the major areas of responsibility in short one- or two-word statements: These areas of responsibility might be product, revenue, capacity, or asset management forecasting. There are three areas that are shared by almost all managers: self-development, forecasting staff/personnel development, and resource management.

For forecast managers, the key areas of responsibility are likely to include forecast evaluation, measurement, monitoring, presentation, and forecaster appraisal and development. Forecast presentation is addressed in Chapter 15. Forecaster performance appraisal/development is equally important, but the methods for achieving success in this area are not restricted to forecast managers. The traditional management literature covers this topic adequately, and it is beyond the scope of this book.

Set Objectives

Once we have determined the purpose and areas of responsibility, the next step is to develop a long-range objective for each area of responsibility. These objectives should be general enough to have long-range significance, and they should contain an indication of the goal that the actual work should accomplish. Some examples are:

- To improve the accuracy of X
- To improve the productivity of X
- To improve managerial and technical skills
- To improve the credibility of the forecasting organization
- To ensure the continuing relevance of X

These objectives are important because they provide the managerial direction and focus that subordinates can embrace and strive to achieve. Forecasters can see how their activities are related to the achievement of organizational objectives. What is implicit in all of these is a striving for improvement that can be translated into actual tasks.

Establish Indicators of Performance

The next step is to define the indicators of performance for the organization. How will the organization know it is making progress toward the achievement of its objectives? What will be the yardsticks? For certain forecasts, one indicator might be the absolute percentage deviation between estimate and actual. For personnel development, indicators might be the demonstrated ability of a forecaster to use a new technique or forecast a new product or service effectively.

Without an understanding of purpose and indicators of performance, we will find it difficult to manage effectively.

Execute

Define Short-Term Goals and Action Plan

With indicators of performance in place, we need to execute a specific short-term goal and action plan for the next 6–12 months. If the goal is to improve the accuracy of a forecast item, a reasonable goal may be to improve the accuracy to within a given percentage, say 5–20%, depending on the item to be forecast. For personnel development, the goal may be to assume responsibility for forecasting revenues within the next 6 months.

Carry Out a Plan for Each Area of Responsibility

Once areas of responsibility are clearly stated, the forecast manager must establish specific activities that can lead to measurable results for the forecasting staff. Because results are evaluated, the action plan needs to be achievable and carried out in reasonable time frames (weeks or months, rather than years).

Evaluate

Develop Objectives for the Forecaster

A primary objective of forecast monitoring is to prevent surprising the company with news about unforeseen exceptions to a forecast. The firm should have sufficient time to evaluate alternative courses of action and not be forced to react to unpredicted, yet predictable events. A second objective of monitoring is to predict accurately a change in the direction of growth. This involves predicting the turning points in the economy and the demand for the firm's products. Quite often forecasters find it difficult to predict a downturn in demand and instead call for an upturn too soon.

It is easy to see why few managers of a business find the exercise of managerial control as challenging as the forecaster does. Forecasters are responsible for a function whose primary output is wholly related to the future environment; being unable to change the environment, the forecaster must instead revise a forecast when it is evident that an original forecast or goal cannot be met. In effect, the forecaster is changing some predetermined goal to approximate expected performance more closely.

Managerial control is a process that measures current performance, based on available information, and guides performance toward a predetermined goal.

The process of forecast monitoring provides the forecaster with an early indication that such changes in forecasts may be required. Through experience, a good forecaster will develop an improved ability to anticipate change and to advise management so that the firm will have time to adjust operations to changing conditions. This, of course, is a valuable attribute in a forecaster.

At a more demanding level, the objectives of monitoring are to predict changes in the rate of growth, to predict the level of growth, and to minimize the impact of forecast changes. The ability to predict any speeding up or slowing down of growth with accuracy helps management to decide on the proper timing of company plans and programs. Accurate predictions of the level of growth—the forecast numbers themselves—allow management to make sizing decisions about investment in facilities, numbers of employees, and appropriate financing arrangements. Last, it is necessary to minimize the internal disruption that results from changing forecasts too frequently. The forecaster could, after all, change a forecast every month so that the final forecast and the actual data are almost identical. However, this does not serve the needs of forecast users. The forecaster must endeavor to minimize the need to change forecasts. The more carefully thought out and thoroughly researched the initial forecast is, the less will be the need to revise it.

Know What to Monitor

There should be a difference between what a forecaster monitors and what the manager monitors. Business forecasters monitor a database that consists of time series and assumptions for customer/geographic segments and product groupings. They are primarily interested in the numerical accuracy of the forecasts and the reliability of the forecast assumptions.

The manager monitors a database that is both more general and more selective. Included in this database are the exceptional cases that forecasters uncover as a result of their detailed monitoring. The manager is primarily concerned with the implications the difference between the initial forecast and the evolving reality will have on the business for which the forecast is made. The manager should know more about that business and should generally be more aware of the significance of forecast changes on business performance than the forecaster.

The specific items that forecasters select to monitor will naturally depend on their areas of responsibility. The indicators that are established in the organization plan are natural candidates for monitoring. After selecting the items to monitor, the forecaster may find the following principles helpful.

- The forecaster should consider monitoring composites or groups of items. Composites often serve as indicators of overall forecast quality and are frequently used as a basis for decision making. They are resistant to individual deviations that may be measurement aberrations and not managerially significant. For example, a forecast of total revenues might be on target, although forecasts of revenues accruing from the sales of a product to residential or business users may need to be adjusted.

- The forecaster should compare the sum of the components of a forecast to the whole to ensure that there is a reasonable relationship between the more stable aggregate forecast and the more volatile bottom-up forecast of many small components. For example, the sum of the individual product forecasts should be compared to a total product-line forecast. In this way, the forecaster can be assured that both upward and downward revisions in the component parts are being made to keep them in reasonable agreement with the total forecast.

- The forecaster should monitor ratios or relationships between different items. The ratio of a given geographic area's sales to the total corporate sales is an example of this approach. Another example is the ratio of sales to disposable personal income.

- The forecaster should monitor time relationships. It may be appropriate to monitor changes or percentage changes over time. The use of seasonally adjusted annual rates is an example, as is the ratio of first-quarter to total annual sales.

- The forecaster should consider monitoring both on a monthly basis, and on a cumulative basis. The sum of the actuals since the beginning of the year should be compared with the sum of the forecasts. This has the advantage of smoothing out irregular, random, month-to-month variations.

- The forecaster should, in all cases, monitor external factors. These are the basic key assumptions about business conditions or the economic outlook. Corporate policy assumptions also need to be monitored.

- The forecaster should monitor user needs. It is possible that budgetary or organizational changes, new or discontinued products, or changes in management will cause changes in the forecast user's needs. Because forecasting is a service function, forecasters need to monitor user needs to be certain that the forecasting service being provided is consistent with evolving business needs. Questionnaires or periodic discussions with users will indicate whether such changes have occurred.

- ■ The forecaster should monitor similar forecasts in several geographic locations. This will help determine whether a pattern is developing elsewhere that may impact the company or area in the near future. Are there areas of the country that generally lead or lag the area? The forecaster may discover that his or her area is not the only area with weak or strong demand; a national pattern may be emerging that needs to be tracked.

The items to be monitored should relate to the purposes and objectives of the organization.

Develop a Measurement Plan

A major aspect of forecast process improvement is forecast measurement or results analysis. For any forecaster, improvement in organizational or staff effectiveness depends on measurement. A forecaster will find it useful to establish a forecast measurement plan to provide indications of overall performance that can be reviewed with upper management. A properly developed plan will show performance trends and highlight trouble areas.

The measurement plan will provide managers with a tool to assist in evaluating both forecasts and forecasters. When a measurement plan exists, forecasters know that they have to explain forecasts that miss the mark. This forces forecasters to structure and quantify their assumptions so that there will be documented reasons to explain deviations from forecasts and actuals.

The goal of a measurement plan is to develop meaningful ways of measuring the performance of the forecasting organization.

More important, adequate documentation enables the forecaster to learn from past mistakes. From reviews of these after-the-fact reports, it can be determined whether the assumptions were reasonable at the time they were made. Which assumptions turned out to be incorrect? Why? Did the forecaster do everything possible to obtain all the facts at the time the forecast was made? Were all sources of information reviewed? Were there any obvious breakdowns in communications? Was the forecast methodology appropriate for the particular problem? The answers to these questions become the information needed to evaluate the forecast and the forecaster.

In reviewing forecasts that were particularly successful, it may be discovered that a forecaster has developed a new method or established new contacts that were responsible for the superior forecast performance. Perhaps the approach can be tried in areas where performance is not as good. The documentation of superior and substandard performance, which will result from the measurement plan, provides the needed inputs to determine areas where improvement in methods or data is required. This document can also be used to support requests for people, data, or other items needed to improve the performance of the organization.

The existence of a measurement plan will also be of value to the users of the forecasts. It will improve their understanding of the limitations that must be placed on the accuracy of the forecasts they receive. For example, suppose the forecaster considers a $+2\%$ miss to be a good job for a given forecast and the measurement plan takes this into account. A user would then be foolish to plan on 0.5% accuracy, which for other reasons may be a desirable accuracy. By providing users with forecasting accuracy objectives, we are in effect providing a range forecast that can help users to scale their plans to differing degrees of forecast sensitivity.

The credibility of the forecasting organization will improve when it is capable of reporting on its own performance.

Develop Indicators of Forecast Accuracy

The development of a measurement plan begins with selection of the indicators that will be used to measure forecast accuracy. Here are four widely used indicators:

- Absolute error = Absolute value of forecast error (Actual − Forecast)
- Percentage error = $100 \times$ (Forecast error)/Actual
- Difference between total past sales and forecast, in percent
- Difference in growth rates

Forecast precision tends to be measured by averages of either absolute or squared forecasting errors. By taking either the absolute values of the errors or by taking the squares of the errors, we eliminate the tendency of negative and positive error to offset one another. The resulting averages then reveal how far away (how distant) the forecasts are from the actual values rather than whether the forecasts tend to be too low or too high.

The most commonly used indicator is the percentage deviation between the estimate and the actual. The actual demand, rather than the forecast, should be used in the denominator. This bases the deviation on actual performance. It is most useful when very large numbers are involved. However, when negative, zero, or relatively low levels of demand for a product are realized, the percentage deviation can become very large and not very meaningful. In those cases, the absolute difference is preferable.

The ratio of the deviation between estimate and actual to the in-service quantity of the series being forecast tends to put the forecast error in perspective. For example, a 100% deviation in forecasting the growth of sales may be only a 1% miss in the total sales. Or a 100% miss in forecasting the growth rate of the GDP may be only a 4% miss in actual GDP. However, when dealing with very large numbers, such as the GDP, a 4% miss can be very significant. In such cases, a useful indicator is the difference between the forecast and actual growth rates (Estimated percentage growth rate − Actual percentage growth rate). For example, the forecast could be for an 8% growth rate while the actual growth rate is only 4%.

The forecast measurement periods are generally monthly, quarterly, annually, long term (5–6 years), or cumulatively. There will probably be different accuracy metrics depending on the use of the forecast. For example, for production planning, a 3-month cumulative forecast versus actual accuracy comparison may be appropriate. For evaluating business plan performance, an annual measure is appropriate. For financial reporting, quarterly and annual intervals apply. To be useful to the manager, the measurement plan should cover all areas for which forecasters have responsibility. The reports of forecast accuracy should include graphs and tables at various levels of reporting (e.g., SKU, product line, geography, and organization level).

Recognize that external or exogenous factors can affect forecast accuracy. For example, the forecast miss can be partially due to forecast error, but it could also be the result of breakdowns in production, inventory or distribution planning or operations. Similarly, competitive actions or changed economic conditions can cause a forecast error. Therefore, the forecast error should trigger a root cause analysis to determine the reasons for the problems so that corrective action can be taken.

Develop Scores for Performance

The score a forecast error receives should not take into account the difficulty of forecasting. If difficulty is taken into account, it will generally not be possible to identify trouble spots; they will be hidden as a result of the scoring system. This means that the score is not a direct indication of the ability of the forecaster.

 To identify trouble spots, it is necessary to have a uniform standard of performance.

Some forecasters will be responsible for forecasting in series or geographic areas that are more difficult to forecast than others. Such forecasts may receive low scores. However, it is better to identify low scores in difficult areas and know the reasons for the low performance than to have uniformly good scores and not know where the performance problems lie. Forecasters can be sure that the forecast users will know from past experience where performance problems are and will not be quickly impressed if they dismiss possible problems simply because forecasts for them seem to score high.

Exhibit 16.1 illustrates the use of revenue weighting to calculate improvement opportunities. The same revenue weighting approach can be used to summarize overall forecasting performance in supply chain forecasting in which quantities of different products cannot be easily or meaningfully aggregated. In this case, the presumption is that lower revenue items are not as important as higher revenue products—which is

EXHIBIT 16.1   Revenue Weighting in Forecasting Performance Measurement

(1)	(2)	(3)	(4)	(5)	(6)	(7)	(8)	(9)	(10)
	Units		MAPE Abs.	1-MAPE Percentage	Price	(3) × (6) Revenue	Revenue Weight	(5) × (8)/100 Accur. × Rev	(8) − (9) Impvnt
Item	Forecast	Actual	Diff (%)	Accuracy	(dollars)	(dollars)	(%)	(%)[a]	Opporty
1	91	70	30	70	250	17,500	21.0	15.1	5.9
2	54	60	10	90	300	18,000	21.6	19.4	2.2
3	48	40	20	80	1200	48,000	57.4	45.9	11.5
Total Wgtd						83,500	100	80.40	19.60

[a] Maximum score for an item = revenue weight (%). In this example, item 3 has greatest improvement opportunity even though the forecast accuracy is higher than item 1. This is because of item 3's high revenue percentage.

generally the case. We can perform a separate analysis for strategic products that are more important than their revenue implies. Graphical plots of forecast accuracy at appropriate summary levels will provide overall measures of forecast accuracy. Because MAPE is used as the metric, it is not possible to observe a bias toward over-forecasting or underforecasting. A companion graph can be provided that indicates the percentage error (plus or minus). In addition, a summary statistic, such as the percentage of forecasts that exceeded actual results, can be added to the graph.

For internal use, a manager may also choose to develop a plan that measures the forecast accuracy of individual forecasters by attempting to incorporate a measure of difficulty into the scoring mechanism. One approach is to review the historical records of forecasts and actuals to determine the average miss and a measure of its variability (e.g., coefficient of variation) for each time series. It may be the case that, the larger the variability, the more volatile and difficult the series is to forecast.

The advantage of this approach is that the current forecaster's accuracy can be compared with others who have predicted the same series. The difficulty is built into the index because difficult areas invariably have larger measures of variability of forecast misses. In practice, the measure of variability used in this process should be resistant to outliers (e.g., downweighting of extreme deviations), so the conventional standard deviation may not always be appropriate.

Because the current miss is divided by a historical variability measure, a large miss will not necessarily penalize a forecaster. This is illustrated in Exhibit 16.2. The forecast miss in a given year in area 1 is 300, and the forecast miss in area 2 is 100. Area 1 had traditionally been more difficult to forecast than area 2, and their variability measures of misses are 300 and 100, respectively. The accuracy ratio in both cases equals 1 and both forecasters receive the same score. The resulting ratios can be graphed or averaged.

This approach measures only the relative forecast accuracy of forecasters. In assessing a forecaster's ability, several other responsibilities are also important. A manager will want to judge a forecaster's ability to sell the forecast to users, documentation of the forecast work, development or testing of new methods, and overall productivity. As a further refinement, Mentzer and Bienstock (1998) recommend

EXHIBIT 16.2 Using Accuracy Ratios to Measure the Forecaster	Measurement	Area 1 (difficult area)	Area 2 (less difficult)
	(Estimate – Actual)	300	100
	Standard deviation of prior forecast miss	300	100
	Accuracy ratio	300/300 = 1	100/100 = 1

the use of multidimensional metrics of forecasting performance in which forecast accuracy is related to business goals, such as inventory and production costs, customer service levels, and profitability.

Reconcile

Select the Best Forecasting Approach

Several approaches to forecasting can improve forecast accuracy. These include top-down versus bottom-up forecasting processes. The bottom-up forecast provides the detail needed to manage operations. However, there is a tendency for the bottom-up forecast to overforecast aggregate demand. The top-down forecasts use aggregate data, which are less volatile and more amenable to a broader range of quantitative forecasting methods. For example, exponential smoothing may be used for lower-level operational forecasts. Regression and econometric methods may be created for product-line and total company forecasts. The bottom-up forecast of units is matched with the top-down forecast, and the reasons for any differences are investigated and reconciled.

 Develop top-down and bottom-up forecasts, and reconcile the two forecasting approaches.

In demand forecasting for the supply chain, bottom-up unit forecasts are multiplied by average prices to develop revenue forecasts. This is compared to and reconciled with an aggregate revenue forecast that uses actual revenues as the data. Use the following guidelines.

- Develop the forecasts and the business plan in parallel with reconciliation at several times. Capacity constraints that would limit demand should be taken into account in developing the final forecast and business plan.

- Use forecast accuracy metrics at various levels of the business to identify improvement opportunities.

- Encourage training in quantitative forecasting techniques and a better understanding of the business and its environment. Also, provide training in the optimal use of the forecasting system.

- Keep historical demands and price data current, and adjust unusual values as appropriate.

- Recognize and reward superior forecasting performance.

Because of the game-playing issues related to sales-force forecasts, one possible approach to improving accuracy is to significantly increase the proportion of the individual's compensation package that is based on forecast accuracy (Gonik, 1978). To ensure that each person is measured against his or her own forecast, people who can override a forecast should have a separate accuracy measure. This diminishes the game playing associated with conflicting goals by influencing compensation adversely.

 Recognition and rewards for improved forecast accuracy should take into account all the people responsible for contributing to the consensus process, not just the forecasting group.

A complication related to this approach is the fact that salespeople may hold back orders until they have an opportunity to forecast them correctly. In addition, the business may take specific action related to a lower revenue forecast (e.g., special deals, promotions, or incentives that result in improving revenues). When the business could have made the forecast in error by taking certain actions, it is not appropriate to penalize the sales force or forecasters for signaling the need for such action. Recognition is usually a simpler but less powerful approach because the specific form it takes is under more control by managers.

When it is possible to obtain sales force forecasting input, the following guidelines should be followed.

- Be sure the right people are assigned to and engaged in the forecasting process. Recognize that all salespeople cannot sell all products (e.g., Asian, European, or South American products may be customized for specific markets). In addition, not all products are equally important. Consider three classes of products. Class A items are high-revenue or strategic items and the forecasts are assigned to salespeople. Class B items are more numerous and less important, and the forecasts are assigned to sales planners or forecasters, who use time series forecasting techniques extensively. Salespeople are asked to identify any unusual factors that will impact sales of these items. Class C items are not sold and will not be sold in a given territory, and no forecast assignments are made.

- For major customers, try to include the customer in the forecasting process. Review the previous forecast-accuracy results with the customer. Review the assumptions and show them plots of history and forecasts and ask for their comments. Try to get them to quantify their input. If possible, get them to enter forecasts directly into the forecasting system; these forecasts still must be reviewed for reasonableness by the sales team and adjusted if appropriate.

- Consider generating statistical baseline forecasts for review and adjustment by the sales force.

- Establish executive review sessions with the sales force to gain insights into the customers, competitors, and product migration plans that could modify forecast assumptions.

- Establish forecast approval points at which sales executives review key forecasts.

- Keep the number of forecast items that the sales team are asked to provide to a reasonable level. (If you ask for too many, you will not get the attention you desire.) Identify items whose demand depends on other related items and consider using regression techniques to forecast dependent items rather than expanding the number of forecasts you ask the sales team to provide. Investigate whether more accurate forecasts of some class B items can be obtained by forecasting the items at a higher organizational level (region or company) where the data are less volatile rather than at an account level.

- Ask for input from the sales force about the language of what it sells and at the level that the salesperson experiences. Have the system translate from the sales-input level to the lower-level component forecasts using typical systems configurations.

Because there are a number of areas to consider, have the forecasters and users identify the characteristics or criteria that will lead to excellence in forecasting (see Chapter 1). Rate current performance of the criteria elements on a six-point scale (0 = Not done, 1 = Done infrequently, 2 = Done some/most of the time, 3 = Always done with opportunity for improvement, 4 = Always done with above-average quality, 5 = Always done with excellence). As a second exercise, rate the importance of the criteria (say from 1 to 10 if there are 10 criteria). Focus attention on the most important items with the lowest performance scores. Alternatively, try to identify areas of greatest potential for improvement.

Integrate Forecasts

The acceptance and proper use of forecasts can improve the forecasting process. To prevent unnecessary disconnects between all the functional organizations that require forecasts, create a process to discuss and reach agreement or consensus on marketing and operational forecasts that are reasonably reconciled. The forecasting needs and requirements of all the organizations are identified so that the agreed-on process meets their critical needs and results in a reconciled business plan.

Create an integrated forecasting process that encourages communication, coordination, and collaboration among marketing, sales, product management, production, distribution, finance, and forecasting organizations.

It is desirable to identify a company forecasting process champion at a reasonably high level to encourage participation by all concerned. If possible, a separate forecasting group should be established to improve the overall forecasting process on an ongoing basis. Within the process, different groups will be accountable for specific forecasts. It should be recognized that the initial forecast created by marketing is unconstrained in terms of the company's ability to meet demand through its operations capabilities. At the subsequent stages of the business-planning process, the forecast used to drive operations must take into account the operational constraints that exist. For example, constrained production capacity information must be fed back to marketing and sales. Also, planned promotional programs must be communicated to production and distribution groups.

 The consensus process must provide feedback to functional organizations.

Support Forecasting Systems

Improved computer systems can also improve the forecasting process. An open systems architecture that allows for all affected groups to interrogate and provide electronic input to the forecasting process is encouraged. Information systems, based on client-server technology or corporate intranets, should be created to share information, make forecast accuracy metrics available online to appropriate groups, and provide descriptions of the products and linkages or translations between the product lines and SKUs. By linking corporate information systems, manufacturing and distribution planning systems, and forecasting systems, it is possible to eliminate the manual transfer of data from one system to another. The benefits are that the information used in the different functional organizations is coordinated by the following.

- Production planners can send out alerts to forecasters or sales teams identifying the most important or critical items for the forecasting cycle so that the forecast providers can focus on these items. A typical supply chain forecast contains so many items that it is not apparent when one or more items are most important at a particular point in time. The factory planners may be making production-capacity-enhancement decisions and require the best forecasts of selected items before investing capital for increased capacity.

- Higher-level summaries of demand history, forecasts, orders, and revenues are available to facilitate developing aggregate forecasts as reasonableness checks on lower-level forecasts.

- The products or SKUs that have been updated or changed since the last forecast cycle can be identified. Because there are potentially several thousands of individual items, it is not practical to expect the users to review all the items every forecast cycle.

- Major customers are encouraged to enter their forecasts online or offline with an upload capability while retaining adequate security and confidentiality. Electronic data interchange (EDI) linkages with suppliers can provide information on parts availability.

- Higher-level management adjustments to the forecast are possible. However, the forecast created at lower management levels needs to be retained in the system. Then, contributors can be measured against their own forecasts.

Get Top Management Involvement

A champion of the forecasting process needs to be identified at a high enough level in the firm to focus attention and resources on improved forecast accuracy. This individual needs to have a stake in the outcome. Operations executives whose organizations depend on accurate forecasts could be excellent champions. The champion does not need to manage the function but does need to have a strong interest in having the function performed well. Management has a tendency to avoid dealing with forecasting issues because these issues have to do with change and chance rather than areas of the business it has greater control over. Once top management recognizes the importance of forecasting, both to the business and operational plans, it will increase its level of support.

The capability to improve the forecasting process is enhanced if the organization can enlist a forecasting process champion.

16.2 AN IMPLEMENTATION CHECKLIST

Managers in industry today expect quantitative analyses in support of forecast-dependent decisions. This section provides guidelines for the implementation of improved forecasting processes. The full implementation checklist is provided in Section 16.4.

Selecting Overall Goals

The first step of the implementation checklist is concerned with selecting overall goals for quantitative modeling. Clearly, it is difficult to be successful in any area of the business without having decided what it is that needs to be done. Because quantitative methods can be used for many areas and the specific requirements of a company will determine which methods are appropriate, it is important for a forecast manager to know the end users' needs.

Before starting the modeling effort, the key point is to determine carefully the specific goals that we hope to achieve.

Elaborate quantitative models are often constructed with the goal of improving forecast accuracy. Because these models can give predictions that are more accurate than official company forecasts using less complex approaches, we should be prepared to work with new models. Managers usually look for more than just a set of forecast numbers. They would like to understand the relationships that exist among the various series of interest and among corporate (internal) data and the external economic and market variables, what the relevant relationships are, and how they have been changing over time. Models also provide management with the ability to explore alternative scenarios. Most likely, optimistic and pessimistic scenarios for economic or market forecasts can be used to assess the demand for a firm's products and services. This helps management generate necessary contingency plans before these are needed. The models can also provide estimates of advertising effectiveness and price elasticities that can be used to assess the impact advertising and pricing strategies may have on revenues. There are numerous business situations in which extremely large numbers of forecasts have to be generated. For example, tens of thousands of forecasts are required to determine the customer-specific requirements for products (SKUs) in inventory and production-planning systems on a weekly basis. To attempt to provide all of these forecasts on a manual, one-at-a-time basis is not practical. Often, stand-alone forecasting systems incorporating statistical forecasting models provide acceptable forecasts for the great majority of cases. The exceptions that warrant additional time and money can then be given the individual attention they deserve.

Forecasters may also want to develop documentation of successful forecasting techniques or models that work well in specific situations. When a request for a special forecast is received, the forecaster can review the documentation to determine the methods that will most likely provide the best results. Unsuccessful attempts should also be noted to avoid the repetition of false starts.

Modeling may also be a way of increasing the productivity of a forecasting organization and reducing overall costs.

A problem that forecasters face is the need to provide substantiation for the forecasts presented to management or regulatory authorities. Good documentation is often required to satisfy reviewers who question the forecasting job that has been done. Forecast tests, stability tests, and simulations are a valuable part of the documentation package. In this regard, the forecaster can also use the forecast test as a criterion for model selection. If a given model's forecast test results in errors that are above the objective set by the forecaster, the model can be rejected.

Obtaining Adequate Resources

Another prerequisite to the successful implementation of quantitative forecasting methods is having adequate resources. First, the forecaster needs to be trained and experienced in the methods that are available for use. Special assistance may be required to gather, verify, input, and process the study data. Access to computers and software is required. Budgetary limitations on salary, equipment, training, and consulting will also need to be determined. If we plan to build or purchase a forecasting system, we can expect expenses and costs for computer-intensive work to build slowly at first but to increase rapidly as the modeling effort intensifies. This aspect of modeling is discussed again under computer-processing considerations.

Defining Data

The forecaster is often faced with a data-collection problem when attempting to build forecasting models. Even when the appropriate independent variables can be identified, it is not always possible to obtain independent projections for these variables. The acquisition of external data has significantly improved over the last several decades as the number of computerized data sources has grown. Many consulting firms and academic establishments are now forecasting a wide variety of national economic/demographic time series, and several also provide regional or localized economic and market forecasts on the Internet.

Many corporations and business firms have research-group staffs that provide forecasts of economic/demographic variables for internal use. These departments provide services to the company's management that enhance and balance the often-conflicting forecasts from outside sources. The advantage of using an internal organization is that company forecasts can be made consistent with the corporate business outlook.

The government is also a data source. Census Bureau publications, the monthly *Business Conditions Digest*, publications of the Department of Commerce, vital statistics data from the Department of Health, and publications of the Federal Reserve Banks are all helpful and are accessible on the Internet. In addition, county and regional planning boards and associations are often interested in economic and demographic projections in connection with funding from the federal government based on population, employment, unemployment, income, and other statistics. Finally, the National Bureau of Economic Research (NBER) provides an analysis of the economic cycles and determines official dates for the beginning and ending of recessions.

Forecasts of independent variables should be carefully reviewed to be certain that they provide a consistent viewpoint.

Forecast Data Management

It is not our intent to recommend ways to create a forecasting system involving system analysis, system design, and system implementation. However, certain data management considerations are worth mentioning. Very early in the forecasting project, the need for standardized naming and filing conventions for data will become obvious. As more and more models are created and new time series added, lack of standard data organization and warehousing can hamper progress significantly. As time goes on, the forecaster will not be able to remember what the various time series names represent and undocumented output will become useless.

When models become part of the everyday forecasting process, it will be useful to establish a database containing the models and the data organized by product and customer/location specific segments. In this way, we are able to maintain a consistent database in which the models and new forecasts are available in a quick, efficient, and cost-effective manner.

Experience also suggests that historical demand file updates and maintenance should come from official corporate information systems. Otherwise, there is no accountability for data integrity, which results in duplication, excess storage for items no longer necessary, and out-of-date product/customer data files.

When establishing a forecasting system, it is advisable to establish password or security conventions and access restrictions. This prevents unauthorized people from accidentally or intentionally using private information, changing forecasting data, or destroying forecasting data. Most database management systems provide a capability for the protection of computerized information.

Selecting Forecasting Software

Statistical software and spreadsheet packages are widely used for the design, modeling, analysis, and implementation of forecasting applications. There are numerous PC-based forecasting packages in use by government, academic, and corporate forecasters. Some of the more popular ones are periodically reviewed by the computer, trade, and professional journals covering forecasting topics (Rycroft, 1999). The forecasters should investigate the availability, quality, and performance of these, and their ability to satisfy their needs in the forecasting environment. No package will be everything to everybody. Also consider the costs before selecting the appropriate software. For example, packages provide economies of scale and considerable savings in time, effort, and cost. The acquisition cost of software may be nominal; however, the investment in learning how to make effective use of it can often be substantial, as the cost of keeping up with its enhancements. The alternative, independent software development, is a major investment that may be beyond the means of most individual organizations.

Training

Hanke (1984, 1989) and Hanke and Weigand (1994) have conducted periodic surveys of business schools in the American Assembly of Collegiate Schools of Business to investigate what business schools are doing to educate forecasters. By comparing survey results over the 10 years, they noted that:

- In 1983, 58% of the schools said they offered a forecasting course at the undergraduate or graduate level; only 47% offered a course in 1993.

- In 1983, 37% of the schools said that they did not teach forecasting because they lacked faculty; this figure dropped to 6% in 1993.

- Regression analysis is still considered the most important technique, but that exponential smoothing has gained in importance.

Most forecasting techniques, no matter how sophisticated, are only as good as the underlying data. However, basic quantitative and **judgmental forecasting** methods should be taught along with studies of best practices. With better data, the value of these techniques will become more evident to management. Managers can absorb and will use the tools that work for them.

Coordinating Modeling Efforts

Quantitative analysis is an endeavor in which two minds are better than one. Members of the group should be encouraged to brainstorm alternative approaches to problem solving. They should also share the results of their work with others, because progress is synergistic.

Periodic conferences are ideal mechanisms to coordinate quantitative methods. Participants should be encouraged to make presentations of their latest work. In this way successes can be transmitted throughout the firm, misconceptions can be corrected, and, equally valuable, methods that have not worked can be discussed. The designation of technical coordinators, whose responsibility it is to assist all model builders, has proved to be a successful way of keeping the implementation project on schedule.

Documenting for Future Reference

The documentation of results is one of the key aspects of any project. It is also the area that is most disliked and easiest to postpone, because it is often done after a forecast has been made. A documentation system such as that covered in the checklist is an effective way of solving the documentation problem. Documentation that takes place while the project is progressing can be planned ahead of time and monitored throughout the project. The establishment of literature, model, data, forecasts, software, and billing files will go a long way toward organizing the project and demonstrating to management the necessary control mechanisms for cost effectiveness.

Presenting Models to Management

The presentation of the model results for evaluation by a company's managers is an important part of the implementation of quantitative methods. Experience has shown that it is best to do this in two different presentations. The first presentation should explain the approach taken, the alternatives considered, and the results from the model when only actual data are used. The purpose of this meeting is to assure management that the methodology is reasonable. Higher management should be encouraged to ask questions so that it understands the strengths and weaknesses of the particular quantitative method or methods presented.

The first meeting should not be one in which higher management is asked to approve a forecast based on a methodology foreign to it. There may be a natural reluctance to accepting the methodology because it is tied in with a presentation that is essentially the selling of a forecast. For these reasons, management's evaluation of the quantitative method should take place in a separate meeting. After gaining an acceptance of methodology, the forecaster can incorporate the model results into the normal presentation of the forecast to management.

The primary concern of management will be the forecast numbers, an assessment of risk, and their implications for the firm's performance in the future.

16.3 USING VIRTUAL FORECASTING SERVICES

There will always be consulting firms and organizations that specialize in market/product forecasting, statistical analysis, data sources, system integration, and training. These firms fill an important niche, providing new, improved, and more sophisticated forecasting services consisting of economic forecasts, demographic projections, long-term trend (futures) scenarios, software systems, computer applications, technical assistance in model building, and training or educational services. With the Internet, more of these services are being hosted through virtual service centers.

Economic and Demographic Forecasts

Until recent years, forecasters faced a great problem in obtaining good forecasts for economic variables for econometric model building. After expending effort in selecting the best variables to include in the model, the forecasters lacked the expertise needed to generate consistent and informed forecasts of the economic variables.

This problem is no longer serious when dealing with national economic variables because forecasts for most of these variables are available from a number of online services or consulting firms. At the local or regional level, however, there are fewer organizations regularly preparing economic/demographic forecasts. The manager should consider the desirability of obtaining consulting help to generate the necessary forecasts if national variables are inadequate. It is best to check within the firm for such advice before looking outside.

Demographic forecasts are also generally available at the national and regional levels but are less so at the local level. Because the funding for an increasing number of government programs depends on the local makeup of the population, local and regional planning agencies are developing expertise in preparing forecasts of births, deaths, migrations, and racial composition. Unfortunately, the forecasts prepared by these agencies may not always be objective, because the projections of increased population can represent additional federal funding.

In many models, the market potential is a function of the size of the population, the number of households, or the age distribution of the population. Therefore, demographic forecasts are increasing in importance and outside consulting help may be required.

Even when economic and demographic forecasts are available, they may not be published when the forecasting organization needs them. Local or regional forecasts are seldom provided as frequently as national economic variables are. Therefore, the need for timeliness and detail may require the assistance of a consultant.

Database Management

Almost all quantitative forecasting methods rely on the availability of computer software. As the scope of the forecasting job expands, the number of data files grows and data management becomes a severe problem. Consultants can sometimes recommend improved, integrated systems for storage and access, which can help reduce costs.

Most large corporations have in-house organizations that develop or acquire the necessary computer programs to use quantitative forecasting methods. However, obtaining the priority to get a program completed when required is not always possible. Fortunately, outside firms offer libraries of application programs as part of a hosted (software rental) service. If a forecasting organization requires a specific kind of program that is not generally available, consulting assistance may be required.

Modeling Assistance

In most cases, only a small number of forecasters in an organization in one location use quantitative methods regularly. When it is not possible to hold periodic conferences among forecasters, outside consulting opinion may occasionally be required to ensure that the analyses being performed are technically sound. This approach may be more desirable than outsourcing technical problems because the forecasters will be increasing their quantitative skills by regularly working with consultants.

Training Seminars

The forecast manager should be concerned about preventing technological obsolescence in the organization. Therefore, some form of periodic training will be required to maintain and expand quantitative skills. For large organizations, this training can be developed internally. For smaller organizations, external consultants can usually provide more cost-effective training.

The company should be cautious in selecting a consultant and in evaluating the services it receives for the monies spent. Most consultants are in business to make money; failure to specify exactly which services are required can be a costly mistake. Generally speaking, the consultant must first learn about the specific industry, the markets it serves, the firm's operations, and the time series to be forecast. This is frequently learned at the expense of the company before any productive work is forthcoming. Be sure to evaluate the consultant's reputation and experience in solving the company's kind of forecasting problems.

The role of consultant in quantitative forecasting methods is to help a company implement methods or techniques that will improve the quality and accuracy of the forecast.

Consultants often sell the use of specific data, forecasts, or application programs. In addition to the per diem expenses for consulting, a firm may find that it has a large bill for computer use generated by the consultant while he or she is becoming familiar with the firm's data. The best way to avoid needless expense is to define the services required very carefully ahead of time and to monitor the consultant's progress with specific indicators of performance.

16.4 THE FORECAST MANAGER'S CHECKLISTS

These checklists can be used as a scorecard to help identify gaps in the forecasting process that will need your attention. It can be scored or color coded on three levels: Green = YES, Yellow = SOMEWHAT, and Red = NO.

Forecast Standards

(See Section 15.7)

Implementation

Step 1. Identify a task or product (What are your needs?)

_____   Are models to be used for short-term or long-term forecasts?

_____   Are models to be used to solve "what if" questions?

_____   Are models to be used to determine elasticities?

_____   Are models needed at all?

Step 2. Priorities (Identify these on the basis of your needs)

_____   Which quantitative techniques are useful?

_____   Should they be implemented?

_____   In what order?

_____   What is the implementation schedule?

_____   How does qualitative analysis fit into total job responsibility?

Step 3. Identification of resources

_____   Is management interest and support available?

_____   Is money available for computer expenses?

_____   Do job responsibilities allow time to meet implementation schedules?

_____   Is adequate support available to maintain files?

_____   Is economic data available for modeling?

_____   Is modeling expertise available for consultation?

Step 4. Database management

_____   Who will enter and update data files?

_____   Who will identify and correct outliers in data?

_____   Will an ongoing program of documentation of outliers be implemented?

_____   Will appropriate time series be base-adjusted, if necessary, on an ongoing basis?

_____   Will seasonally adjusted data be created and updated periodically?

_____   Will data be maintained at the local, area, or company level?

Step 5. Intracompany coordination of modeling techniques

_____   How many individuals in the company will be using quantitative techniques?

_____ Can intracompany communications through seminars (and so on) reduce the redundancy and increase the effectiveness of quantitative modeling?

Step 6. Documentation of modeling work for future references

_____ Will modeling work be documented for future reference by others engaged in quantitative analysis?

_____ Will folders be organized for different aspects of modeling work?

- **Literature:** for publications about work in the modeling field, including trade journals and textbooks on mathematics, statistics, and economics; literature from vendors; modeling studies done by others; and so on

- **Models:** about types of models developed, any changes and reasons for the changes, including information on statistical tests, estimation of parameters, forecast tests, and simulations

- **Data:** about types and sources of data, as well as explanations of adjustments and transformations

- **Forecasting:** containing records on forecasts, forecast errors and monitoring information, and any analyses of forecast errors

- **Software:** about available computer programs

- **Billing and Related Expenses:** about costs related to modeling work

Step 7. Presentation of modeling work for evaluation

_____ What kind of feedback on modeling results should be sent to higher levels of management?

_____ How should this be done, and how often?

Software Selection

Step 1. Identify needs

_____ What level and detail is being forecasted (product, customer, geography, time granularity)?

_____ Are the end-user needs well understood?

_____ Are models to be used to solve forecasting problems?

_____ What are the available sources of data?

_____ Are staffing and staff qualifications adequate?

Step 2. Establish goals and objectives

_____ Have you established your goals and objectives for the forecasting process?

_____ What are the strengths and weaknesses of the information systems?

_____ Are there requirements for both hardcopy and online forecasting output?

_____ Have you set up a planned approach to implementation?

Step 3. Determine functional requirements

_____ Have you determined the scope of the system in terms of number of forecasts, size of historical file, system interfaces, and hardware/software performance criteria?

_____ What is the environment under which the system is expected to work?

_____ What are the time and cost factors related to installing the forecasting system?

_____ How are support issues for the system going to be handled?

Step 4. Establish selection criteria

_____ What are the program features and capabilities required to support the forecasting process?

_____ Have the reporting and export functions been identified?

_____ Have performance and maintenance standards been established?

Step 5. Review products

_____ What type of systems will be reviewed (mainframe, PC, client-server, intranet, other)?

_____ Have you established and prioritized a list of requirements and options?

_____ What features, modeling, and reporting capabilities are available?

_____ Can you identify pros and cons of each system under review?

_____ Are purchase price, implementation/support time, and costs provided by vendor?

Step 6. Evaluate systems

_____ Have systems been reviewed based on established criteria?

_____ Have you established a short list of potential vendors that fit your needs?

_____ Is there a clear set of evaluation standards prepared for the vendor presentations?

_____ Can the system be customized, and by whom and at what cost?

_____ Are there options to develop a system in-house?

Step 7. Check references

_____ Have you checked functionality against your requirements?

_____ Can the vendor provide user references in your industry/area?

_____   Do vendors provide adequate system documentation?

_____   Have you established implementation, training, and support schedules?

_____   Can you test the system with live data from your own company?

_____   Can you review a vendor's operational system in another company?

Step 8. Acquire the system

_____   Have you developed a purchasing recommendation?

_____   Can you provide a time, cost, and implementation schedule?

_____   Have you established performance criteria with a vendor?

_____   Are contracts and payment schedules in place?

SUMMARY

Improving the overall forecasting process can result from a number of well-planned activities. Preparing information systems supports the forecasters as well as multiple users and allows for integrated databases so that manual data transfer is eliminated. The systems should be able to provide online as well as reports of forecast accuracy and allow for higher-level management adjustments and results reporting at each level where a forecast was entered. The system should allow for direct input of forecasts by customers and of material availability by suppliers. Measuring forecast accuracy is important in identifying areas for improvement, understanding accuracy capabilities, and to developing credibility with users. Two generic approaches can be pursued: (1) Measure only the forecast and make no allowance for difficulty (this approach identifies trouble spots) and (2) measure the forecaster by taking into account the relative difficulty of the forecasting environment. Monitoring forecast performance consists of activities designed to prevent surprise for a company by highlighting the need for a change in the forecast. These activities include monitoring composites or groups of items, the sum of the parts to the whole, ratios of related items, monthly and cumulative results, company and external factors, results in other locations, and user needs. Reconciling forecasting approaches includes top-down and bottom-up forecasting with sales force and/or customer input. Develop the forecast at the same time the business plan is developed with periodic reconciliation; training in forecasting methods, the business environment, and the forecasting systems; and recognition and reward. Forecast integration refers to the need to encourage communication, collaboration, and coordination among all the organizations involved with forecasting including the forecasting organization, marketing, sales, production, distribution, finance, and product management. It involves understanding the needs of all the organizations and developing a process to facilitate information sharing and consensus on the final forecasts that are used in business planning and operations.

We recommend a disciplined approach for implementing new forecasting methods or process improvements. A specific methodology should be selected for on-the-job implementation; deadlines should be established and the resources that will be made available to complete the project should be specified. An implementation plan should indicate the methods to be implemented and indicators of progress to ensure that the plan does not die from lack of follow up. Considerations that should be incorporated into the plan are highlighted in a sample manager's checklist included in this chapter. Training courses improve the likelihood that new techniques are implemented; however, the value of training is often dissipated because of lack of specific on-the-job reinforcement. Making the implementation of new methods as routine as any other job requirement is the surest way to achieve success. Managing an organization that uses quantitative methods requires above-average technical competence; a background in management is also recommended. Forecasters and managers alike may find that a jointly compiled checklist will help in the technical evaluation of forecasting models. (A sample checklist is included in this chapter and Chapter 14.) The purpose of such checklists should be to establish standards for the forecasting organization. The forecaster should use the checklists in the preparation of the forecasts and have them available for subsequent review by the forecast manager if an exceptional circumstance makes this advisable.

The philosophy of forecast evaluation is one in which primary emphasis is placed on the process rather than the numbers. If the forecaster has meticulously followed a proper forecasting process, the end result will be as good a forecast as can be developed. If not, a manager may need to find a better-qualified forecaster.

REFERENCES

Gonik, J. (1978). Tie salesmen's bonuses to their forecasts. *Harvard Business Rev.* May–June, 116–23.

Hanke, J. (1984). Forecasting in business schools: A survey. *J. Forecasting* 3, 229–34.

Hanke, J. (1989). Forecasting in business schools: A follow-up survey. *Int. J. Forecasting* 5, 259–62.

Hanke, J., and K. Weigand (1994). What are business schools doing to educate forecasters? *J. Business Forecasting* 13.3, 10–12.

Mentzer, J. T., and C. C. Bienstock (1998). *Sales Forecasting Management*. Thousand Oaks, CA: Sage.

Rycroft, R. S. (1999). Microcomputer software of interest to forecasters in comparative review: Updated again. *Int. J. Forecasting* 15, 93–120.

CASE 16A DEMAND FOR ICE CREAM
(CASES 1A, 3A–8A, 10A, AND 13A–15A CONT.)

An entrepreneur has asked for a comprehensive forecasting strategy that can be used on an ongoing basis. Use the previous cases and the end-of-chapter checklists as a guide to construct a flowchart of a forecasting process for ice cream demand.

(1) Describe the critical data preparation issues surrounding this particular product.

(2) What are the key requirements for executing the modeling steps for this particular product?

(3) How can you effectively evaluate the forecasting performance of this product in its particular environment?

(4) How would you reconcile the modeling results to derive a single-number forecast for this product?

CASE 16B DEMAND FOR APARTMENT RENTAL UNITS
(CASES 6C, 8B, AND 13B–15B CONT.)

The owner wants to implement an ongoing forecasting process that helps him improve the efficiency of the apartment complex and has asked you to put a report together.

(1) Use the previous cases and the end-of-chapter checklists as a guide to construct a flowchart of a forecasting process for apartment rental unit demand.

(2) Describe the critical data preparation issues surrounding this particular product.

(3) What are the key requirements for executing the modeling steps for this particular product?

(4) How can you effectively evaluate the forecasting performance of this product in its particular environment?

(5) How would you reconcile the modeling results to derive a single-number forecast for this product?

CASE 16C ENERGY FORECASTING
(CASES 1C, 8C, 11C, AND 13C–15C CONT.)

It is time to pull together everything that you have learned from the previous cases into a comprehensive forecasting implementation strategy.

(1) Use the previous cases and the end-of-chapter checklists as a guide to construct a flowchart of a forecasting process for energy demand.

(2) Describe the critical data preparation issues surrounding this particular product.

(3) What are the key requirements for executing the modeling steps for this particular product?

(4) How can you effectively evaluate the forecasting performance of this product in its particular environment?

(5) How would you reconcile the modeling results to derive a single-number forecast for this product?

CASE 16D U.S. Automobile Production (Cases 1D, 3D, 4D, 6D, 7D, and 13D–15D cont.)

From the previous studies, you recognize that, in this environment, it must be extremely difficult to forecast the demand for domestic automobiles. Based on your experience from the previous cases, you are asked to summarize your findings and make recommendations for an ongoing forecasting process.

(1) Use the previous cases and the end-of-chapter checklists as a guide to construct a flowchart of a forecasting process for automobile demand.

(2) Describe the critical data preparation issues surrounding this particular product.

(3) What are the key requirements for executing the modeling steps for this particular product?

(4) How can you effectively evaluate the forecasting performance of this product in its particular environment?

(5) How would you reconcile the modeling results to derive a single-number forecast for this product?

CASE 16E Demand for Chicken (Cases 8E, 12B, and 13E–15E cont.)

Your last assignment is to assist one of the principals in the firm with an implementation of an ongoing forecasting process that will create long-term trend forecasts for various meat products.

(1) Use the previous cases and the end-of-chapter checklists as a guide to construct a flowchart of a forecasting process for chicken demand.

(2) Describe the critical data preparation issues surrounding this particular product.

(3) What are the key requirements for executing the modeling steps for this particular product?

(4) How can you effectively evaluate the forecasting performance of this product in its particular environment?

(5) How would you reconcile the modeling results to derive a single-number forecast for this product?

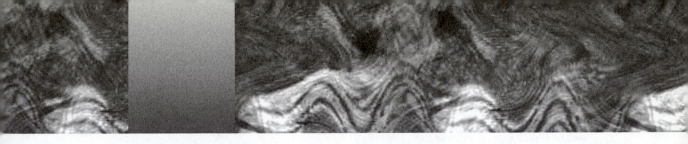

GLOSSARY

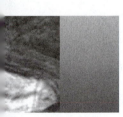

A

ABC analysis an arrangement of activities, such as total dollars spent for purchase of particular items, in descending value, so that resources can be allocated according to the relative importance of each activity

abandonment the final stage of a **product life cycle**

accuracy how close the forecast is to the actual value

aggregate forecasting and replenishment (AFR) the market-level forecasting of consumer demand and the supporting supply process for a defined set of products. The aggregation of data across products, markets, and time produces a high-level plan requiring an operating environment that is responsive to detailed fluctuations, primarily through inventory intensive supply processes

Akaike criterion designed to evaluate models of different complexities

algorithm a computational procedure containing a finite sequence of steps; a set of rules that specify a sequence of actions to be taken to solve a problem

analysis of variance (ANOVA) a technique in which the total variation of a dataset is assumed to be influenced by different causes and the variation due to each cause is separated from the total variation and measured. For business data, ANOVA is used to decompose the time series into trend, seasonal, and irregular components

ANOVA table table summarizing the total variation of the data about its mean in terms of the variation due to the regression fit and the unexplained variation; for example, a two-way table summarizes trend and seasonality

artificial intelligence software that can be programmed to assume capabilities such as learning, adaptation, and self-correction

artificial neural network (ANN) a software implementation of **artificial intelligence**

assets things of value that include good will from an acquisition; good-will balances should be amortized

autocorrelation the association or mutual dependence between the values of the same time series at different time periods or lags

autocorrelation coefficient a measure of the autocorrelation between two values of a time series a distance k apart; it is dimensionless

autocorrelation function the plot of **autocorrelation coefficients** versus the lag k

autocorrelation plot a plot of the sample **autocorrelation coefficients**

autocovariance coefficient a measure of the covariance between two values of a time series a distance k apart

autocovariance function the plot of **autocovariance coefficients** versus the lag k

autoregressive integrated moving (ARIMA) process process that can be generated by summing or integrating the stationary ARMA process d times

autoregressive model the current value of the process expressed as a finite, linear aggregate of previous values of the process and a **white-noise process**

B

back order the receipt of an order for a product when there are no units on hand in inventory

backshift operator a notational device for expressing ARIMA models in a compact form. It shifts a series backward (i.e., lags it) k periods

baseline forecast a forecast based on "business-as-usual" assumptions that does not include sales-force overrides or management judgment

batch forecasting the opposite of online or real time; often refers to programs that run at a fixed later time

Bayesian estimation a mathematical formulation, using Bayes's theorem, by which the likelihood of an event can be estimated taking explicit consideration of certain contextual features (such as amount of data, nature of decision, etc.)

Bayesian information criterion (BIC) criterion designed to evaluate models of different complexities; also known as the Schwarz information criterion

benchmark a fixed point of reference; a standard for comparison; an outstanding example, appropriate for use as a model

bias the difference between the actual and its expected value (forecast)

Box-Jenkins method a three-step process for building time series models: identification, estimation, and diagnostic checking

Box-Ljung test a test for randomness by calculating a sample statistic based on the residual autocorrelations; also known as the Q criteria test

buffer stock *see* **safety stock**

business cycle the cyclical movement in economic data attributed to economic expansions and contractions

C

calendar effect the tendency of data to perform differently at different times of the year, for example during the Easter holiday

causal forecasting model a quantitative forecasting method that relates a time series value to other variables that are believed to cause its pattern

classical decomposition a technique that expresses variation into unobservable components, such as trend, seasonality, calendar effects, and irregular

Cochrane-Orcutt transformation a procedure used to reduce or eliminate serial correlation in residuals

coefficient of determination a measure of goodness-of-fit for a regression model. It is the percentage of the total variation about the mean value of the data that is explainable by performing a linear regression; also known as R^2 statistic

collaborative forecasting and replenishment (CFAR) a collaborative effort between suppliers and retailers to improve the accuracy of sales forecasts and reduce inventory

combining forecasts an approach creating a single forecast from forecasts derived from multiple models or techniques

comparison values calculations in a two-way **ANOVA table** to assess the extent of any systematic nonadditivity

Conference Board a not-for-profit organization that creates and disseminates knowledge about management and the marketplace to help businesses strengthen their performance and better serve society

Consumer Confidence Index a survey based on a representative sample of 5000 U.S. households. The monthly survey is conducted for the **Conference Board**

Consumer Price Index (CPI) a measure of the average change in prices over time relative to a base period of goods and services purchased by households

correlation matrix an array of pairwise ordinary product moment correlation coefficients; used in linear regression models to assess the strengths of relationships among variables

correlogram a display of sample autocorrelations used in analyzing the time-dependent properties of time series; a sample autocorrelation function

Collaborative Planning, Forecasting, and Replenishment (CPFR) an industry initiative energized by the **VICS** organization that is focusesd on improving the partnership between manufacturers and distributors and retailers through shared information

continuous replenishment (CRP) the practice of partnering between distribution channel members that changes the traditional replenishment process from distributor-generated purchase orders based on economic order quantities to the replenishment of products based on actual and forecasted product demand

cross-price elasticity a measure of the percentage change in the demand for good A as a result of a given percentage change in the price of good B

cross-sectional data samples or data that are taken from an underlying population at a particular time

cycle a unit of economic expansions and contractions; cycles are usually irregular in depth and duration

cycle time the length of time between placing two consecutive orders, that is, the time required manufacturing a product or producing a service from the release of the order to its completion

D

damped trend exponential smoothing a form of exponential smoothing in which the trend is damped by the volatility of the data; more volatility leads to greater dampening

data pieces of quantitative information, for example, dollar sales of a product, numbers of building permits issued, and units of raw material on hand

data mining the process of data analysis of large databases aimed at finding unsuspected relationships, which are of interest or value to the database owner

data series *see* **time series**

data value a single, atomic piece of data that cannot be subdivided and still retain any meaning

data warehouse a database for query and analysis, as opposed to a database for processing transactions

database a collection of data organized for ease of use

Database Management System (**DBMS**) software used to access the data stored in a database and present multiple views of the data to end users

decomposition method an approach to forecasting that regards a time series as consisting of a number of unobservable components, such as trend, cycle, seasonality, and irregular

degrees of freedom (**df**) a number used in statistical **significance testing** to quantify the number of effective independent quantities in a calculation

Delphi method a qualitative forecasting method that obtains forecasts through a group consensus

demand the quantity of goods that buyers are ready to purchase at a specific price in a particular market at a given period of time

demand planning the strategy for meeting the independent demand for all products at all locations

dependent demand the demand for one component derived from the demand for another component or final product; also called internal demand

dependent variable the variable that is determined or explained by one or more explanatory factors in a regression model

deterministic model a mathematical model that, given a set of input data, produces a single output or a single set of outputs

deviation the difference between an observation and a fixed value

diagnostic plot a **scatter diagram** of the comparison values against the residuals in a two-way **ANOVA table**

differencing the mathematical operation of subtracting a lagged version of a **time series** from the time series in order to remove or reduce the seasonal and/or trend effects in the series

discrete choice models models that mimic the buying behavior of individuals where the individuals have to make a choice from among several discrete alternatives. Usually a planned experiment used in market research

Distribution Resource Planning (**DRP**) a plan to manage the tactical needs of the supply chain for managing finished goods inventory and determining manufacturing requirements as well as the daily need to determine what inventory should be deployed to the distribution center (DC)

distributor a business that does not manufacture its own products but instead purchases and resells these products; such a business usually maintains a finished-goods inventory

dummy variable a binary variable whose values are 0 or 1, used to quantify seasonality effects, isolated, or even qualitative effects; also called an indicator variable

Durbin-Watson (**DW**) **test** a test for autocorrelation in the residuals from a regression model

E

econometrics an extension of regression analysis that, along with economic assumptions, allows for the mutual dependence among all variables in a number of related equations

economic indicator an economic time series that is assumed to have a reasonably stable relationship to the average or whole economy; used to identify turning points in the level of general economic activity

economic lot size the number of units produced at one time to minimize the combined costs of setups and carrying inventory over a specified planning period, usually 1 year

economic order quantity (**EOQ**) a type of fixed-order quantity that minimizes the total inventory costs in a constant-demand inventory model. It determines the amount of an item to be purchased or manufactured at one time

Efficient Consumer Response (**ECR**) a strategy, established by the grocery industry, in which the retailer and supplier trading partners study methods to work closely together to eliminate excess costs from the supply chain and better serve the consumer

elasticity describes the responsiveness of changes in demand in prices or any of the other variables; see also **own-price elasticity** and **cross-price elasticity**

enterprise resources planning (**ERP**) describes next-generation **MRP** II systems that have evolved to include functionality for enterprise and supply chain management

estimation using statistical techniques to determine the estimated values of the parameters in a model; part of the model-building process

event any occurrence that is expected to impact the sales or order forecast, for example, promotion, price reduction, inventory control tactics, and plant closings

ex ante forecast a forecast made by incorporating predicted values of the independent variables into the forecast period of the model; also known as postsample forecasting

ex post forecast a forecast made by incorporating actual values of the independent variables into the forecast period of the model; also known as postsample forecasting

excess inventory any inventory in the system that exceeds the minimum amount necessary to achieve the desired throughput rate at the constraint or that exceeds the minimum amount necessary to achieve the desired due date performance

expert system a computer program that uses knowledge and reasoning techniques to solve problems normally requiring the abilities of human experts; software that applies humanlike reasoning involving rules and heuristics to solve a problem

exponential smoothing a forecasting technique based on taking weighted sums of past data in order to smooth or forecast a time series

exponentially weighted moving average (EWMA) a set of weights applied to the past history of a time series in order to generate a forecast

external data source for data outside of one's business or industry

external demand *see* **independent demand**

F

fall-off the stage in a **product life cycle** when sales begin to decline

F test a statistical test that measures the overall effectiveness of a fitted regression relationship

feedback a return of part of the output of a machine, process, or system to the computer as input for another phase, especially for self-correction or control purposes

fill rate percentage of product shipped (or received) out of the total order quantity

fit period time period over which data are used to estimate model coefficients

flowchart diagrammatic representation of the operations involved in an algorithm or automated system. Flow lines indicate the sequence of operations or the flow of data, and special symbols are used to represent particular operations

forecast error the difference between the observation when it comes to hand (actual) and its forecast

forecast evaluation a process used to evaluate the quality and value of a forecast

forecast function the function that provides the forecasts at origin *t* for all future lead times; also known as forecast profile

forecast horizon the period over which the forecast is created; also known as the prediction horizon

forecast performance an assessment of the **accuracy** and reliability of a forecast

forecast profile *see* **forecast function**

forecast test the formal analysis of forecast accuracy using models

forecasting accuracy a measure of how close the forecast is to the actual value

forecasting competitions studies performed with various techniques to compare the forecast performance of models

forecasting model a representation of reality used for making projections

forecasting period a time frame outside the **fit period** for which model projections are made

forecasting process a systematic procedure for producing and analyzing forecasts

forecasting simulation *see* **sliding simulation**

forecasting support systems (FSS) computer-based set of procedures that supports a forecasting process

format the particular arrangement or layout of data on a medium, such as the screen or a diskette

forward buy the purchase of a larger-than-normal quantity due to an anticipated shortage or price increase

G

goodness-of-fit a measure of how well the total variation about the mean value of data is explained or fit by a statistical model

H

Henderson averages measures used to calculate the trend-cycle component in X-11-based seasonal adjustment procedures

heteroscedasticity variability in data that is not constant or not homogeneous (nonconstant variance)

heuristic an approach based on commonsense rules and trial and error rather than on comprehensive theory

hierarchical approach an approach, used in numerous technologies, including machine vision, process control, networking, databases, and planning, in which the scope of work is arranged in hierarchies that establish priorities and appropriate routings

hierarchical database database architecture in which data are arranged in the form of an inverted tree structure in which no data value has more than one parent

Hildreth-Lu procedure a procedure used to reduce or eliminate serial correlation in residuals

histogram a bar graph on which the X axis is divided into ranges of data values and the Y axis shows the frequency or percentage of the total number of observations for each range

Holt-Winters model an exponential smoothing model of level, trend, and seasonal smoothing equations; also known as Winter's method

I

impulse a **dummy variable** that takes on a nonzero value at some time and is zero elsewhere

independent demand demand originating from external sources such as direct customer orders; also called external demand

independent variable variable used in a regression relationship to explain or predict values of the dependent variable

Index of Industrial Production (IP) a production index, developed by the U.S. Federal Reserve, that measures real output and is expressed as a percentage of real output in a base year, currently 1997

indicator variable *see* **dummy variable**

input-output table table quantifying relationships among sectors of the economy or industry

intercept the coefficient associated with the constant term in a regression relationship

internal data source of data from within one's company or industry

internal demand *see* **dependent demand**

inventory turns the number of times that an inventory cycles or turns over during a year. A frequently used method to compute inventory turnover is to divide the average total inventory level into the annual cost of sales. For example, this method applied to an average inventory of $3 million with an average cost of sales of $21 million shows that inventory turned over seven times

invertible model a model that generates weights when applied to the past history of a time series in order to generate a forecast and that dies out the further back in time one goes

irregular unusual or rare events arising in a time series

J

judgmental forecasting a technique that relies on human judgment from individuals or groups to make a forecast

L

lag the number of periods or shifts by which a time series is offset

lag structure the dynamic relationship between an output and an input to a system; it can be represented by a **transfer function**

lead time the time frame over which a forecast is required. In inventoried items, it is the period that begins when an order to replenish stock is placed with the factory and lasts until the order is delivered into stock

lead-time demand the number of units demanded during the lead-time period

lead-time demand distribution the probabilistic distribution of demand that occurs during the lead-time period

leading indicator measure that identifies changes in a business cycle or other elements of economic activity that it is assumed to precede in time

least-squares method a method of estimation in which a formula is derived by minimizing the sum of the squares of the differences between the actual data values and the formula approximation

linear filter a linear operation that transforms a **white-noise process** (random shocks) to the process describing the observed time series

long-term forecasting a process that supports strategic planning with forecasts from 2 to 5 years or longer

lot sizing determining the number of units to purchase or manufacture at one time

M

macroeconomics an area of economics concerned with the study of the whole economy and its interrelationships. Key variables studied are GDP, personal income, employment, price levels, and other indicators of economic performance

manufacturer the producer of a good or a service

manufacturing lead time the time required to manufacture a product, generally assuming all needed materials are in stock

manufacturing resource planning (MRP) a computerized method for planning the use of a company's resources, including scheduling raw materials, vendors, and production equipment and processes; includes financials, manufacturing, and distribution management

market share sales of a business as a percentage of the total sales of the market it actively serves; *see also* **relative market share**

materials rquirement panning (MRP) a computerized method whose function is to schedule production and control the level of inventory for components with dependent demand; MRP was the first phase in the development of MRP II

median polish a smoothing procedure used for isolating large outliers in a two-way table analysis

metric a verifiable measure stated in either quantitative (e.g., 97% inventory accuracy) or qualitative (e.g., as evaluated by customers, the company is providing above-average service) terms

min-max system a replenishment system in which inventory level falling to the min (minimum) triggers an order of whatever quantity will reach the max (maximum) level

mixed ARMA model a **model** that includes both autoregressive and moving average terms

model a simplified representation of reality; usually represented in terms of mathematical equations

model building an iterative process for developing a model using prior information and data

monitoring a procedure for checking residuals from a model to see whether specific forms of model inadequacy have occurred

moving average an arithmetic average calculated for a moving sequence of data

moving average process the current value of the process expressed as a linear combination of current and previous shocks

multiple linear regression a statistical technique for relating a variable of interest (the dependent variable) to one or more independent variables

multiplicative error an error term in a model that is multiplied in the model equation rather than added

multiplicative seasonality seasonal fluctuations that widen as the average annual level of the data increases

multiplicative trend a trend component that is multiplied in a model equation rather than added

multivariate model a model involving multiple instances of a dependent variable

N

neural network a processing architecture derived from models of neuron interconnections of the brain. Unlike typical computers, neural networks are supposed to incorporate learning (rather than programming) and parallel (rather than sequential) processing

new product the percentage of a business's current-year sales accounted for by products introduced during the 3 preceding years. New products differ from product-line extensions and model changes by requiring one or more of the following: relatively long gestation periods, major changes to the manufacturing facilities, separate promotional budgets, and separate product managements

noise in general, any unwanted disturbance superimposed on a useful signal and tending to obscure information content

nondeterministic time series *see* **statistical time series**

nonnormality not normally distributed; *see also* **normal distribution**

normal distribution a distribution of values taking on a symmetric bell-shaped curve centered on a mean that equals the median. Approximately 68% of the distribution falls within ±1 standard deviation from the mean. The standard normal distribution has a mean of 0 and a standard deviation of 1

O

observation the value in a **time series** or cross-section. Any data in a database

order forecast forecast of anticipated orders created by combining **POS** data, causal information, and inventory strategies to generate specific ordering requirements to support the sales forecast. Actual numbers are time-phased and reflect inventory objectives by product and receiving location. The short-term portion of the forecast is used for order generation, whereas the longer-term portion is used for material, production, and financial planning

order quantity the minimum size order a customer must make

ordinary correlation coefficient a measure of linear association between two variables; also known as the product moment correlation coefficient

ordinary least squares (OLS) a fitting procedure used in regression analysis, based on the criterion of minimizing the sum of squared residuals

origin time at which a forecast is made for a future **lead time**

own-price elasticity a measure of the responsiveness of the quantity demanded of a good to a change in its own price

P

parameter a fixed, unknown numeric characteristic of the population; the focus of statistical estimation

parsimony the principle, espoused by John Tukey, that a forecaster should employ the smallest number of parameters in a practical application of a model

Pegels classification the grouping of exponential smoothing techniques that gives rise to 12 **forecast profiles** for trend and seasonal patterns

performance measurement a procedure to monitor model or forecast accuracy in terms of multiple metrics

periodicity repeating time periods, such as hours per day, weeks per month, and months per year

pipeline the combined manufacturing production processes required to fabricate, transport, and deliver a single product to retail

point-of-sale (POS) retail sales data collected at stores; also called scan data

postsample forecasting *see* **ex post forecasting** and **ex ante forecasting**

prediction horizon *see* **forecast horizon**

prediction intervals *see* **probability limits**

price elasticity the degree of quantity demanded for a product in response to a change in its price

probability limits a way of expressing the accuracy of forecasts; may be calculated for any set of probabilities, say 50, 90, or 95%. The realized value of the time series, when it eventually occurs, will be included within these limits with the stated probability; also known as prediction intervals

procurement lead time the time required from the point of recognizing a need for an item to receiving it; often refers to the longest **lead time** among those for all needed materials for a given product; allows process optimization and the ability to predict potential problems

product life cycle the steps through which a product passes during its life: introduction, growth, maturity, saturation, and decline

product moment correlation coefficient *see* **ordinary correlation coefficient**

production planning deciding which manufacturing output (production plan) will most efficiently satisfy sales or forecasted sales while meeting general business objectives of profitability, return on assets, customer service, and so forth

projection techniques procedures or algorithms of a quantitative or qualitative nature used to forecasts future periods

Q

Q criteria test *see* **Box-Ljung test**

quick response (QR) a general term for such programs as **VMI** and **ECR;** a management approach to making the manufacture and supply of products to retail faster and more efficient, allowing lower inventories, faster turns, and higher in-stock service of retail, particularly in general merchandise

R

R^2 statistic *see* **coefficient of determination**

R&D cost total research and development cost, including both process and product. If shared with another business, these costs include this business's share of the cost. Process

R&D expenses should be future-oriented expenses; thus, maintenance-engineering costs should be included under manufacturing and not R&D

random series a time series whose autocorrelations are zero at all nonzero lags

regression a statistical technique relating the variability of a key (or dependent) variable to one or more explanatory (or independent) variables. Simple linear regression determines a straight line between two variables; polynomial regression fits a polynomial; and multiple regression expresses the relationship between a dependent variable and several independent variables. Regression models are usually fit by the **least-squares method**

regression equation an algebraic relationship that describes the mean value of a **dependent variable** in terms of fixed values of one or more **independent variables**

relational database a database organized and cross-referenced in rows and columns that provides greater flexibility than a **hierarchical database** in answering ad hoc queries; *see also* **relational database management system (RDMS)**

relational database management system (**RDMS**) a software application that manages a structured collection of information, automatically maintaining defined data relationships that can be accessed by simultaneous users to update or review the data

relative market share the market share of the business divided by the combined shares of the top three competitors

relative price the average selling price for the business's products relative to the average selling price of its top three competitors' products. For example, if the business's prices average 5% above its competitors' prices, report 105%

replenishment lead time the time from the moment it is determined that a product should be reordered until the product is available for use or sale

residual a deviation from a fit, given as the difference between the actual and predicted values in a regression equation

resistant method a technique that safeguards against unusual values

robust method a technique that safeguards against departures from classical modeling assumptions

robust regression a method of regression analysis in which extreme values are downweighted; this results in an analysis in which regression coefficients (and the fitted equation) better represent the bulk of the data

S

safety stock inventory maintained in order to reduce the number of **stockouts** resulting from higher-than-expected demand during lead time; also known as buffer stock

sales forecast a projection of future retail sales for a given time period and location, created by combining **POS** data, seasonality, causal information, and planned events

sampling measuring at regular intervals the output or variable of a process to estimate certain characteristics of the process

scan data *see* **point-of-sale (POS)**

scatter diagram a plot of corresponding pairs of points of two variables to depict a linear or nonlinear relationship

Schwartz information criterion *see* **Bayesian information criterion (BIC)**

seasonal adjustment an analytical procedure designed to remove and adjust for seasonality in data

seasonal indexes a set of factors, either additive or multiplicative, used to quantify the seasonal impact in a time series

seasonal inventory inventory built up in anticipation of a peak seasonal demand (e.g., costumes for Halloween)

seasonality regularly recurring or systematic yearly variation in time series

serial correlation the effect in time series due to data being sequentially related

service level the average number of **stockouts** allowed per year

shock *see* **white noise process**

significance testing a statistical procedure used to test whether a statistical hypothesis holds true

simple linear regression an algebraic relationship between a dependent variable and a single independent variable

stock-keeping unit (SKU) the lowest level of a demand unit tracked in a forecasting and inventory planning system

sliding simulation the use of a mathematical model by a computer program to envision process-design scenarios with real-time visual and numerical feedback; also known as forecasting simulation

standard deviation a statistical measure of the dispersion of a set of values around the mean; the square root of the **variance**

state space models a general mathematical framework for representing discrete-time linear systems, now generally used to describe exponential smoothing and ARIMA time series models

stationary process a stochastic process that remains in statistical equilibrium about a constant mean level

stationary time series a realization from a stationary stochastic process that can usually be described by its mean, variance, and autocorrelation function

statistical time series a time series whose future values can be described only in terms of a probability distribution; also known as nondeterministic time series 000

stochastic process a model that describes the probability structure of a time series

stockout the condition when available inventory is not sufficient to satisfy demand

structured process a systematic procedure described in terms of inputs, intermediate steps, and outputs

supply chain all the activities involved in delivering a product, from the raw material through delivery to the customer, including sourcing raw material and parts, manufacturing and assembly, warehousing and inventory tracking, order entry and order management, distribution across all channels, delivery to customer, and the information systems necessary to monitor all these activities

Supply Chain Management (SCM) a set of processes that produces plans or sets of time-phased numbers representing the best estimate of what demand or supply will be at a given time

survey data information gathered through a sampling process involving questionnaire design, data collection, data processing, and analysis

T

t **test** a statistical test used extensively in regression analysis to test the hypothesis that individual coefficients are significantly different from zero

time series a collection of ordered observations over time (e.g., monthly sales figures for a number of years); regarded as a sample realization from a **stochastic process;** also known as data series

transfer function the linear operator (equation) that transforms a **white-noise process** into a process for the observed time series

transformation a mathematical operation to enhance the analysis, display, and interpretation of data

trend the prevailing tendency or directed movement of data over time; as applied to a straight line, it is often referred to as slope

U

unbiased the quality of being free from bias or prejudice; in forecasting, either an objective qualitative procedure or the mathematical properties of an estimator; statistically, this occurs when the expected value of an estimator of a parameter equals the parameter

V

variable a quantity that can assume any of a given set of values

variance a measure of dispersion of a set of data around the mean; calculated by summing the squares of the differences between each value and the mean and then dividing by the number of values. A larger variance indicates a larger dispersion of values

Vendor Managed Inventory (VMI) the practice of retailers' making suppliers responsible for determining order size and timing, usually based on receipt of **POS** and inventory data; its goal is to increase retail **inventory turns** and reduce **stockouts;** also known as vendor-managed replenishment (VMR)

vendor-managed replenishment (VMR) *see* **Vendor Managed Inventory (VMI)**

Voluntary Interindustry Commerce Standards (VICS) a nonprofit organization focusing on the improvement of product and information flow throughout the supply chain

W

white-noise process a sequence of independent random variables from a fixed distribution, usually assumed to be normal and have mean zero and a constant variance; also known as shock

Winters method *see* **Holt-Winters model**

X

X-11 a computer program developed by the U.S. Census Bureau to provide seasonal adjustment of time series

X-11-ARIMA a computer program developed by Statistics Canada that significantly enhanced the accuracy of seasonal adjustment from the U.S. Census **X-11** program

X-12-ARIMA the most current enhanced X-11 program developed at the U.S. Census Bureau

INDEX

A

Accuracy, 161–193
  in ARIMA models, 482, 484
  bias and, 162–163, 164–166
  in econometric forecasting, 260, 264
  evaluation methods, 166–169
  exponential smoothing and, 305, 306
  fixed-origin simulations and, 555
  forecast end product and, 550–551
  of forecasters, 585
  forecast presentations and, 562
  holdout samples and, 552–554
  indicators of, 583–584
  measurement periods and, 584
  measures of, 169–173
  normality and, 98–99
  outliers and, 77–78, 378
  precision and, 162, 163–166
  qualitative methods compared, 32
  recognition/rewards for, 587
  RMSE and RMPSE, 171–173
  time horizons and, 50–52
  tracking tools, 176–183
Additivity, 379–381
Advanced planning system (APS), 324
Aggregate performance, 168
(AIC) Akaike information criterion, 479
Analysis-of-covariance models, 442
ANOVA models
  day of the week effect, 135–136
  regression models, 397, 398
  residuals and outliers, 379–380
  seasonality and trend, 131–134
  uses of, 134, 135
ARIMA models, 465–510. *See also* Box-Jenkins methodology
  adequacy, validating, 484–486
  advantages/drawbacks, 466
  autocorrelation, 469–477
  causal forecasting, 358

  checklist for, 502–505
  fitting models to data, 478–484
  linear filter model, 467–468
  main features of, 478
  nonseasonal, 477–478, 486–494
  product life cycle, 26
  reasons for use, 466
  seasonal, 494–496, 496–502
  strategy for building, 468–469
  uses of, 465, 466
  vs. other techniques, 33
  X-11/X-12 programs, 215–221
ARIMA models, forecasting with, 511–540
  backshift operator, 537–538
  checklist for, 531
  in compact form, 537–538
  forecast error, 539–540
  forecast probability limits, 528–531
  forecast variance, 539–540
  nonstationary time series, 519–525
  order of, 520

ARIMA models, forecasting with, cont.
seasonality, 525–527
stationary time series, 512–519
ARIMA *(p, d, q)* models, 520–525,
530, 539
ARMA process. *See* autoregressive-
moving average (ARMA) process
Artificial neural nets (ANN), 39–40
Assumptions and rationale
consensus on, 19
credibility and, 546
measurement plans, 582–583
in presenting final forecast,
547–548
supporting, 561–562
Assumptions of methods
data characteristics and, 50
in inferential statistics, 98–99
linear regression, 361, 378–381, 407
normality, 98–99
randomness, 371–374
testing for specificity, 547–548
Asymptotic efficiency, 115–116
Autocorrelation analysis
detecting autocorrelation, 93–96,
137–138, 374
objective of, 88–89
residuals in ARIMA models, 485
serially correlated data, 88–96
Autocorrelation coefficient, 89–90,
92–93, 471, 472, 485–486
Autocorrelation correction, 407–415,
433
Autocorrelation functions (ACFs)
decaying/truncating, 473–474, 475,
487, 498
interpreting, 469–477
nonseasonality, in ARIMA models,
489–494
ordinary correlograms, 92–93, 470,
489–494, 497–498
seasonality, in ARIMA models,
477, 494–495, 497–498
Autoregression model, 469
Autoregressive (AR) process,
470–474, 514–515, 518
Autoregressive integrated (ARI)
model, 520
Autoregressive-moving average
(ARMA) process
ACF and PACF, 476, 498–499
general form of, 468
mixed, 474–476
noise and, 513–515
origin of, 512

parametric model, 519
purpose of, 467–468
stationary time series, 512–513
Averaging forecasts, 545

B

Backshift notation, 476, 502
Backshift operator, 470, 537–538
Bias, 162–166, 169–170, 176–183, 557
BIC (Schwarz Bayesian information
criterion), 479
Bisquare estimator, 448, 450
Bisquare weighting function, 222, 224
Black box. *See* linear process/linear
filter model
Bottom-up approaches, 586
Box-Jenkins methodology. *See also*
ARIMA models
appropriate use of, 469
causal forecasting, 358
parsimony, 468, 474, 479
stages of, 469
theoretical foundation of, 34
vs. econometric method, 35
Box plots, 85–86, 101, 129, 130,
376–377
Business cycles, 16, 236–237, 238,
249–253, 265–267
Business forecasting, 235–275. *See
also* econometric analysis
economic indicators, 236–247
elasticities, 253–260
regression models, 451–455
using turning points, 247–253

C

Calibration period. *See* fit period
Causal forecast models, 357–390
function of, 358
linearity, achieving, 389–390
linear regression as, 395
model building strategy, 358
multiple linear regression models,
363–369
regression models, 359–363
residual patterns, 369–378
validating assumptions, 378–381
Census seasonal adjustment method,
215–221
Central Limit Theorem, 99
Central tendency, 76–77, 78
Centre for Forecasting, Lancaster
University, 15
Checklists
for ARIMA forecasting, 531

for ARIMA models, 502–505
for final forecast number selection,
546–550, 563–564
for forecast managers, 596–600
for implementing forecasting
process, 590–594, 596–600
purpose of, 543
for regression models, 450–451
Chi-squared test, 485–486
Cochrane-Orcutt procedure, 411–414
Coefficient of determination. *See*
R^2 statistic
Coincident indicators, 237–240, 246
Collaboration, 20
Collaborative Planning, Forecasting,
and Replenishment (CPFR), 334
Collinearity, 429–430, 431–432
Combining forecasts, 40, 545
Communication, 20, 32, 39, 588–589
Composite indicators, 243–244, 581
Computer software selection, 593,
595
Computer systems architecture, 589
Conference Board, 243, 245
Confidence intervals, 99–102, 403,
404, 501, 502
Consensus process, 589
Consistency, 17, 114, 115
Consultants, 596
Consumer confidence, index of,
124–125
Consumer Price Index (CPI), 15,
258–260
Consumer sentiment, index of,
487–490
Cooperation, 19, 593–594
Correlation. *See* autocorrelation;
regression
Correlation, lag 1, 409
Correlation, serial, 88–96, 408–411
Correlation coefficient
autocorrelation coefficient, 89–90
in linear regression, 74–75, 401,
407
outliers and, 430–431
robustness, 112–113, 425–426
Correlation matrix, 75–76, 429–430
Correlograms
in ARIMA models, 474, 485,
486–494, 497–502
autocorrelation, 374, 409
interpretation of, 92–93, 138
nonstationarity, 137–139
of seasonal data, 95–96, 97,
497–498, 499–502

Correlograms, ordinary, 92, 486–487, 489–494, 497–498, 499–502
Correlograms, partial, 487, 488–489, 491–494, 497, 499–502
Council of Supply Train Management, 322
Credibility, 546–552, 561, 562
Cross-correlation coefficients, 95
Cross-elasticity, 235, 257–258
Cross-functional business processes, 332–333
Cross-sectional data, 80–83
Cumulative frequency distributions, 83
Cumulative miss, 180
Cumulative sum tracking signal (CUSUM), 180–181
Current Employment Statistics (CES), 69–70
Current level, 287–292, 300, 318–320
Current seasonal index, 288–292, 318–320
Current trends, 288–292, 318–320
Customer collaboration, 39
Customer inputs, 559–560, 587, 589
Cycles. *See also* business cycles; time series analysis; trend-cycle
   as abstract, 34
   defined, 121
   quantification of, 121
   stochiastic cycle, 479, 482, 483
   two-way tables, 156–159
Cycle time, 325, 327. *See also* manufacturing lead times

D

Daily volume, 127–128
Dampening. *See* trends, damped
Data. *See also* historical data
   final analysis and review, 546–547
   judging quality of, 15–17
   limited, forecasting with, 55
   locating sources, 14–15, 592
   randomness assumption, 371–374
Data, external, 14–15, 17
Data, internal, 14, 17
Database management, 595
Data management, 592–593
Data summaries, 76–83, 84–88
Data warehouse, defined, 14
Day of the week effect, 135–136
Decision trees, 30–31
Decomposition methods, 33–34, 156–159. *See also specific methods*
Degrees of freedom, 396

Delphi method, 29
Demand
   determinants of, 254
   dimensions of, 329–331
   future, time integration and, 548
   multilevel forecasting, 330
   place segments of, 330, 331
   price and, 255–256, 257
   theory of, 254
Demand, dependent, 325–326
Demand, independent, 325, 326
Demand forecasting. *See also* disaggregated product-demand forecasting; supply chain
   accuracy, need for, 161–162
   changes in demand, 12–13
   defined/characterized, 2–4
   econometric systems and, 35
   elasticities, 253–260
   guidelines, 18–20, 586–587
   importance of, 253
   integrated system for, 328–337
   key concepts, 332
   for new products, 38
   Pareto's law, 562
   price elasticity and, 254–258
   retailers' role in, 39
   simple smoothing, 53–55
   turning-point analysis, 249–253
   unusual variation in, 215–216
Demand management, 325–326, 332–333, 337–346
Demand management system architecture, 329
Demand theory, 13
Demand variability analysis, 334–337
Demographic data, 17
Demographic forecasts, 595
Depth of simulations, 554, 556, 557
Deterministic models, 31–32, 33, 156
Diagnostic checking, 18, 549–550
Differencing. *See also* first differences; second differences
   in ARIMA models, 486, 519–520
   econometric models and cycle forecasts, 253
   nonstationarity, 139–142, 519–520
   seasonality, 197–200, 525–526
   transformations and, 142–145
   uses of, 139
Differential intercept coefficient, 441
Disaggregated product-demand forecasting, 321–355
   automated, 337–346
   checklist for, 346–347

integrated system framework, 328–337
   for supply chain, 322–328
Discontinuities, 442–443
Dispersion. *See* variability (spread)
Distribution resource planning (DRP), 324, 326–328, 347–348
Documentation, 561, 591, 594
Dummy variables (indicator variables), 308, 407, 434, 440, 441–445
Durbin h statistic, 409
Durbin-Watson statistic, 408–409, 433
Dynamic simulations, 556
Dynamic systems modeling, 32–33

E

Econometric analysis, 235–275
   overview, 34–35, 236
   applications of, 236
   business forecasting and, 260–265
   defined, 235
   pitfalls/limitations, 263–265
   product life cycle, 26
   recursive systems, 262–263
   seasonal adjustment and, 253
   theoretical foundation of, 34
   types of models, 262
   uses of, 236, 261–262
   vs. other techniques, 33, 35
Economic forecasting
   accuracy of, 2
   consumer sentiment, index of, 487–490
   cycles, analyzing/predicting, 121
   differencing and, 139–143
   input-output analysis, 32
   seasonal variation, 123–124, 141, 200, 202
   virtual forecasting services, 595
Economic indicators, 236–247. *See also* coincident indicators; lagging indicators; leading indicators
   classification of, 237
   composite indicators, 243–244
   consumer price index, 15, 258–260
   macroeconomic demand, 236
   selection criteria, 245–247
   sources of, 14–17, 245
   turning-point technique, 247–253
Efficient Consumer Response, 323
Elasticities, 253–260
Environmental variables. *See* variables, external
Error-minimization criteria, 344

Errors. *See also* forecast errors
  absolute *vs.* squared, 168–169
  fit errors, 47–48, 479–482
  mean absolute error (MAE),
    184–185, 186–187
  mean absolute percentage error
    (MAPE), 78
  mean error (ME), 169–170
  mean percentage error (MPE),
    169–170
  mean squared error (MSE), 55
  median absolute percentage error
    (MdAPE), 78, 170–171, 544
  moving average percentage errors
    (MAPE), 500, 503
  multiple error measure, 557
  multiplicative error, 299
  one-period ahead forecast errors,
    137
  percentage errors (PE), 500
  relative error measures, 161, 175
  root mean squared error (RMSE),
    171–173
  root mean squared percentage error
    (RMPSE), 171–173
  turning-point errors, 177–178
Estimation/estimators, 102–105,
  113–116, 478–484
Evaluation stage, 12, 543, 544,
  580–586
Event adjustments, 38, 307–309
Execution stage, 12, 578, 579
Exploratory data analysis (EDA),
  67–116
  correlation matrix, 75–76
  data summaries, 84–88
  defined, 68
  M-estimation method, 103–105
  robustness, 102
  scatter diagrams, 71–74
  time plots, 69–71
Exponential decay, 292
Exponential growth patterns,
  296–298
Exponentially weighted averages,
  282, 284, 288
Exponentially weighted moving averages, 520
Exponential smoothing, 279–320
  overview, 279–280
  automated systems, 337–346
  damped and exponential trends,
    292–301
  defined/characterized, 280
  evaluation procedures, 305–306

prediction limits for models,
    298–301
  in product life cycle forecasting, 36
  seasonal models, 301–307
  smoothing levels and constant
    change, 287–292, 318–320
  smoothing weights, 281–286
  types of techniques, 286–287
  vs. other techniques, 33
Exponential smoothing, simple
  assumptions behind, 52
  current level, 290–291, 343
  defined, 49, 283
  equation, 344
  methodology, 282–286, 290–291
  uses of, 53
  as a weighted average, 53–55
Ex post forecasting, 393, 433
Extrapolation, 430

F
Federal Reserve Board, 16–17,
  120–121, 492–44
Federal Statistics Web page, 15
Filtering. *See* linear process/linear
  filter model
Final forecast number, selecting,
  543–575
  "best-fit" number, 550
  checklist for, 546–550, 563–564
  credibility, establishing, 546–552
  documentation supporting, 561,
    591, 594
  management, gaining acceptance
    from, 560–562
  multiple methods for arriving at,
    549
  preparing forecast scenarios,
    544–545
  presentation of, to users, 562
  probabilities and risk, 550–551
  reconciling sales force and
    customer inputs, 559–560
  simulations, 552–559
First differences
  defined, 91
  of log-transformed data, 142,
    496–497
  nonstationarity, 139–141, 522
  of transformed data, 142–143
Fit errors, 47–48, 479–482
Fit period
  pruning, 558
  simulations and, 552, 553,
    558–559

single *vs.* multiple, 558
  vs. test period, 166–167
Focus Forecasting, 343
Forecast cycle, defined, 331
Forecast errors
  absolute values of, 170–171
  analysis of, 547–548
  ARIMA models, 515–516, 539–540
  autocorrelation, 179
  bias and, 162–163, 164–166, 181
  calculating, 55–56
  cumulative, monitoring, 181–183
  defined, 47, 165
  disaggregate product-demand
    forecasting, 336–337
  distribution of, 67
  influence of, 54, 55
  magnitude of, 583–584
  naïve forecast *vs.* simple
    smoothing, 55–57
  nonrandom, 183
  precision and, 163–166
  prediction intervals and, 100, 101,
    179
  probability limits and, 528, 529
  regression models, 433
  scatter of, measuring, 184–185
  signal tracking, 180, 183,
    184–188
  squared values of, 171–173
  summarizing, 544–545
  variance of, 179, 540
  vs. fit errors, 47–48, 167
Forecasters
  assessing ability of, 585–586
  items monitored by, 580–582
  objectives for, 580
  roles/responsibilities of, 10, 20,
    578–579
  training, 19, 593, 596
Forecast horizon, defined, 331, 553
Forecasting
  ambiguities/uncertainties, 19
  "best practice," 333
  creating a structure for, 11–18
  defined, 2
  key stages in, 11–12
  nature of, 3–4, 10
  purpose of, 2–4, 321
  as a structured process, 20
Forecasting models. *See also*
    specific models
  overview, 6–10, 34
  defined, 6
  diagnostic checking, 549–550

evaluating, 17–18
validity of, 19
Forecasting process, implementing, 577–603
  checklist for, 590–594, 596–600
  PEER methodology, 577–590
  virtual forecasting services, 594–596
Forecasting techniques, 27–63
  overview, 27, 40–42
  fit *vs.* forecast errors, 47–48
  judgment and, 50–52, 378
  life-cycle perspective, 35–37
  market research, 37–38
  moving averages, 42–47
  multimethod approaches, 52
  neural nets for, 39–40
  new product introduction, 38
  promotions/special events, 38
  sales force composites, 39
  selection of, 28–35
  weights, 48–50, 52–58
Forecast miss. *See* forecast errors
Forecast numbers. *See* final forecast number, selecting
Forecast package, 561–562
Forecast profile (forecast function), 43, 118–119, 517–518
Forecast support systems (FSS), 328–329
Forecast users
  acceptance, 560–562
  credibility and, 546–552
  information distribution, 329
  needs of, 332–333, 581–582
  types of, 10–11
  understanding forecast limitations, 583
Frequencies, tabulating, 80–83
Frequency distributions, 82–83, 85–86
FSDJ (Dow Jones Industrial Average of Stock Prices), 71
F statistic, 404, 405, 406
F statistic, incremental, 404, 406–407

G

Game-playing, 587
GDP, time plot of, 237, 239
GNP. *See* GDP
Goals, setting, 579, 580, 590–591
Goodness-of-fit, 167–168, 393, 399, 404–405, 479
Granularity, 70–71, 330–331
Group dynamics, 29

H

Henderson curves, 219
Heteroscedasticity, 375
Heuristic methodology, 343
Hildreth-Lu procedure, 413–414
Hinges, defined, 86
Histograms, 83, 98
Historical analog, 29
Historical data. *See also* time series
  autocorrelation analysis, 67
  cycle patterns, 250
  documentation of, 547
  evaluating sources of, 17
  long-range projections, 40–52
  smoothing of, 41–47
  weighting, 48–50
Holdout samples, 166, 305, 552–554. *See also* test period
Holt-Winters' method, 288, 291–292, 296–298, 301–309
Huber M-estimator of location, 104–105
Huber M-estimator of standard error, 104–105
Huber weights, 447, 448, 449

I

Improvement opportunity items, 562
Income elasticity, 258
Indicators. *See* economic indicators
Indicator variables. *See* dummy variables (indicator variables)
INDPRO index, 16–17, 120
Industrial production, index of, 492–494
Information sharing, 589
Initialization period. *See* fit period
Inner fences, defined, 81
Input-output model, 33
Integrated moving average (IMA) model, 520
Integrating forecasts, 588–589
Intention-to-buy anticipation survey, 33
International Institute of Forecasters, 15
International Journal of Forecasting (IJF), 166
Internet
  as data source, 14–15
  M3 competition, 166
Interquartile difference (IQD), 79–80, 86
Interval forecast, 100. *See also* prediction intervals
Intuition, 18–19

Inventory, 323, 325, 326, 327, 347–348. *See also* disaggregated product-demand forecasting; distribution resource planning (DRP); Supply Chain Management (SCM)
Inventory, projected, 348
Inventory turns, 323
Invertibility: ACF and PACF, 476
Invertibility requirement, 538
Irregular fluctuations, 124–126, 133–134, 156–159, 513–514. *See also* white noise
Item-level performance *vs.* aggregate performance, 168

J

Judgment, 50–52, 378, 551–552
Judgmental forecasting, 593
Judgmental overrides, 333

K

Kolmogorov-Smirnov test, 486

L

Lack-of-fit criterion, 434
Ladder charts, 176–177
Lagged, defined, 90–92
Lagged dependent variables, 444
Lagging indicators, 237, 241, 242
Lags, 44, 467–468
Leading indicators
  defined, 237
  limitations of, 247
  origin of, 236–237
  reverse trend adjustment, 244–245
  uses of, 237, 240–242
  vs. other techniques, 33
Leading indicators, index of, 243, 244
Lead time. *See* cycle time; manufacturing lead times
Lead time of the forecast, 554, 556–557
Least-squares assumption, 362, 408, 440, 448
Least-squares estimators, 116, 446–447, 448, 450, 478. *See also* M-estimation method
Least-squares fit, 401–402
Life cycles. *See* product life cycles
Linear correlation (association), 74–75, 401–402
Linearity, achieving, 389–390
Linear process/linear filter model, 467–468, 511, 518–519

Linear regression, multiple
checklist for, 450–451
confidence intervals and F tests, 404
correlation coefficients, 401
creation of, 363–369
defined, 359, 363
log-linear models, 366–369
multicollinearity, 431–432
normality assumption, 403
transformations, 366–369, 438–439
with two explanatory variables, 365–366
Linear regression, simple, 359, 361, 401–402, 404–407, 414–415
Linear regression analysis, 391–427.
See also regression models
autocorrelation correction, 407–415, 433
causality and correlation, 401
creating and interpreting output, 395–402
graphing relationships, 391–395
inferences about parameters, 402–407
prediction intervals, 414–415
summary statistics, 393–394
uses of, 391
variation, 396–397
Linear relationship, defined, 395
Linear trend line fit, 395
Livingston Survey, 260
Location. *See* central tendency
Logarithmic transformations
in ARIMA models, 490–491, 496–497
exponential growth patterns, 296–298
with nonstationary data, 142–145
regression models, 366–369, 438–439
in turning point analysis, 251–252, 253
variance and, 379
Long-range forecast methods, 32
l-step-ahead forecast, 517–518, 539

M

Macroeconomic demand analysis, 236
Macroeconomics, 2, 235
Management as forecast users, 10, 560–562, 590, 594
Managerial control, 580
Manufacturing lead times, 331.
See also cycle time

Manufacturing resource planning (MRP), 324
Market intelligence, 333–334
Market planning, 5–6
Market potential, defined, 13
Market research, 32, 37–38
Market share, estimating, 11
Markov process, 514
Master production scheduling (MPS), 324
M-competition, 40, 166, 341, 545
Mean absolute deviation (MAD), 79
Mean absolute error (MAE), 170–171, 184–185, 186–187
Mean absolute percentage error (MAPE), 78, 170–171, 344
Mean error (ME), 169–170
Mean percentage error (MPE), 169–170
Means
as "best" estimator, 98–99
confidence intervals and, 100–101
defined, 76–77
outliers and, 77–78, 79, 105
tapered mean, 222, 224
trimmed, 78, 105
Mean squared error (MSE), 55
Median absolute deviation (MdAD), 79, 112–113
Median absolute percentage error (MdAPE), 78, 170–171, 544
Median polish, 379
Medians, 77, 78, 85, 100–101, 112–113
Median smoothing, 47
Memory patterns, 95
M-estimators, 103–105, 447–450
Mixed autoregressive-moving average (ARMA) process, 474–476
Money supply, 139–143, 490–492
Monitoring forecasts, 183–188, 528–531, 580–582
Months for Cyclical Dominance (MCD), 219
Morphological research, 30
Moving average (MA) process.
See also autoregressive-moving average (ARMA) process
in ARIMA model, 472–474, 494–496, 513–514
defined, 468
invertibility, 476
probability limits for, 528–530
Moving average percentage errors (MAPE), 500, 503

Moving averages, 33, 44–47, 207–209, 217–218
Moving averages, centered, 203, 219
Moving averages, simple, 42–43
Moving averages, weighted, 11, 48–50, 222
Moving median smoothing, 46–47
m-period ahead forecasts, 43, 49, 167
Multicollinearity, 431–432
Multi-equation causal models, 9
Multimethod approaches, 52
Multiple error measure, 557
Multiple linear regression models.
See linear regression, multiple
Multiple-tier charts, 129, 130
Multiplicative error, 299
Multiplicative trend, 299
Multiseries comparisons, 168

N

Naïve forecast1 (NF1), 48, 55, 57, 173–175
Naïve_3 technique, 302, 303
National Bureau of Economic Research (NBER), 236–237
Neural nets, 39–40
Noise. *See* irregular fluctuations; white noise
Nonrandom patterns, 374–376, 486
Nonstationarity
ARIMA models, 512, 518–519, 519–525
components of, 513
correlograms, 137–139
defined, 137
differencing, 139–142
log transformations, 142–145
Nontraditional approaches, 102–105
Normal distribution, defined, 98–99
Normality
overview, 96, 97–102
assumptions of, 97
confidence intervals and, 99–101
mean, effect on, 98–99
prediction intervals and, 99–102
Q-Q plots for detecting, 87–88, 103–104
in regression, 402–403
transformations and, 144
variance, effect on, 98–99
Normal probability plots, 87–88, 143–145, 466
Notre Dame—Business Forecasting Course, 15

O

Objectives, setting, 579, 580
One-period ahead forecast
    accuracy and, 167
    in ARIMA models, 515–517
    defined, 43
    exponential smoothing, 284, 306
    weighted moving average, 49
One-period ahead forecast errors,
    47, 137
Ordinary least squares (OLS), 362,
    408, 440, 448
Origin of a simulation, 554–556
Outer fences, defined, 81–82
Outliers. *See also* robustness
    in ARIMA models, 482–484
    in data with high irregular
        components, 133–134
    defined, 77
    in demand management models,
        340, 342
    determining outlier boundaries, 102
    error minimization methods, 344
    inner and outer fences, 80–81
    interpolation, 307–308
    MdAD and, 79
    means and medians, effect on,
        77–78, 79
    moving median smoothing and, 47
    Q-Q plots for detecting, 103
    regression models and, 430–431,
        444–445
    resistant smoothing in time series,
        221
    robust/resistant methods and, 102,
        112–113, 425–426, 446–447
    special event adjustments,
        307–309
    UMdAD and, 80
    visual inspection for, 342
Overfitting, 430, 484, 485
Own-price, 235, 257

P

Panel consensus, 28–29
Parameter definition, 12, 20, 30
Parametric model, 519
Pareto's law, 562
Parsimony, 468, 474, 479, 527
Partial autocorrelation functions
    (PACFs), 469–477, 487, 488–495,
    497–498
P_CMAVG_t, 203, 204–205
Pearson product moment correlation
    coefficient (r), 74–75, 401

PEER methodology
    defined, 12, 577–578
    for demand forecasting, 18–20
    "evaluate," 580–586
    "execute," 579
    "prepare," 578–579
    "reconcile," 586–590
Pegels diagrams, 118–119, 286–287,
    298–299
Percentiles, 85–86
Percent miss, 180
Percent residual effect, 128
Performance
    aggregate *vs.* item-level, 168
    evaluating, 55, 57, 547, 580
    forecast presentations and, 562
    indicators of, establishing, 579
    standards of, 584–586
Periodicity, 95–96, 97, 486
Periodogram analysis, 486
Period (time) granularity, 330–331
Personal income time plot, 237, 239
Planned receipts, defined, 348
P_MAVG_t, 43, 49
Point-of-sale (POS) data, 211–215
Policy information, 360
Policy variables. *See* variables,
    internal
Precision, 99, 162, 163–166,
    170–173, 395–397
Prediction intervals
    calculating, 99–102
    causal regression with, 414–415
    defined, 99, 100, 414
    as early warning signals, 180–183
    empirical, 178–179
    for forecast errors, 179
    as percent miss, 180
    seasonality and exponential
        smoothing, 302–303
    for time series models, 178–179
    uses of, 179
Prediction-realization diagrams,
    177–178
Preparation stage, 11, 12–17, 578–579
Presentation of forecasts, 562, 594
Pressures analysis, 247–249, 265–267
Price elasticities, 255–257
Probability limits, 528–531
Probability plots, 143–145
Process of forecasting. *See* forecasting
    process
Product demand. *See* demand
    forecasting; disaggregated
    product-demand forecasting

Product development, 36
Product hierarchy, 330
Product introduction, 29, 36, 38
Production planning, 39
Product life cycles, 35–37
Product moment correlation
    coefficient, 74–75, 401
Projection techniques, 2, 9
Promotions
    effects of, 38
    exponential smoothing, 307–309
    influence on time series, 126–127
    moving averages and, 47
    patterns characterizing , 212–213
    vs. seasonality, 126–127
Psi (ψ) coefficients, 467
P-term centered moving average
    (P_CMAVG_t), 203, 204–205
P-term moving average (P_MAVG_t),
    43, 49
P_WMAVG_T, 49–50

Q

QR (Quick Response), 323
Qualitative methods, 28–32, 36, 41
Quantile-quantile plots (Q-Q plots),
    86–88, 87, 103
Quantitative methods
    overview, 41
    deterministic, 31–32, 33
    group coordination on, 593–594
    intuition and, 18–19
    probability statements and, 528
    statistical (stochastic), 31, 33
    validation, importance of, 19
Questionnaires, 29

R

R^2 statistic, 397–400, 405
Random errors, 102, 179, 467, 468,
    518–519
Randomness, testing for, 371–374
Random samples, defined, 99
Rank correlation test, 372
Rates of return analysis, 70–71
Ratio-to-moving average method,
    202–205
Reasonableness, 73, 546, 549–550,
    581
Recalibrating *vs.* updating, 559
Recessions, 236, 237, 244, 245, 546
Reconciliation stage, 12, 18, 578,
    586–590
Recursive systems, 262–263
Regression: history of term, 360

Regression coefficients, 362, 404–407
Regression curve, defined, 360
Regression models, 427–464. *See also*
    linear regression
    assessing adequacy of, 428–434
    causal forecasting, 359–363
    checklist, 450–451
    defined, 359
    dependent variables, 363, 364
    exchanging X and Y, 400–401
    explanatory variables, 364, 365,
        393, 431–432
    intercepts, 442, 453, 454
    leading indicators and, 245
    linear association in, 401–402
    nonstationary time series, 523–524
    normality, 144, 402–403
    outliers, 430–431, 444–445
    overfitting, 430, 484, 485
    probability distributions, 403–404
    qualitative variables, 440–445,
        451–455
    random errors and, 179
    residuals, 436, 437–439, 445–446
    robustness, 112–113, 362, 425–426,
        446–450
    slopes, 442
    time series as variables, 429
    transformations, 144
    variables, selecting, 434–440
    vs. other techniques, 33
Regression parameters, 361, 428
Regression SS, 397
Relative error measures, 161, 175
Relative frequency distributions,
    82–83
Reliability, 543, 544
Residual analysis, 369–378. *See also*
    fit errors; forecast errors
    in ARIMA models, 482–484, 485
    autocorrelation in, 88, 408–409,
        413
    correlograms, 485, 487
    graphical aids, 376–377
    median polish, 380–381
    M-estimators, 447–450
    patterns in, 374–378
    randomness assumptions and,
        371–374
    in regression models, 436,
        437–439, 445–446
    seasonal adjustment, 253
    sequence of, 445
    serial correlation, 409–411
    straight-line trend, 143

    two-way tables, 158–159
    uses of, 369–370
Residual series, 73
Resistant smoothing, 221–225
Resistant statistics, 9, 77–78, 80, 106
Resources, obtaining, 591
Retail sales, 39, 237, 240
Revenue, defined, 256
Revenue forecasting
    overview, 4–5
    basis of, 393
    cycle forecasts, 250–253
    forecast presentations and, 562
    price elasticity and, 256–257
Revenue repression factor, 257
Revenue weighting, 584–585
Rewards for accuracy, 587
Risk, 550–551, 561
Robust correlation coefficient,
    425–426
Robust correlations, 429–430
Robustness, 9–10, 102, 106
Robust regression, 362, 440,
    444–450
Role-playing, 30
Rolling simulations, 341, 555–556,
    558
Root cause analysis. *See* demand
    variability analysis
Root mean squared error (RMSE),
    171–173
Root mean squared percentage error
    (RMPSE), 171–173
Rule-based forecasts, 40
Runs-of-signs test, 372–374

S

Safety stock, 326, 327
Sales and operation planning (S&OP),
    324
Sales force composites, 39, 560
Sales force inputs, 11, 559–560,
    587–588
Sales forecasts, 6, 13, 560, 587
Sales quotas, 560
Sample standard deviation, 79, 99,
    399, 400
Sample variance, 77–78, 98–99, 145
Sampling distribution, 99
Scale statistic, 80
Scaling of MdAD and IQD, 80
Scatter diagrams, 71–74, 91–92,
    437–439
Scatter plot matrix, defined, 428
Scenarios, preparing, 544–546

Seasonal adjustment
    in ARIMA models, 476–477,
        496–502, 525–527
    census method, 215–221
    defined, 45
    econometric regression, 253
    exponential smoothing, 301–307
    future seasonal factors, 217
    monthly/quarterly data, 215–221
    moving averages, 44–45
    ratio-to-moving average method,
        202–205
    in regression models, 443–444
    resistant smoothing in, 221–225
    starting values and, 345
    trend adjustment and, 526–527
    uses of, 201–202, 215
Seasonal ARIMA models, 525–527
Seasonal component, defined, 129
Seasonal decomposition
    additive *vs.* multiplicative,
        200–201, 205–215, 218, 301,
        303–307
    of monthly data, 205–208
    objective of, 200
    of quarterly data, 208–211
    of weekly point-of-sale data,
        211–215
Seasonal indexes, 206–207, 208,
    288–292, 318–320
Seasonal-irregular (SI) ratios, 220
Seasonality, 195–234
    as abstract, 34
    additive, 287, 301, 304–306, 308,
        338–339
    in ARIMA models, 494–496
    correlograms and, 95–96, 97,
        137–138
    defined, 45, 122, 196
    detection methods, 225–227
    differencing, 141, 142, 197–200
    exponential smoothing and,
        286–287, 288–292, 318–320
    factors influencing, 196
    graphical displays of, 129–130
    ladder charts and, 176–177
    moving averages and, 44–45
    multiplicative, 287, 301, 304–308,
        338–339
    Pegels diagrams, 118–119, 287
    trend and, 119, 121–124, 128–136,
        379
    two-way tables, 130–132, 156–159
    types of, 286, 287
    vs. promotion scheduling, 126–127

Second differences, 140–141, 145–146, 522

Simulations, 552–559

Skewed data distributions, 87–88

SKUs, 324, 331

Sliding simulation, 341

Smoothing, 41–47, 222–225, 299–300, 343–344. *See also* exponential smoothing

Smoothing weights, 285, 299–300, 343, 344–345

Software selection, 593

Special events, 38, 307–309. *See also* promotions

Spread. *See* variability (spread)

Spreadsheets, 40–41, 130–132, 148, 156–159, 185–188

SRC index of consumer confidence, 124–125

Standard deviation, sample, 79, 99, 399, 400

Standard deviation of the location estimator, 104–105

Standard deviation of the sample values, 78–79

Standard error. *See* standard deviation, sample

Standardized variables, 74

State-space models, 298, 299

Stationarity, 136–145, 476, 512–519

Statistical (stochastic) methods. *See also* econometric analysis; summary statistics
overview, 31, 33
analysis procedure steps, 96
automated, for demand forecasting, 337–346
"best" estimators defined, 998
regression coefficients, 404–407
robust estimators, 102
time series decomposition, 33–34
for trend detection, 146–148
types of, 31, 33–35

Statistics, resistant, 9, 77–78, 80, 106

Stem-and-leaf displays, 84–85

Steps, defined, 81

Stochiastic cycle, 479, 482, 483

Stochiastic model, 512

Stochiastic processes, 512, 513. *See also* linear process/linear filter model

Stock prices, 373

Straight-line projection. *See* trends

Structural analysis, defined, 261

Structured process, 20

Summary statistics, 33, 76–83, 401–402, 404–407. *See also* means; medians; trimmed means

Sum of squares (SS), 396, 397–398

Supply chain, 329, 352–353, 586

Supply Chain Management (SCM), 322–328, 330–331

Supply train, integrated, 322–323

Supply train, traditional, 322–323

Surveys, 15, 29–30, 33

T

Tapered mean, 222, 224

Tapered smooth , 224

TCSI decomposition (Shiskin X-11), 33

Technological forecasting, 30

Test period, 166–167, 557–558

Tier plots, 129, 130

Time horizons, 50–52, 331, 553, 557–558

Timeliness of data, 17, 553–554

Time-phased replenishment plans, 347–348

Time plots, 69–71, 199

Time series, defined, 4, 117

Time series, random, 93–96

Time series, trending, 45–46, 50–51, 119

Time series analysis, 117–159. *See also* seasonality; stationarity; trends
administrative decisions and, 126
association between series, 95
assumptions of, 33–34
autocorrelation, 92–95
collinearity, 429
correlograms, 92–93
cyclical: TPA forecasts, 249–253
day of the week effects, 135–136
decomposition, 33–34, 128–136
describing correlation in, 88–96
differencing, 139–142
empirical, 467–468
first differences, 91
fluctuations, 124–126
homogeneous, 139
lagged, 90–92
linearity, achieving, 389–390
nonstationary, 519–525
periodic cycles and, 96, 97
in product life cycle forecasting, 36
purpose of, 118
as regression variables, 429–430, 432–433

rule-based forecasts and, 40

scatter diagrams, 71–74

seasonality, 121–124

with seasonal periodic component, 527

short, dealing with, 559

simple moving averages, 42–47

stationarity, 136–145, 512–519, 519–525

trading-day patterns, 127–128

trends, 120, 121, 145–146

uses of, 34, 117

variability, 156–159

weekly patterns, 126–127

weighted moving averages, 49–50

white noise and, 513–514

Top-down perspective, 586

Total SS (sum of squares), 396, 397

Tourism forecasting, 8–9, 254

Trace forecasts, 554

Tracking signals, 180–181, 183, 188

Tracking tools, 176–183

Trading-day (TD), 127–128, 202–203

Training, 19, 593, 596

Transformation. *See also* logarithmic transformations
defined, 42, 74
linearity, achieving, 389–390
linear regression, 366–369, 379, 446
nonstationarity and, 141, 142–145
normality and, 98
scatter diagrams and, 73–74
square-root, 142–143, 439
variance-stabilizing, 490

Treasury yield curve, 121, 122

Trend-cycle
defined, 121
in ratio-to-moving average method, 203, 205
in regression models, 429, 444
turning points, 247–253
X-11 and X-12 programs, 218–220

Trending time series, 45–46, 50–51, 119

Trend profiles, 118–119

Trend projections *vs.* other techniques, 33

Trends
as abstract, 34
ARIMA models, 521–522, 526–527
change, explaining, 547
choosing methods in demand forecasting, 342
classifying, 145–146

Trends, cont.
  collinearity and, 429–430
  correlograms and, 95–96, 97,
      137–139
  current trends, exponential smooth-
      ing, 288–292, 318–320
  cyclical component of, 121
  defined, 45, 120–121
  detecting, 146–148, 184
  granularity, 70–71
  model selection, 339–340
  in regression models, 429–430
  in residuals, 375–376
  reverse trend adjustment of leading
      indicators, 244–245
  scatter diagrams for analysis of,
      73–74
  seasonality and, 121–124, 128–136,
      379, 526–527
  straight-line, 143, 146, 522–525
  time plots for analysis of, 69–71
  two-way tables, 131–132, 156–159
  types of profiles, 286
Trends, damped
  damped trend technique, 287
  declining/downward, 146, 292
  in demand forecasting, 342
  exponential smoothing, 292–297,
      300–301
  uses of, 301
Trends, exponential, 146, 293–298,
    299–300
Trends, linear, 293, 296–298,
    299–300, 301
Trend smoothing. *See* exponential
    smoothing; smoothing
Trigg tracking signal, 183
Trimmed means, 78, 105
T statistic, 400, 405–406
Tukey, John W., 67, 68, 98
Turning-point analysis (TPA), 201,

247–253, 265–266, 580
Two-way tables, 130–132, 135–136,
    156–159

U

Unbiased estimators, 114, 115
Unbiased interquartile difference
    (UIQD), 80
Unbiased median absolute deviation
    (UMdAD), defined, 80
Unemployment, 241, 242
Unified framework, 341–342
Updating *vs.* recalibrating, 559

V

Validation, 549–550
Validation period. *See* test period
Validity testing, 544
Variability (spread)
  defined, 78
  identifying sources of, 156–159
  log transformations, 497
  measures of, 78, 79, 80, 105
  in tabulating frequencies, 80–81
Variables, categorical, 443
Variables, dummy. *See* dummy
    variables (indicator variables)
Variables, endogenous *vs.* exogenous,
    262
Variables, external, 359
Variables, internal, 359
Variables, qualitative, 451–455
Variables, standardized (defined), 74
Variables, transformed. *See* transfor-
    mation
Variance
  in ARIMA models, 529–530,
      539–540
  irregular component of, 134, 135
  log transformations and, 379

nonconstant, 375
  residual variance, 159
  seasonality and, 134, 135
  in trend analysis, 147–148
  trend and, 134, 135
Variance, sample, 78–79, 98–99, 145
Vendor Managed Inventory (VMI),
    323
VICS (Voluntary Interindustry
    Commerce Standards), 334
Virtual forecasting services, 594–596
Visionary technological forecasting, 30

W

Warning signals, 180–181
Waterfall chart, 503
Weekly patterns, 126–127, 213–214
Weighted averages, 53–55, 281
Weighted least squares, 447–448
Weighted moving averages, 11,
    48–50, 222
Weighted regression, 399
Weighted sums, 53
Weights
  in ARIMA models, 467
  artificial neural nets and, 39
  averaging and, 545
  choosing, 55
  exponentially decaying, 281, 282,
      283
  with recent history, 48–50
  smoothing weights, 281–286,
      299–300, 343, 344–345
  unit forecast miss, 562
Whiskers, defined, 86
White noise, 467, 486, 513–514, 526
Winsorizing. *See* trimmed means

X

X-11/X-12 programs, 215–221